Interdisciplinary Evolution Research

Volume 11

Roma Tre University
Rome, Italy

Augusta Gaspar, Centre for Psychological Research and Intervention
Catholic University of Portugal
Lisbon, Portugal

Benedikt Hallgrimsson, Department of Cell Biology and Anatomy
University of Calgary
Calgary, Canada

Misato Hayashi, Primate Research Institute
Kyoto University
Inuyama, Japan

Lydia Hopper, Lester E. Fisher Center for the Study and Conservation of Apes
Lincoln Park Zoo
Chicago, USA

John Hunter, Department of Comparative and Digital Humanities
Bucknell University
Lewisburg, USA

Antonis Iliopoulos, Department of Philosophy
Emory University
Atlanta, GA, USA

Mary Lee Jensvold, Fauna Foundation
Carignan, Canada

Carl Knappett, Department of History of Art
University of Toronto
Toronto, Canada

David Leavens, Department of Psychology
University of Sussex
Brighton, UK

David Morrison, Sequence Alignment and Phylog. Networks, Systematic Biology
Uppsala University
Uppsala, Sweden

Roland Mühlenbernd, Leibniz-Zentrum Allgemeine Sprachwissenschaft (ZAS)
Berlin, Germany

April Nowell, Department of Anthropology
University of Victoria
Victoria, Canada

Mark Pagel, School of Biological Sciences
University of Reading
Reading, UK

Elena Pagni, Department of Psychology

Federal University of Juiz de Fora (UFJF)
Juiz de Fora, Brazil

Anna Prentiss, Department of Anthropology
University of Montana
Missoula, USA

Timothy Racine, Department of Psychology
Simon Fraser University
Burnaby, Canada

Eugenia Ramirez-Goicoechea, Department of Social and Cultural Anthropology
National University of Distance Education
Madrid, Spain

Cristian Saborido, Dept Logica, Despacho 2.25
UNED, Fac Filosofia Edif Humanidade
Madrid, Spain

António J. Santos, William James Centre for Research
ISPA - University Institute
Lisbon, Portugal

Peter Saunders, Department of Mathematics
King's College
London, UK

James A. Shapiro, Department of Biochemistry and Molecular Biology
University of Chicago
Chicago, USA

Chris Sinha, Department of Cognitive Science
Hunan University
Changsha, China

Sune Vork Steffensen, Department of Language and Communication
University of Southern Denmark
Odense, Denmark

David Suárez Pascal, Department of Evolutionary Biology
UNAM, National Autonomous University of Mexico
Mexico City, Mexico

Franceso Suman, Pikaia -The Evolution Portal
University of Padua
Padova, Italy

Sandra Swart, History Department
Stellenbosch University
Stellenbosch, South Africa

Ian Tattersall, Division Anthropology, Central Park

The peer-reviewed Springer book series ***Interdisciplinary Evolution Research*** provides a scholarly platform for rigorous scientific, epistemological, and educational investigations into biological, sociocultural, and technological evolution. Volumes focus on well-defined evolutionary problems that benefit from inter- and transdisciplinary analysis with the goal to develop unified jargon and methodology thereby enabling the crossing of disciplinary boundaries.

Book proposals The Series welcomes proposals for anthologies and monographs in the following 5 areas of investigation. 1. Molecular, Biological and Ecological Evolution 2. Symbolic (Behavioral, Cognitive, Social, and Communicative) Evolution 3. Primate, Hominin, and Human Evolution 4. Cultural and Technological Evolution 5. Epistemology, Education, and Origins of Evolutionary Sciences.

Books can focus on theories, topics, data-analysis or methodologies.Inquiries & proposals can be sent to the Series Editor Nathalie Gontier; nlgontier{@}ciencias. ulisboa.pt using the form available on the right.

Iain Davidson

Art or Scribbles? In the Eye of the Beholder: The Evolutionary Emergence of Visual Communication

Springer

Iain Davidson
University of New England
Department of Archaeology, Classics and History
Armidale, NSW, Australia

ISSN 2199-3068 ISSN 2199-3076 (electronic)
Interdisciplinary Evolution Research
ISBN 978-3-031-90457-8 ISBN 978-3-031-90458-5 (eBook)
https://doi.org/10.1007/978-3-031-90458-5

This Springer imprint is published by the registered company Springer Nature Switzerland AG
The registered company address is: Gewerbestrasse 11, 6330 Cham, Switzerland

To Helen
whose long record of support for me has
given me focus
and guided me away from errors.
With all my love

Foreword by Adam Brumm

Everything Iain Davidson has written in this book about the evolution of art is probably wrong. This is not a fault of the book or its author—it is the nature of archaeology. Iain is an archaeologist who, perhaps more than any other, is aware of the inadequacy of the traces left to us from the past, and more so the stories archaeologists invent to explain them. He may have spent more time than anyone thinking about what the first marks left by humans we would recognise as art, looked like, and why they were made. This is intellectually demanding work, and I am humbled he asked me to write the foreword to a book that presents the fruits of that thinking. It follows I should qualify my opening sentence. We can never know how and when art began because we were not there to see it happen. Any effort to do so is therefore flawed. In the absence of evidence, we can only theorise. Iain holds himself to the same strict standards he holds other archaeologists, so he will be the first to admit that his contributions to this matter are, in all likelihood, incorrect. In my view, however, what he has to say about the origin of art in *Art or Scribbles* is more likely to be *less* wrong than anything on this subject written before it. And this is the highest compliment I can think of that can be said of the work of an archaeologist.

Why Is the *Mona Lisa* So Famous?

My 11-year-old daughter asked me this recently and I struggled to answer the question to her satisfaction. I too was intrigued by this problem when I was about her age. It made no sense to me why Leonardo da Vinci's painting was considered to be the greatest piece of art ever created—what was so *good* about it?

Art or Scribbles proposes a theory that can help us understand how and why the *Mona Lisa* became so famous. But it also shows us what strange anomalies such celebrated works really are in the context of art during the vast majority of human history. In particular, it tells us just how unlikely it is that an ice age version of da Vinci's masterpiece could have come to be without writing and modern technology.

Consider the spectacular hunting scene depicted in the rock art at the cave of Leang Bulu' Sipong 4 in Maros-Pangkep, Sulawesi, an artwork Iain discusses in this book. My team discovered this cave painting in 2018 and subsequently dated it to at least 48,000 years ago (see postscript for further details). It was painted on the rear wall of a high-level cave passage. The entrance to the cave is a small hole located several metres above the ground and requires a challenging climb up a nearly vertical cliff-face to reach. The location of this Indonesian painting recalls the ice age cave art of France and Spain, which is also in difficult-to-reach places far from the public eye.

It is possible the scene in Leang Bulu' Sipong 4 was lost to memory after the artist died and lay unseen in the cave until we rediscovered it in 2018. But Iain's way of thinking suggests another scenario. The location in which the artist chose to produce this visual record of a narrative hints at the sort of arcane and secretive knowledge that is communicated through oral traditions and rituals and constitutes the fuel on which power relations in small-scale human societies (e.g., foragers) tend to operate. In Iain's vision, the story the artist had in mind when he or she painted this composed scene in this place was told to others within their lifetime, and this story endured in the cultural beliefs of the artist's society, perhaps long after their death.

The story lived on because that is how the "art-system" as Iain characterises it has worked throughout much of human history. Some people within the local foraging community, probably senior elders, knew about the art and enhanced their status by determining who could see it and hear the story told about it and/or participate in rituals associated with the telling. The number of people alive at any point in time who knew about the painting and its story was probably very few. However, so long as some individuals benefited from controlling access to both, the presence of this art was remembered. Eventually the story of the painting will have changed as the people who learned it from the artist died. But it is likely an oral narrative of some kind was continually associated with this artwork because (1) the painting persisted in the cave and could be seen, (2) the figures portrayed in the scene are interacting with each other in a way that "speaks for itself," and (3) controlling knowledge of the painting and the associated story gave some people in the community power.

Iain's work suggests the scene at Leang Bulu' Sipong 4 could have had an important role in the lives of the Pleistocene foragers who resided near the cave over hundreds of generations or more.[1] Knowledge of this artwork and its story may have spread to neighbouring groups of foragers and to unrelated peoples elsewhere in Sulawesi, perhaps even to adjacent islands. However, the art was fixed in a particular part of the landscape that few (if any) outsiders could visit. The art's story

[1]One support for this is the observation that these near-inaccessible caves were still the focus of special activities in much more recent times. For example, Iron Age people buried their dead in these places, and Muslims have long gone into them to venerate stalagmites they believe are mythical beings (doing so is known in Indonesian as *bertapa*, 'to seek supernatural power'). These recent visitors to the caves are unrelated to the people who made the art. However, like the earlier inhabitants they see the caves as important. (We do not know if the Pleistocene art was ever a part of modern rituals as the rock paintings only entered the written record in 1952.)

would also change as it moved outside its cultural context and eventually be lost altogether.

The theory proposed in *Art or Scribbles* suggests that art worked in this way for some 50,000 years if not longer, until the invention of writing and the development of more advanced means of communication (e.g., colour photography) enabled knowledge of individual artworks and their stories to be told to many people.

Seen in this context, the answer to my daughter's question is clear. The fame of the *Mona Lisa* has less to do with its artistic brilliance (although it is a superb painting) and more with the rise of the Fine Art industry over recent centuries. The popularity of the painting has also increased exponentially with the power of twentieth-century print media and the capacity for stories written about it to reach a mass audience. Media coverage of the 1911 theft of the *Mona Lisa* may even have been what brought the first global renown to a painting that up until then had been fairly obscure.[2]

Nevertheless, if we think about art in the way Iain suggests, then it is also evident that the *Mona Lisa* and the hunting scene at Leang Bulu' Sipong 4 are more similar to each other than they first appear. The existence of something as complicated as the former is inconceivable without the "art-system" that seems to have already been in play when the latter was produced in Sulawesi tens of thousands of years ago.

Most notably, the experience of looking at either artwork compels us to contemplate its meaning, to imagine what the makers *meant* by these images. This behaviour has probably been a key characteristic of what it is to be human for a very long time.

In that sense, it could be argued that neither of these paintings are as important for our understanding of the human story as the first marks made by our ancestors that would fall within the rubric of art, but outside the "art-system" as Iain has defined it.

It is possible there was a stage in our evolution when hominins made marks we would now interpret as art but at a time preceding the emergence of modern human cognition and language. So this would have been before hominins even had the capacity to be storytellers, let alone to tell stories about particular pictures.

If those markings still exist, they are somewhere in Africa where our genus evolved.

I hope to live long enough to witness the enormity of their discovery.

---

[2] https://www.britannica.com/story/why-is-the-mona-lisa-so-famous

Postscript

In 2014 my team of Indonesian-Australian archaeologists published a paper that put forward the claim that cave art on the Indonesian island of Sulawesi dates to at least 40,000 years ago. That argument was based on uranium-series dating of calcite growths that had formed on top of paintings in the Maros-Pangkep caves.

Prior to that, the oldest known rock art was in Europe and dated to 40,000 years ago. Indeed, western scholars had long looked upon the ice age art of France and Spain as the paragon of early human artistic creativity. The Sulawesi art therefore presented a challenge to this standard view that the first "true" art arose in Europe.

In 2019 we published another paper about the Sulawesi cave art. This time we reported the discovery of the oldest surviving example of a narrative composition—a composed scene—in art. At a Maros-Pangkep cave site, Leang Bulu' Sipong 4, we dated a painting that seems to depict therianthropes (part-human, part-animal figures) hunting warty pigs and dwarf buffalo. Uranium-series analysis suggests this artwork is at least 43,900 years old. At the time of writing, we have another paper in press that pushes back the age of this scene by about 4000 years, while another cave art scene that also depicts humans interacting with an animal is older still.

I first saw parts of the hunting scene in Leang Bulu' Sipong 4 as blurry images WhatsApp'd to me by an Indonesian colleague directly from this cave in 2018. At that time the figures representing the therianthropes went unnoticed. However, even in the pixellated iPhone snaps I could see what looked like long, thin painted lines coming from the chest area of one of the dwarf buffalo figures; I thought these might be depictions of spears. Later we realised there are also small, stick figure-like humans with animal heads on the other end of those spear-like lines—the hunters.

When I first visited the site to see this painting, I was bemused by something I had read in one of Iain Davidson's most recent papers: "Contrary to the popular image beloved of cartoonists who show stick figure humans hunting herds of animals, scenes were not an important part of Paleolithic cave art, or so it has been said…."

Those cartoonists had been right all along, I thought.

Iain was one of the first scholars to see the implications of the Sulawesi rock art dates and the importance of the early use of scenes in this art. He has written about this art in several papers and now in this book. I like to think our discovery has changed even this longstanding authority's view of art—as it has changed mine.

Griffith University
Nathan, Australia
June 2024

Adam Brumm

Acknowledgments and Admissions

All books are written under a contract from the publishers. Discussion with the editors produced these words from them:

> the Publisher is bound by the Author's moral rights when making edits to the original work or creating derivative works. This means that the intellectual and personal interests of the Author must be taken into account and *the author's original work cannot be falsified or distorted* (my italics).

I am happy with that reassurance. I now state that:

> Except as otherwise acknowledged in references, this book is my original work and cannot be used in a way that falsifies or distorts what I have written. So far as I know.

I have many people to thank going back to my time as an undergraduate in the 1960s. Most of the people from that period of time know who they are, and I have thanked them personally or in print before, or they have died.

I must begin, though, with special thanks to my valued colleagues who wrote words of support to the publishers before I began writing. I thank you all from the bottom of my heart. April Nowell and I now go back quite a long way, and we have edited two volumes together, one of them on art. She has been a friend as well as a colleague to the extent that it is difficult to know how to express my gratitude adequately. Thank you.

Adam Brumm passed through my Department in Armidale to consult one colleague while both were studying for their PhDs, and he has subsequently followed in the footsteps of another senior colleague in Sulawesi with spectacular results. He is now the leading authority on the earliest rock art and scenes in the world and, in the process, has thought deeply about the interactions between people and animals as shown in the art. We are back where my own research started, only he has gone further. It is a privilege that he consented to write a foreword for the volume. Thank you.

Martin Porr and I crossed (intellectual) swords about matters little related to art at a time when we lived on opposite sides of the world. Since he came to Australia, he has raised the intellectual debate in Australian rock art studies, particularly through giving due weight to the impacts of colonization and to Indigenous knowledge. It took me much longer than him to realize the extent of our indebtedness to Aboriginal knowledge. I thank all three.

On the subject of Aboriginal engagement, I have written this book while living on Anaiwan land, and I have worked with Aboriginal people on their Country in many different parts of Australia. I thank all those who have helped me learn about their Country. I have been privileged to work with Isabel Tarragó and her mother Mrs Hansen of the Wangkamadla people since 1981. Isabel's husband Ray was a Valencian from the town 125 km north of the city where I studied bones for my PhD. We had a memorable field trip in 1982 to Mrs Hansen's traditional Country on the edge of the Simpson Desert and we toured many of the places of significance to her as she explained them to her daughter. We were honored to have her tell us about the art sites on the edge of the Mulligan River (which now gives its name to Isabel's people whose Native Title has been acknowledged there). Together with Isabel's daughter, Avelina, I have been lucky enough to participate in several trips up and down the Simpson dunes to the site of Nganganterra, which turns out to have been so important in my telling of the story in Chap. 9. My work would not have been possible without those experiences and those long-standing friendships. Isabel and Avelina read Chap. 9 when I had written it and gave their approval.

I also want to make a special mention of Anne and Mike Eastham. They both taught art at Wigan Technical College, and Anne studied bird bones in Iberia (Eastham, 2021) while I was researching the bones of other animals. In 1972 at a campsite south of Valencia, over a memorable dinner of chicken cooked with juniper berries, they tried to convince me that my lack of skill at producing a drawn likeness of anything was not an impediment to taking art seriously. I hope they understand how much I have benefited from that conversation and others.

Bill Noble and I wrote many things together, including a book, arguing about the importance of language in human evolution. Many of the things I write about here are an outcome of that fruitful collaboration. Phil Barnard and Dick Byrne also made a huge difference to my life by inviting me to participate in a research workshop in Budapest over several weeks in 2003. We worked out a model of the evolution of human cognition (Barnard et al., 2007), mostly due to Phil, which has guided the way I look at art in human evolution. I was lucky enough that Piers Kelly took up a Fellowship in my Department at about the time I was getting heavily involved in this writing. He has rapidly become a friend and has offered many wise words about art and communication from his experience in linguistics and the study of message sticks. My friend Peter McClenaghan, with whom I and our spouses shared a memorable trip to the art gallery in Toledo Cathedral, selflessly volunteered to read the whole manuscript, because he is deeply read on the art of Caravaggio. His comments were invaluable for the development of important aspects of the work. Finally, my life partner, Helen Arthurson, also read the whole manuscript and made important comments, some of which prevented me from going off in the wrong direction. All that I have written in the last 33 and 1/3 years has been improved by her comments.

Among those still living at the time of writing, I thank (again, if necessary) the following people if I have got something right (and probably several others whose names should be here too). I am also certain in the knowledge that you, the reader, understand that it is my fault, not theirs, if I have managed to get something into print which does not follow their advice.

In alphabetical order of surnames:

A

Steve Ahoy
Karim Alizadeh
Jesús Altuna
Maxime Aubert

B

Anna Belfer-Cohen
Federico Bernaldo de Quirós
Ruth Blasco
Rudie Botha
Carolyn Boyd
Heather Burke

C

Matt Cawood
Betsy Chapoose
Jean Clottes
Nick Conard
Meg Conkey
Pearl and Gordon Connelly
Nick Cook
Güner Coşkunsu
Elisabeth Culley

D

Francesco d'Errico
Bruno David
Livio and Patricia Dobrez

E

Phil Elsegood
Jordi Estévez

F

Valérie Feruglio
Danae Fiore
Warren Fish
Roslyn Frank

G

Pascal Gagneux
Clive Gamble
Alejandro García Moreno
Jason Gibson
Manuel Gonzáles Morales

H

Chris Henshilwood
Wilfred Hicks
Thomas Higham
Jillian Huntley

J

Jacques Jaubert
Peter Jarman
Antonieta Jiménez

K

Ramona Koval
Emma Kowal
Lynne Kelly

L

Michelle Langley
Bob Layton
Darrell Lewis

M

Leila McAdam
Ian McNiven
Ana Belén Marín-Arroyo
Mark Moore
Oscar Moro Abadía
Frances and Howard Morphy
Ken Mulvaney
Fred Myers

N

It is odd that there are no names beginning with N in this list, although I have probably published more often with co-authors whose names begin with N than anyone else. I have thanked them elsewhere.

O

Jim O'Connell
Sue O'Connor

P

John Parkington

Q

Michael Quinnell
Jeff Quilter

R

Mal Ridges
Rebekah Roberts
June Ross
Lynette Russell

S

Pat Shipman
Anne Solomon
Jerry Spangler
Jim Specht
Lawrence Straus
Steve Sutton

T

Paul Taçon
Alandra Tasire
Noreen Tuross

U

Sean Ulm

V

Helga Vierich
Xon Vila
Valentín Villaverde

W

Lynley Wallis
Graeme Ward
Lloyd Weeks
Kurt Wehrberger
Peter White
Polly Wiessner

Finally, a word about chronology. As an archeologist I decided long ago that it was absurd to give dates as anything other than dates before present or years ago—and Radiocarbon dates have been calculated as years before 1950 since it was realized that nuclear weapons testing could have had an effect from about that date. That is three quarters of a century ago, well beyond acceptable ranges of uncertainty about Radiocarbon dates nowadays. I tried out my proposal to use "years ago" on a few people and said that I would refer to Leonardo da Vinci as living between 571 and 504 years ago. The problem is that those dates have no fixed reference. I calculated them in 2023, but this book will probably be published in 2025 and you might be reading it at any time after that. So, it makes sense to use the conventional dates of Leonardo's lifetime as 1452-1519. But those are dates in what some call AD. Some good (I think) people decided to change that denomination from AD to CE so as not to offend other religions. I would have thought that it makes no difference at all whether you say the dates are in the years of "your lord" or within an era defined by that same "lord." 1450 would be 2203 years after the foundation of Rome, but why should we choose Rome? Muslims would say that 1450 was 853 AH (Hijri Years), Chinese 4146, and Hebrews 5210. Wikipedia lists 31 different calendars (https://en.wikipedia.org/wiki/1450), so why choose a particular one? I have chosen to use dates which I was brought up to call AD without any letters and adding a date in years ago. Any earlier dates, I will always use only years ago. As a result, Julius Caesar died more than 2270 years ago, the great Greek sculptor Praxiteles is known to have lived more than 2500 years ago, Buddha was born about 2900 years ago, and literature began to be written down in Mesopotamia 4800 years ago.

References

Barnard, P. J., Duke, D. J., Byrne, R. W., & Davidson, I. (2007). Differentiation in cognitive and emotional meanings: An evolutionary analysis. *Cognition & Emotion, 21*(6), 1155–1183. https://doi.org/10.1080/02699930701437477

Eastham, A. (2021). *Man and bird in the Palaeolithic of Western Europe. Man and Bird in the Palaeolithic of Western Europe*. Oxford, UK: Archaeopress.

Competing Interests The author has no competing interests to declare that are relevant to the content of this manuscript.

Preface and Introduction

Background

More than 30 years ago, I was in the bush north of Boulia, Queensland, Australia, recording archeological sites with colleagues. Early in the morning of 8 August, 1989, a little bit south of Maggie's Creek, we had just finished inspecting a bush grave with a timber surround, probably from the nineteenth century. As we returned to our vehicle, an 82-year-old man, whom I will call Bill, a grazier, stopped his Land Cruiser to ask if we needed help, because that is the protocol in the sparsely populated country around there. He is dead now, but at the time he was one of Australia's largest landholders who grew beef on some of Australia's richest grazing land. He told us what we already knew, because we had been recording it: there were examples of Aboriginal art nearby in the rock shelters on one of the properties his family leased from the State Government. "Not that I would call them art, so much as scribbles," he said. The statement is one that has occupied me for those 35 years and continues to be provocative. This book is about the meaning of that statement: what is the difference between art and scribbles, and why?

The book you hold in your hands is about art: what art is; how art arose; how art became part of the behavior of humans worldwide when it is almost completely absent among all other animals; and how archeological art became art as it is defined today. I am writing primarily about visual art but most of what I argue applies in some way to other forms of art as well. The conclusion I will reach is that art is not a characteristic of an object, but a relationship between that object and a person who interacts with it (see Table 1). In the case of visual art, I mean the person who made it and the one who sees it—the seer. As Esther Pasztory (2005), then Professor of Pre-Columbian Art History and Archeology, describes it "'Art' does not reside in objects; it is society that decides what is and what isn't art." This book explores that idea, noting that the judgments of society begin with the statements of the individuals who produce the art, but continue with those of all who see it—the critics. The nature of the relationships depends on the cultural backgrounds of the producer and

Table 1 Key thoughts for the book

Author	Source	Statement
Iain Davidson	This book	Art is not a characteristic of an object, but a relationship between that object and a person who interacts with it
Esther Pasztory	2005 book	"'Art' does not reside in objects; it is society that decides what is and what isn't art"
Alice Procter	2020 book	The seers also work within conventions, and they often find it strange when anyone presents a different history and status of what they see
Pierre Bourdieu	1968 book	The producer always works within the conventions and mental considerations that they have as a result of living in a particular culture
Oodgeroo Noonuccal (Kath Walker)	1970 poem "The Past"	Let no one say the past is dead The past is all about us and within
Brett Whiteley	Letter to Australian painter Lloyd Rees	"These little pictures signed in the corner L Rees, seemed to have filtered and sometimes dredged the whole of Europe through them…"
Iain Davidson	This book	Why do some people see something and call it art, but others see the same thing and say it is only scribbles?
Fred Myers	Title of 1982 paper about fieldwork in Central Australia with Pintupi people	"Always ask"

the seer. For some people particular marks are art, for others they are scribbles. The question is: "Why the difference?"

How do people look at art, and do archeologists do it differently? I have been researching around the edges of art since the early 1970s, squirrelling away chance remarks among my "real" archeological work and culminating in this book. You will read a number of those anecdotes that I stored in my mind over the years, and my thoughts about them influenced everything else I did. They appear here leading to some statements about both archeology and art that I think will be of interest to people interested in either of them. One of my friends was kind enough to comment on chapters as I wrote them and towards the end he asked who I was writing for. For now it is enough to say: "This book is really for people of all sorts including those who study art."

This is a personal account that I want to be readable and is intended for anyone who has any interest in art. It is aimed at a broad readership of people who want to know more about the evolution of art as well as of anyone interested in what they call art who may want to extend their interests in directions other than those they are usually most familiar with. The book crystalized into this form as I was writing it, but chiefly as a result of a presentation I made to the Center for Academic Research and Training in Anthropogeny (CARTA) at the University of San Diego in 2020. I hope you enjoy reading it—there is no point if no one gets any pleasure from it, especially when we think about art.

What I have written here is a result of a lifetime of archeological work during which I: studied the bones of animals from the 15 thousand-year-old Upper Paleolithic cave site of Parpalló near the present coast in eastern Iberia (Davidson, 2012) where pictures of the same species were painted and engraved on plaquettes of stone; excavated remains of other animals that were painted on the walls of the Cueva del Niño in the hills west of there (Davidson, 2022; García Moreno, 2022; Marín-Arroyo et al., 2022); taught and studied the art of Australian Aboriginal people (Ross & Davidson, 2006), and with others excavated and recorded art sites south of Cloncurry in Northwest Queensland (Davidson et al., 2005); I was lucky enough, as a result, to be among the small number of international scholars invited to enter Chauvet Cave in France and advise on the research into its wonderful art (Clottes & Geneste, 2012); I worked with others on the evolution of cognition from the earliest times and how that view relates to art (Barnard et al., 2007, 2016; Noble & Davidson, 1996); and I visited the site of Blombos and inspected some of the remarkable discoveries there (Henshilwood et al., 2002, 2004, 2009, 2011) and elsewhere in South Africa, including the Drakensberg (Vinnicombe, 1976). Like anyone else, I have benefitted from visiting art and art museums around the world, but particularly in London, Madrid, Barcelona, Paris, Avignon, Florence, Sydney, Melbourne, Brisbane, Canberra, Perth, Darwin, Ieramugadu (formerly Roebourne), Armidale, New York, Vancouver, Christchurch, Bendigo, Hamilton (Victoria), and Darwin. I take photographs. All of this has contributed to my culture.

I began to think seriously about art in 1980 when I first wrote about that which was excavated at Parpalló in the Valencian Community about 100 years ago. The art was incidental to my other archeological research, but perhaps it was central to my thinking. I treated it briefly in my PhD and at greater length in a conference paper in December 1980, taking into account my studies of bones of the animals that were represented in the art of that site. One of my earliest students in 1974 wanted me to run a course on European Cave Art, and in my youthful arrogance I refused on the grounds that there was nothing useful written about it. I apologize to her. There wasn't, but that would have been a good start to teaching people. It would have been fun building something back then.

The Development of the Argument

This book sorts out the difference between the judgments of whether marks are part of art or just scribbles (Chap. 1). As I thought about art at different stages of my career, I came to the unsurprising conclusion that art was a very important part of human behavior; indeed, it may be one of the defining characteristics of humans. So, against some opposition from well-meaning but very knowledgeable friends, I decided that art was too important to be left to people who are interested in art

(Chap. 2); most art historians are not interested in earlier histories of art, and some would confine what they call art to the last 350 years or less.

We need to understand how people with different approaches see art, using the one word but seeing different things. There is some disagreement about whether archeological art can be included with later art. I canvas various definitions, and why people define things, in Chap. 3. I introduce the concept of the art-system and borrow from others the ideas that art is lifelike, involves unquestioned beliefs, is beautiful, uses skill and acute observation, and creates pleasure. In Chap. 4, I discuss at length one of the popular definitions of art that, within the art-system, it is a form of communication by visual means. "A picture is worth a thousand words," they say, but art is only a form of visual communication because people use words when they are communicating, not just pictures. The pictures are not enough.

I recount an anecdote claiming that the Internet revolution was the second great revolution in human history. I suggested then that it was actually the sixth such revolution. The production of visual images was the second revolution, which, unlike all the others, generally does not seem to have a substantial literature. I hope people will see this book as one such contribution. We can show that people share several semantic roles with chimpanzees; those same roles show that in stone tool making our ancestors left a material product which persisted and could be interpreted by others; when those ancestors made art as a result of those same roles, some of the art could only be recognized by the definition of things the people had in mind—what I call "conceptual rather than physical products." This is a key to the whole analysis.

In Chap. 5, I explore the six revolutions in communication that I identified in the previous chapter. There is an important visual element to all of the major changes in communication, but it is important to realize that there is no unbroken thread from language to the "selfie" (Fig. 5.3). In each of the revolutions in communication, there were significant changes in the relationships between the artist and the audience as a result of the visual and material elements in communication. New revolutions are built on earlier ones but not in a continuous process. I call the people in that audience seers, because I am writing about their (your/our) experience of a visual product even though I am claiming that the experience is embedded in the stories that artists tell.

In the second part of the book starting with Chap. 6, I turn to "meaning" and how archeologists go about understanding it. I begin by exploring the concept of dual expression—the thing itself and the idea associated with it—in the context of my simple history of communications. I develop this concept by beginning with another anecdote—about people observing an exhibit of paintings by Picasso and Braque and wishing they knew what the works meant. What emerged among humans was an expectation of meaning, which is generally not present among other animals, and the great mystery of art is that those meanings are almost always indeterminate. That is how we can understand the common reaction to art expressed by those people at the exhibition. We expect a meaning, but we do not know what it is. Much of the rest of the history of art is about reducing that indeterminacy. Different elements of the visual representation of storytelling also involve different relationships with the numbers of people who can know the story, where they must be, and the time it takes to tell them. Different means of representation also have different relationships with power, but that is about the sociology of the relationships, rather than about the art alone.

I contrast the requirements to recognize something archeological as art, and what sorts of art there can be (foreshadowing the three types of Art—Art 1, Art 2, and Art 3—I elaborate from John Robb in Chap. 3). I look at the central theoretical feature about the need for storytelling and the expectation of storytelling when looking at art. I also consider different ways people look at scenes in art as a means by which viewers can access the storytelling involved.

In Chap. 7 I point out that the indeterminacy in mark-making means that seers of art can come up with different ideas associated with a single object and I consider some of the early objects that have been called Art. They often required storytelling for seers to understand them at the time and I show how this interacts with the evolution of cognitive ability. We need to avoid seeking to understand them only in our terms and not in the context of the time of their creation. I then turn to what I call "Seven Types of Ambiguity" (with acknowledgment) which characterize art of different types, different times, and different cultural and social contexts. These have an important relationship to the expectation of meaning and its indeterminacy. I suggest that the emergence of human cognition involving art was more complex than anyone has shown.

The suggestion of art historians (people who are interested in art) is that Fine Art began in the eighteenth century (just under 300 years ago) and that they could drop the "fine" from that phrase. Looking at the situation, what actually became fundamental at that time was the commerce in art, and the establishment of the role of "artist" as a profession.

This exploration leaves me in a position to elaborate in Chap. 8 a complex model of the ways in which different characteristics of art changed during the course of human evolution. I define and discuss seven states of human evolution under the headings of "Creating the expectation of meaning," "Ritual and the importance of distance," "Overcoming the tyranny of place," "Fixing the story," "Freeing the story from institutions of control," "Overcoming the tyranny of time," and "Let me tell you what you see." As the final heading implies, art generally is about different ways of seeing the world, sometimes with the assistance of important people who conducted rituals, sometimes by making important things visible to everyone in the society, sometimes by making seers notice things they had not seen, and sometimes by making us aware of the way our perception works.

Chapter 9 draws all these threads together and explores how they apply to the art of Sahul. Although Aboriginal Australians at contact with Europeans did not use writing, printing, or the telegraph, they spread a set of complex beliefs over most of Australia, and much of that was made possible by encoding beliefs in rock art. I explore the implications of this, including the fact that there was relatively little commerce in art anywhere in Sahul.

There is a concluding chapter. In this I summarize the argument in the whole book and present some observations about the implications for an understanding of art generally. This is certainly a book for archeologists and for people interested in any art, but it also says something to all of us about what art is and what it does.

Discussion

I am an archeologist, but I really wanted to avoid detailed archeological arguments—I hope I have done that. There is an anecdote in Chap. 6 about a referee (probably #2) who suggested my work would be improved if I included in the bibliography some works by art historians (people who are interested in art). Rather than taking the advice to just add works to the bibliography, I read some art history, and I discovered some alarming things, mostly about the art that art historians are prepared to accept, to be interested in. I realized that the answer to the question my friend had posed after reading most of the book was that I have written for precisely those people who are interested in art—maybe even for art historians. What I do with this book, then, is to move away from a restricted view of art and write in such a way that you, the reader, can see that it is appropriate to give the name "art" to all things archeologists and art historians call art.

Remember, much of art history concentrates on art in the last couple of hundred years in a rather small part of the world. Pasztory's (2005, e.g., Ch 15) work is saturated with the idea that art is restricted by the ways art historians have used the idea, and that there is much art in societies where Western Europeans would not have expected to find it. As archeologists, we have been rather shy, on the one hand, about admitting that we are straightforwardly studying art, and, on the other, brazen in not getting to grips with the issue of why we call it art. I hope you will find that I come to terms with a way of looking at art so that anyone can call it art. And I have not included in the bibliography anything that I have not read and does not contribute to my argument in some way.

There are artists who look at things or think of things and produce something in consequence. We tend to say that the object they produce is a work of art, but in fact the art starts from the relationship between the artist, the object, and its subject. Most importantly it exists because of what the producers say to the people who see it. Those people together make up the society that Esther Pasztory wrote about. In this book, I define how that set of relationships works. The important point is that the producer always works within the conventions and mental considerations that they have as a result of living in a particular culture (Bourdieu, 1968). Importantly, what they produce contributes directly to that particular culture, and sometimes it changes the culture. Pierre Bourdieu's image of products embedded within a more-or-less fixed culture is better seen as part of a circular process whereby producers work within their set of cultural values, but what they produce is part of the next stage of cultural values. The next set of producers works with the earlier set as part of the new culture. As we will see, the eighteenth-century English painter, Thomas Gainsborough, broke the rules that the equally English, Joshua Reynolds, had set, and everyone afterwards followed him; Pablo Picasso and Georges Braque changed the way artists see, and everyone afterwards understood that seeing was what mattered. I think that now I understand a lot more about how and why that happens.

The seers also work within conventions, and they often find it strange when anyone presents a different history and status of what they see. Procter (2020) showed the influence of a rather unreflective set of colonial attitudes on the way art has been presented in museums. This cultural context has been fundamental from the very beginning of the appearance of things that result from such artistic relationships. Since 1972, John Berger has influenced large numbers of seers because his original television series, called *Ways of Seeing*, reached a large number of people and he turned it into a very accessible book in the same year (Berger, 2008). He spoke about what we see in the art made between 600 years ago (and the dominance of perspective representation) and the recognition of the implications of the photograph and its domination of the visual environment we all share. He went on to describe how this depends upon the conventions we have learned about our perceptions. That is an example of someone showing how art has contributed to our culture by changing the way we see the world. "What you saw depended upon where you were when. What you saw was relative to your positions in time and space. It was no longer possible to imagine everything converging on the human eye as on the vanishing point of infinity" (Berger, 2008, 18). This book is in that tradition, but extends the story to take into account archeological aspects of art. I will show how the emergence of art in the past influenced the ways we make and look at art today.

About halfway through writing this book, I happened upon an article in a daily newspaper by Zora Simic (2023) reviewing a book by Claire Dederer about "Monsters" in art (I have accessed her earlier work Dederer, 2017), monsters because of their personal behavior towards others. There was a quotation from the book by Dederer (which I have not accessed) which goes: "Consuming a piece of art" involves "two biographies meeting: the biography of the artist, which might disrupt the viewing of the art, the biography of the audience member, which might shape the viewing of the art." Of course, Dederer's point was that many famous artists were or are monsters, but then so, too, are some of the people in that "audience." The statement is very similar to the point of view I have been expressing now for some time (Davidson, 2020, 2023). We have to deal with all four of these elements if we are going to understand what is actually happening.

In her poem "The Past" Oodgeroo Noonuccal (Kath Walker) (1970), Aboriginal poet from Stradbroke Island, wrote:

> Let no one say the past is dead
> The past is all about us and within.

I will present an archeologist's sense in which the lessons of the past are all around us and that those lessons come to us through the art of the longer time period that archeologists study.[1] Of course, those two perspectives—mine and Oodgeroo's—

[1] Tempting though it might be, we could discuss statements in late 2024 about slavery by elected and unelected leaders of the UK to the effect that "none of us can change the past." The enslavement of people has been pervasive, but that would take us away from the main theme. I could argue that the income from the trade in people was an important element of the ability of people in such positions to purchase art 300 years ago, but again it would divert us.

come together in Australia. I will consider a longer past that is contained within all of art and how art has been part of society since humans became human. The lesson is that producers and observers—seers—have used art to make society. I use the slightly unusual name of "seers" because I want you to notice whenever you read about them that you are in the role of seeing, in such a way that you will not just pass over any more common word I might use instead. I will explore how I think that has worked in Australia towards the end of the book.

Structure

A feature of this work is that at the end of each chapter I have written a small Essay attached to relevant images. These images are part of my own cultural background, and the Essays describe something of my relationship to the objects. In presenting these images, the reader has the opportunity not only to enjoy some art works that have been important to me but should be able to appreciate the influences that got me to the position that I adopt in the book. They are, therefore, an essential and accessible part of the whole argument. There is, in Chap. 1, a quotation from a letter written by the Australian artist Brett Whiteley to another Australian artist Lloyd Rees as he lay on his deathbed. Whiteley described how … "these little pictures signed in the corner L Rees, seemed to have filtered and sometimes dredged the whole of Europe through them…" (as in Fig. 1). These images and their accompanying Essays are elements of the whole of art through which my views have been dredged. All of us, artists and seers, have different experiences, and to that extent different cultural backgrounds. So, our own past is all around us. These personal Essays represent my past through which I acquired and internalized a knowledge of art.

In the next few pages, I will spell out some of the more fundamental aspects of the outline I sketched in the section on the development of the argument. I begin the ten chapters with some strange contrasts, and various oppositions continue to be a characteristic of art. Why do people have different opinions about art? Why do some people see something and call it art, but others see the same thing and say it is only scribbles? In order to answer these questions, we need two discussions. First, what constitutes that experience and those relationships we call art? This book challenges the standard definitions of art such that by the end of the book, my primary emphasis in the definition is the relationship between people and the art object, rather than just the object itself. This view has important consequences when we try to incorporate archeological art into the narrative of the rest of art. Second, we need to consider what might be called the sociology of art—how do different groups of people interact with each other using their understandings of what art is? How do people come to think that art communicates, and does it always communicate the same thing to everyone?

In considering communication: what is the relationship between communication through art and other forms of communication? I will show that all forms of communication—especially writing—have some sort of visual element often involving

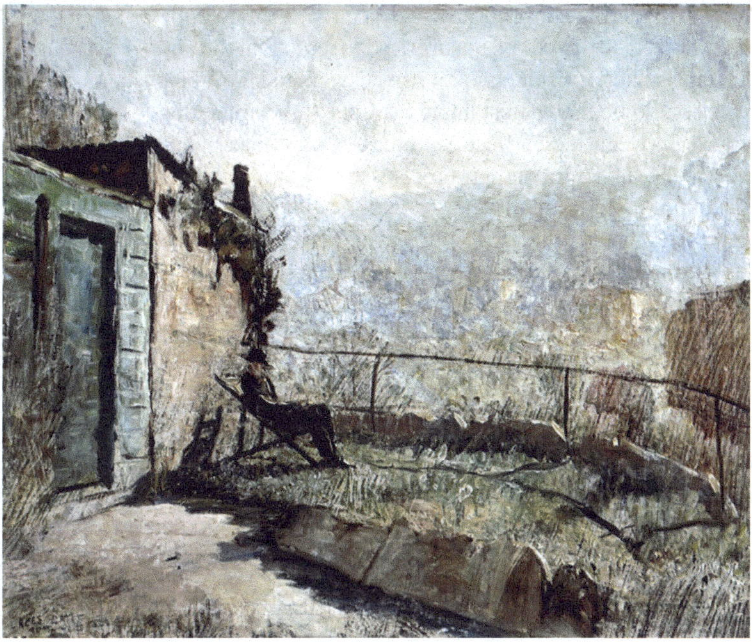

Fig. 1 Lloyd Rees Spring Sunshine. (New England Regional Art Museum, Armidale and © Alan and Jancis Rees/Copyright Agency, 2025)

symbols. What difference does it make to communication, first, that there is a visual element, and then, that the visual element persists so much longer than the words of speech? What is the implication of persistence for the way people think about art, and how they think about communication? Does the art itself communicate to people or does their own interaction with it have as a result that they think particular things? Can we say that art does NOT communicate, but people have learned to say that their thoughts have been communicated to them when they see something in the relationship? We will end up understanding the implications of that for the expression that "A picture is worth a thousand words."

The beginning of the book also addresses questions concerning art and symbolism: How does art relate to symbols, especially if shared conventions and arbitrariness are both features of symbols? Does the nature of the interaction between art and symbols mean that the art can stand by itself to communicate to people—does the art have agency of its own? Or is it essential that, before an observer can understand how art communicates, they must understand the conventions and arbitrariness that people accept within their culture and for whom it is a symbol? Do people derive meanings from the conventional nature of symbols—including that special, esoteric, meaning of symbols as representing some sort of belief system? Does this explain the complex relationships people have with the meanings of the things we call art? Are conventions unchanging, or can we see how the gradual shift of conventions can lead to changed relationships between art and society? Do people from societies with different conventions have different relationships between themselves and the art objects? You will be unsurprised to learn that the answer is usually "Yes."

In the later chapters of the book, I use answers to all of the questions in the early part to address some more substantive features of the archeology of art. I begin with the contrast between the earliest examples of what has been called art (Culley & Davidson, 2021) and a second group consisting of those objects, now mostly found on walls and in galleries. Only the second group is called art by people who define themselves as interested in art. The first group is archeological, but there is a bigger group of visual images now known primarily to archeology—how are the early examples and the more recent ones related? The second group ends up with what some like to call Fine Art. I was introduced to this distinction between art and Fine Art by reading Larry Shiner (2001), who borrowed it from Paul Kristeller (1951) (and many others). The bigger group—the first—appears not to be very important to those who become experts on Fine Art, and I want to change that. How are they related? And how is it that all four categories, which I will define using numbers—earliest archeological art (Art 3), other archeological art (Art 2), Fine Art (Art 1), and art that the mavens do not regard as Fine Art—are called art by somebody? Such questions are often posed by those who study Fine Art as if they are the guardians of the real nature of what Art is, as opposed, they say, to what is mere "craft," or perhaps "graffiti." This book turns the problem around and asks what factors led from the earliest art to the protected confines of the Fine Art world, and which other categories were given too little attention in the process; in grazier Bill's word, as if they were mere scribbles.

Along the way, I consider the importance of scenes—special combinations of images, first found in rock art and elsewhere—derived from the recent book April Nowell and I edited about *Making scenes* (Davidson & Nowell, 2021). In Chap. 6, I give the example of scenes as a way in which people can move from producing the expectation of meaning (however indeterminate) in the marks they produce towards reducing the indeterminacy of that meaning. Scenes give the appearance of letting people read a story from what is shown, whether or not they have been told the story by the producer. This leads to a discussion of the patterns of the way scenes have been used in art and their relation to the stories told using art.

I then work out a framework by which we can consider the various changes in the emergence of art at different stages of human evolution and history. I discuss the implications of these changes first for archeological cases outside Australia and then for the understanding of the way art was used throughout Australia.

The archeohistory of Australia involves a series of changing symbolic relationships between people and their environments. This chapter considers the 65 thousand or so years people—already symbol using humans, not least because art appears to be earlier in Sulawesi—have lived in Sahul as requiring processes by which peoples in different parts of Australia and New Guinea worked out how to live with each other in the driest and most variable continent on earth.

The final chapter takes the well-known saying that "a picture is worth a thousand words," and links it to all I have discussed before. In this chapter I set up a discussion of the ways in which pictures have meanings that might be different for different people, and the way meaning becomes a property of art. The chapter presents a case for the relationship between these two visions, and how, in considering them, we can understand much more about art in the history of humans being people.

In this book, then, I take several chance remarks about art and explore their implications in order to understand how they contribute to an understanding of the relationships among the earliest art, rock art, archeological art, and the other things called art. Some people write about the history of European art as if it were somehow the most important evidence of human creativity. On the contrary, the essence of art is the cultural context in which people produced it. In many (perhaps all) places people produced art. In all places, what they produced shared common features about the interaction between artists and their cultural contexts.

Some of those interactions, maybe all of them, include associated rituals. More generally, the interactions include the back story of the work that relates the person who produced it to the people who see it, the subjects (material or mental) the producers have in mind and the objects they produce for the seers at the time, and those other seers (ourselves) who may not know the original stories. Those objects are too often seen as the only "thing" that is art, whereas I argue, the art is in the whole system of relationships. The fact that the material object is interpretable by seers is made easier if it represents a scene. People see the world through the materiality and persistence of the object and the mentality of the associated stories. Only when people from different cultural contexts view the material object does anyone express a view that some are not art, just scribbles. Sometimes those things that look like scribbles are the ones where the meaning to the artist and the seers depends on their shared cultural knowledge.

Bourdieu was right to show that we need to consider the work of art in its whole cultural context, but with the certainty of an early Greek philosopher, we should acknowledge that that cultural context is constantly changing as a result of the way people use art.

Essay: What This Is All About (Anthropomorph and Arcs, Selwyn Ranges, Queensland)

This book is ALL about relationships and how we use things we call art to define aspects of those relationships and how these interact with cultural context. Context is everything, or, at least, everything depends on the cultural expectations that people have.

As we were walking to an archeological survey location south of Cloncurry, Queensland, my students and I first spotted a rock shelter halfway up a cliff that we have previously called Site 3 (Davidson et al., 2005). We said that we would investigate it after we had done our work. One student, however, had other ideas, and went up and saw this wonderful image (Fig. 2 Left). He told us about it, and it looked to us like a stylized human figure with a headdress. The local graziers had not seen it, though the rock shelter is only a few kilometers from their homestead, and closer still to the camp they used while mustering cattle. They were neighbors of the grazier, Bill, whose comment gave me the title for this book. In terms of cultural context, I have no doubt that this image deserves to be called art. But it is very possible that, for grazier Bill and his family, even an image as magnificent as this would be more likely to fall into his category of scribbles. Bill's family could trace their ancestry back to a time in the 1860s when, 900km due east of Cloncurry, people broadly related to the artists had killed his great-grandfather—newly arrived as an immigrant from Scotland—during the theft of the land from its Aboriginal inhabitants. We can only speculate what cultural values his family brought to their thoughts on the cultural achievements of Aboriginal people?

At the time, we thought that the appropriate Aboriginal organization to grant consent for fieldwork was the Kalkadoon Tribal Council (KTC) in Mount Isa, and I

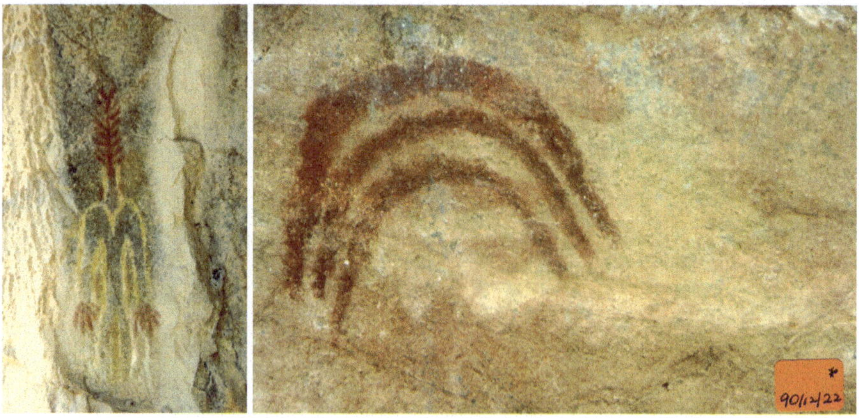

Fig. 2 Left, Anthropomorph from Site 3 in NW Queensland. Right, Nested arcs from Site 2. The nested arcs are superimposed on a hand stencil. (Photos by Iain Davidson, with permission from Pearl Connelly and the Mitakoodi Native Title organization)

subsequently took the Chairman of the KTC, Ken (now deceased), to see Site 3. Ken was a wonderful photographer, and he took a photo of this image. Some months later he rang me in Armidale to ask for my permission to use his photograph in a book he was publishing to raise funds for the KTC (Isaacson, 1995). I was astonished. I needed (and wanted) to work with the KTC to do my work. Aboriginal people have always identified with land they call their Country (Rose, 1996; Sutton, 1995), and I was showing this image to Ken because it was in the Country I thought KTC had authority over (though, since then, Native Title has sorted out where boundaries might reasonably be claimed). Here was the Head of the organization from whom I sought consent, asking me to use his own photograph of an image from his own country. When I told him that I found this paradoxical, his response was simply that I had found it (though actually, the student had as part of my research team). Although it was in "his" Country (at least he was the head of the Aboriginal organization that operated there), and it was his photo, he said it was polite to recognize my contribution by asking my permission. As Fred Myers (1982) showed in another Aboriginal context, it is always polite to ask.

Our original published description of this image was as follows (Davidson et al., 2005):

> The third image provides some comfort for an anthropomorphic interpretation. The use of distinctive colours allows the viewer to recognise a convention of 'modular representation' which picks out particular features of the image for separate treatment (plate 13.5). So this image has a red-branched 'headdress', red hands, and red patches on either side of the face and under the arms. The outline is yellow, but there is evidence for over-painting of the image, so that an earlier outline was cream. ... The important point here is that the careful dissection of the imagery through the identification of the modularity and the recognition of variation within the modules, combined with some iconic links, allows the recognition that the north-west central Queensland anthropomorphs are not just iconic. At very least the schematisation is symbolically opaque—but there are no referents to the motivation of the symbolism except through the recognition of a link between human and animal. We are confident that there was once a motivated symbolism of this sort, told in stories or sung in songs at ceremonies ....

The same student was also one of the people who reported the site we called Site 2 in the same analysis. He and another student had been assigned a particular piece of survey work, but they saw rock shelters on the side of the valley and climbed up to them. There were several spectacular paintings in them which I later showed to Jean Clottes, who was then the leading authority on cave art in France. Clottes (2000) described his perception of the visit to Site 2 in the company of Alf Barton from the Kalkadoon Tribal Council. The image on the right (Fig. 2 Right) was in a small shelter together with a stencil of a shell pendant (Fig. 9.17) and some other paintings including one of the figures that appears to have a head-dress. This image, which I say falls into the general category "parallel or nested arcs," might fall into the category "scribbles" for the grazier, Bill. It turns out that images with nested arcs, often with more of them than this, are found in rock art right across Australia, at least as far west as you can go into Murujuga, on the west coast of the continent.

Rather sneakily, I included it with images in a seminar presentation at which Jean Clottes and David Lewis-Williams were among the participants (Conkey et al.,

1997; Davidson, 1997). Clottes and Lewis-Williams were then working through their argument about shamanism represented in the rock art of the world (Clottes & Lewis-Williams, 1998; Lewis-Williams, 2012), an argument that rested, among other things, on the identification of images known as "entoptics" apparently seen by people experiencing what are sometimes called "visions" in altered states of consciousness. Among the entoptics is the "catenary entoptic," so named because the curve seems to be similar to the resting position of a chain supported at each end. I asked David whether he had noticed the "catenary entoptic" among my pictures. He had not. He thought I had shown pictures of boomerangs. The point here is that the interpretation of the art depends on the cultural context through which it is being viewed—boomerangs are represented quite differently in the region. For me the context was plain—in the presence of Clottes and Lewis-Williams, these curved lines were probably an example of the catenary entoptic that might have been seen by people in altered states of consciousness. For Lewis-Williams, not suspecting I was imbued with knowledge of that context, they were merely Australian images, and hence likely boomerangs. Context is everything, or, at least, everything depends on the cultural expectations that people have.

References

Barnard, P. J., Davidson, I., & Byrne, R. W. (2016). Toward a richer theoretical scaffolding for interpreting archaeological evidence concerning cognitive evolution. In T. Wynn & F. Coolidge (Eds.), *Cognitive models in Palaeolithic archaeology* (pp. 45–67). Oxford University Press.

Barnard, P. J., Duke, D. J., Byrne, R. W., & Davidson, I. (2007). Differentiation in cognitive and emotional meanings: An evolutionary analysis. *Cognition & Emotion, 21*(6), 1155–1183. https://doi.org/10.1080/02699930701437477

Berger, J. (2008). *Ways of seeing*. Penguin UK.

Bourdieu, P. (1968). Outline of a sociological theory of art perception. *International Social Science Journal, 20*(4), 589–612.

Clottes, J. (2000). *Grandes girafes et fourmis vertes. Petites histoires de préhistoire*: La maison des roches, editeur.

Clottes, J., & Geneste, J. M. (2012). Twelve years of research in Chauvet Cave: methodology and main results. In J. McDonald & P. Veth (Eds.), *A companion to rock art* (pp. 583–604). Blackwell.

Clottes, J., & Lewis-Williams, J. D. (1998). *The shamans of prehistory. Trance and magic in the painted caves*. Harry N. Abrams.

Conkey, M. W., Soffer, O., Stratmann, D., & Jablonski, N. G. (Eds.). (1997). *Beyond art: Pleistocene image and symbol*. The California Academy of Sciences.

Culley, E. V., & Davidson, I. (2021). Art, Sign, and Representation. In N. Gontier, A. Lock, & C. Sinha (Eds.), *The Oxford handbook of human symbolic evolution, second edition* (pp. 331–368). Oxford University Press.

Davidson, I. (1997). The power of pictures. In M. Conkey, O. Soffer, D. Stratmann, & N. G. Jablonski (Eds.), *Beyond art: Pleistocene image and symbol* (Vol. 23, pp. 128–158). The California Academy of Sciences.

Davidson, I. (2012). What a carry on? Portable art and changes of symbolic meaning. *L'art pléistocène dans le monde / Pleistocene art of the world / Arte pleistoceno en el mundo, Actes du Congrès IFRAO, Tarascon-sur-Ariège, septembre 2010, N° spécial de Préhistoire, Art et*

Sociétés, Bulletin de la Société Préhistorique Ariège-Pyrénées, LXV-LXVI, Book p. 268–269, CD: p. 1559–1570.

Davidson, I. (2020). Marks, pictures and art: Their contribution to revolutions in communication. *Journal of Archaeological Method and Theory, 27*(3), 745–770. https://doi.org/10.1007/s10816-020-09472-9

Davidson, I. (2022). Prefacio. In A. García Moreno (Ed.), *La Cueva del Niño (Aýna, Albacete). Estudios en el 50 aniversario de su descubrimiento* (pp. 17–48). Instituto de Estudios Albacetenses Don Juan Manuel.

Davidson, I. (2023). Humans making history through Continuities and Discontinuities in Art. *Cambridge Archaeological Journal, 33*, 637–654. https://doi.org/10.1017/S0959774323000057

Davidson, I., Cook, N. D. J., Fischer, M., Ridges, M., Ross, J., & Sutton, S. A. (2005). Archaeology in another country: Exchange and symbols in North West Central Queensland. In I. Macfarlane, M.-J. Mountain, & R. Paton (Eds.), *Many exchanges: Archaeology, history, community and the work of Isabel McBryde* (pp. 101–128). Aboriginal History Inc.

Davidson, I., & Nowell, A. (Eds.). (2021). *Making scenes: Global perspectives on scenes in rock art*. Berghahn Books.

Dederer, C. (2017). What do we do with the art of monstrous men? *The Paris Review*. Retrieved from https://www.theparisreview.org/blog/2017/11/20/art-monstrous-men/

García Moreno, A. (Ed.) (2022). *La Cueva del Niño (Aýna, Albacete). Estudios en el 50 aniversario de su descubrimiento*. Instituto de Estudios Albacetenses Don Juan Manuel.

Henshilwood, C. S., d'Errico, F., Vanhaeren, M., van Niekerk, K., & Jacobs, Z. (2004). Middle Stone Age shell beads from South Africa. *Science, 304*(5669), 404.

Henshilwood, C. S., d'Errico, F., & Watts, I. (2009). Engraved ochres from the Middle Stone Age levels at Blombos Cave, South Africa. *Journal of Human Evolution, 57*(1), 27–47. Retrieved from http://www.sciencedirect.com/science/article/B6WJS-4WDNBPR-1/2/bdfae7761fb2b33c23791223f8922970

Henshilwood, C. S., d'Errico, F., Yates, R., Jacobs, Z., Tribolo, C., Duller, G. A. T., et al. (2002). Emergence of modern human behavior: Middle Stone Age engravings from South Africa. *Science, 295*(5558), 1278–1280. https://doi.org/10.1126/science.1067575

Henshilwood, C. S., d'Errico, F., Niekerk, K. L. V., Coquinot, Y., Jacobs, Z., Lauritzen, S.-E., et al. (2011). A 100,000-Year-Old ochre-processing workshop at Blombos Cave, South Africa. *Science, 334*, 219–222.

Isaacson, K. (1995). *All one people. This land Australia*. Aebis Publishing.

Kristeller, P. O. (1951). The modern system of the arts: A study in the history of aesthetics part I. *Journal of the History of Ideas*, 496–527.

Lewis-Williams, J. D. (2012). Rock art and shamanism. In J. McDonald & P. Veth (Eds.), *Rock art companion* (pp. 17–33). Blackwell.

Marín-Arroyo, A. B., Davidson, I., & García Moreno, A. (2022). Análisis tafonómico y arqueozoológico de la fauna de mamíferos y la microfauna de la Cueva del Niño. In A. García Moreno (Ed.), *La Cueva del Niño (Aýna, Albacete). Estudios en el 50 aniversario de su descubrimiento* (pp. 179–214). Instituto de Estudios Albacetenses Don Juan Manuel.

Myers, F. (1982). Always ask: resource use and land ownership among Pintupi Aborigines. In N. Williams & E. Hunn (Eds.), *Resource managers: North American and Australian hunter-gatherers* (pp. 173–196). Westview Press.

Noble, W., & Davidson, I. (1996). *Human evolution, language and mind*. Cambridge University Press.

Oodgeroo Noonuccal [Kath Walker]. (1970). The past. In *My People* (p. 93). Jacaranda Books.

Pasztory, E. (2005). *Thinking with things: Toward a new vision of art*. University of Texas Press.

Procter, A. (2020). *The Whole Picture: The colonial story of the art in our museums & why we need to talk about it*. Cassell.

Rose, D. B. (1996). *Nourishing terrains: Australian Aboriginal views of landscape and wilderness*. Australian Heritage Comission.

Ross, J., & Davidson, I. (2006). Rock art and ritual: An archaeological analysis of rock art in arid Central Australia. *Journal of Archaeological Method and Theory, 13*(4), 305–341. https://doi.org/10.1007/s10816-006-9021-1

Shiner, L. (2001). *The invention of art: a cultural history*. The University of Chicago Press.

Simic, Z. (2023, 11 May 2023). Friday essay: cancellation or conflicted joy – Grappling with the work of our 'art monsters'. *The Conversation*. Retrieved from https://theconversation.com/friday-essay-cancellation-or-conflicted-joy-grappling-with-the-work-of-our-art-monsters-202748

Sutton, P. (1995). *Country. Aboriginal boundaries and land ownership in Australia*. Canberra: Aboriginal History Inc.

Vinnicombe, P. (1976). *People of the Eland. Rock paintings of the Drakensberg Bushmen as a reflection of their life and thought*. University of Natal Press.

Contents

Chapter 1
Art or Scribbles

Abstract In the course of human evolution, art and evolving cognition were bound together. Leonardo da Vinci called art a *cosa mentale*. The book will emphasize that what made a difference to the explosion of art after the Middle Ages, in addition to perspective, paint, paper and canvas, was the use of scenes that could be interpreted without the words of the artist, as well as the expansion of commerce.

The study of the evolution of art requires us to acknowledge several elements. This chapter introduces the object that is recognized as art, as well as the idea that in addition to that object we, you the reader and me the writer, have to acknowledge that there is a process of production that involves the artist, the subject of the art, the stories that the artist tells about it, as well as the seers of the art. The chapter goes on to emphasize the importance of the context within which people get to experience and comment on the process that results in art. It is in this context that some refer to an object as art, while others see the same object and call it scribbles. The chapter finishes with a discussion of the way the processes of art are intertwined with the complex sets of relationships among people.

The Essay at the end of the chapter discusses two paintings called *Las Meninas*, that by Velázquez and one of the series, inspired by the Velázquez painting, by Picasso. I use the Velázquez painting to set out the sort of things that I include in an art-system, setting the painting and its subject into some of the context of its time, but also the context in which I got to know it in the 1970s.

> There really is no such thing as Art. There are only artists. (Ernest Gombrich *The Story of Art* 1995 p. 15)

For every work that gets called art, there was an artist, who produced the work: Gombrich was right. Ernest Gombrich first published his guide to the *Story of Art* in 1950, which once made a great impact on the art world. It is risking fate to claim he was wrong, but this book is about his proposition that artists were more important than art but that is not enough. I want to emphasize the stories that artists told about the relationship between their work and the ideas they had or the things they saw. Those stories were for an audience who could see the work, whom I choose to call "seers."

I. Davidson, *Art or Scribbles? In the Eye of the Beholder: The Evolutionary Emergence of Visual Communication*, Interdisciplinary Evolution Research 11,
https://doi.org/10.1007/978-3-031-90458-5_1

Thus, every object that might be called a work of art had an artist, a subject which was revealed by the stories, and a viewer who saw the work and heard the stories. That is the starting point of the evolution of art, and of every art experience that anyone has ever had or has now. Gombrich was trying to move people away from the idea that the only thing of importance is the work itself, even for those who are interested in art objects. I want you to join me in discovering how that is part of the story, but only a part.

This book is about how art arose, and how it became what we call art today. I decided to write it as I was thinking about several chance remarks I heard people make at different times from different parts of the world and with different personal experiences and expertise. Together they force me to address several issues about art and it turns out that these issues sum up the whole achievement of our species: what is art and why do all people seem to be involved in some form of it? Those questions can only be understood if we also look at how art came into being. We use the word art because that is what we are familiar with wherever we live in the world. There is a big literature about how we might want to call many things art but they once went by another name in further different places and times (Kristeller, 1951).

The big question becomes how, in the course of history, did stuff that is so widespread turn into the stuff that artists, critics, curators, philosophers or other interested people call art today and what was the impact? Are all of the things and relationships that get called art really the same sort of thing? Those are the central questions this book seeks to address. I am an archeologist, and archeologists ask how things come in to being as we try to write what I call archeohistory—the narrative of past events that includes inferences from archeological evidence. My question is how art came in to being and became part of the history of relationships among people. In addition, as a human being, I am interested in the art that surrounds all of us in the world and in the social divisions that are essential to the way we produce it and appreciate it. This book is about all of these things.

The distinction between art and scribbles, with which I began, is part of a widespread set of attitudes to art. Some people know art when they see it; they equally know what is "not-art". In this way they make judgments about people who have views different from their own beliefs. This is very similar to attitudes about culture: there can be different understandings of the meaning of the concept, and unless people clarify these understandings, they give rise to many conversations at cross purposes (Davidson, 2016). Many people believe that their own position about an issue is the only one with any virtue, even in their own group of people. "Art is what I say it is, whatever you may think." I will discuss various understandings about the concept of art in Chap. 3, after introducing some of the fundamental ideas behind the way I treat art in this book. Various authors have looked at the concept of "art" in historical terms to show that the modern usage is relatively recent (and correct!). I will put that history into context. One argument is that there is a distinction between art and craft. In the late nineteenth century, the popular understanding of this difference was one reason why it was so difficult for people to accept the existence of Cave Art in Spain and elsewhere (Moro Abadía, 2006). Another important argument was that outside the west people made many things that might be art during or for rituals. That led to the initial exclusion of non-Western art from art museums and art galleries of Europe and America. Once the work could be seen without its attendant rituals and beliefs, the argument went, people might move it from the ethnographic museum to the art museum. The most obvious example was the bronzes and other art from Benin which the British captured by military force in 1897 (Hicks, 2020; McDougall, 2008). I will address in a later chapter the fundamental question of the content of archeological museums such as the British Museum, and whether we should see the objects in the museum as art or something else. Alice Procter (2020) has written an excellent account of the multiple meanings of displays in Museums and the unhappy role of colonialism in the creation of the collection in museums. The case for restitution of the art of Africa to its places of origin is, by now, overwhelming (Adjei & LeGall, 2024).

Most importantly in the history of art, Pablo Picasso, visiting the museums of Paris, saw some Iron Age Iberian sculpted heads from his native Spain, as well as masks ripped from their ritual context and placed in ethnographic museums in Paris. William Rubin (1980, 59–99) has also given an account of Picasso's use of masks. Picasso used stylization of masks in his painted art at the beginning of the twentieth century, most famously in the revolutionary *Les Demoiselles d'Avignon* and his the portrait of Gertrude Stein is Fig. 3.23 on p. 97.

The Iron Age bronze in Fig. 1.1 is not one of the figures Picasso drew on but does exemplify some of the effective simplicity of sculpture from that period, and, in particular, the stylization of facial features. The figurine is in the Museum of Archeology in Valencia, http://mupreva.org/cat/2899/en. I will argue, we should use the same name, art, without separating craft from it, nor excluding it if we know it was involved with ritual. Archeologists have been, by turns, over generous and over cautious in their recognition of ritual as an element of what we find. Curiously, there is a whole set of other rituals when people place art in art museums and galleries and others go to see it. These include the deference with which the museum is approached, the special building within which it is housed, generally the special clothing people wear when visiting, the hushed tones of the audience, the discreet relationship between interpretation and image, the fact that the museum houses art divorced from its contexts of production or use, and so on (see also Duncan, 1995, 2005). But we tend to ignore those sorts of ritual in our own behavior.

Art historians got it wrong when they argued that art (by which they meant Fine Art, but they generally dropped the "fine") was invented in the eighteenth century. First it is a niche argument within art history, second it is very Eurocentric, and third it depends on two other arguments: that, for example, Picasso's masks only became art when he got rid of the ritual element of them; and, that the social system within which the art was produced had been altered by the expansion of commerce. These things seem to be related.

In the first place, I will argue, art was always there, inside and outside Europe, even when it was involved with ritual—and particularly because of the rituals associated with the relationship between modern seers and the works they see, from whatever time period. I think that what happened in the eighteenth century was that scenes made it possible to paint pictures with stories that were unrelated to ritual (i.e. Fine Art), or to classical or biblical stories, and that stripping ritual from a specific context of the art meant that commerce in pictures opened up. Technology of the paint (oil paint) and of the support (painting on the newly available paper and on canvas) made that commerce even more attractive.

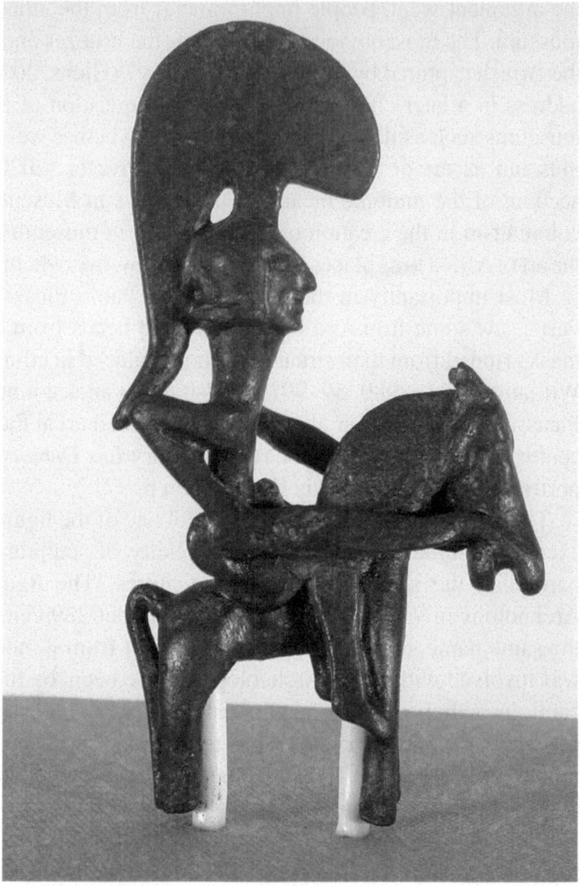

Fig. 1.1 Iron Age bronze figurine of warrior from La Bastida de les Alcusses (Moixent, València). (Image from Museu de Prehistòria de València (with permission))

The second point is that art might be considered in a different way when commerce becomes dominant (but tell that to Van Gogh). Importantly, though, it makes very little difference to the work of art whether the producer is funded by a patron and makes works to contract or gets paid by the buyers in the marketplace. Larry Shiner (2001, 35–37) pointed out that Leonardo signed a contract to paint one of his famous works, and that he was told what the painting should include, and the color of the clothing of "Our Lady". That may be true, but it does not mean that he did not produce a work of art any less than if he had had to put the work up for auction. And the fact that he was a painter of genius can be discerned even though the subject was given him. Somewhere in there, there is art, but two things stand out. First, my remarks still leave us short of understanding what art is, and second, we know a lot about the social context of production of the work. But suppose we did not. Suppose all evidence of that social context had been destroyed, would we still be able to discern that the painter had NOT produced a work of Fine Art. In other words, is a work an example of Fine Art because of some characteristic of the work, or because of what we know of the way it was produced? The choice between these two options may be in the attitudes of the person who sees it. To paraphrase another set of circumstances: we will decide what is included in art, and the circumstances in which it is included. I hope this book helps.

One of the principal determinants of the distinction between art and scribbles is the nature of person who is making the distinction. Both Desmond Morris (2013, Ch. 3) and John Matthews (2013, 311–352) have analyzed "scribbles" by chimpanzees to show that they were far from random. The central point of discussion in this book is that each person has their own particular history and attitudes derived from that history. The attitudes that people show are less about the marks they see than about their own story, and more about their own beliefs, derived from their education and experience, than about the story of the marks said to be art. Attitudes and stories are both central to any understanding of what art is, but they are also central to many of the examples that some might call scribbles. In particular, much art was always involved in ritual at the time of its production: cave art of Europe (Arias, 2009) (Fig. 1.2); art of the beginnings of agriculture in Western Asia—considered as likely interpretations at Göbeki Tepe, in Turkey (Peters & Schmidt, 2004) (Fig. 6.2) and at 'Ain Ghazal, in Jordan (Rollefson et al., 1992) (Fig. 1.3); temple art of Angkor Watt (Fig. 1.4); Church art of Europe (Gombrich, 1995, Chs 9–10) (Fig. 1.5); Benin bronzes (Hicks, 2020) (Fig. 1.6). I could provide hundreds of examples from around the world.

If I understand it correctly, many art historians and other people who are interested in art would deny that any of these examples are art at all, because they are from before the eighteenth century, or from Asia or Africa. The possible exception is the eighteenth century image from Toledo, a combination of sculpture and painting.

Russell McDougall (2008, 75) pointed out that "it is now customary for art historians to locate the European 'discovery' of African art with the masks favoured by Fauvism and Cubism", in the manner of Picasso as I described earlier. The Benin

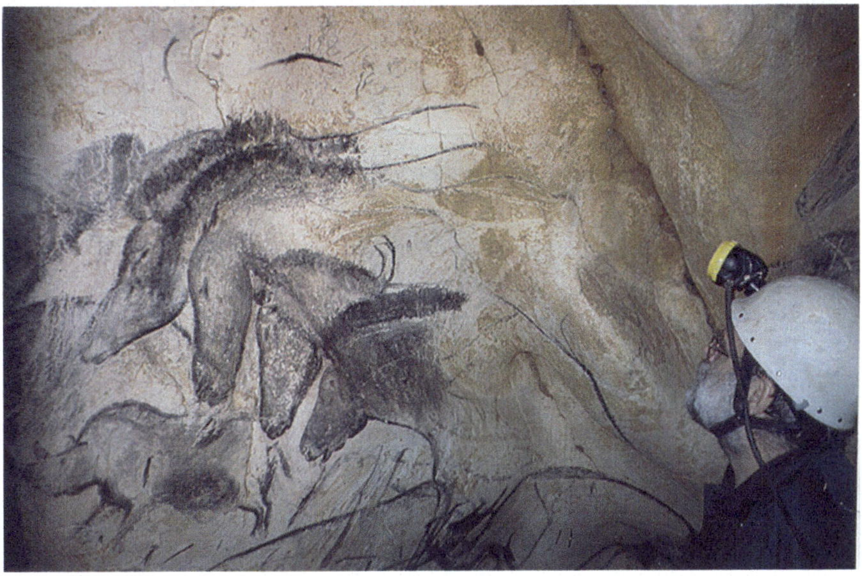

Fig. 1.2 Horse panel at Chauvet Cave being admired by the author 2002 (Photo Jean Clottes)

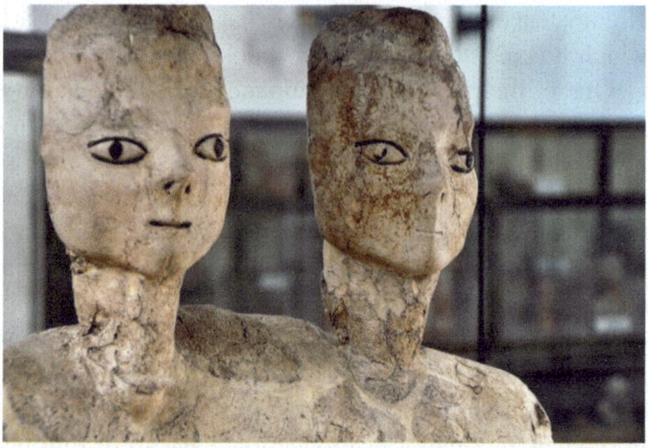

Fig. 1.3 Figurines from the Neolithic site of 'Ain Ghazal, near Amman, Jordan. In the Amman Museum. (Photo Iain Davidson)

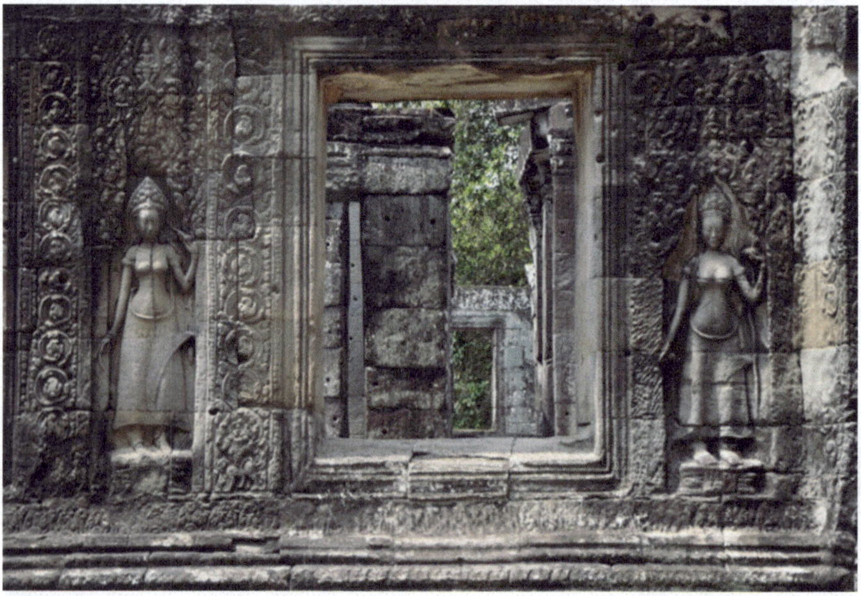

Fig. 1.4 One of the corridors of the Preah Khan at Angkor, a temple established by Jayavarman VII in the late twelfth century. As told to me by Roland Fletcher, the figures are either *apsara* or *devatas* both of whom are part of the iconography of Hinduism and Buddhism in Angkor. The wall decorations are also part of the iconography. The holes in the stone further back are fittings to attach metal panels around doors. (Photo by Roland Fletcher, used with permission)

bronzes "entered the European canon of beauty almost a decade earlier, soon after the British sacking" of Benin, described in 1903 by Roth (1968 (1903)).

Let me put into context the story about the grazier, Bill, who was unconvinced that he should recognize as art the images on rocks in the land he and his family grazed. He was the great-grandson of a European pastoralist who migrated from Britain to Queensland in the middle of the nineteenth century. His ancestor was, therefore, one of those Europeans who felt they had a right to introduce cattle to a

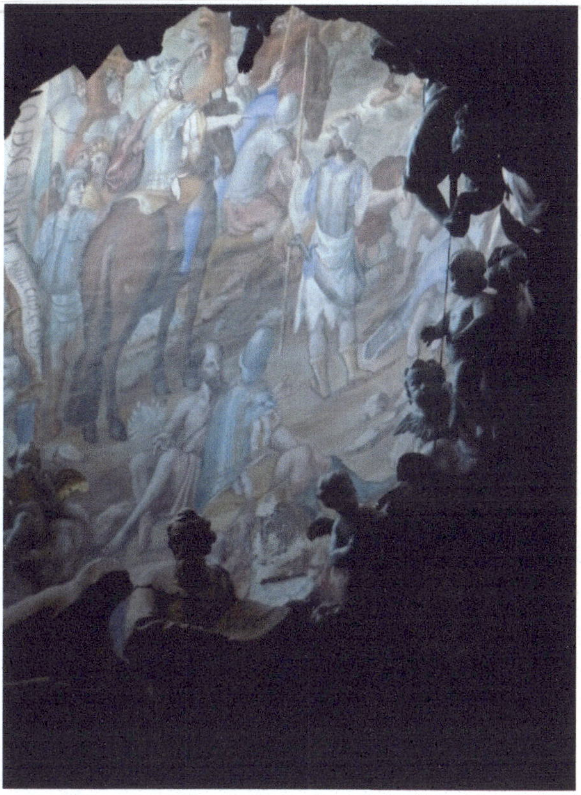

Fig. 1.5 Cathedral art, Toledo Cathedral. *El Transparente* window, Narciso Tomé, 1721–1732. (Photo Iain Davidson)

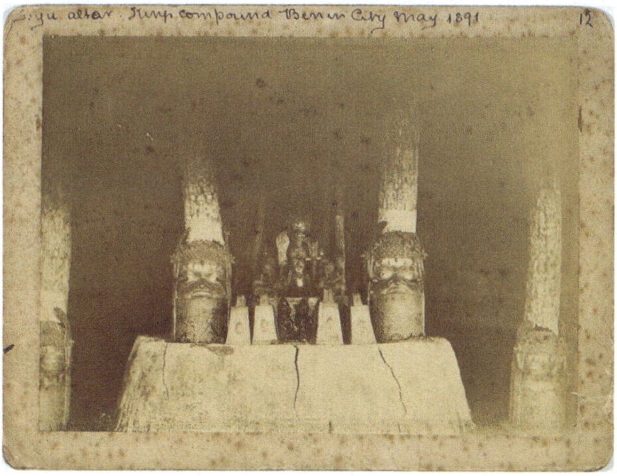

Fig. 1.6 Earliest known photograph of Benin bronzes on display in 1891, in an ancestral shrine. This picture was taken 6 years before the sacking and looting of Benin City by the British Navy. (Photo by Cyril Punch, published in Roth, 1968 (1903)). Wikimedia Commons, Public Domain

new continent and graze them for their own profit. He was killed by Aboriginal people in circumstances that we need not examine closely to realize that his death arose from a conflict over use of the land and its Indigenous people. Presumably similar conflicts had occurred for similar reasons in the Highland Clearances in the ancestor's native Scotland. That assessment of the causes results from my engagement in research into the interactions between the Aboriginal people of Queensland and the Europeans who now have power there, as a member of the research team into the Native Mounted Police (Barker et al., 2020; Burke et al., 2022; Davidson et al., 2020). Figure 1.7 shows ruins of one set of Police Barracks established near Boulia in NW Queensland.

Bill was being helpful when he gave me and my colleagues advice, and he could see that the paintings were made by the Aboriginal people on whose land his family now lived and made money. I do not know, but I have more than an inkling, that his attitudes to the Aboriginal owners of the land owed something to the history of his forebears who came to Australia from nineteenth century Europe. The conflict between Bill's ancestors and the Aboriginal owners of the land, and the particular attitudes Bill's family derived from it, are not the main point of this story. It is relevant, though, that people derive their attitudes to art from their own personal education and experience—Bill's story shows how that can happen. Other stories and chance remarks that I shall introduce are part of a common tradition that talks about some images as art and others as scribbles.

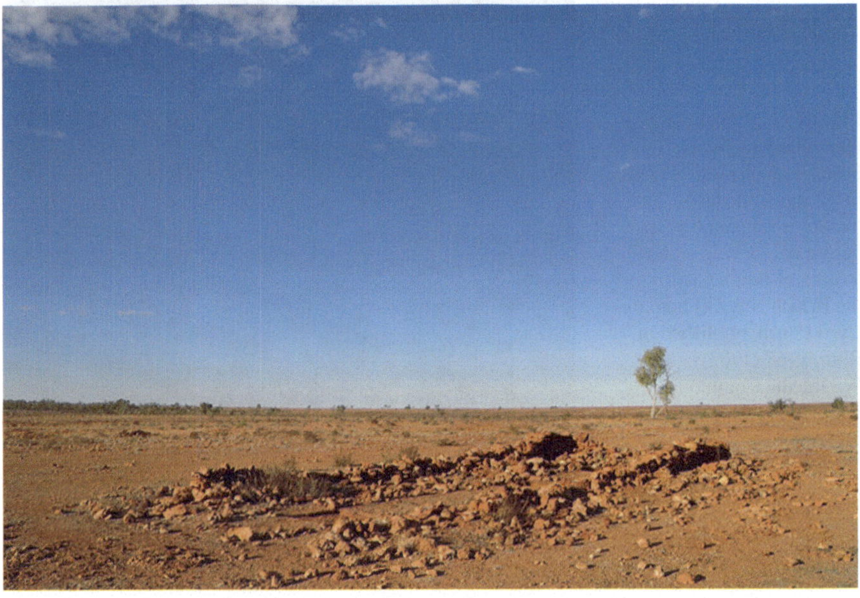

Fig. 1.7 Ruins of Native Mounted Police Barracks, Boulia, NW Queensland. (Photo Iain Davidson)

Art seems to be one of those things that provokes different attitudes from different observers. For example, the Australian painter Brett Whiteley remarked, writing to his Australian inspiration, Lloyd Rees (see Fig. 5.22), as he lay dying:

> … these little pictures signed in the corner L Rees, seemed to have filtered and sometimes dredged the whole of Europe through them. They contained nature AND ideas, they contained naturalism but seemed also very invented, and the adventure of them was that they showed the decisions and revisions that had been made while they had been painted.

What does this tell us about art in any of its manifestations? Most importantly: mark-making by itself is not all that art is. The object that is called art includes a story about what the producer was doing when they produced it. That story depends, in turn, on the producer's personal experiences—their personal history—at the time of making the art. Whether the object is viewed as art by a seer depends on that seer's personal history at the time they in turn see it. The seers might, for example, be professional art critics, or members of a culture with completely different values, or tourists. Each work of art is different for different artists and for different seers, as it was "dredged … through" the experience of artist and seers.

In the example of Picasso's use of masks for some of his most famous paintings, he could be said to have dredged the pictures through his experience of the archeological finds from his home country, and through the ethnographic museums of Paris. Likewise, it would appear that Mark Rothko's art was dredged through his experience of visiting the archeological site of Pompeii and seeing its 2000-year-old paintings (Fig. 1.8). The story is recounted in many places (Harris, 2015, 26–31; Street, 2021, 13–16).

There is a well-known picture of children looking at their phones while sitting on an art museum bench (Fig. 1.9). In the photo the painting *The Night Watch* by Rembrandt van Rijn can be seen on the wall behind the children. The photo seems to be a story about how children nowadays are obsessed with what is on the screen of an electronic device—whether that be phone, tablet, or computer—at the expense of the beauty around them. That is an interesting story to be explored about the power of one sort of visual message rather than another. I am very grateful to Gisjbert van der Wal both for his permission to use his great photo and for his free

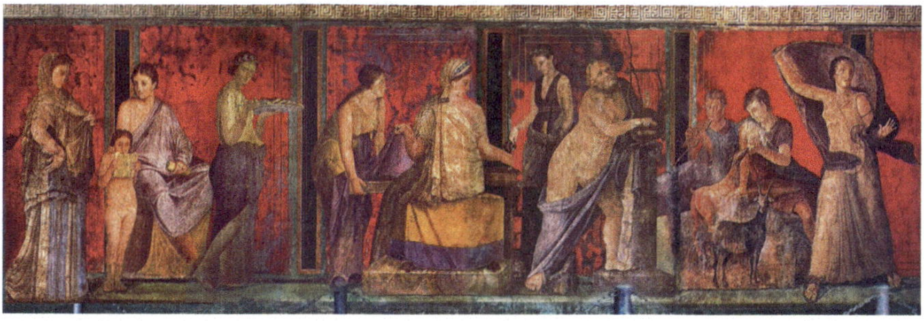

Fig. 1.8 Villa dei Misteri #1, Pompeii, Italia (2009) photographed by Wolfgang Rieger. It is thought to depict the initiation of a woman into a cult devoted to Bacchus, the god of nature, fertility, and wine. (Public Domain, Wikimedia Commons)

sharing of this story about the work the children were doing. He has published the story at the Flickr link given in the caption.

In the context of the argument in this book, the photo is interesting for another reason. The students were not doing the wrong thing but, rather, finding the answers to questions about the meaning of the painting they had been brought to see, questions that had been set by the Museum application they had downloaded onto their phones. You can see that the children are listening to something on their phones, not just reading text messages. The children would have read in the app (which I downloaded on 4 August 2022 through PlayStore) that the painting of *The Night Watch* was Rembrandt's "largest and most famous painting … made for one of the three headquarters of Amsterdam's civic guard" and that "The nickname Night Watch originated much later, when the painting was thought to represent a nocturnal scene."

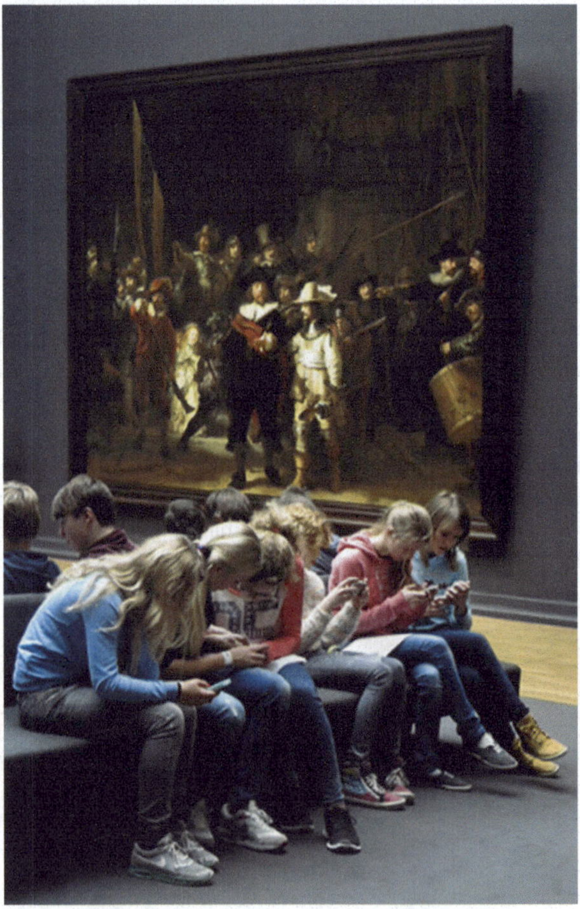

Fig. 1.9 Children reading about the interpretation of Rembrandt's *The Night Watch* in the Rijksmusem in Amsterdam. (Photo Gijsbert van der Wal with permission. https://www.flickr.com/photos/gijsvanderwal/15893868835)

In addition to the artist, teachers and museum curators have their own sets of ideas about what students should be able to say about a painting. In this way the elders channel the children's understanding of meaning through the app, and ultimately the attitudes of all of us. At issue, then, is the nature of the art as the artist produced it and spoke about it, but also the nature of the way individuals are educated into understanding it. In Whiteley's view, the way Rees painted could be considered to reflect what he learned from the art of "the whole of Europe." Whiteley, too, had travelled Europe and seen much of the art that Rees had seen, so he could understand the "dredging." Whiteley himself studied the art of Vincent Van Gogh to get insight into, perhaps, the way an artist works with a troubled mind. He died of an overdose in 1992. He also commented that Van Gogh invented "dot-dash drawing. The dot ● came from Seurat and the dash ■■■■■ from Pisarro and Impressionism" (Whiteley, 1983).

In the context of chance remarks that had an impact on my thinking, one reviewer of an earlier paper commented: "It may be clarifying to introduce a brief reflection … adding to the bibliography some texts referring to the creation of the concept of art (e.g., Shiner 2003)." Shiner (2001, xv) was admitting being puzzled by his youthful visit to both the Art Institute in Chicago and the Field Museum in the same city. One had famous paintings, the other had "endless cases full of African, Oceanic and Native American artifacts." When, later, he went to college in the same city, some of the African masks had "miraculously migrated" from the Field Museum to the Art Institute. Guided by an anthropologist pointing out that the African masks were used in ritual and that there was no category of art in the languages of their makers, Shiner found the concept of "art" problematic.

Shiner discovered a clue to his own attitude to art in the work of Paul Kristeller (1951) who acknowledged that the idea goes back even further, at least to Robin Collingwood (1938, 5–7), a scholar who had experience in archeological art. Kristeller wrote about the history of classical authors who discussed something like arts or aesthetics but in ways that seem quite alien to the general understanding of art in the twenty-first century and in fields other than European classical or mediaeval discourse. Kristeller (1951) joined in the argument that modern attitudes to aesthetics emerged in Europe of the eighteenth century at the same time as "the term 'Art,' with a capital A and in its modern sense, the related term 'Fine Arts' (Beaux Arts)." As he said, these attitudes involved "taste and sentiment, genius, originality and creative imagination." I would like to know how the performance of ritual involving such objects (as, for example, in the Benin Bronzes) apparently negates these attributes?

For what it is worth, Rembrandt also died before the eighteenth century when Kristeller (1951) and others would have it that art was "invented." By that definition, the painting *The Night Watch*, whatever the social context of its production, was not art—something that is not immediately obvious. The art historian John-Paul Stonard (2021, 288), for example, compared Rembrandt to the earlier Italian artist Titian (Tiziano Vecelli) saying: "Rembrandt … sought to rival Titian's rough brilliance of style, and succeeded, even surpassing the great Venetian painter in the range of effects of light and atmosphere he could emulate in paint." The eighteenth century will prove not to be about the invention of art so much as about the expansion of the market in art.

Critics make things into art. The point here is that it is people of all sorts, not just art historians, who decide whether something is or is not art. To be sure, there are people, for example those who write as academics, or write books or for newspapers, who set themselves up not as artists but as professional commentators on art. They have every right to be called critics. Some other people may not earn their living writing criticism of what they see in order to earn the title, but they are being critics when they make the judgment. Every person who says "I don't know whether it is art, but I know what I like" is being a critic. When art historians decide that something is art rather than craft they are making such a judgment. Oscar Moro Abadía (2006) has shown that the pervasiveness of the distinction in the nineteenth century was critical in the declaration (at that time) that cave art was not possible because people who were labelled "primitive" were deemed to practice craft, not art. Archeologists have been a little meek on this point: we do not need the approval of professional art critics to tell us that what we study is art. As a critic myself and as an archeologist, I can tell you, first hand, that some cave art is up there with the best (Fig. 1.2).

But what is also inescapable is that we can argue from the contexts of such cave art that ritual was involved in its making and in people seeing it (for example, June Ross and I showed that the art of Central Australia meets these criteria in Ross & Davidson, 2006). Does that exclude it from being art? Shiner was right to be puzzled by the distinction and the elision of meanings. When talking about the contemporary Aboriginal art of the Western Desert of Australia, Fred Myers (2002, 1) wrote about "the processes through which the paintings have entered the international art market as a result of the imaginings of diverse people." The imaginings of diverse people are necessary for things to be regarded as art. To be regarded as Fine Art, perhaps, the art market is necessary. Indeed, Shiner (2001, 75) contrasted seventeenth century lack of libraries, concerts and museums with the situation in mid eighteenth century England where all of these had "appeared across Europe along with the rise of a distinct market and public for the Fine Arts and the new concepts of Fine Art, artist, and the aesthetic." The issue is not so much whether art came into existence in the eighteenth century so much as whether the social circumstances existed in which it could achieve a value that would grow and be available to more people. The issue of the history of art, therefore, can either be about the social relations of production—the replacement of patronage by the market and museum—or it can be about whether the artists and producers had the skill to imagine how to represent what they saw or what they thought. How people came to imagine anything is an issue in human evolution. One of the mysteries that I will address in this book is how, relatively early in their evolution, some of our ancestors came to produce things that can be singled out as art.

We are going to look at the way archeological art emerged and changed among people whose ancestors had had no concept or practice that resembled either art or Fine Arts. The relevance of Kristeller and Shiner is that they were concerned primarily with the distinction between Art and "Fine Arts" and how it emerged in the eighteenth century. I am saying that scholars and others were misled in the eighteenth century into thinking that the important thing was "Fine Art," because it is absurd to exclude artists from before that time, such as Leonardo and Michelangelo

Buonarotti, for example, or Albrecht Dürer, Jan Van Eyck, Johannes Vermeer, Rembrandt, and Velázquez from the category of Art, whether Fine or not. We can understand both art and scribbles better if we acknowledge that misdirection and instead seek to understand what it involved. Then the question is how far back and how far away we choose to include works of art. The Art Insitute of Chicago, for example, had an exhibition of Ancient Roman Art (https://www.artic.edu/highlights/19/ancient-roman-art). Many things might make the journey from Field Museum to Art Insitute if we did not decide on one or other location because of knowledge of where or when they were produced and the social context of the production. It is a separate question how such things might enter the art market as a result of the imaginings of people.

Shiner (2001, xvi) recognised that:

> One of the lessons I learned in my struggle to understand what happens when African or Native American ritual objects are transformed into art in the modern sense is that institutions such as the museum, the concert hall, and the literature curriculum have been central to constituting things as art. I soon came to the conclusion that we could no longer do either aesthetics or the history and criticism of music, art, and literature in traditional ways. I became something of a pest, constantly taking my fellow academics to task for not prefacing their eager praise of African or Native American or Chinese "art" with some sort of disclaimer. These colleagues were understandably irritated since they sincerely believed they were engaged in a worthy, even complimentary, pursuit and wanted to get on with their aesthetic appreciation and analysis. Although most of them were prepared to admit that the meaning of "art" might differ from culture to culture or that it had changed some in our own culture since the eighteenth century, they didn't view cultural and historical differences as having significant consequences in the way I did.

How did art emerge from the relationships between people and the things they produce and the ideas of producers? How do people manipulate those relationships in their society? If we were not part of the society, let us say it was African, Native American or Chinese (and I apologize if you do belong to one of these groups), how do we outsiders work out the ways in which those relationships were used to make history? Is it acceptable to look at what they did in their society with our cultural expectations? What if the question was the other way around, and we were African, or Native American or Chinese, how would we work out how art was made in the west? You may immediately conclude that your culture and my culture, your art and my art are quite different from each other. That makes it difficult to be dogmatic about what we are talking about.

I will suggest that "cultural and historical differences" have a completely different set of "significant consequences" from those Shiner imagined he identified. From its very beginning, and everywhere, art has always been embedded in its cultural context—the contexts of the people who made it as well as those of people who see it. Morphy (1998, 24), writing about Australian Aboriginal art, spoke about the interaction between "the motivations of the artist" and the "expectations of the

European audience" throughout his book. It is the "museum, concert hall and literature curriculum" and other places with curriculum, such as art schools and degrees, that engender their particular, culturally specific, attitudes to aesthetics that are exceptional as a context for art. Their attitudes do not represent the way most art in most places and most times has been produced. The usual cultural contexts are ritual, purpose or utility. Even when the art is displayed in the museum, performed in the concert hall, or studied in the literature curriculum, ritual, purpose and utility are intrinsic to those contexts. People who see paintings or sculptures in a museum are learning the cultural context of consuming the art: as I have described in this chapter, visitors to the museum are acknowledging the ritual of museum visitation (Duncan, 1995, 2005)—such as hushed voices, special times and places, and other features familiar from the generalizations about ritual described by Roy Rappaport (1999). In addition, listeners in the concert-hall observe the accepted practices for the particular type of concert—such as silence rather than applause at the end of the movement of a symphony, throwing flowers at opera, demands for encores at the end of chamber music, the rituals of the mosh-pit in popular music. They have not always done so (Shiner, 2001, 130–132). Students of the literature curriculum may be an exception because reading is an individual activity but may still involve quasi-ritual practices such as using a particular chair or reading at a particular time. It is possible to consider the use of newspapers as objects of habitual behavior at particular times of day, or of books read by reading groups as examples of ritualized behavior (though many of the rituals do not involve reading). Alberto Manguel (1997, 110–114) described the publication of a newspaper *Le Aurora* in 1860s Cuba that described the political conditions under which cigar-makers were working and exposed them to great writers at the same time. As "barely 15 per cent" of these workers could read the paper, the publisher, Martínez, decided to select individuals who could read, and the tobacco workers were happy to pay for the selected ones to read the newspaper aloud as they worked. For that situation, Manguel refers to the "reading ritual," and, given that it was a communal experience, it seems likely that "ritual" was the right description. It seems difficult to show that western "Fine" Art is devoid of ritual—except perhaps where the experience is individual, as in reading.

The history of art seemed to confuse "social and institutional changes" of view when western critics (or artists) wanted to "turn ritual works into 'art'." They seemed to define the boundaries of art so tightly that ritual, purposeful or utilitarian works in non-western society were excluded from it. In reality, gatekeepers of western art were substituting the ritual, purpose or utility of the original producers for others within their own cultures. In their definition, art only existed if the rituals of the original producers were stripped from the works or if the objects were appropriated away from the situation in which they were produced and valued for their aesthetics. But the process merely replaced the conventions of the producers with those of the society that appropriated the works. The rituals that attended the production and display of the Benin Bronzes were replaced with the rituals of observing the same bronzes as art in the British Museum.

Lots of people begin their books by saying they are about art, but they are all slightly different from each other. Sometimes the books are about the definition of art (Collingwood, 1938), or the way philosophers think about it (Davies, 2006).

Sometimes they are about specific examples of art (Clottes, 2001) or about the history of things that people call art (Gombrich, 1995). Some books are about the place of art in the evolution of human behavior (Sachs-Hombach & Schirra, 2013) or the role played by art in historical events (Procter, 2020). Many books seek to justify the use of the word "art" to describe archeological art (Bradley, 2009), many more just assume that we know what we mean by the word and proceed almost without discussion (Dissanayake, 1992).

In this book, you will read about what different people at different times have thought about art. I will discuss some important elements of art as we know it today and how those elements may have first occurred in the past as humans evolved. I have been concerned about the presence of scenes in art (Davidson & Nowell, 2021) and how among early examples of art sometimes there were scenes, sometimes not. Elisabeth Culley (2021) explored the different contexts in which different types of scenes were prevalent. I will argue that scenes were essential for the emergence of art—people and objects—as a category that could be understood separately from ritual and other contexts, whether or not they were produced with ritual. For others, the most important questions about art have been about how different sorts of art have different functions in society (Robb, 2017). What leaps to my attention from these (and other) examples is that an object is not art by itself, but that art consists of relationships between people, things, and ideas in quite specific ways. We will discuss those relationships at length in this book.

There are two elements of those relationships which are intertwined but can be considered separately: they are culture (Davidson, 2016) and stories (Minc, 1986; Nowell, 2023). Both, of course, are complex ideas and, as with art, there is a common understanding which makes it seem almost unnecessary to talk about the possible variations of meaning. Both also have quite specific roles in the story we are going to unravel here. We need to think of culture coming to have that name through the combined attitudes of people in societies. In different cultures people produce works they are content to call art. Part of the way they reach such contentment is through the stories they tell others about the works they call art. Art is a product of cultural context, but it also creates cultural context.

The conclusion of this work is that art is a part of the way in which people relate to each other within cultures and tell stories that are important to them about things they can see or think about. Art and stories together set out (and reflect) the elements of the cultural values people share with others. In these relationships people produce art—the relationships between the producers and their material product are the art—and there is a contemporary audience of people I call seers. Some of the material survives for later people to see—people who are not part of the relationships, are ignorant of the stories and are outside the culture. All of this means that the relationships which define the art can be used by members of a society to relate to others and to the circumstances in which they live. In addition, over time, people who saw the art—the things and the relationships—used it as part of a contribution to the way history happened over thousands of years. Over time, indeed, the stories changed and so did the symbolism involved in the art and the stories. In the end, the pictures inspire thousands of words to be written about them—they do not replace the words. This is their story.

1.1 Essay 1 Art or Scribbles? (Las Meninas by Velázquez and by Picasso)

I saw the Velázquez and the Picasso paintings called *Las Meninas* in my two favorite museums in Spain. The original (Fig. 1.10 Left) is in the Prado Museum which is a few hundred meters north of Atocha, then the main train station in Madrid. It was there that mobile phones set off terrorist bombs in 2004. In the 1970s when I was a student in Iberia, I sometimes dropped a friend at the station, drove up to the Prado Museum, found an easy car park, went into the Museum, paid homage to one or other of my favorite painters and paintings, which were mostly oil paintings on canvas, and was out of the other side of Madrid within an hour. The Velázquez was kept in a darkened room with light from the window on the right-hand side of the painting, as in the painting, and with far too many of us crowded into the room to gaze on the masterpiece. The absence of Velázquez from Clark's BBC television synthesis on *Civilisation* (Clark, 1969) offended my Spanish friends in the 1970s as I will describe in Chap. 3. Michel Foucault (2002, 3–16) wrote descriptively about the painting throughout the first chapter of his book, and was certain "without any doubt" (p.9) that the figures seen in the mirror at the back of the room represent the King and Queen, who were the subject of the painting.

There are, straightforwardly, four elements in any work of art by any definition. No single painting more characterizes the four elements I have defined for any work

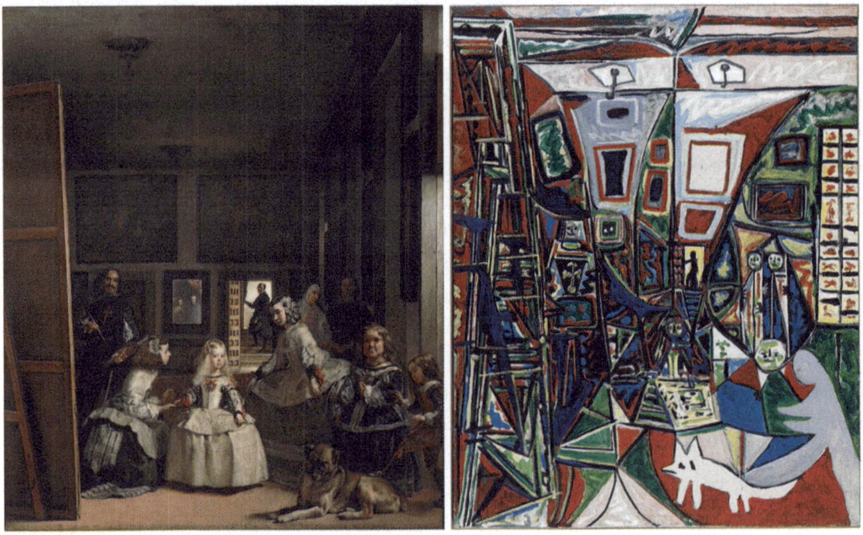

Fig. 1.10 Left: Diego Velázquez (1599–1660). *Las Meninas* in El Prado, Madrid (in the public domain from Wikimedia Commons). Right: Pablo Picasso. *Las Meninas*. Cannes, 2nd October, 1957. Oil on canvas 161 × 129 cm. Museu Picasso, Barcelona. Gift of Pablo Picasso 1968 MPB 70.465 Photo Fotogasull. © *Succession Picasso/Copyright Agency, 2025*

of art. There is (1) a person who makes marks as a painter (or surfaces as a sculptor)—the producer, who in the fullness of time will be an artist. In this case Velázquez, painter to the seventeenth century court of the Spanish royal family of Philip IV, painted the canvas, but "Velázquez" (in an early example of a selfie) is also probably the person shown producing the painting on the easel in his canvas. There is also (2) a person who observes that process—the observers of the canvas, including you and me after the event, and in this case the Meninas and others who were observing what "Velázquez" was producing on the easel. Then, there is (3) the mark itself—the product, which nowadays some would say is art—and for the canvas shown in this painting the product is the picture of an artist producing a painting on an easel. We have little idea what that painting is. In fact, (4) the subject of the painting is the artist doing that painting with Las Meninas and others in his studio. What we see in the Velázquez canvas is a work of art by Velázquez, the artist, of a painter, "Velázquez," producing a picture. Velázquez the artist is invisible—though it is generally thought that he was both producing the picture we are looking at and representing himself as a painter in a self-portrait.

The picture we see, by Velázquez the artist, represents the infant princess, and she is undoubtedly one of the subjects of the painting we see. We may or may not know the subject that the painter is painting—but it was probably not the princess. There is a reflection in a mirror at the back of the room, and it appears to represent Philip IV and his wife, but we cannot be sure that the image in the mirror is of these two people looking at the act of producing or of the image of them on the canvas—the king and queen as the painted objects. It would have been cute to represent them both twice in the mirror, once of them as the subject of the painting in the scene, once of them as the object of the painting in the scene. Foucault (2002, 17), likewise, expected that the image in the mirror could be more complex.

We take it for granted that "art" is "of" something, so that the subject is a fourth element—that which the marks represent in some way, probably iconically—but as in this masterpiece we know Velázquez was painting himself ("Velázquez") producing a painting, that was the subject of his work. We do not necessarily know what subject "Velázquez" was painting, though some have speculated that his subject was the king and queen seen reflected in the mirror, at the back of the room. The recursivity of this painting exemplifies the point that art-as-we-know-it was more complex than "just" representation. To add another layer of complexity, some commentators add that the reflections in the mirror hark back to (were "dredged through") the Jan van Eyck *Arnolfini Portrait* of the fifteenth century (Fig. 7.27). That painting on oak, was, in turn, an early example of the use of oil paint to produce a portrait on a wooden panel (Gombrich, 1995, 236–9). In the van Eyck painting there is also a mirror in the back of the room which shows something. For what it is worth, the Velázquez painting was done in the cultural context of the court of the seventeenth century and undoubtedly contributes to the cultural context of most art critics, including, as it happens Clark (1969), Foucault (2002), Picasso and ourselves.

Picasso's 58 studies of *Las Meninas* are in the museum that celebrates his works in Barcelona (http://www.museupicasso.bcn.cat/en) which is a collection of mediaeval palaces in the mediaeval quarter of the city. In terms of the questions in this book, it would be possible for grazier Bill to look at the Picasso version of the Velázquez painting and argue that it is nothing more than scribbles (assuming that his judgment was based on the nature of the painting). Whenever I stopped in Barcelona, I visited the Picasso Museum. The important lesson I learned during those visits is that Picasso trained in the skills of an artist from his earliest years in school, and then at night classes in Barcelona. As we all know from his early paintings on display in the Museum, he was capable of producing works in the style of many of the greats before he set out on his experiments with different ways of seeing and representing things. The Picasso here (Fig. 1.10 Right) seems to be that painted about 2 October 1957 (https://www.blogmuseupicassobcn.org/2015/08/the-chronology-of-lasmeninas-of-picasso/?lang=en/. Accessed 4.04 pm 24 October 2021).

Picasso said to his friend Jaime Sabartés (with grammar as it is in Wikipedia): (https://en.wikipedia.org/wiki/Las_Meninas_%28Picasso%29. Accessed 4.06 pm 24 October 2021)

> If someone want to copy Las Meninas, entirely in good faith, for example, upon reaching a certain point and if that one was me, I would say..what if you put them a little more to the right or left? I'll try to do it my way, forgetting about Velázquez. The test would surely bring me to modify or change the light because of having changed the position of a character. So, little by little, that would be a detestable Meninas for a traditional painter, but would be my Meninas.

By his emphases, Picasso made the background and the ceiling full of light and the light coming in from the right now has the appearance of a bank of windows (or stage lights). Among the human figures, Picasso only emphasized the artist and the princess in this painting. The other figures are less important, but the schematic silhouette at the door is as clear as in the original painting. But we are inclined to say that the Picasso image on the right, if it was not an example of scribbles, is certainly changed. Each seer can make their own judgment of whether it is detestable.

References

Adjei, S. K., & LeGall, Y. (Eds.). (2024). *Fifteen colonial thefts: a guide to looted African heritage in museums*. Pluto Press.

Arias, P. (2009). Rites in the dark? An evaluation of the current evidence for ritual areas at Magdalenian cave sites. *World Archaeol, 41*(2), 262–294. https://doi.org/10.1080/00438240902843964

Barker, B., Wallis, L. A., Burke, H., Cole, N., Lowe, K., Artym, U., et al. (2020). The archaeology of the 'Secret War': the material evidence of conflict on the Queensland frontier, 1849–1901. *QAR, 23*, 17–41. https://doi.org/10.25120/qar.23.2020.3720

Bradley, R. (2009). *Image and audience. Rethinking prehistoric art*. Oxford University Press.

Burke, H., Wallis, L. A., Davidson, I., Cole, N., Barker, B., & Hatte, E. (2022). The shape of absence: community archaeology and the heritage of the Queensland native mounted police, Australia. *J Community Archaeol Herit, 9*(2), 120–133.

Clark, K. (1969). *Civilisation*. John Murray, London.

Clottes, J. (Ed.). (2001). *La Grotte Chauvet. L'art des origines*. Seuil.
Collingwood, R. G. (1938). *The principles of Art*. Oxford University Press.
Culley, E. V. (2021). 12. A comparison of "scenes" in parietal and non-parietal Upper Paleolithic imagery: formal differences and ontological implications. In I. Davidson & A. Nowell (Eds.), *Making scenes: global perspectives on scenes in rock art*. Berghahn Books.
Davidson, I. (2016). Stone tools: evidence of something in between culture and cumulative culture? In M. N. Haidle, N. J. Conard, & M. Bolus (Eds.), *The nature of culture* (pp. 99–120). Springer.
Davidson, I., Burke, H., Sullivan, L., Wallis, L. A., Artym, U., & Barker, B. (2020). Cultural conflict in text and materiality: the impact of words and lead on the Northwest Queensland colonial frontier, Australia. *World Archaeol, 51*(5), 724–740. https://doi.org/10.1080/00438243.2020.1711153
Davidson, I., & Nowell, A. (Eds.). (2021). *Making scenes: global perspectives on scenes in rock art*. Berghahn Books.
Davies, S. (2006). *The philosophy of art*. Wiley Blackwell.
Dissanayake, E. (1992). *Homo Aestheticus. Where art coimes from and why*. University of Washington Press.
Duncan, C. (1995). The art museum as ritual. *Art Bull, 77*(1), 10.
Duncan, C. (2005). *Civilizing rituals: inside public art museums*. Routledge.
Foucault, M. (2002). *The order of things: an archaeology of the human sciences*. Routledge.
Gombrich, E. H. (1995). *The story of art* (16th ed.). Phaidon.
Harris, C. D. (2015). *Cinnabar: the symbolic, seductive, sublethal shade of Pompeii*. Brandeis University, Graduate School of Arts and Sciences.
Hicks, D. (2020). *The brutish museums: the Benin bronzes*. Pluto Press London.
Kristeller, P. O. (1951). The modern system of the arts: a study in the history of aesthetics part I. *J Hist Ideas*, 496–527.
Manguel, A. (1997). *A history of reading*. Flamingo Books.
Matthews, J. (2013). Seven spots and a squiggle: the prehistory of pictures. In K. Sachs-Hombach & J. R. J. Schirra (Eds.), *Origins of pictures. Anthropological discourses in image science* (pp. 311–352). Halem.
McDougall, R. (2008). Chapter 4. The making of *Great Benin*: Felix Norman Roth and Henry Ling Roth. In R. McDougall & I. Davidson (Eds.), *The Roth family, anthropology and colonial administration* (pp. 73–92). Left Coast Press.
Minc, L. D. (1986). Scarcity and survival: the role of oral tradition in mediating subsistence crises. *J Anthropol Archaeol, 5*, 39–113.
Moro Abadía, O. (2006). Arts, crafts and Paleolithic art. *J Soc Archaeol, 6*(1), 119–141.
Morphy, H. (1998). *Aboriginal art*. Phaidon Press.
Morris, D. (2013). *The artistic ape: three million years of art*. Red Lemon Press.
Myers, F. (2002). *Painting culture: the making of an aboriginal high art*. Duke University Press.
Nowell, A. (2023). Oral storytelling and knowledge transmission in upper paleolithic children and adolescents. *J Archaeol Method Theory, 30*(1), 9–31.
Peters, J., & Schmidt, K. (2004). Animals in the symbolic world of pre-pottery Neolithic Göbekli Tepe, South-Eastern Turkey: a preliminary assessment. *Anthropozoologica, 39*(1), 179–218.
Procter, A. (2020). *The Whole Picture: the colonial story of the art in our museums & why we need to talk about it*. Cassell.
Rappaport, R. A. (1999). *Ritual and religion in the making of humanity*. Cambridge University Press.
Robb, J. (2017). 'Art' in archaeology and anthropology: an overview of the concept. *Camb Archaeol J, 27*(4), 587–597.
Rollefson, G. O., Simmons, A. H., & Kafafi, Z. (1992). Neolithic cultures at 'Ain Ghazal, Jordan. *J Field Archaeol, 19*, 443.
Ross, J., & Davidson, I. (2006). Rock art and ritual: an archaeological analysis of rock art in arid Central Australia. *J Archaeol Method Theory, 13*(4), 305–341. https://doi.org/10.1007/s10816-006-9021-1

Roth, H. L. (1968 (1903)). *Great Benin: its customs, art and horrors*. Barnes and Noble.
Rubin, W. (Ed.). (1980). *Pablo Picasso: a retrospective*. The Museum of Modern Art.
Sachs-Hombach, K., & Schirra, J. R. J. (Eds.). (2013). *Origins of pictures. Anthropological discourses in image science*. Halem.
Shiner, L. (2001). *The invention of art: a cultural history*. The University of Chicago Press.
Stonard, J.-P. (2021). *Creation: a fully illustrated, panoramic world history of art from ancient civilisation to the present day*. Bloomsbury Publishing.
Street, B. (2021). *How to enjoy art: a guide for everyone*. Yale University Press.
Whiteley, B. (1983). *Another way of looking at Vincent Van Gogh, 1888–1889: paintings, drawings, photographs, silkscreen prints, etchings and sculptures created over a period of fifteen years, 1968–1983*. R. Griffin.

Chapter 2
Art Is Too Important to Be Left to Those Who Are Interested in Art

Abstract The chapter begins with an exposition of my credentials for engaging with it. I give a brief account of some of my experiences with the archeology of art in which I learned of the importance of the context of the art and its likely association with ritual. Recognizing that art is important for thinking, I try to show why it is too important to be left to those who "are interested in art." I further foreshadow how art changes the seers who see it and is changed by being seen. It is complicated by the fact that most of the art of the past is symbolic, but we will see that the important part of symbols is the people who see them. I end with brief words on surveys of art history and show how the processes of art in the past have generally been glossed over.

The story of the discovery of the origins of art in Europe has been told many times (very usefully by Bahn, 1998; David, 2017). Although there are many new discoveries, for our purposes there are three important facts: 1) bones engraved with images of animals were recovered from several sites in France by the mid-nineteenth century (Fig. 2.1), (2) many of them known from a synthesis of finds published by the French Edouard Lartet and his rich British banker friend Henry Christy (Lartet & Christy, 1875).

The existence of (3) cave art in Europe was claimed in 1880 when Marcelino Sanz de Sautuola announced that, he and his daughter Maria had found paintings of bison, a horse and a deer on the ceiling of the cave of Altamira in northern Spain (for a good account of this and the controversies surrounding it, read Lawson, 2012, 48–53, 253–261). The claim was disputed, dismissed by those who had not seen it, and embroiled in the controversy over the opposition between arts and crafts and the abilities of "primitive" people (Moro Abadía & González Morales, 2004a, b, 2005). The art in Altamira was recognized by Cartailhac and Breuil early in the twentieth century (Cartailhac, 1902; Cartailhac & Breuil, 1906) together with that in other caves known by then.

By the 1970s I knew that art of fisher-gatherer-hunters dating back more than 14 thousand years had been found in four caves in a small region of eastern

I. Davidson, *Art or Scribbles? In the Eye of the Beholder: The Evolutionary Emergence of Visual Communication*, Interdisciplinary Evolution Research 11,
https://doi.org/10.1007/978-3-031-90458-5_2

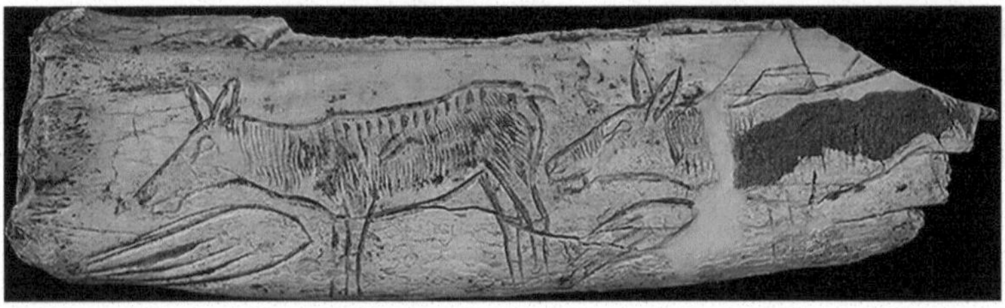

Fig. 2.1 Engraving on bone from the Upper Paleolithic site of Le Chaffaud, France, MAN No. 30361 (© MAN/Loïc Hamon in email from helene.bendejacq@culture.gouv.fr)

Spain—Parpalló (Gandía, País Valenciano) (originally reported by Breuil in 1914, as reported by Pericot 1942, 16, n.4), Volcán del Faro, Les Mallaetes and Cueva del Niño (Aýna, Albacete) (Fig. 2.2). I described these in my PhD most of which was published in 1989 (Davidson, 1989b).[1]

In my research at that time, I was trying to understand how human behavior produced the Late Paleolithic economy and my starting point was identification and analysis of the excavated animal bones that were principally remains of food. I was not much concerned with art, although these sites had produced examples of it, but I later decided the relationships entailed in the art were an important part of that economic story. This was the period of the Late Pleistocene which ended with the rise of sea-level and loss of coastal plains. Agriculture was introduced to Spain somewhat later from other countries with new plants and animals (e.g., Fernández et al., 2013; Tejedor-Rodríguez et al., 2021).

In my research in Iberia, I wanted to see whether there was evidence of earlier changes in behavior that might have been considered as indigenous precursors to agriculture using the plants and animals native to the Peninsula. It seemed to be unrelated to art, even though I knew of Parpalló as an art site. I was pleased to find that I could not identify any, and subsequently formed an argument that the nature of the management of resources in some regions depended on the way people used art to manage their relations with the plants and animals around them (Davidson, 2012a). Art and environment were closely tied together in the beliefs and practices of the people for whom they were fundamental.

The most remarkable of the sites I studied was the cave of Parpalló (Gandía, País Valenciano) (Pericot García, 1942), which, in addition to abundant animal bones, also contained a sequence of images of the same species of animals—red deer, spanish ibex, horses and cattle (Fig. 2.3) engraved or painted on small plaquettes of stone—that have now been dated between 27 thousand years ago and 13 thousand years ago (Davidson, 2012b). The paintings and engravings on the plaquettes have been described in an exhaustive study by Villaverde Bonilla (1994).

[1] Following the restoration of the official use of Valencian language, Les Mallaetes is now known as Malladetes (Barx, País Valenciano) (Villaverde et al., 2021), Volcán del Faro (Cullera, País Valenciano) is also known as Cova de l'Hort de Cortés (Aura Tortosa et al., 2020).

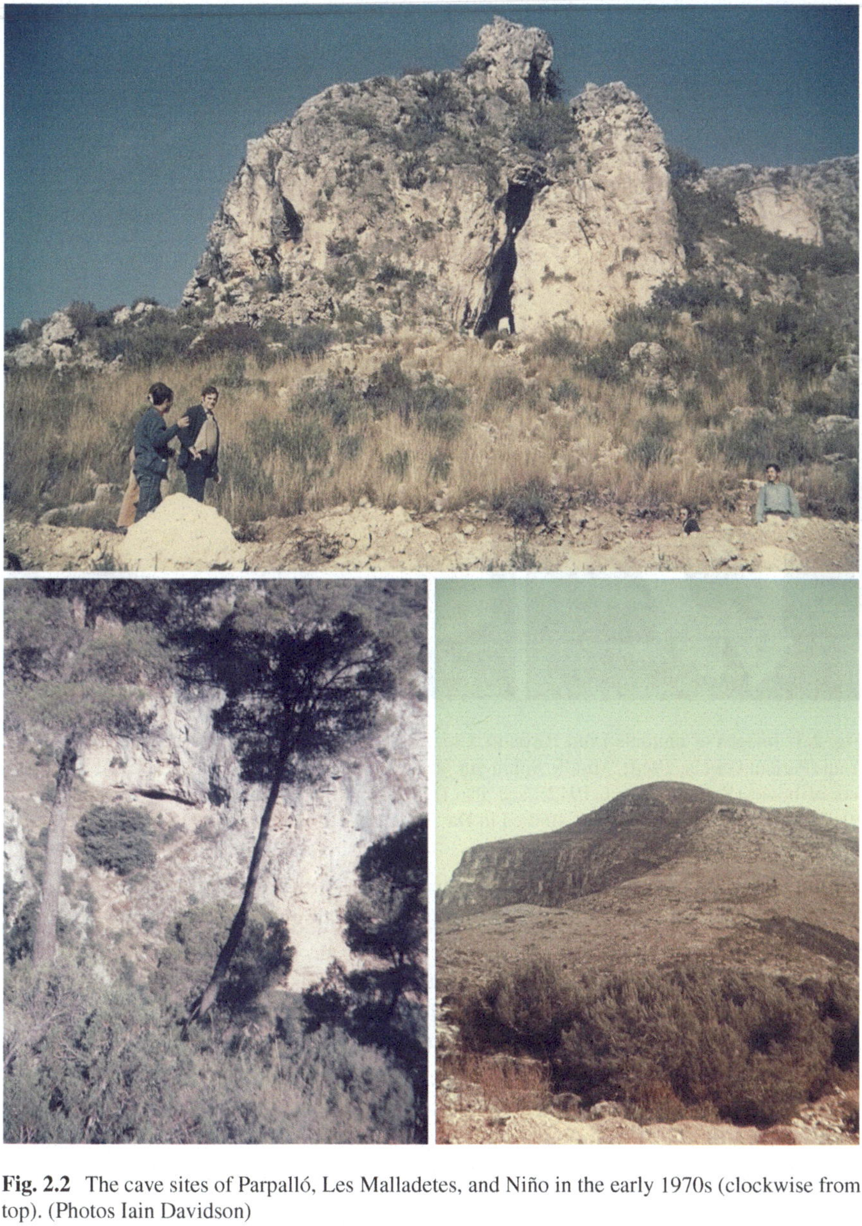

Fig. 2.2 The cave sites of Parpalló, Les Malladetes, and Niño in the early 1970s (clockwise from top). (Photos Iain Davidson)

My interpretation involved seasonal movement between suitably located places, together with fission and fusion of social groups that enabled them to manage the way in which people guaranteed food from the resources in their environment (Davidson, 1989b). As part of the contextualization of the sites in the landscape I defined likely behaviors involved when people moved between sites as they

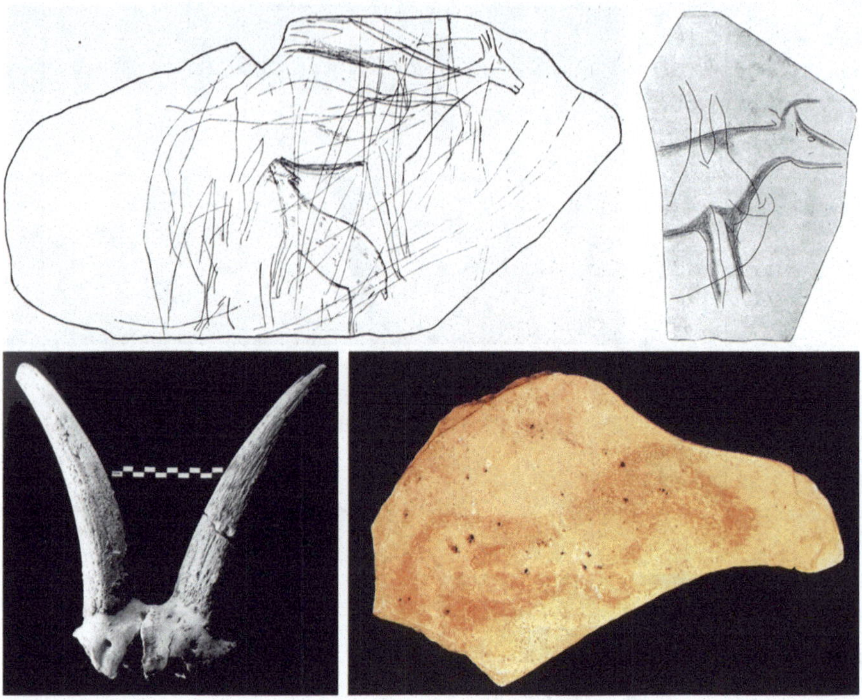

Fig. 2.3 Images of animals from Parpalló. Clockwise from top left: Engraving of suckling deer from (Pericot García, 1942), Middle Solutrean, see also (Villaverde Bonilla, 2021, 226); fore quarters of bovid (Pericot García, 1942), see also (Davidson, 2005); painted equid (Pericot García, 1942); horn cores of spanish ibex studied in Davidson (1989b). (Photo Iain Davidson)

exploited the environment. One important element of that argument emphasized the likelihood that there was a ritual concentrated at a single location, Parpalló, which involved the art produced on plaquettes (Davidson, 1989a). Together with other students, we excavated one of the small sites in the hills, the Cueva del Niño (Aýna, Albacete), where deer, horses, ibex, cattle (Fig. 2.19 Left), and possibly a snake were painted on the cave walls (Almagro Gorbea, 1971). The site was not included in my 1989 publication but is now published (García-Moreno, 2022; García-Moreno et al., 2021), including new analysis of the paintings, which include cave art inside the cave and later Levantine art in the rock shelter outside it (Garate Maidagan & García-Moreno, 2022; Mateo Saura, 2022). It is still the only site in Iberia with art of both styles.

I was excited to compare the distribution of different species of animals in the bones and in the art so that I could try to understand whether they both varied in response to changes in the environment or if people made choices that went beyond the simple extraction of food as it was available in the changing environment. The relationship between the art and the economic remains would be an indicator of the way in which people thought about their environment and economy and made appropriate choices. Without the art we would know much less about the thinking of those people and how it changed. By analyzing these materials, I showed how animals, art, human relationships, ritual, and material culture were all related so that archeologists were not simply analyzing sequences of stone and bone tools but that these were part of people's culture. This could be shown because there was a change 20 thousand years ago and I argued that it represented what I called "a real cultural change" (Davidson, 2012b, CD 1563).

By January 1981 I was convinced that I needed to know more about the relationships between art and meaning and that I would dedicate part of my research to that. To do so, I needed to know how art is used in modern societies, and I needed to begin that work by speaking to people who know about it. There was none more knowledgeable at the time than Patricia Vinnicombe who was an expert on the paintings of the Drakensberg in southern Africa (Fig. 2.4) (Vinnicombe, 1976), and on the engravings of the Burrup Peninsula near Dampier (now generally known by its original name of Murujuga) (Vinnicombe, 1976, 2002). I later decided that the sort of ethnoarcheology of art that I had thought I could do was actually impossible (Davidson, 1995) because art is about the use of symbols, and they have an arbitrary relationship with the things they refer to. The symbols are determined by the conventions of the people who made the art, and it is impossible for anyone to understand the nature and significance of the conventions if we are people from outside the system.

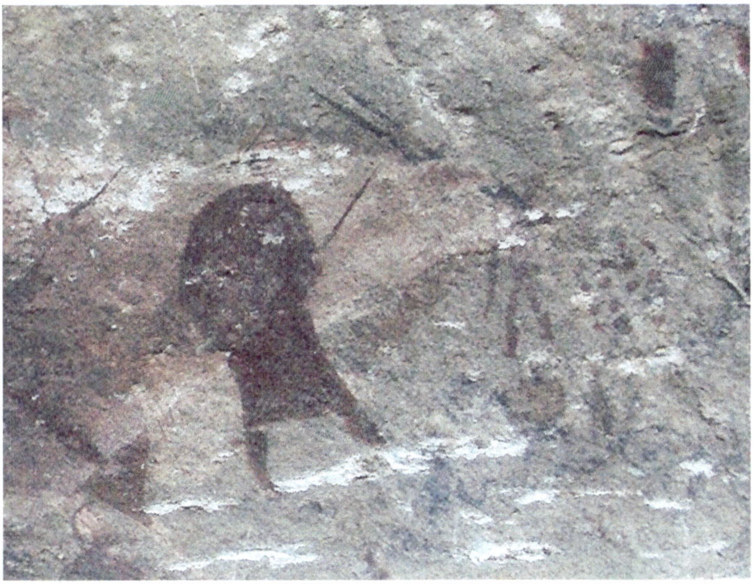

Fig. 2.4 Superimposed paintings from Main Caves, Giant's Castle, Maloti-Drakensberg National Park, KwaZulu Natal Province, South Africa. The site is mentioned in the book *People of the Eland* by Patricia Vinnicombe. (Photo Iain Davidson)

I showed the first draft of my conference paper to Vinnicombe, which included the thought that "rock and cave art is too important to be left to people who are interested in art." She advised me to leave that sentence out. I did. She may have been wise. I did not publish a version of the sentence until 2020, but it is an important point for this book to justify that remark. There are several things implied when I use that statement which can be exemplified by the rock art panel in Fig. 2.4—one that Vinnicombe included in her study. The painting shows, among other things, an eland facing right which has a grid of red dots near the end of its nose, and another figure, also facing right, which might be a person dressed in a caross—a cape of eland skin—and carrying a spear. There are other small figures, mostly facing left, in red (near the nose of the "eland") and black (lower down), and some seem to be images of people.

For "people who are interested in art," I venture to suggest, this is an image about eland, and perhaps about people carrying spears and others dancing. For the archeologist, the first task is to disentangle the different times at which artists painted images on the rock (e.g., Russell, 2000) and then, perhaps, to decipher what they were depicting. Vinnicombe had access to the manuscripts in which Orpen reported the testimony of Bushman Qing about the beliefs associated with the Drakensberg paintings (Vinnicombe, 1976, 176–7). The issue here is that there are various archeological interpretations of art associated with eland which run from Vinnicombe's (1976) "thinking strings" to Lewis-Williams's emphasis first on "medicine men" (Lewis-Williams et al., 1982) then on to "sorcerers" and eventually to "shamanism" (Lewis-Williams, 1985) that involves shamans bleeding from the nose in altered states of consciousness (Lewis-Williams, 2002). Are those red dots really representations of blood emerging from the nostrils of the eland? Only when archeologists have worked out what is represented, aided in this case by detailed historical accounts written down in the nineteenth century (Vinnicombe, 1976), can anyone begin to find ways to understand what the painters were thinking when they represented these things in this way, and why they painted them. Not all of archeologists' attempts reach a wider public, sometimes because they record meticulous detail.

In the words of the historian, Ann McGrath (2015, 2), "archaeologists have tended to publish their findings as scientific reportage around distinctive sites rather than as peopled, connected histories in a contextualised landscape." The statement is true, in that archeologists (myself included) have been unwilling to reach beyond the available data to write stories that the general public can consume—but it is not and has never been true that we have ignored "contextualized landscapes." The absence of archeologists' stories for the general public is part of the reason why the narrative of archeohistory is dominated by people who are not archeologists, such as physiologist/geographer Jared Diamond (1998), primatologist/evolutionary psychologist Robin Dunbar (2004), historian Yuval Harari (2014), historian Billy Griffiths (2018), genomicists such as David Reich (2018) and philosophers such as Kim Sterelny (2021). There are exceptions, though, such as Clive Gamble (2013), or Pat Shipman (2015), both also authors of several technical and other discursive works in archeology, and Thomas Higham (2021), an archeologist whose primary

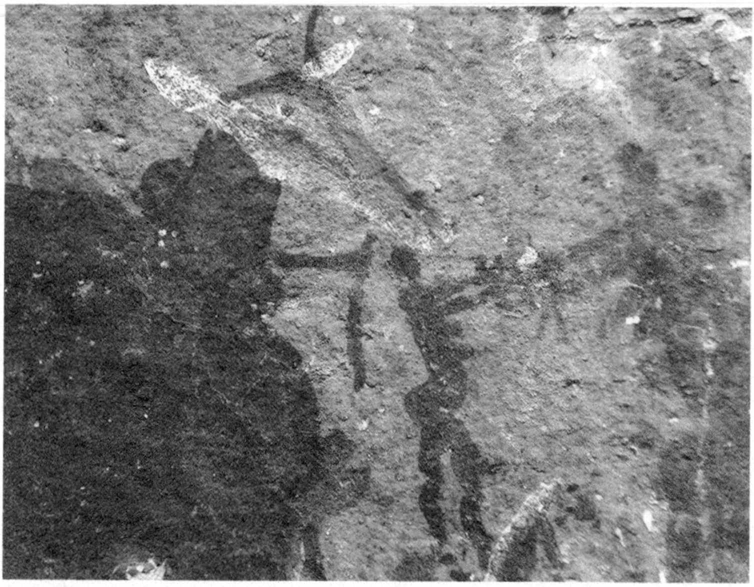

Fig. 2.5 Eland head from Main Caves, Giant's Castle, Maloti-Drakensberg National Park, KwaZulu Natal Province, South Africa. The site is mentioned in the book *People of the Eland* by Patricia Vinnicombe. (Photo Iain Davidson)

work has been with radiocarbon dating. It is also part of the reason why the texts by people who are interested in art are so cursory about early art.

Archeologists have not ignored contextualized landscapes in the Southern African rock art, because the archeohistory cannot be written without understanding the full set of contexts of the painting (e.g., Parkington & Rusch, 2003). In the case of the South African paintings and their implications, the narrative has been explored by Lewis-Williams (2002) including the possible implications beyond South Africa of important ritual leaders who can be called "shaman" (Clottes & Lewis-Williams, 1998). For modern archeologists to tell stories about the set of images is difficult without acknowledging the "scientific reportage" that archeologists must undertake to reach an interpretation. The narrative cannot be written by someone who does not take into account the processes of reaching those archeological conclusions—including, in this example, serious questioning of the appropriateness of the interpretation of shamanism (e.g., McCall, 2007; Solomon, 2013). The art is the source of a story about the archeohistory of past people provided we also consider the way in which the imagery was built up over time. Just because the image of the eland is beautifully produced, as are others in the caves (Fig. 2.5), the story is not necessarily best told by "people who are interested in art" indulging in the claim that the paintings have moved from the realm of rock art to art. Artists or art historians may say the processes are irrelevant, but when we analyze them, we understand the importance of the painters and the paintings in both space and time.

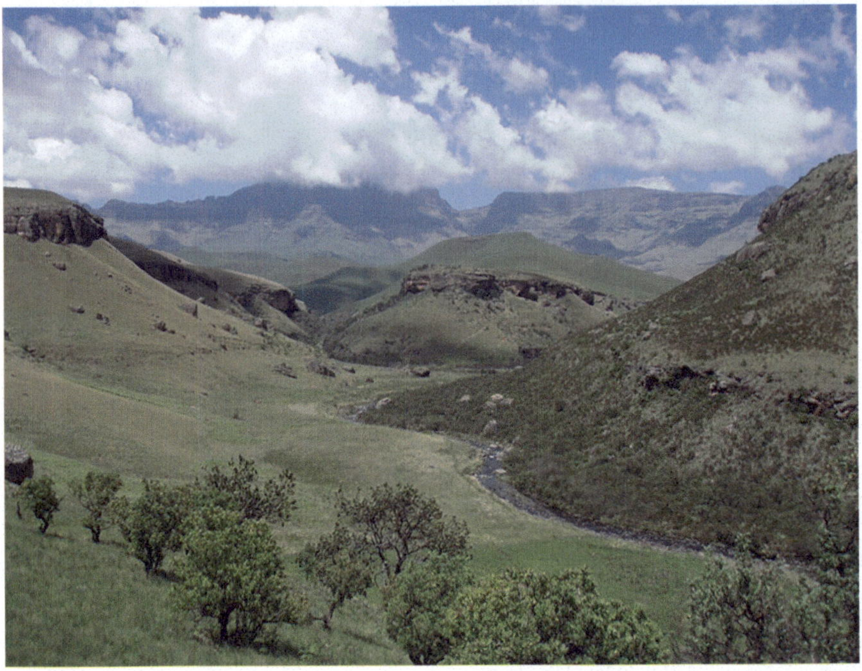

Fig. 2.6 Location of Main Cave, Giant's Castle, Ukhahlamba Drakensberg Park, North Cave (Centre). (Photo Iain Davidson)

Finally, the panel itself does not exist independently from the other panels in the cave (Figs. 2.4 and 2.5 are in the same cave), nor from the location of the individual site (Fig. 2.6) and its situation in relation to other sites in the region. As a general principle, one part of considering this wider context is to set the art into the framework of other examples of art. Elsewhere, my colleagues and I showed the multiple layers that need to be built up in setting the art into context by outlining a study in North West Central Queensland (including the art in the first Essay of this book) (Davidson et al., 2005). Only by considering the several layers is it possible to see that the paintings (such as the one in Fig. 2 Left) are part animal and part human.

The study of the archeology of art, therefore, includes the archeologist's "scientific reportage" before it becomes an object of study for "people who are interested in art." The interest derives from the scientific reportage, which, in the South African example, would include the whole corpus of paintings in the caves, the contexts of the caves relative to each other, as well as the ethnohistorical documentation available about those caves. Only through that process is it possible for "people who are interested in art" to avoid stereotypes about people they imagine were responsible for the art.

We will deal in detail with the question of what art is, and the importance of aesthetics for art, in the next chapter, so our task here is to determine which people are interested in art. What that really means here is that we need to look at some surveys of art history and the stereotypes they bring to three categories of art: first, the earliest art and its evolution, including rock and cave art; second, other archeological art; and third, art outside the collection (or canon) of western art. Then we will consider why rock and cave art are important.

2.1 People Who Are Interested in Art: How Do they See the Earliest Art and Its Evolution?

As an introduction to the question of stereotypes, we might consider the statement by the prominent art critic Robert Hughes (1991, 139) who commented on the Henri Matisse painting *Dance* 1909 (TV program available online at https://www.youtube.com/watch?v=nIhc8qsJXIA) that it was "one of the few entirely convincing images of ecstasy made in the 20th century." It is much copied and seared into the eyes of most people who have them open (Fig. 2.7: there is also a copy on the walls of Margaret Olley's *Yellow Room* in Fig. 3.6). Figure 2.7 is from the wall of the Restaurant Covadonga in Cienfuegos in Cuba. Leaving aside the differences between the original and the copy, the point for this book is that Hughes went on to say of the painting: "that circle of twisting stamping maenads takes you right back down the line to the red-figure vases of Greece and beyond them to the caves. It tries to be as old as dance itself."

In the same place, Hughes described Matisse's *Music* (Fig. 2.8) of the following year and the same atmosphere, as "music at its origins—enacted ... by five naked cavemen, pre-historical, almost pre-social." In this Hughes displayed his stereotypical bias about "cavemen" without any hint of knowledge of the evidence.t

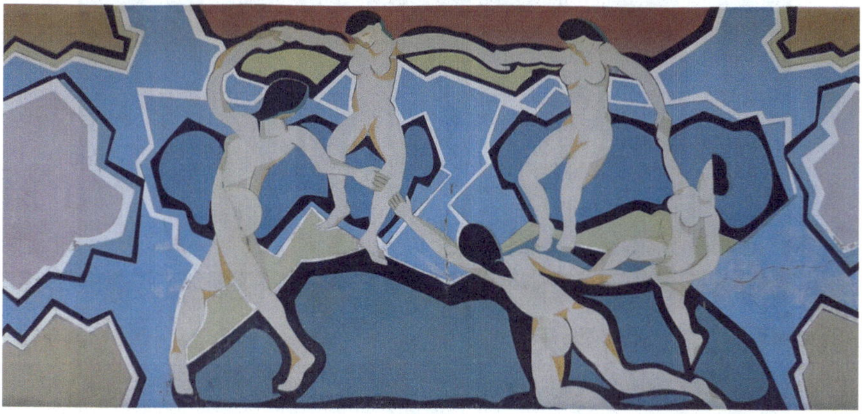

Fig. 2.7 *Dance* (1909) by Henri Matisse, reproduced with changed background, on the wall of the Restaurant Covadonga in Cienfuegos, Cuba. (Photo taken from a moving vehicle, May 2013 by Iain Davidson)

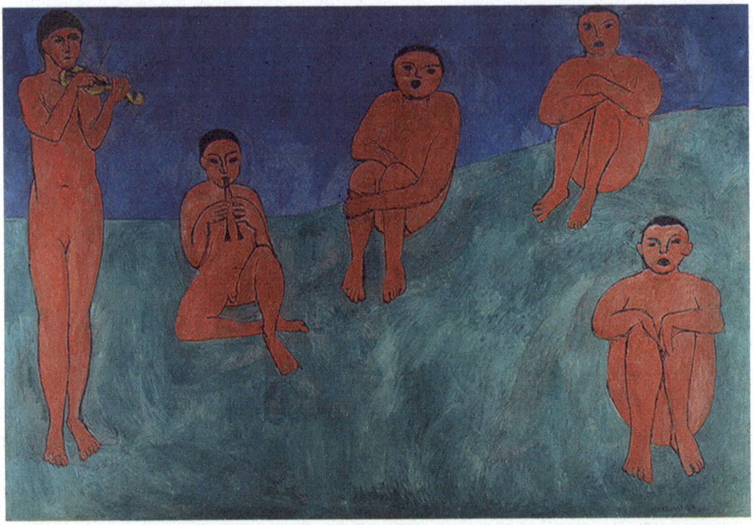

Fig. 2.8 *Music* by Henri Matisse (1910). (State Hermitage Museum, Saint Petersburg, Russia, PD-US. https://en.wikipedia.org/w/index.php?curid=26957974. Public Domain in the USA)

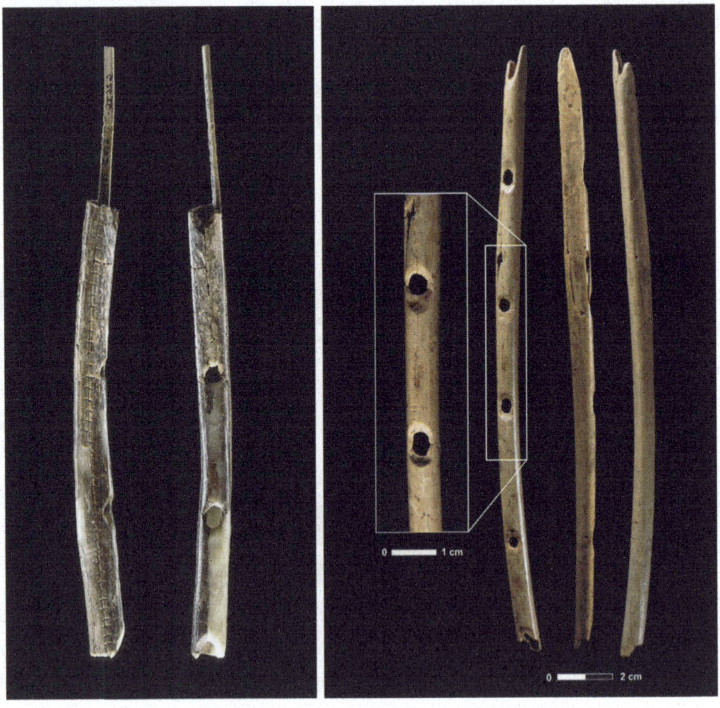

Fig. 2.9 Left Flute from Geissenklosterle, SW Germany, made on ivory (University of Tübingen, Photo: J. Lipták). Right Flute from Hohle Fels, SW Germany, made on the radius of a griffon vulture (University of Tübingen, Photo: M. Malina). They are more than 35 thousand years old. (Conard et al., 2009 with permission)

In fairness, in 1910, the paintings of Altamira had been acknowledged for only 8 years, and, until 2012, no one knew that one of them might be dated to about 36 thousand years ago (Pike et al., 2012). We are dealing, though, with a comment by an art critic—a person "who is interested in art"—in 1981 and by then it was well-known that there were hardly any images of people in the art of "the caves," neither dancing nor playing music, nor, in general, hunting. Now we know that, although there are finds of the earliest musical instruments from Germany dated more than 35 thousand years ago (Conard et al., 2009) (Fig. 2.9) as well as (much) later cave art with scenes of music-playing and dancing (Dubey-Pathak & Clottes, 2021), music and dance are almost completely absent from cave art in Europe (Davidson & Nowell, 2021b, 7).

In some other countries, the localization of scenes of music-making in some regions but their absence from regions nearby (Brusgaard & Akkermans, 2021, 137; Davidson & Nowell, 2021b, 7) reinforces the point that conventional choices, not "primitive" ones, were being made about what to represent on cave walls from among the many activities of human societies nearly 40 thousand years ago. The present book is concerned with the choices made in producing the oldest art as well as its relations with modern art.

Attacking the problem from the other end, art historian Whitney Davis (1993, 327–28) pointed out that many textbooks of art history began their first chapters with the same pictures from Altamira and the other well-known example of cave art at Lascaux as if these were enough to represent 30 thousand years of pictorial art. In the case of Altamira, the story about the enthusiastic child rushing out of the cave claiming she had seen bulls (David, 2017, 19) (Fig. 2.10) seems to have been

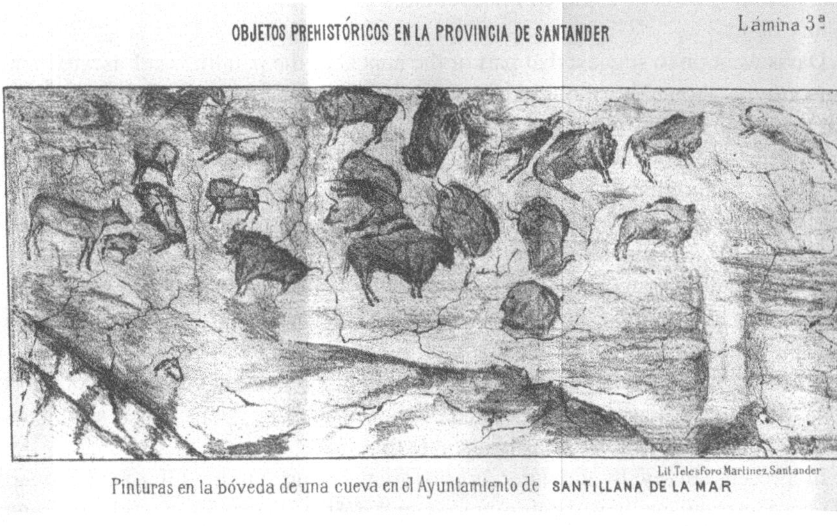

Fig. 2.10 Original sketch of the painted ceiling of Altamira published by Sautuola in 1880 (obtained from the Biblioteca Pública de Cantabria by Alejandro García-Moreno). This is, therefore, the earliest published image of cave art

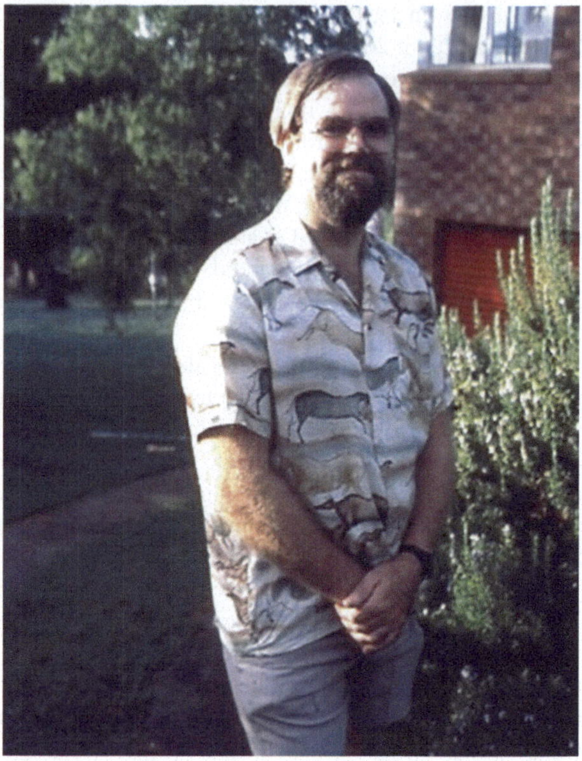

Fig. 2.11 The author in a shirt printed with images from Lascaux, 1983–4

exaggerated (Lawson, 2012, 51) but remains one of the foundation stories of the history of art.

Davis went on to suggest that part of the appeal of the paintings at Lascaux, now known to be about 21.5 thousand years old (Ducasse & Langlais, 2019) was first that there was a good story—another foundation story but which, in this case, was probably true—about how they were discovered by young children looking for their dog—and second that they can be seen, although not in the flickering light of limited range inside the cave, in reproductions in the same manner as "Egyptian tomb paintings, Greek pedimental reliefs, … or the Sistine Chapel ceiling." In all likelihood they can no longer be seen by people, though some of us have been lucky enough to see them in the site (Aujoulat, 2004). None of the art historians who have used the images of Altamira and Lascaux mention that 15 thousand years passed between the first image at Altamira and the bulk of the paintings at Lascaux (a taste of the paintings at Lascaux can be seen on the shirt in Fig. 2.11). It was a long and slow beginning and not necessarily a single tradition.

In a recent survey, by a painter, art historian and art curator John-Paul Stonard (2021), these images have been replaced in the first chapter by well-documented images from recent finds of archeological art. The earliest representational paintings then known (Fig. 2.12 which is of a painting 1.5 thousand years more recent than the older image used by Stonard) date from 45.5 thousand years ago and were

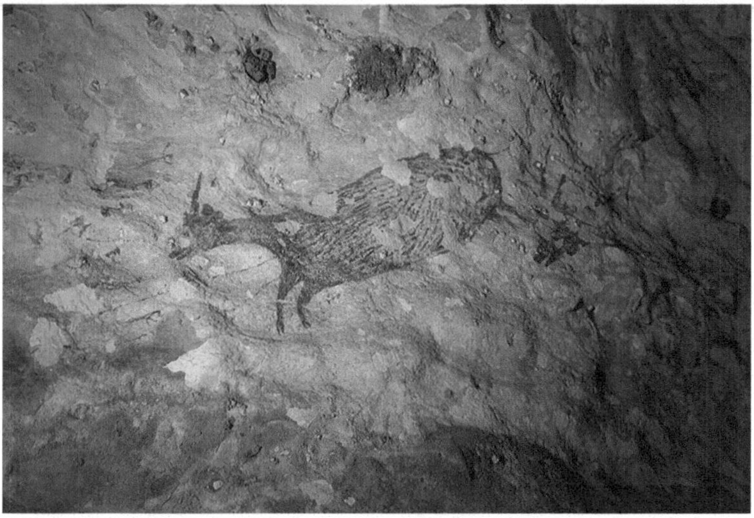

Fig. 2.12 Anoas and human-like figures, Leang Bulu' Sipong 4 (Aubert et al., 2019). (Photo credit Ratno Sardi, permission from Adam Brumm and Maxime Aubert)

found in Sulawesi (Brumm et al., 2021), but older paintings from Sulawesi have now been published (Oktaviana et al., 2024); the mammoth ivory Hohlenstein Stadel figure (Wynn et al., 2009) (Fig. 2.14) dated to approximately 39–41 thousand years ago (Kind et al., 2014); and the horse panel at Chauvet Cave dated between 38 and 33 thousand years ago (Quiles et al., 2016) (Fig. 1.2). It was a good idea to choose more recently discovered images, but some of the art in Sulawesi is 9 thousand years older than the oldest art at Altamira; the Hohlenstein Stadel figure is about 4 thousand years older; and the Chauvet Cave horses might be the same age as Altamira.

Stonard (2021, 4) went on to discuss the motivation for the carving of the Hohlenstein Stadel figure: "the finished carving may have had magical properties, or may even have been worshipped as a god—or perhaps it was carved simply to be admired as an image." Archeologists have run the gamut of "explanations" of the earliest rock and cave art from "art for art's sake" (see e.g., Culley & Davidson, 2021; Moro Abadía & González Morales, 2005), through totemism (discussed later in this Chapter) (e.g., Davidson, 2017), hunting magic (for an account of the way such interpretations have been superseded see Conkey, 1999) and expressions of sexuality (now comprehensively debunked by Nowell & Chang, 2014), to structural expression of the values of society with either two principals (Leroi-Gourhan, 1982) glibly compared with male and female, or four (Descola, 2010), and have found them all wanting. It is worth saying something about the once-fashionable accounts by André Leroi-Gourhan (Davidson, 2017) and more recently Philippe Descola (Davidson, 2017): anything can be divided into two or four parts. As a teenager my friends and I used to play a game "It is either this thing, or it is not". Assigning significance to such divisions is much more problematic. Their meaning, as Descola

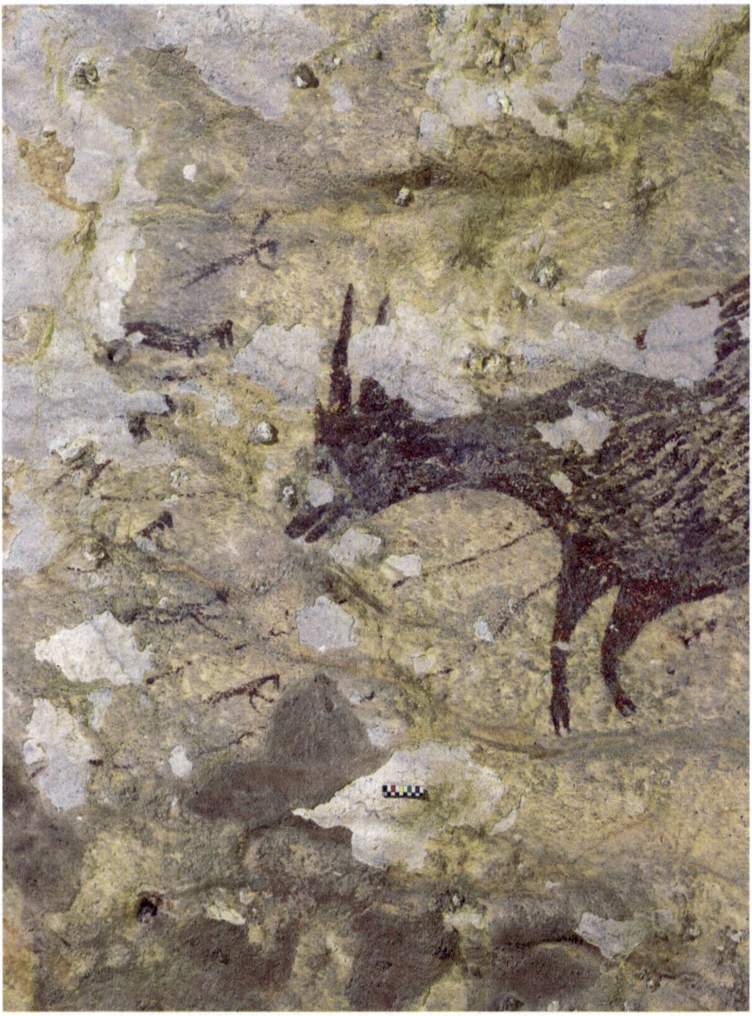

Fig. 2.13 Details of human-like figures to the left and in front of the anoa shown in Fig. 2.12. See comments by Adam Brumm in Foreword. (Photo credit Ratno Sardi, permission from Adam Brumm and Maxime Aubert)

showed, can only come from the storytelling of modern people about such divisions. It is misleading, probably quite wrong, for "people who are interested in art" to grab one or other of these stereotypical, generally unreflective, "explanations" as if it were the last word on that earlier art, without understanding the scientific (and often unscientific) processes or the modern storytelling involved in stating those conclusions. In this case it can only give the wrong impression about the "primitiveness" of the past, an impression that derives primarily from attitudes of modern

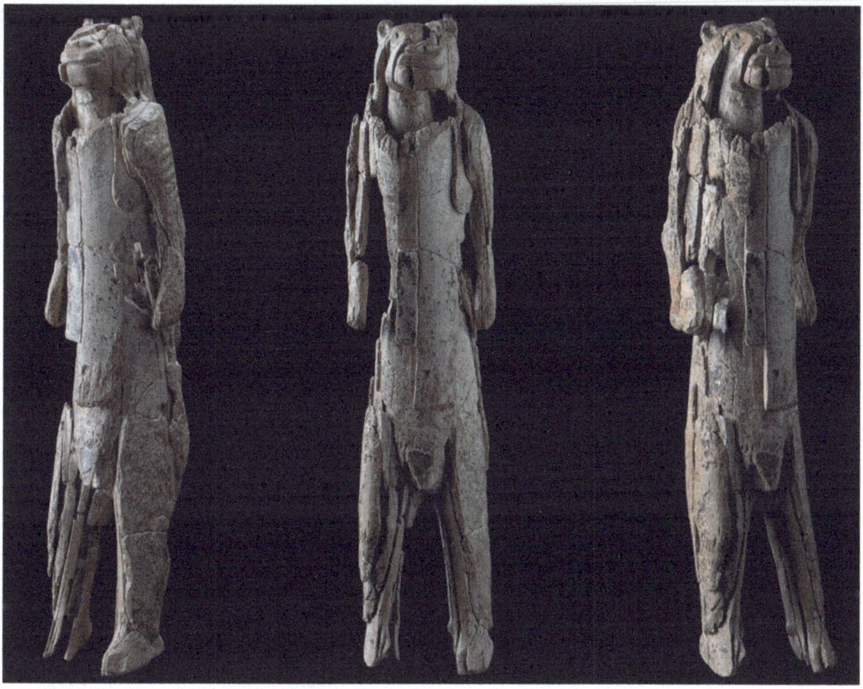

Fig. 2.14 Hohlenstein Stadel figure dated around 41–39 thousand years ago, reconstructed by Kind et al. (2014). ©Landesamt für Denkmalpflege im Regierungspräsidium Stuttgart/Yvonne Mühleis and Museum Ulm. (Photos by Claus-Jochim Kind)

specialists about something other than the past towards the people from such a past 30 thousand years ago as well as prejudices that have very little to do with the evidence of the art and other archeology. Better to use the evidence of the art and its other archeology as an indication of the spatial contextualization of the people who used the sites (Conkey, 1980) or the art (Gamble, 1982) whether the studies are 40 years old or only six (Fritz & Sauvet, 2018). By using these analytical conclusions, we can build up a picture of whether either the producers or the seers were primitive or not, although that is not a judgment we wish to make nowadays. If we must make judgments about humans from the time of the beginning of art, they should arise from analysis by archeologists, not from prejudices of art historians.

The figures from Sulawesi are remarkable enough because their discovery and ancient dating was so unexpected. After all, getting to Sulawesi involved a sea-crossing that was one of the earliest documented in the world, some crossings very early that might be explained as incidental, others later and "purposive" (Shipton et al., 2021) to the extent that it was, once, suggested that crossing the sea in watercraft was one of the best markers of the modernity of humans (Davidson & Noble, 1992). These images of anoas, warty pigs, people and hand stencils have been found

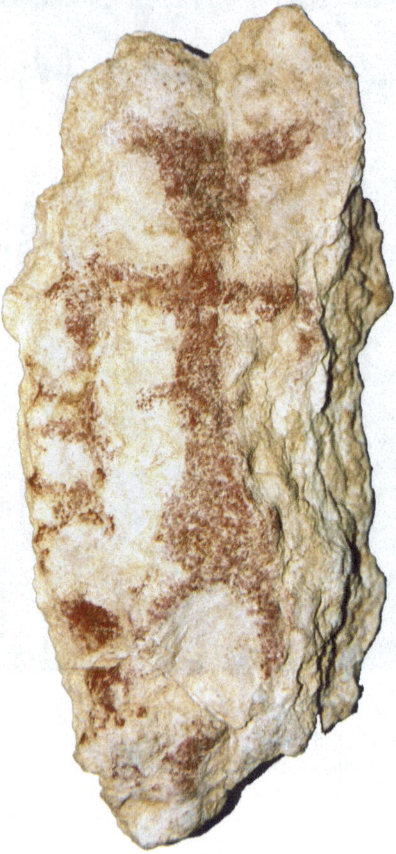

Fig. 2.15 Painting of probable anthropomorph, from Grotta Fumane, Verona (Italy), dated from 40 to 37.5 thousand years ago (Marín-Arroyo et al., 2023). (Image and permission thanks to Marco Peresani)

in several caves in the Maros region of Sulawesi and have produced reliable dates much older than any for representational art in any other region of the world. While the Leang Bulu' Sipong 4 image was once the oldest painting site known in the world—at least 43.9 thousand years old (Aubert et al., 2019)—that title is now held by an image of a warty pig from Leang Todongnge, illustrated by Stonard, which is at least 45.5 thousand years old (Brumm et al., 2021) and from Leang Karampuang is 51.2 thousand years old (Oktaviana et al., 2024). We may not have heard the last word about this.

The anoa shown in Fig. 2.12 seems to be being attacked by human-like figures to the left of the animal or in front of it (Fig. 2.13). This means that the sites of Maros in Sulawesi provide not just very early examples of art, but, unlike the art from the region where the next oldest art is concentrated, Western Europe, this is an unequivocally early representation of a scene involving interaction between human-like creatures and animals (see Davidson & Nowell, 2021a).

In addition, these are not just any human-like creatures. They seem to be humans represented with non-human heads. Although he does not make the point about the art in Sulawesi, Stonard also illustrates the most famous art example—a carving in mammoth ivory—that appears to combine the characteristics of a human (upright bipedalism) with those of an animal (generally described as a European lion) (Fig. 2.14), although a recent study has suggested instead that it represents a bear (Clifford & Bahn, 2020). There is no such "explaining away" for the figures at Leang Bulu' Sipong 4.

It turns out that, when you go looking, there are several examples of figures that combine human and animal characteristics elsewhere in the world (Taçon & Chippindale, 2001). Those would include the probable early painting from Grotta di Fumane (Fig. 2.15) (Peresani, 2019) across the Alps from Hohlenstein Stadel as well as the creatures in the painting from Leang Bulu' Sipong 4 and the much later ones we described in North West Central Queensland (Davidson et al., 2005).

The point here is that there is an argument to be presented about the implications of such human/animal composites for the way people thought at the time, just because we can only imagine that the images were created by humans. There is no written account of this but there are aspects of human thought that must follow from the existence of these images (Wynn et al., 2009). Tom Wynn and Fred Coolidge (p. 76) argue that the Hohlenstein-Stadel figure required its makers to use "basic ontological categories; taxonomies and multi-sensorial models; and abstract concepts." Some of this argument requires that the identifiable regions of the brains of distant ancestors had similar associations to those neurologists can identify among modern people. It may be true that ancient brains and modern ones prove to have similar sets of associations, but that should not be a starting assumption. Instead it should be a product of analysis. It can only really be discovered by the analysis of archeological materials without assuming that something about people's brains had already appeared as if by magic.

On ontological categories, the authors suggest that the category "tool" cannot be identified earlier than 1.5 million years ago long before the conceptual separation of "animate and inanimate objects"—between animals and things. This would be true whether or not it was appropriate to translate "ontology" as "colonialism … culture … alterity … otherness … [or] reality" (Davidson, 2022). Importantly, this is reflected in another approach which recognizes that evolving people must have demonstrated several stages in the perception of things—subjects in the real world—before they could produce other objects as representations of ideas that existed only in the imagination (Barnard et al., 2016; Davidson, 2019). Before you can carve or paint a creature that combines the appearance of a human in one part with the appearance of another animal in another part, it is necessary to conceptualize the existence of parts, then that both humans and other animals can be classified as having distinct parts, and finally that part of one can be substituted (in some way) for the part of the other that we now recognize as equivalent. There is a lot of thought needed to get there. The therianthropes of Leang Bulu' Sipong 4, Fumane, and, perhaps, of the Hohlenstein Stadel figure required conceptualization that parts from different things—human body, animal head—could be combined. In addition, they both required the conceptualization that people could adopt the characteristics of part of another animal; then it required the conceptualization that that way of

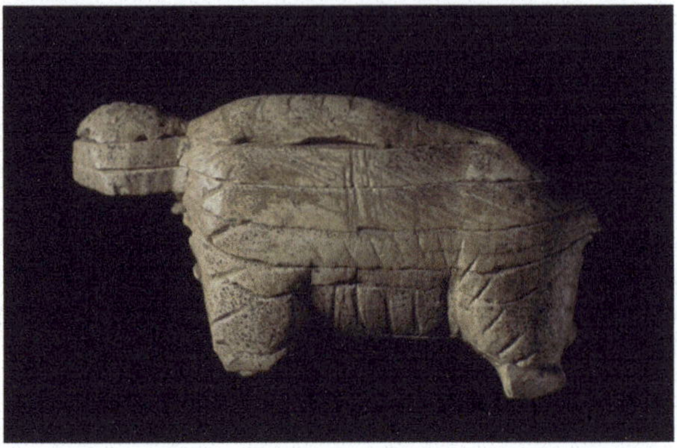

Fig. 2.16 Carved Mammoth Ivory figure of Mammoth from Geissenklosterle, near Blaubeuren, Germany. While some other figures have crossed lines this figure is incised with parallel lines on belly and limbs. (Copyright Landesmuseum Württemberg, P. Frankenstein/ H. Zwietasch)

constructing the world could be represented in either two (the painting) or three (the sculpture) dimensions.

Reasoning of this sort is much more revealing than the assertions of art historians using old stereotypes from early archeologists about magic or religion in early human art. Such reason is a part of the early history of art that might escape anyone not willing to undertake the step-by-step scientific argument—the scientific reportage that leads to peopled, connected histories. Not only does this tell us something about human thinking in the past by the producers of the art but also something about people from one discipline thinking about the studies of the past by people from another discipline who wrote a long time ago. We do not need to engage in speculation about magic or worship, or whether it was with a lion that the creature combined with human characteristics or a bear that was copied with no demonstration of the recognition of the concept of "part." In addition to the important argument by Wynn and Coolidge, these composite human/animals must be considered as symbolic—they represented something by convention using choices that seem to us arbitrary—even if, as seers we are not privy to the conventions that gave the symbols meaning. And consider, for a moment, the left shoulder of the Hohlenstein-Stadel figure. Whether this is a figure of a composite animal, symbolic because of the combination, or a figure of an animal such as a bear, none of these can be said to have parallel marks on the shoulder. In the case of Hohlenstein-Stadel, the marks were made by repeated scoring and they are very like parallel marks on other objects, not least those on the other figurines from Swabia which are more generally shown as cross-hatched sets of parallel lines (Fig. 2.16).

Such marks symbolized something: they were both arbitrary and conventional and only derived their meaning from the stories that were told about them or the knowledge of such stories in the community. Earlier patterned lines have been described, but in the earlier cases, the lines did not seem to be made by repeated scoring. Those were more likely to be accidental marks, not made by deliberate cutting, but selected by the excavators (Davidson, 1990; Davis, 1993). I was once challenged by a referee about the attachment of stories and songs to early images with the assertion that it was ridiculous to contemplate that there were songs associated with these figures. My response is that generally when I say there were stories, those might have been delivered in song. It is impossible to contemplate that such markings on carved figures could be given meaning such that they were repeated on several objects without some story-telling by the producers. As I wrote previously (with April Nowell), between such simple marks and these "there was a great distance of travel and an equally great symbolic journey" (Davidson & Nowell, 2021a).

Through these simple examples, we can see that the technical work of scientific analysis to disentangle archeological evidence is not straightforward. Observation and analysis can reveal important insights about the way we understand the relations between producer, subject, object and the seer of the art in the present, and then plausibly in the past. The subject may at first have existed only in the mind of the producer but comes to signify something about the way the producer saw the world through the story about the object. Today we have the object and we are seers, and through experiments we can reconstruct something of the relationship between object and subject. The stories that give meaning to the object can only be reconstructed through archeologists understanding the landscape of contextualization of similar objects.

So, the business of interpreting the art of the past looks as if it is complicated and requires scientific reportage and clarity of thought. And ideas about it change through time. It may be best to adopt my original line that the art is too important to be left to people who are interested in art, who cannot be expected to keep up with the scientific reasoning that leads to reportage about the past because it is not their job. It can allow archeologists to work out how artists saw the world and to understand how art fitted into that world.

2.2 Early Art in Surveys of Art History

Archeologists who study art have the great advantage that the subject is constantly changing as new discoveries are made. At the same time the discipline changes so that the interpretation of new finds reflects the theoretical currents in the wider world. Sometimes the whole story has changed because of new discoveries, sometimes new theoretical approaches to art have made archeologists think again about what it is they are studying.

By the same token, artists are constantly creating new art, and art historians are constantly describing the art of the past according to different points of view. The literature on the history of art is so vast that the whole book could be consumed by an analysis of the ways in which our perceptions of the world—including the world of archeological art—have been changed by the ways artists have seen it. But I am an archeologist, so I have been selective. I will look at a limited number of texts and consider their treatment of the art of the distant past.

Helen Gardner's *Art Through the Ages* (1926), was first published in 1926 or 1927 by which time she was working at the Art Institute of Chicago, the same place that contributed to Shiner's puzzlement years later, as I wrote in the previous chapter. Gardner's book was the foundation of what has been called the "Bible of Art History" and has been described as the "benchmark text" for art history (Kader,

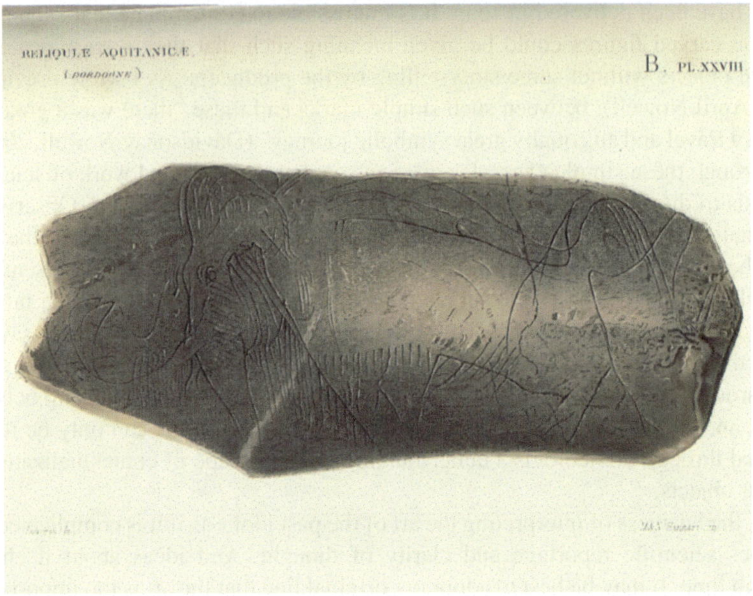

Fig. 2.17 "A portion of the outside layer of a Tusk of an Elephant, most probably a Mammoth, with the outline of a hairy Elephant cut on its outer surface". From Lartet and Christie (1875, Plate XXVIII)

2000). The book was revised three times in her lifetime, and subsequently others took over the editorship so that it has now reached at least 16 editions. Because the frequent editions after her death retained her name in the title, she represents the history of art history. The 1948 third edition, published just after her death, has been said to have "tried to contextualize her book within the emerging disciplines of prehistorical anthropology [sic], and non-European art mediated by historical facts and aesthetic theory" (Kader, 2000, 166). This edition (Gardner, 1948, 33) emphasized the techniques by which images were made, and their impact, including the mammoth shown in Fig. 2.17 in the following terms:

> An infuriated mammoth charging forward. There is a largeness, a striking direct statement of a few essentials, expressed by a line so sure that it is convincing and so economical that it incorporates all details without specifically stating them.

Gardner's intention was to write a single volume book surveying all art, but, so far as "prehistorical anthropology" was concerned, she was hamstrung in part by the slow accumulation of knowledge of the earliest art. Lascaux had only been discovered in 1940, but, other than press reports during wartime, the earliest published account appeared in the same year as Gardner's third edition (history of Lascaux taken from Lawson, 2012, 340–343). She published illustrations of seven items on bone or antler, and only attributed two of Breuil's paintings of cave art, one from Font de Gaume, the other from Altamira. She failed to recognize the chronological distinction between cave art and the later Levantine art paintings of eastern Spain, but in 1948 that chronology and distinction was not really understood (e.g., López-Montalvo, 2018).

The work has been revised and reprinted many times since and it is now subdivided into different topics in separate volumes (e.g., Kleiner, 2016). Gone are the details of the techniques of representing animals using engraved lines, and the work considered a sculpted antler from La Madelaine (Fig. 33A in Gardner, 1948) the same site as the source of Fig. 2.17, with an account of the technical achievement of carving a bison from reindeer antler (Kleiner, 2016). It seems to be true of Gardner, during her lifetime, that she placed emphasis on "aesthetics over historical fact" (Kader, 2000, 170), being concerned with the artistic principles of how an effect was achieved without forcing a view that these might exemplify the historical accumulation of principles. She does not seem to be concerned with the theoretical question of how these principles might arise, say, as aspects of what makes representation possible. They might, for example, either be a property of the magic of translating reality into another medium, or a property of the limitations of human perceptions. A full account of the scientific reportage and the archeohistorical contextualization might be something we can approximate on condition that the origins of art are not concealed in the considerations of what became dominant in later art. We might even make some judgments about the relative importance of a biological origin of art or a purely cultural one (Davies, 2006).

Gombrich's great work *The Story of Art* was first published in 1950 2 years after the last revision that Gardner herself was responsible for and likewise revised (into 16 editions) and reprinted many times before the author's death in 2001. By 1995, Gombrich conformed to the stereotype that Davis (1993) commented on, with pictures of Altamira and Lascaux at the beginning but then the book diverged into a discussion of the parallel feelings of Gombrich himself and what he imagined were those of the painters as part of a claim that these were "the oldest relics of that universal belief in the power of picture-making." For Gombrich (1995, 39–43), in art, the image that had been created invited a respect equivalent to that you might feel for another person—one of the central mysteries of art. While Gombrich's language about the creators of the early art of the caves was full of stereotypes about behaviors of (mostly modern) people he called "primitive" that involved magic or identification with wild animals, he was also at pains to show that similar practices and beliefs could be paralleled "within the prescribed rites and customs of our [western, white, male] lives." He got close to recognizing the importance of cultural differences between societies in acknowledging that "the whole story of art is not a story of progress in technical proficiency, but a story of changing ideas and requirements." In these pages, I agree on the importance of changing ideas, but give greater weight to the cultural differences among artists and their ideas. Gombrich (1995, 48–49) illustrated the relationship between the work of art and the stories entailed by illustrating a model of a house made by the Haida of Vancouver Island in western Canada and repeating the associated story. Figure 2.18 shows a bronze sculpture by the Canadian First Nations artist Bill Reid of Spirit of Haida Gwai from Vancouver Island. "The work depicts a canoe filled with thirteen entangled ethereal beings, the majority of whom are non-human and of mythical Haida origin: the Raven (*Xhuuya*), the Wolf, the Eagle, the Frog, the Bear Family, the Beaver, the Dogfish Mother, and

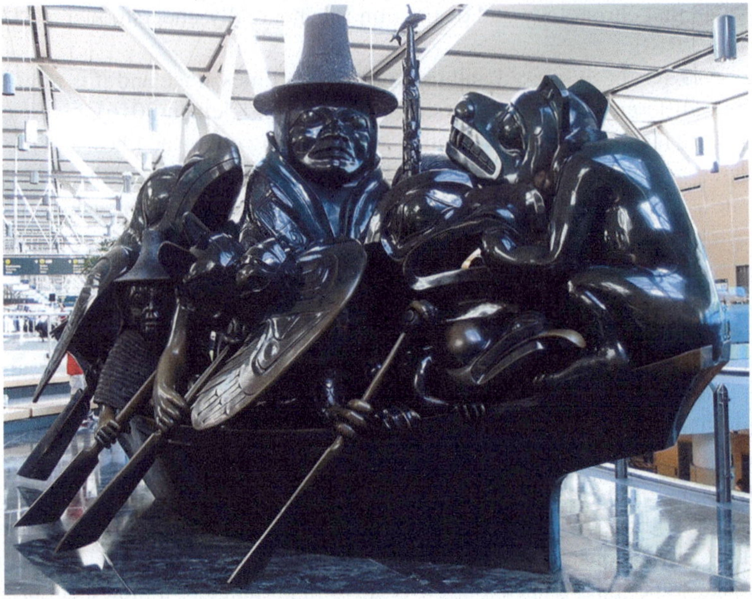

Fig. 2.18 Bronze sculpture in Vancouver Airport, Canada, by the Canadian First Nations artist Bill Reid of Spirit of Haida Gwai on Vancouver Island. The work depicts a canoe filled with 13 mythical Haida beings, including the Raven (*Xhuuya*), the Wolf, the Eagle, the Frog, the Bear Family, the Beaver, the Dogfish Mother, and the Mouse Woman (*Quaganjaat*). (from https://live.staticflickr.com/8296/7880338734_58a63d2755_o.jpg. Photo: Guilhem Vellut/CC-BY 2.0 Deed)

the Mouse Woman (*Quaganjaat*). Three humans are aboard: the Bear Mother, the Chief (who sits at the centre holding his staff), and the "Ancient Reluctant Conscript," as Reid has described him, who paddles humbly alongside all the others. The sculpture has been seen as both a portrait of Reid himself and as a depiction of the condition of life on Earth as seen through his eyes. Of it he wrote, "The boat goes on, forever anchored in the same place." (from https://www.aci-iac.ca/art-books/iljuwas-bill-reid/key-works/spirit-of-haida-gwaii/).

The important point here is that the work of art entails the object, as most commentators admit. It is a special act to view it separated from the artist and the story they tell about it—often without knowing either. Equally the seer of the object is entailed in the whole work of art because the story involves conventions, details and traditions that can best be understood within the shared culture of the artist/storyteller and the seer. In the case of the of *Spirit of Haida Gwai* sculpture, for many people, particularly Canadian, it is remembered as having featured on the Canadian $20 note between 2004 and 2012. The stories once in the public domain expand with the viewing of the work of art—as other sights and experiences are "dredged through" this work. For many people passing through the airport it is simply a magisterial construction which implies story, artist and context without the seer actually knowing much about them.

In going through this argument about the "art" of the "primitives," Gombrich also revealed much about his own attitudes to the story of art. Art is lifelike; it is not just about decoration; it involves beliefs we accept by custom and without question; it is beautiful; it uses skill to render lifelike images that involve acute observation; it may involve pleasure. What he omitted was the complex cultural contexts within which his own judgments were embedded.

An analysis of materials in other art and other museums shows that the politics of art and judgments about it are contingent on prevailing attitudes in the society of those who collect and display art (Procter, 2020). Not least of these attitudes was the idea that what people in Western Europe and North America valued was, in some sense, where everything was going. It once appeared to be, for example, that colonialism was the natural order of things, though now we can see otherwise, so that anything that seemed outside that normal progress must have another name—such as "craft." Others have grappled with this issue in a variety of different ways—Nelson Goodman (1976) sought to develop a theory of symbols; Pasztory (2005, 4) opted for the "abolition of the concept of art" which she replaced with "a cognitive interpretation of things;" Dana Arnold (2020, 31), though brief, decided to "see how the various ways of writing about art can change the way in which we see the object and think about its history;" for Stonard (2021, 1) the images of art are about "the encounter … with ourselves." Each of these people "interested in art" has emphasized that the important point is how people—the artist, the original seers, they the critics, or you and me, waiting to be told about the art—see the object, and the effects that seeing has on us as seers. How people are tied in to the conventions involved in recognizing symbols; how it is human brains that interact with images, "all equally rational and irrational" (Pasztory, 2005, 4); how the image itself is not unvarying but depends on the ways in which we know about it, in particular through our expectations about progress or evolution; and finally how we ourselves are changed by seeing the object.

All of these are attitudes to art that we will emulate in one form or another, but, as demonstrated in the foundation texts of art history by Gardner herself and by Gombrich, the art that then was conceived as prior to all other art was diminished by labelling its artists as primitives or in other similar fashions. For Shiner, whose curiosity arose from the distinction between Fine Art and Ethnographic Art, it would seem the solution came from separating off the (very) late emergence of Fine Art in a small part of the world. It is the function of this book to show that not all the seers were the same, that the human mind was evolving, and part of the way that was achieved was through the images and the stories that came to be told about them. Some say that art has been around for 100 thousand years—it has certainly been around for 50 thousand—and the expectations of artists and seers have changed in that time (and some would say that the species has too). It has changed in different ways in different places and at different times, and so have the expectations of artists and seers, but it was all art. It is just that there have been some special circumstances that led to a specialized bit of it being called "Fine" Art which in other places is called "High" Art. The history of this process includes the interactions at different times between the producers, the objects they produce, the stories they tell about the subjects and how that makes them produce objects in a particular way, and how our cultural embeddedness enables us to construct relationships between producer, subject, object, stories and ourselves.

2.3 Essay 2 Art Is Too Important to Be Left to Those Who Are Interested in Art (Niño Main Panel, and Kapoor and the Guggenheim Bilbao)

Early art, and the appearance of art on rocks around the world, is a fundamental part of the evolution of art to become an important characteristic of the lives of the people who made it and of the people in their societies who are consumers of art. The art has changed to allow multiple different perspectives through isolation or combinations, including by taking photographs. The work is the same, but the context is radically transformed.

Some works of art are obviously just that. The panel of paintings in the Cueva del Niño, Aýna (Albacete, Spain) where I excavated in 1973 (Davidson & García-Moreno, 2013) was originally described by Martín Almagro Gorbea (1971) and clearly represents several species of animals commonly hunted during the Upper Paleolithic—deer, horse, ibex—and they resemble other paintings and engravings from that period. As with those other examples of art on cave walls, there is no representation of the landscape. The radiocarbon dates (using a method that would not be chosen if there were better samples) were around 27 thousand years old (Garate Maidagán & García Moreno, 2011; García-Moreno et al., 2021).

It is possible to analyze the panel to show that the structure was not haphazard (Davidson, 2021). From our point of view, as seers of the art, the individual representations of animals now constitute a single panel, at some other time they may not have. It is worth pointing out that one attempt to date the representations in the panel using stylistic comparisons with Parpalló and elsewhere suggested that the paintings were from several different periods (Fortea Pérez, 1978)—but if the Diego Garate and Alejandro Garcia dates are to be believed, the art of Niño was produced before Parpalló was first occupied. The dating of the art and its relationship to other aspects of the society of the painters depends on a wider understanding of the behavior of people at different times in this upland part of eastern Iberia.

While I was in Spain, Ann Eastham was studying the bird bones from Iberian Paleolithic sites (Eastham, 2021) while both she and her husband Mike were also studying Paleolithic art (Eastham & Eastham, 1979, 1995; Eastham, 2012). I learned about art from her and her husband who both taught about art in England and studied Paleolithic art. I learned several things from them: (1) it is not true that I cannot draw, just that I **do** not draw—skill at drawing derives from doing it; (2) many images (such as those at Niño) can be drawn with one sweep of the arm from a single position; (3) it is often the case that images look different depending on the position of the illumination. I now also realize that we often judge skill at drawing if an image fits within norms of expectation about representation.

The context of the images at Niño is that evidence was left on and in the surfaces in front of the panel and through archeological excavation we recovered evidence of posts placed into holes in that surface (Davidson & García-Moreno, 2013) with much associated evidence of burning (Davidson, 1981). Without archeological excavation and recording, that context does not exist any more. As I wrote in 1981: "It does not require much imagination to speculate on the possibility that these

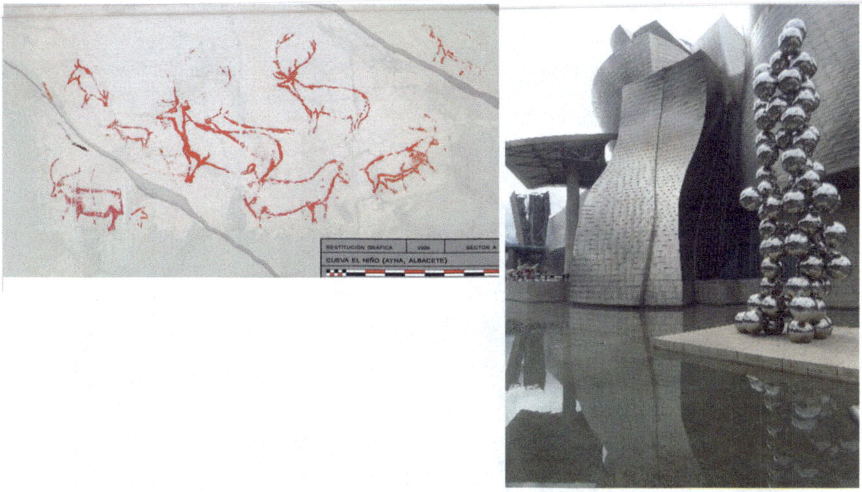

Fig. 2.19 Left: tracing of paintings from Cueva del Niño, Aýna, Albacete (Courtesy Diego Garate). Right: Anish Kapoor *Tall Tree and the Eye* and Guggenheim Bilbao. (Photo Iain Davidson)?

features represent an association between the paintings, structures and fire which could, doubtless, be backed up with ethnographic parallels to suggest ritual activities. Needless to say, there would be no justification for such flights of fancy."

The Guggenheim Museum at Bilbao is one of a number of buildings around the world designed by Frank Gehry, and possibly the most famous. It was opened in 1997 and is credited with generating huge income for the Basque country in Iberia—the Bilbao Effect—and it was described by architect Philip Johnson as "the greatest building of our time" (Tyrnauer, 2010). But the building itself is only part of the fame, for it is surrounded by other works of art both in terms of the Anish Kapoor installation in the steel spheres that constitute the figure *Tall Tree and the Eye* (Fig. 2.19 Right) or the Louise Bourgeois *Maman* (Fig. 2.20), a casting of a spider, which is on the other side of the pool from the Kapoor. The pool itself is often the source of moisture for a fog sculpture by Fujiko Nakaya.

> Kapoor's monumental *Tall Tree and the Eye* (2009), recently installed outside the museum alongside outdoor works by Louise Bourgeois, Daniel Buren, Jeff Koons, and Fujiko Nakaya, consists of 73 reflective spheres anchored around three axes. This illusionistic work continues the artist's examination of complex mathematical and structural principles embodied in sculptural form. The mirrored surfaces of the orbs reflect and refract one another, simultaneously creating and dissolving form and space. Images of the surrounding city, including the Nervión river, Buren's sculptural intervention on La Salve Bridge (Arcos rojos/Arku gorriak, 2007), and the museum itself, are cast into dynamic suspension. Kapoor reminds us of the instability and ephemerality of our vision-and by extension of our world. https://www.guggenheim-bilbao.eus/en/the-collection/works/tall-tree-and-the-eye (Accessed 4.29pm 24 October 2021).

Whatever we think of this explanation of the meaning of the installation, for our purposes here, it is important that the Kapoor installation exists in a cultural context

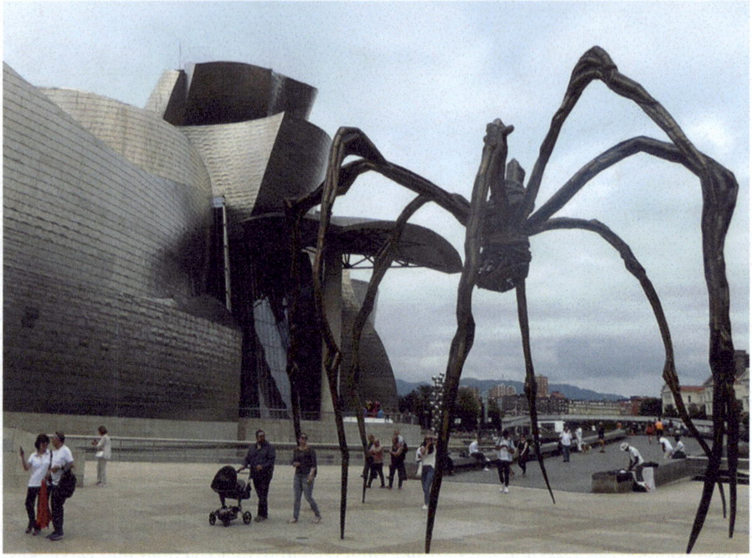

Fig. 2.20 *Maman* by Louise Bourgeois at the Guggenheim Museum, Bilbao. The Kapoor *Tall Tree and the Eye* from Fig. 2.19 can be seen behind it, with the Koons tulips to its left. (Photo Iain Davidson July 2018)

of the titanium walls of the Museum, the water in which installation and museum walls are reflected, and a view to the Buren installation on the La Salve bridge. These are all part of the photograph in the right frame of the figure. *Maman* is just out of the photograph on the left, and a Koons sculpture of tulip flowers to the left (Fig. 2.20).

At the time this photograph was taken, there were many people in the forecourt. Few were looking at the works of art. "Works of art" are the productions by artists, of objects, that can be seen by seers in the context in which they now exist and have some relationship with the subject the producer saw, but that is not all. In this instance, we might ask whether the individual sculpture or the building are the works of art, or the juxtapositioning of them. Or, perhaps, my photographs are the work of art, as are the hundreds of thousands taken by other tourists.[2]

Early art, and the appearance of art on rocks around the world, is a fundamental part of the evolution of art to become an important characteristic of the lives of the people who made it and of the people in their societies who are consumers of art. The context and archeology of the art at Cueva del Niño shows that the relationships among the producers and the seers of the images of animals were somehow restricted, however familiar the animals were as part of the lived lives of the artists and seers. By the same token, the Guggenheim Museum and its setting have created

[2]A small sample can be seen at https://flickr.com/search?text=Anish%20Kapoor%20Bilbao (accessed 17 November 2024).

a context where the producers of the art have disappeared (or become little more than famous names) and we, the seers, think of the artistic object as special although we are often hard put to nominate what subject it might relate to. Nevertheless, the art has changed to allow multiple different perspectives through isolation or combinations, including by taking photographs. The work is the same, but the context is radically transformed.

References

Almagro Gorbea, M. (1971). La cueva del Niño (Albacete). La cueva de la Griega (Segovia). Dos yacimientos de arte rupestre recientemente descubiertos en la Península Ibérica. *Trabajos de Prehistoria, 28*, 9–62.

Arnold, D. (2020). *Art history: A very short introduction*. Oxford University Press.

Aubert, M., Lebe, R., Oktaviana, A. A., Tang, M., Burhan, B., Hamrullah, et al. (2019). Earliest hunting scene in prehistoric art. *Nature, 576*(7787), 442–445. https://doi.org/10.1038/s41586-019-1806-y

Aujoulat, N. (2004). *Lascaux. Le geste, l'espace et le temps*. Seuil.

Aura Tortosa, J. E., Vadillo Conesa, M., Miret i Estruch, C., Verdasco Cebrián, C., Jiménez Barredo, F., Jordá Pardo, J. F., & Soler Mayor, B. (2020). La Cova de l'Hort de Cortés-Volcán del Faro (Cullera, Ribera Baixa, País Valenciano). Datos para la discusión de las ocupaciones presolutrenses. *Munibe Antropologia-Arkeologia*, (71), 23–39.

Bahn, P. G. (1998). *The Cambridge illustrated history of prehistoric art*. Cambridge University Press.

Barnard, P. J., Davidson, I., & Byrne, R. W. (2016). Toward a richer theoretical scaffolding for interpreting archaeological evidence concerning cognitive evolution. In T. Wynn & F. Coolidge (Eds.), *Cognitive models in Palaeolithic archaeology* (pp. 45–67). Oxford University Press.

Brumm, A., Oktaviana, A. A., Burhan, B., Hakim, B., Lebe, R., Zhao, J.-X., et al. (2021). Oldest cave art found in Sulawesi. *Science Advances, 7*(3), eabd4648. https://doi.org/10.1126/sciadv.abd4648

Brusgaard, N. Ø., & Akkermans, K. A. N. (2021). 9. Hunting and Havoc. Narrative Scenes in the Black Desert Rock Art of Jebel Qurma, Jordan. In I. Davidson & A. Nowell (Eds.), *Making scenes: Global perspectives on scenes in rock art* (pp. 134–149). Berghahn Books.

Cartailhac, E. (1902). Les cavernes ornées de dessins. La grotte d'Altamira, Espagne. Mea culpa d'un sceptique. *L'anthropologie, 13*, 348–354.

Cartailhac, E., & Breuil, H. (1906). *La caverne d'Altamira à Santillana, près Santander (Espagne)*. Monaco.

Clifford, E., & Bahn, P. (2020). The bear necessities. A new view of the so-called 'lion man'. *Curr World Archaeology, 100*, 24–29.

Clottes, J., & Lewis-Williams, J. D. (1998). *The shamans of prehistory. Trance and magic in the painted caves*. Harry N. Abrams.

Conard, N. J., Malina, M., & Münzel, S. C. (2009). New flutes document the earliest musical tradition in southwestern Germany. *Nature, 460*, 737–740.

Conkey, M. W. (1980). The identification of prehistoric hunter-gatherer aggregation sites: The case of Altamira. *Current Anthropology, 21*(5), 609–630.

Conkey, M. W. (1999). A history of the interpretation of European 'palaeolithic art': Magic, mythogram, and metaphors for modernity. In A. Lock & C. R. Peters (Eds.), *Handbook of human symbolic evolution* (pp. 288–349). Blackwells.

Culley, E. V., & Davidson, I. (2021). Art, sign, and representation. In N. Gontier, A. Lock, & C. Sinha (Eds.), *The Oxford handbook of human symbolic evolution* (2nd ed.). Oxford University Press.

David, B. (2017). *Cave art*. Thames & Hudson.

Davidson, I. (1981). *Late Palaeolithic economy in eastern Spain*. (PhD). Cambridge University.
Davidson, I. (1989a). Freedom of information: Aspects of art and society in western Europe during the last ice age. In H. Morphy (Ed.), *Animals into art* (pp. 440–456). Unwin Hyman.
Davidson, I. (1989b). *La economía del final del Paleolítico en la España oriental*. Diputación Provincial.
Davidson, I. (1990). Bilzingsleben and early marking. *Rock Art Research, 7*, 52–56.
Davidson, I. (1995). Paintings, power, and the past: can there ever be an ethnoarchaeology of art? (Review of *Ancestral Connections: Art and an Aboriginal system of knowledge.* [H. Morphy]). *Current Anthropology, 36*(5), 889–892. Retrieved from http://links.jstor.org/sici?sici=0011-3204%28199512%2936%3A5%3C889%3APPATPC%3E2.0.CO%3B2-2
Davidson, I. (2005). The painting and the tree: symbolism in the Upper Palaeolithic. A tribute to a great Basque scholar. *Munibe (Antropologia-Arkeologia), 57*, 197–205.
Davidson, I. (2012a). Symbolism and becoming a hunter-gatherer. *L'art pléistocène dans le monde/Pleistocene art of the world/Arte pleistoceno en el mundo, Actes du Congrès IFRAO, Tarascon-sur-Ariège, septembre 2010, Symposium « Signes, symboles, mythes et idéologie… ». N° spécial de Préhistoire, Art et Sociétés, Bulletin de la Société Préhistorique Ariège-Pyrénées, LXV-LXVI*, Book: p. 292–293, CD: p. 1689–1705
Davidson, I. (2012b). What a carry on? Portable art and changes of symbolic meaning. *L'art pléistocène dans le monde/Pleistocene art of the world/Arte pleistoceno en el mundo, Actes du Congrès IFRAO, Tarascon-sur-Ariège, septembre* 2010, *N° spécial de Préhistoire, Art et Sociétés, Bulletin de la Société Préhistorique Ariège-Pyrénées, LXV-LXVI*, Book p. 268–269, CD: p. 1559–1570.
Davidson, I. (2017). Paleolithic art. In J. Jackson (Ed.), *Oxford bibliographies in anthropology*. Oxford University Press.
Davidson, I. (2019). Evolution of cognitive archaeology through evolving cognitive systems. In K. A. Overmann & F. Coolidge (Eds.), *Squeezing minds from stones* (pp. 79–101). Oxford University Press.
Davidson, I. (2021). 1. Scenes and non-scenes in rock art. In I. Davidson & A. Nowell (Eds.), *Making scenes: Global perspectives on scenes in rock art* (pp. 16–31). Berghahn Books.
Davidson, I. (2022). Review of Oscar Moro Abadía & Martin Porr (ed.). 2021. Ontologies of rock art: Images, relational approaches, and Indigenous knowledges. Abingdon: Routledge. *Antiquity, 96*(389), 1349–1351. https://doi.org/10.15184/aqy.2022.90
Davidson, I., Cook, N. D. J., Fischer, M., Ridges, M., Ross, J., & Sutton, S. A. (2005). Archaeology in another country: Exchange and symbols in North West Central Queensland. In I. Macfarlane, M.-J. Mountain, & R. Paton (Eds.), *Many exchanges: Archaeology, history, community and the work of Isabel McBryde* (pp. 101–128). Aboriginal History Inc.
Davidson, I., & García-Moreno, A. (2013). La excavación arqueologica de la Cueva del Niño (Aýna, Albacete) de 1973: secuencia estratigráfica y materiales. *Al-Basit, 58*, 91–117.
Davidson, I., & Noble, W. (1992). Why the first colonisation of the Australian region is the earliest evidence of modern human behaviour. *Archaeology in Oceania, 27*, 135–142.
Davidson, I., & Nowell, A. (2021a). Epilogue. Is there more to scenes than meets the eye? In I. Davidson & A. Nowell (Eds.), *Making scenes: Global perspectives on scenes in rock art* (pp. 327–331). Berghahn Books.
Davidson, I., & Nowell, A. (2021b). Introduction. Behind the scenes—did scenes in rock art create new ways of seeing the world? In I. Davidson & A. Nowell (Eds.), *Making scenes: Global perspectives on scenes in rock art* (pp. 1–15). Berghahn Books.
Davies, S. (2006). *The philosophy of art*. Wiley Blackwell.
Davis, W. (1993). Beginning the history of art. *The Journal of Aesthetics and Art Criticism, 51*(3), 327–350.
Descola, P. (Ed.). (2010). *La Fabrique des images. Visions du monde et formes de la représentation*. Somogy et Musée du Quai Branly.
Diamond, J. (1998). *Guns, germs and steel. A short history of everybody for the last 13,000 years*. Vintage.

Dubey-Pathak, M., & Clottes, J. (2021). 8. Music and dancing scenes in the rock art of Central India. In I. Davidson & A. Nowell (Eds.), *Making scenes: Global perspectives on scenes in rock art* (pp. 122–133). Berghahn Books.

Ducasse, S., & Langlais, M. (2019). Twenty years on, a new date with Lascaux. Reassessing the chronology of the cave's Paleolithic occupations through new 14C AMS dating. *PALEO Revue d'archéologie préhistorique*, (30–1), 130–147.

Dunbar, R. I. M. (2004). *The human story*. Faber & Faber.

Eastham, A. (2021). *Man and bird in the Palaeolithic of Western Europe* (pp. 1–160). Archaeopress Publishing Ltd.

Eastham, A., & Eastham, M. (1979). The wall art of the Franco-Cantabrian deep caves. *Art History, 2*(4), 365–387.

Eastham, A., & Eastham, M. (1995). Palaeolithic images and the Great Auk. *Antiquity, 69*, 1023–1025.

Eastham, M. (2012). Pleistocene prelude: the implications of the way things are drawn by Ice Age Draftsmen in Europe. *Time and Mind, 5*(3), 299–315. https://doi.org/10.2752/175169712X13376094321417

Fernández, E., Thaw, S., Brown, T., Arroyo-Pardo, E., Buxó, R., Serret, M., & Araus, J. (2013). DNA analysis in charred grains of naked wheat from several archaeological sites in Spain. *Journal of Archaeological Science, 40*(1), 659–670.

Fortea Pérez, J. (1978). Arte Paleolítico del Mediterraneo español. *Trabajos de Prehistoria, 35*, 99–149.

Fritz, C., & Sauvet, G. (2018). The role of art in Prehistoric societies. *Quaternary International, 491*, 1. https://doi.org/10.1016/j.quaint.2018.09.037

Gamble, C. S. (1982). Interaction and alliance in Palaeolithic society. *Man: Journal of the Royal Anthropological Institute, 17*, 92–107.

Gamble, C. S. (2013). *Settling the Earth: The archaeology of deep human history*. Cambridge University Press.

Garate Maidagan, D., & García-Moreno, A. (2022). Capitulo 9. El arte rupestre paleolítico de la Cueva del Niño. In A. García-Moreno (Ed.), *La Cueva del Niño (Aýna, Albacete). Estudios en el 50 aniversario de su descubrimiento* (pp. 243–302). Instituto de estudios Albacetenses.

Garate Maidagán, D., & García Moreno, A. (2011). Revisión crítica y contextualización espacio-temporal del arte parietal paleolítico de la Cueva del Niño (Aýna, Albacete). *Zephyrus, 68*, 15–39.

García-Moreno, A. (Ed.). (2022). *La Cueva del Niño (Aýna, Albacete). Estudios en el 50 aniversario de su descubrimiento*. Instituto de estudios Albacetenses.

García-Moreno, A., Cubas, M., Davidson, I., Garate, D., López-Dóriga, I., Marín-Arroyo, A. B., et al. (2021). El Niño Cave (Aýna, Albacete, Spain): Late Middle Palaeolithic, Rock Art and Neolithic Occupations from Inland Iberia. *Proceedings of the Prehistoric Society, 87*, 1–9. https://doi.org/10.1017/ppr.2021.14

Gardner, H. (1926). *Art through the ages: An introduction to its history and significance* (Vol. Vol. 1). Harcourt.

Gardner, H. (1948). *Art through the ages: An introduction to its history and significance* (3rd ed.). Harcourt.

Gombrich, E. H. (1995). *The story of art* (16th ed.). Phaidon.

Goodman, N. (1976). *Languages of art: An approach to a theory of symbols*. Hackett Publishing.

Griffiths, B. (2018). *Deep time dreaming: Uncovering ancient Australia*. Black Inc.

Harari, Y. N. (2014). *Sapiens: A brief history of humankind*. Random House.

Higham, T. (2021). *The world before us. How science is revealing a new story of our human origins*. Penguin Books.

Hughes, R. (1991). *The shock of the new*. Thames and Hudson.

Kader, T. (2000). The Bible of Art History: Gardner's "Art Through the Ages". *Studies in Art Education, 41*(2), 164–177.

Kind, C.-J., Erbinger-Rist, N., Wolf, S., Beutelspacher, T., & Wehrberger, K. (2014). The smile of the Lion Man. Recent excavations in Stadel Cave (Baden-Württemberg, south-western Germany) and the restoration of the famous Upper Palaeolithic figurine. *Quartär, 61*, 129–145.

Kleiner, F. S. (2016). *Gardner's art through the ages: The western perspective* (Vol. 1). Cengage Learning.

Lartet, E., & Christy, H. (1875). *Reliquiae Aquitanicae: Being contributions to the archaeology and palaeontology of Périgord and the adjoining provinces of southern France*. Williams and Norgate. Online at https://www.forgottenbooks.com/en/books/Reliquiae_Aquitanicae_1000698892

Lawson, A. J. (2012). *Painted caves. Palaeolithic rock art in Western Europe*. Oxford University Press.

Leroi-Gourhan, A. (1982). *The dawn of European art. An introduction to Palaeolithic cave painting*. Cambridge University Press.

Lewis-Williams, J. D. (1985). The San artistic achievement. *African Arts, 18*(3), 54–59, 100.

Lewis-Williams, J. D. (2002). *The mind in the cave*. Thames and Hudson.

Lewis-Williams, J. D., Bardill, P. N., Biesele, M., Yearwood, S., Clegg, J., Davis, W., et al. (1982). The economic and social context of Southern San Rock Art [and comments and reply]. *Current Anthropology, 23*(4), 429–449. Retrieved from http://www.jstor.org/stable/2742269

López-Montalvo, E. (2018). Hunting scenes in Spanish Levantine rock art: An unequivocal chrono-cultural marker of Epipalaeolithic and Mesolithic Iberian societies? *Quaternary International, 472*, 205–220. https://doi.org/10.1016/j.quaint.2018.03.016

Marín-Arroyo, A. B., Terlato, G., Vidal-Cordasco, M., & Peresani, M. (2023). Subsistence of early anatomically modern humans in Europe as evidenced in the Protoaurignacian occupations of Fumane Cave, Italy. *Scientific Reports, 13*(1), 3788. https://doi.org/10.1038/s41598-023-30059-3

Mateo Saura, M. Á. (2022). Capitulo 10. El arte rupestre levantino de la Cueva del Niño. In A. García-Moreno (Ed.), *La Cueva del Niño (Aýna, Albacete). Estudios en el 50 aniversario de su descubrimiento* (pp. 303–329). Instituto de estudios Albacetenses.

McCall, G. S. (2007). Add shamans and stir? A critical review of the shamanism model of forager rock art production. *Journal of Anthropological Archaeology, 26*(2), 224–233. Retrieved from http://www.sciencedirect.com/science/article/B6WH6-4MBCBR1-1/2/6a34da6703f4893845d7723c2dd0a74b

McGrath, A. (2015). Deep histories in time, or crossing the great divide. In A. McGrath & M. A. Jebb (Eds.), *Long history, deep time. Deepening histories of place* (pp. 1–32). Australian National University Press.

Moro Abadía, O., & González Morales, M. R. (2004a). 1864–1902: El reconocimiento del arte paleolítico. *Zephyrus, 57*, 119–123.

Moro Abadía, O., & González Morales, M. R. (2004b). Towards a genealogy of the concept of "Paleolithic mobiliary art". *Journal of Anthropological Research, 60*(3), 321–340.

Moro Abadía, O., & González Morales, M. R. (2005). "El arte por el arte": revisión de una teoría historiográfica. *Munibe (Antropologia-Arkeologia), 57*, 5–15.

Nowell, A., & Chang, M. L. (2014). Science, the media, and interpretations of Upper Paleolithic figurines. *American Anthropologist, 116*(3), 562–577.

Oktaviana, A. A., Joannes-Boyau, R., Hakim, B., Burhan, B., Sardi, R., Adhityatama, S., et al. (2024). Narrative cave art in Indonesia by 51,200 years ago. *Nature, 631*(8022), 814–818. https://doi.org/10.1038/s41586-024-07541-7

Parkington, J., & Rusch, N. (2003). *Cederberg rock paintings*. Krakadouw Trust.

Pasztory, E. (2005). *Thinking with things: Toward a new vision of art*. University of Texas Press.

Peresani, M. (2019). Grotta di Fumane (Fumane in Valpolicella, VR). *Le ricerche preistoriche dell'Università di Ferrara", in Annali dell'Università di Ferrara-Museologia Scientifica e naturalistica, 15*, 59–64

Pericot García, L. (1942). *La cueva del Parpalló*. Consejo Superior de Investigaciones Científicas.

Pike, A. W. G., Hoffmann, D. L., García-Diez, M., Pettitt, P. B., Alcolea, J., De Balbín, R., et al. (2012). U-series dating of Paleolithic art in 11 caves in Spain. *Science, 336*(6087), 1409–1413. https://doi.org/10.1126/science.1219957

Procter, A. (2020). *The Whole Picture: The colonial story of the art in our museums & why we need to talk about it*. Cassell.

Quiles, A., Valladas, H., Bocherens, H., Delqué-Količ, E., Kaltnecker, E., van der Plicht, J., et al. (2016). A high-precision chronological model for the decorated Upper Paleolithic cave of Chauvet-Pont d'Arc, Ardèche, France. *Proceedings of the National Academy of Sciences, 113*(17), 4670–4675. https://doi.org/10.1073/pnas.1523158113

Reich, D. (2018). *Who we are and how we got here: Ancient DNA and the new science of the human past*. Oxford University Press.

Russell, T. (2000). The Application of the Harris Matrix to San Rock Art at Main Caves North, Kwazulu-Natal. *The South African Archaeological Bulletin, 55*(171), 60–70.

Shipman, P. (2015). *The invaders. How humans and their dogs drove Neanderthals to extinction*. Belknap Press.

Shipton, C., O'Connor, S., & Kealy, S. (2021). The biogeographic threshold of Wallacea in human evolution. *Quaternary International, 574*, 1–12. https://doi.org/10.1016/j.quaint.2020.07.028

Solomon, A. (2013). The death of trance: Recent perspectives on San ethnographies and rock arts. *Antiquity, 87*(338), 1208–1213.

Sterelny, K. (2021). *The Pleistocene social contract: Culture and cooperation in human evolution*. Oxford University Press.

Stonard, J.-P. (2021). *Creation: A fully illustrated, panoramic world history of art from ancient civilisation to the present day*. Bloomsbury Publishing.

Taçon, P. S. C., & Chippindale, C. (2001). Transformations and depictions of the First People: Animal-headed beings of Arnhem Land, N.T., Australia. In K. Helskog (Ed.), *Theoretical perspectives in rock art research* (pp. 175–210). Instituttet for sammenlignende kulturforskning.

Tejedor-Rodríguez, C., Moreno-García, M., Tornero, C., Hoffmann, A., García-Martínez de Lagrán, Í., Arcusa-Magallón, H., et al. (2021). Investigating Neolithic caprine husbandry in the Central Pyrenees: Insights from a multi-proxy study at Els Trocs cave (Bisaurri, Spain). *PLoS One, 16*(1), e0244139.

Tyrnauer, M. (2010). Architecture in the age of Gehry. *Vanity Fair*, August 2010. Retrieved from https://www.vanityfair.com/culture/2010/08/architecture-survey-201008

Villaverde Bonilla, V. (1994). *Arte Paleolítico de la Cova del Parpalló*. Servei d'Investigacio Prehistorica, Diputacio de Valencia.

Villaverde Bonilla, V. (2021). 15. Scenes in the Paleolithic and Levantine Art of Eastern Spain. In I. Davidson & A. Nowell (Eds.), *Making scenes: Global perspectives on scenes in rock art* (pp. 224–239). Berghahn Books.

Villaverde, V., Sanchis, A., Badal, E., Bel, M. Á., Bergadà, M. M., Eixea, A., et al. (2021). Cova de les Malladetes (Valencia, Spain): New insights about the early upper Palaeolithic in the Mediterranean Basin of the Iberian Peninsula. *Journal of Paleolithic Archaeology, 4*(1), 5. https://doi.org/10.1007/s41982-021-00081-w

Vinnicombe, P. (1976). *People of the Eland. Rock paintings of the Drakensberg Bushmen as a reflection of their life and thought*. University of Natal Press.

Vinnicombe, P. (2002). Petroglyphs of the Dampier Archipelago. *Rock Art Research, 19*, 3–27.

Wynn, T., Coolidge, F. L., & Bright, M. (2009). Hohlenstein-stadel and the evolution of human conceptual thought. *Cambridge Archaeological Journal, 19*(1), 73–83. https://doi.org/10.1017/s0959774309000043

Chapter 3
Art, Rock Art or Not Art

Abstract In this chapter, I introduce the concept of the art-system, defined by the producer and the stories they tell, the seers, and the object that the producer makes, and the seers see. I introduce four types of definition: ostensive, lexical, operational and functional, and I also extend the categorization of different types of art used by John Robb and call them Art (a generic term), Art 1 (mostly what some specialists think of as "Fine Art"), Art 2 (Archeological Art) and Art 3 (the earliest stuff that gets called art, out of which other archeological art emerged). In addition, I discuss criteria from Gombrich (art is lifelike, involves unquestioned beliefs, is beautiful, uses skill and acute observation, and creates pleasure) and a dictionary definition (creativity, skill, imagination, beauty, and emotional power). All these criteria are related to the concept of aesthetics. I show how that has been resolved into a choice between "I like it" and "it is beautiful" and opt for the personal "I like it" because it requires all people involved in "art," whether they are producers or seers to be agents in their own emotional reactions.

3.1 Talking About Art

The time has come to talk about art. I ended the previous chapter with a grand vision of a set of relationships that are present in all art. I am concentrating on visual art, especially painting, but the principles apply to all sorts of art, such as architecture, dance, literature, music, photography, poetry, etc. These relationships in what I choose to call the art-system involve artists—the producers (1)—and the things that get called art (2)—the objects artists produce. Fundamental to that set of relationships are the stories (3) artists tell about the subjects they turn in to art and the people who see it (4). The seers do not necessarily hear the stories but see the art that results from them, so they are not an audience. That is how I am referring to them—as seers. For convenience I use this slightly strange term to apply even when the art I am discussing includes music or literature where the visual is not the most important aspect.

I. Davidson, *Art or Scribbles? In the Eye of the Beholder: The Evolutionary Emergence of Visual Communication*, Interdisciplinary Evolution Research 11,
https://doi.org/10.1007/978-3-031-90458-5_3

There is, of course, a group of people who make up a crucial fifth element of the system and they are the observers of images—also seers—who are sufficiently separated from the act of creation not to be hearers and readers of stories from the time of production. They may be like us, those who appreciate art in some way, despite the fact that we have some part missing. For example, I know very little about the Spanish royal court in the seventeenth century, but I can still appreciate both Diego Velázquez's *Las Meninas* and Picasso's. The argument of this book is that what we call art is part of the whole system of relationships among these five elements. We should avoid thinking that only the object is the work of art. Art is in the system of relationships which includes people—producers and seers—as well as the object that is called the "work of art" and, indeed, its subject. This is a reason why Soffer and Conkey (1997) and April Nowell (2006, 2017), amongst others, write about visual cultures—to understand what is going on, we need to consider more than just the object, and to take into account the cultural context in which it was produced. It is also the reason why much digital art and even art produced by Artificial Intelligence (the fad of the moment) is also part of an art-system. But I have a *caveat* that it may not be the art object that some would wish.

It has been reported that Rembrandt's painting *The Night Watch*, shown in Fig. 1.9, is being restored by removal of varnish and other treatments. As part of the process, panels are being created to replace those from the edges of the painting which were removed when the work was cropped in 1715 (nearly 310 years ago) so that it could be displayed in the Amsterdam Town Hall (Goldstein, 2021). These new panels are being created by applying Artificial Intelligence to panels derived from a copy made in the seventeenth century by Gerrit Lundens. This ingenious use of modern technology might be considered as enhancing the Rembrandt as it has been displayed for the last 300 plus years, but we should be quite clear that it will no longer be a work straightforwardly BY Rembrandt. The reconstructed panels will not be his. The art-system has been expanded to include additional producers and takes the emphasis away from the "work of art" as an object by itself.

I also want to consider the criteria that are part of the assessment of art across societies, criteria which were smuggled in by Gombrich (1995, Ch. 1) and are key parts of most definitions of what art is: Art is lifelike, involves unquestioned beliefs, is beautiful, uses skill and acute observation, and creates pleasure. Gombrich wrote about them long ago—and was indeed reflecting his own unquestioned attitudes to art—but his ideas are still current among many later people who are interested in art. Thus, Denis Dutton (2009) offered a dozen aspects of art which might occur in some combination or other: Direct pleasure; Skill and virtuosity; Style; Novelty and creativity; Criticism; Representation; Special focus; Expressive individuality; Emotional saturation; Intellectual challenge; Art traditions and institutions; Imaginative experience. This list appears to be more comprehensive, yet there is clearly a substantial overlap with Gombrich (as we might expect).

I suggest that there are important exceptions to these criteria. Lifelike or representational images—a notion that goes back to the Athenian philospher Plato (e.g., Dutton, 2009, 31–2)—are not the only types of art but art frequently involves non-representational lines, and in the case of ancient or modern artists it may involve all sorts of things that have nothing to do with likeness or representation. By the same token, since the establishment of the importance of perspective and *chiaroscuro*, the use of light and shade, in the Renaissance, what constituted likeness was transformed, at least in parts of Europe, and those two techniques became part of what was taken for granted about how representation was achieved. But much art, at least since the Impressionists in the mid-nineteenth century, has involved the questioning of beliefs about representation which made possible a questioning of what likeness was, how it was achieved, and then whether likeness itself was important. Art, it turns out, is not only about beauty and pleasure—you would think the market might not respond to ugliness. Pablo Picasso's *Guernica*, surely one of the best known and widely appreciated images of the mid-twentieth century, is not straightforwardly a likeness of anything and could be criticized because it does not involve beauty or pleasure. Picasso made many sketches for *Guernica* which were not, however, part of its public exhibition before its travel to Spain after the death of the dictator, Francisco Franco.

The philosopher Arthur Danto (2013, ix–xii) looked to the early twentieth century for the death of this lifelike or representational art that appealed to Plato:

> … Plato's definition of art remained in place, with little to argue about until it stopped capturing the seeming essence of art. … Historically it happened with the advent of Modernism … Plato had had an easy run, from the sixth century BC until AD 1905–7, with the so-called Fauves—Wild Beasts—and Cubism. … Marcel Duchamp found a way of eradicating beauty in 1915, and Andy Warhol discovered that a work of art could exactly resemble a real thing in 1964, though the great movements of the 1960s …made art that was not exactly imitation. Oddly, sculpture and photography shifted the center of artistic self-awareness in the seventies. After that, everything was feasible. Anything went, leaving it uncertain whether a definition of art is any longer possible. Anything cannot be art.

Arnold (2020, 102–4) broadening the definition, emphasized that art is representational, but that it has only ever been an illusion created by the skills of putting marks on a support (e.g., paint on canvas) or creating surfaces by modifying a material to create an object to which a name is given (e.g., Michelangelo's *David*). Arnold had previously acknowledged that art is an illusion when disentangling some of the complexities of Velázquez's *Las Meninas* (Fig. 1.10), an illusion with the intent of "creating a realistic effect" (p. 22). Whitney Davis (2017) began his book with an example of how a single discontinuous line, by artist Tony Urquart, can be the framework of an illusion that was transformed into a picture of a fence crossing a prairie in winter (near London, Ontario) by the addition of very few other lines.

Consider the view of the southern Iberian town of Antequera in Fig. 3.1 Top, etched before printing by Lionel Lindsay (Holden, 2003). I photographed the same view in Spain in 2005 (Fig. 3.1 Bottom). Lindsay was representing the scene as he saw it in 1929 that was unchanged in major elements in 2005. He may have added the person riding the donkey. He exaggerated the height of the *Convento del Carmen* in the middle of the image, foreshortened the *Real Colegiata de Santa María la Mayor* to its right, and made the mountains in the background more dramatic. Representation is important, but the exaggerations make Lindsay's image—rather than my photograph—into a work that exists in an art-system. Gombrich's word

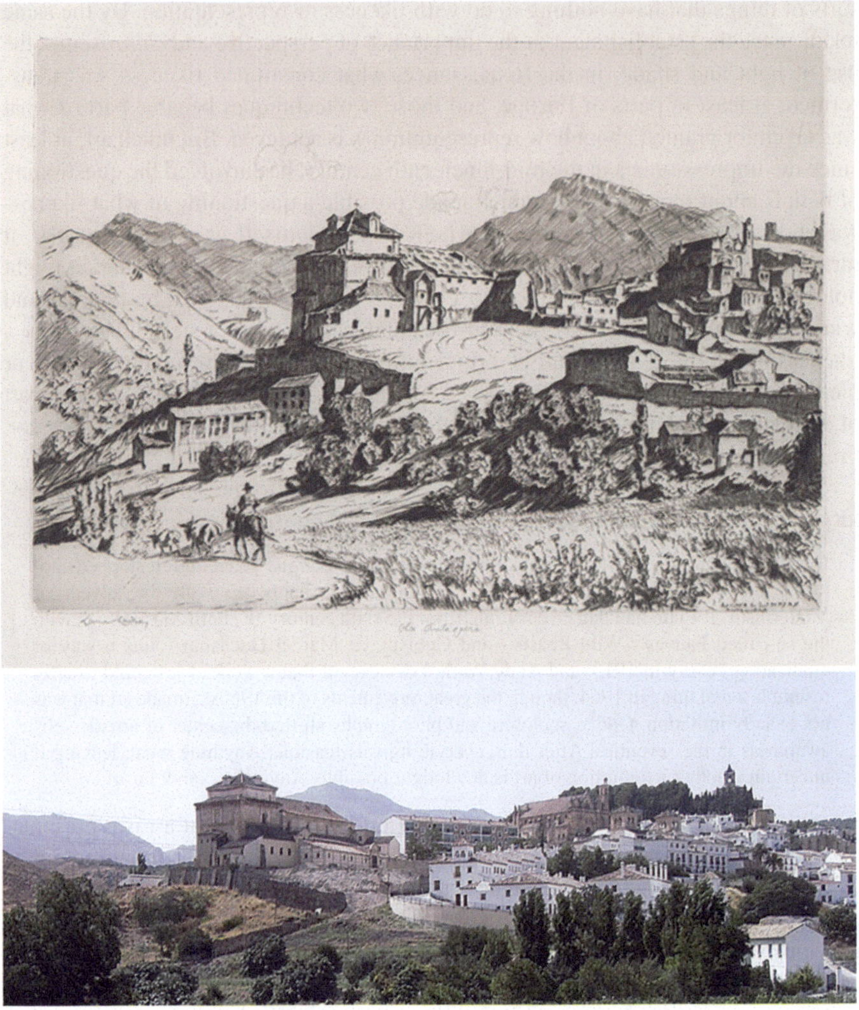

Fig. 3.1 Top: *Old Antequera*, Andalusia, Spain. Drypoint etching 1929 by Lionel Lindsay courtesy of the National Library of Australia (nla.obj-152556951). (© *Alan and Jancis Rees/Copyright Agency, 2025*). Bottom: Photo of the same place in 2005 by Iain Davidson

"lifelike" may be better than "representational". Foucault (2002, Chap. 3) discussed "Representing" at length through an exploration of the fictional character of Don Quixote, pointing out that he, in turn, lived through the appearance of reality he acquired from other works of fiction. In reality, all creations in the art-system take on a life of their own in this way, generating new narratives for the seers.

For the etching *Old Antequera*, the art-system includes the likeness, but it also includes the whole storied history of the etching: a print was given to the Armidale Teachers' College by Howard Hinton who amassed the wonderful collection of pictures that now are the core of the New England Regional Art Museum (Baguley &

Kerby, 2021); in the Armidale Teachers' College the collection inspired individuals who taught all over Australia; it once adorned an office in the administrative buildings at the University of New England where I fell in love with it having been enchanted by the reality of southern Spain decades before. The art-system, thus, ramifies to include many aspects of what I recognize as my cultural values, in much the same way as the generation of new narratives within other art-systems.

A simple reason to prefer the word "lifelike" over "representational" is given by the example of religious art. Ikons and other religious images of people (and their imaginary counterparts) were generally painted long after the lives of those people—if they ever existed except in the stories that are told about them. (I will discuss, in Chap. 7, the confusion between ikons as religious images and the semiotic meaning of the word icon. It may be that the semioticians, probably Charles Peirce, chose the word badly). Figure 3.2 shows banks of ikons for sale in Sofia, Bulgaria which diminishes, perhaps, the impact of each individual work.

Some time around 700 years ago on a wall in a religious building, Duccio di Buoninsegna painted a pair of Old Testament heroes, one of whom was Jacob. Jacob's story is told in a memorable myth that is foundational within three sets of unquestioned beliefs, Judaism, Christianity, and Islam. In what sense can the images in Duccio's painting (https://provincedesienne.com/2019/09/04/duccio-di-buoninsegna-jeremias/ downloaded 9 Feb 2025) be said to be representational? Even if we were to accept the historicity of the tale, how is anyone to know whether the historical Jacob looked anything like the painting that was produced several

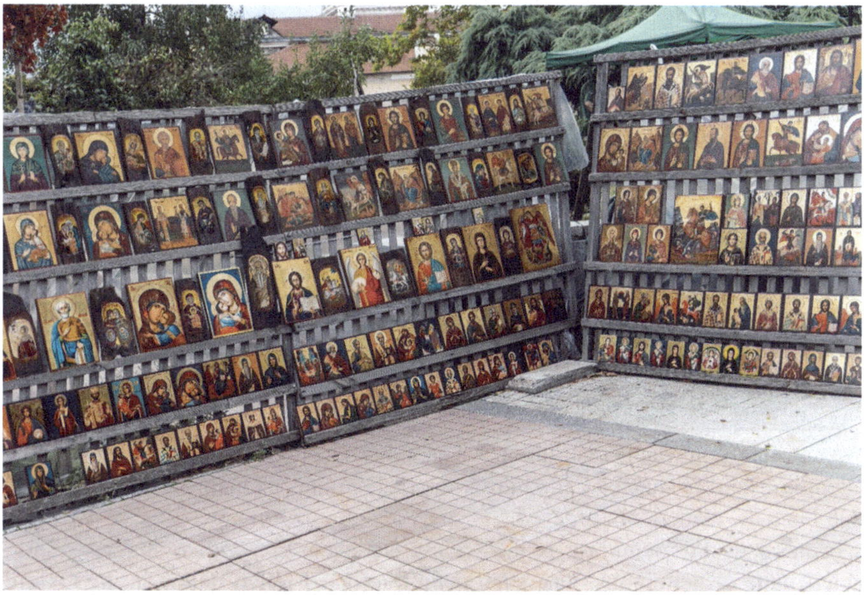

Fig. 3.2 Walls of ikons for sale at a street market in Sofia, Bulgaria, 19 September 2024. (Photo by Anna Soffía Óskarsdóttir with permission)

thousand years after the story was supposed to have happened and more than 2000 km away? The sole authority would have to be that Duccio was known to be capable of producing lifelike images, but even he could probably not do so at such a long time after the (supposed) event.

Duccio's unquestioned belief in the stories of the Old Testament meant that he had an idea of Jacob as a person, so he painted a person. He connected his image to the story in the Bible by painting a cartouche over the image of a person that contained a fragment of text that would have been known to the seers at the time. Those seers were embedded in the same culture as Duccio through the telling of the story in the religious ceremonies of the time. Now, 700 years later, many seers—tourists visiting Siena—are probably embedded in cultures less dominated by those religious stories, often in countries thousands of kilometers from Italy and further from the lands where the stories were originally told. What is the art? It could be just the picture, or it could be the whole set of interactions within the cultural context of production, seeing and knowing—the art-system.

Then, we must evaluate the skill involved in art, because skill is also involved in almost all products of human manufacture. Indeed, skill is almost a definition of those "crafts" which are so often contrasted with art (for example Ingold, 2000a).

3.2 Can Anyone Define What Art Is?

How then do we go about defining art? Some would avoid the question altogether, such as by choosing to talk about visual cultures instead of art (Davis, 2022; Nowell, 2017; Soffer & Conkey, 1997) (or, indeed, about the art-system) but that may do no more than avoid the problem. Others would state clearly that "art is the creative production of novelty" (Ingold, 2000a, 350). Elisabeth Culley and I (Culley & Davidson, 2021, 331) have outlined some characteristics of the things that get called art.

> Art historians have traditionally chronicled visual and tangible objects that were made by humans—including paintings, drawings, monuments, and sculptures—and that embody extraordinary technical skill and aesthetic qualities. Unique but realistic depictions have been favored, and finely executed images spanning tens of thousands of years were uncritically subsumed within the same domain of interest … . 'Art' was also explicitly distinguished from decorative crafts, utilitarian items, and other artefacts of daily life … as intrinsically valuable for its aesthetic appeal. 'Real' art was self-motivated; produced and consumed for its own sake … .

We may question whether definition is a useful concept, though clearly many people attempt it. In this chapter I am aiming to reach an understanding of what art is so that I and the readers can proceed.

Central to any definition is the widespread view that a dictionary is a suitable source of definitions. One such definition that I found online in the Oxford English Dictionary (https://www.oed.com/dictionary/art_n1?tab=meaning_and_use#136737394) is definition 1.8.a: "The expression or application of human creative skill and imagination, typically in a visual form such as painting or sculpture, producing works to be appreciated primarily for their beauty or emotional power," with the caveat that "Although this is the most usual modern sense of art when used without any qualification, it has not been found in English dictionaries until the 19th cent. Before then, it seems to have been used chiefly by painters and writers on painting." For most specialists a dictionary is only a starting point—hence this discussion. I will have something to say about elements of this definition in what follows: creativity, skill, imagination, beauty, and emotional power. I note that missing from the criteria I took from Gombrich are the longstanding requirement for art to be lifelike, involve unquestioned beliefs, and to involve acute observation. I would include his requirement for it to create pleasure with the dictionary definition of the condition of emotional power, such that the dictionary definition extends the types of emotion that may be elicited by art.

For some, such as philosophers, art must be defined so that we understand that we are all talking about the same thing (Davies, 2015). That seems laudable, though, of course, it may be a good idea to work out who is included in "we." All of the individuals who are included in that "we" may have different definitions. As my conversation with grazier Bill has shown, where some see art others see scribbles—and the argument of this book is that what is said to be art depends not on some objective criteria but on the social perceptions of things artists tell us about. For many, art of all sorts must include concepts of beauty, so any discussion that includes the possibility of art must include some definitions of that ethnocentric word, beauty. Not for nothing we say that "beauty is in the eye of the beholder." Gombrich (1995), of course, just took it for granted that beauty was a property of art, or rather of the work artists produce. In fairness, he did contrast a painting by Peter Paul Rubens of his beautiful son (in about 1620) with a drawing by Albrecht Dürer of his aged and weather-beaten mother (pp. 15–17) (in 1514), and a tender drawing by Picasso of a hen and her chicks (from 1940–41, p. 26) with his arresting drawing of a strident rooster (from 1938, p. 27), showing that beauty may be a matter of judgement rather than of absolute standards. He could have included, perhaps, one of Picasso's sketches for *Guernica*. Under those circumstances it is not clear that beauty is a property of the image alone, so much as something that might arise in the relationships between producer, subject, image and seer. The "less beautiful" of the images in these pairs might qualify under the criterion of emotional power, as would Picasso's sketch for *Guernica*. Some of Goya's paintings were far from beautiful (see Fig. 3.25).

All art is a version of something else: it may be a representation of something that the artist sees but it can never represent the living thing—even a dead shark in a tank of formaldehyde (just supposing that anyone considers it is an example of art) is not a representation of the real thing, unless we say that it represents a real dead shark—but it IS a real dead shark. Geometric marks might represent something—a thing or an idea—but they will do so by convention and with agreed symbolism. There is, in fact, an element of arbitrariness within the constraints of convention that makes all art symbolic in some way (see Culley & Davidson, 2021).

Stephen Davies (2016, 24) wrote that "a definition should identify the elements that all art must have, such that anything possessing them thereby is art." This definition supposes that "art" is a label that can be attached to some (but not all) of the things people create, the things we see or appreciate with other appropriate senses. Of course, by using those words I am using two concepts that apply to works of art—they are created, and they can be appreciated by human senses. For what it is worth, pigeons have been trained to discriminate between the creations by different people: music by Johann Sebastian Bach from that by Igor Stravinsky (Porter & Neuringer, 1984); and paintings by Claude Monet from those by Picasso (Watanabe et al., 1995) with some intriguing extensions of their abilities. But they have never created something that would fit into the art-system. As always with such animal experiments, it is not clear what exactly the animals were responding to, and we need to take into account that honeybees can also perform the visual task (Wu et al., 2013). With humans we can question artists, as well as those who respond to art: with the animals, the attempts to work out answers are equally frustrated by the lack of communication between animals and humans. I think it is true that it is only in the one-sided relationship between humans and other animals that the animals have been persuaded to create art, sometimes by careful training, sometimes by the humans being the ones who decided that the animals had created something that could be called art, sometimes by both (we discussed the possibility of primate art in Noble & Davidson, 1996, 78–79). We will return to this.

There are some obvious exceptions to my generalization, some very well-known: the dams and lodges of beavers (Fig. 3.3); the nests of all birds (Fig. 3.4); and the constructions by bower birds (Fig. 3.5).

There are probably other examples, but I suspect in all examples, individual animals each create something—in some cases by learning as they develop in a community of other creators, that might almost lie within a definition of the species—but do not innovate new creations outside their characteristic productions (Healy et al., 2008).

Most obviously, birds of almost all species make some sort of nest, though there may be differences between the sexes in the relative contributions, and consideration of the organization of labor and materials suggests that some complexity of cognition is generally required (Mainwaring et al., 2014). In the case of weaver birds, the nests are made by elaborate knotting and mostly made by the males (Tello-Ramos et al., 2024) (Fig. 3.4).

Tim Ingold has compared the knotting of the naturally-sourced fibers of weaver bird nests with the looping of string baskets and bags young women learn to create among accomplished women of the Telefol from New Guinea (Ingold, 2000a). In both cases, Ingold argues, the skill with which the fibers are combined is learned by individuals as their bodies develop and grow, and not according to instructions or rules that could be applied by people outside the society. As with another well-known example of comparison between the products of animal behavior and of human behavior (McGrew, 1987), the similarities are strong(ish) when a particular category of production is analyzed, but the ape comparisons miss the point about the whole context of productions and use, and the similarities die there. On the human

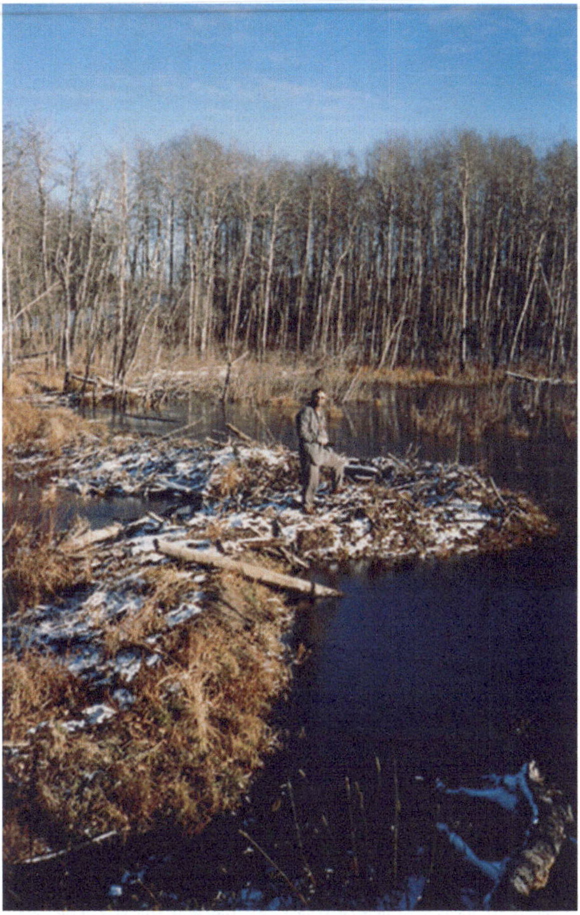

Fig. 3.3 The author standing on a dam of logs felled by beavers. Alberta 1989. (Photo in possession of the author by David Lubell and Mary Jackes)

side of the comparison the simple tools required more complex learning and manufacture (for example, a digging stick required the prior making of stone tools) and were embedded in a life full of ritual and art (Davidson, 2007, 80; Noble & Davidson, 1991, 242). By the same token the *Bilum*, or string bag, that the Telefol women make, not only requires that the strings be made (often by young children) before they are used for manufacture but is symbolically embedded in the cultural productions of the society. The *Bilum* has many forms, is used in ritual and enmeshed in the production and maintenance of social relations (MacKenzie, 2004). Women sometimes planted particular species for dye, often made dyes from plants, and decorate designs onto the bags using those dyes. Marking *bilums* using dye is one of the arenas for novel production (Andersen, 2015). In both cases of the comparison between animal and human production, although similarities can be described, the overwhelming difference is in the social, cultural and ritual context and variation of the human productions, which, in the New Guinea case, also involve richly decorated house boards (Craig, 1967). Point to point comparisons show some similarity,

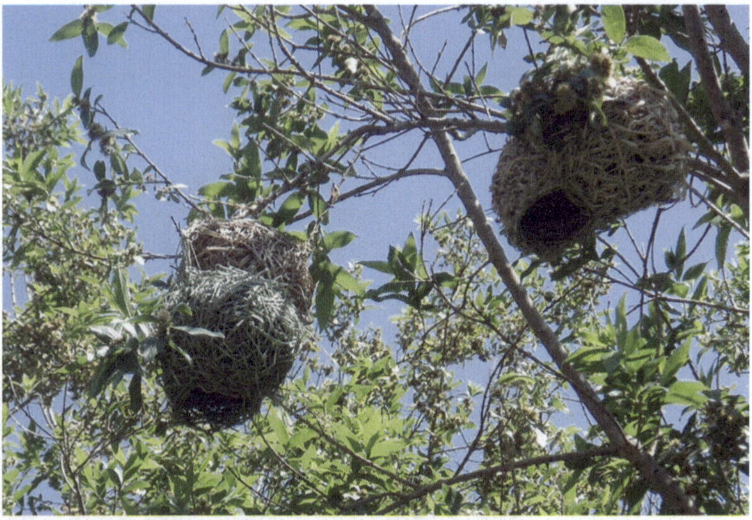

Fig. 3.4 Weaver bird nests in KwaZulu Natal, October 2006. (Photo Iain Davidson)

but as Ingold argued previously about beavers (Ingold, 1983), the idea of creative production of novelty does not arise in the animal cases.

Bower birds create a path between lines of upstanding grass and collect a variety of things to display around the path. The grass structure is relatively unvarying within a species but different between species (Madden, 2001). The collected objects are often in shades of blue, but they vary. Classically they may include blue drinking straws and milk bottle caps, but they have also been shown to include stone artefacts, or animal bones from archeological sites (Solomon et al., 1986). The bird does not produce those objects, just assemble collections around the bower. Those shown in Fig. 3.5 include broken glass of different colors, bones, bullet cases and beer-can ring pulls. The display is not a nest and has meaning—universally thought to be to attract a mate—but not on the basis of an analysis of the detailed contents. The display appears to vary from bird to bird, from generation to generation, though the content of the collections appears to be almost irrelevant. As with almost all animal research particular aspects are seen to be increasingly complex, as more is known (Madden, 2008).

Different types of definition of art, then, might not help very much if they are confined to descriptions of the state of a thing, without taking into account its contexts in evolutionary or developmental terms.

Consider for a moment three main types of definitions (Table 3.1): "Ostensive" definition ("Here is an example of…"); "Lexical" definition ("We can replace this word with the phrase…"); and "Operational" definition ("We name the thing that has all of the following characteristics…"—Davies's definition). When we point to a painting by Margaret Olley (Fig. 3.6) and say: "That is an example of art," we are using an ostensive definition. Olley's copies of paintings by Henri Matisse are an indication that she has dredged her picture (and her life) through the art of Europe

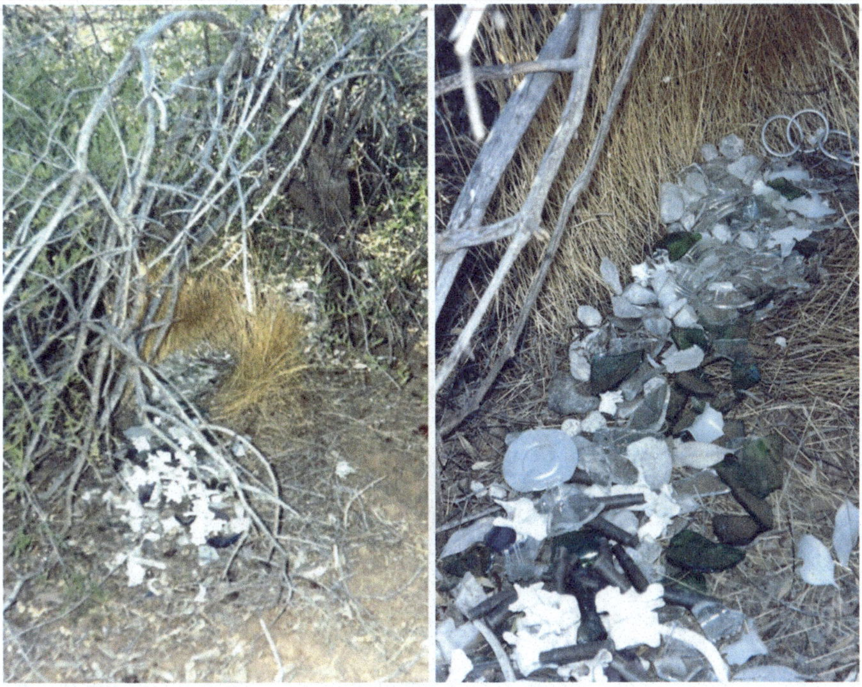

Fig. 3.5 Bower on Mount Tracy Station, near Cloncurry (Left) with detail of the collections including shell cases, bones, broken glass, tin lids and ring pulls from beer cans (Right). (Photo Iain Davidson, 1991)

(and New York). That does not make her painting into art by itself but is an indication of the process of acquiring the creativity that Brett Whiteley pointed to. The problem with ostensive definition is that knowing that one thing is an example of art does not enable us to identify that another thing is also art, that a sculpture or any other work is also art.

Ingold (2000b) offered a lexical definition in suggesting that the word "art" could be replaced by "image" or "depiction". As I have discussed previously (Davidson, 2013, 22–23), the idea has some merit, but when you scratch the alternative words—image or depiction—their definitions take us back to questions of likeness or representation. We could choose to use those words instead, and we would avoid the uncertainties of the word "art," but we would lose something at the same time. The earliest things that have been called art do not include likeness or representation, but involve non-depictive marks, from sites like Trinil, in Java (Joordens et al., 2015) at 500 thousand years ago (Fig. 3.7) or Blombos, in South Africa, 100–75 thousand years ago (e.g., Culley & Davidson, 2021) (Fig. 3.27 Right). At some stage, our ancestors learned that marks could be used not just to make patterns but to make "images," "representations," and art. The patterns of Trinil or Blombos might have been used in the same way as other art, but for them to achieve this, there must have been stories attached to the marks. Such stories might have depended on

Table 3.1 Categories of definition

	Simple gloss of meaning	Art
Ostensive	This is an example of…	Margaret Olley's *The Yellow Room*
Lexical	We can replace this word with the phrase …	The word "art" could be replaced by "image" or "depiction"
Operational	We name the thing that has all of the following characteristics…	Art is lifelike, involves unquestioned beliefs, is beautiful, uses skill and acute observation, and creates pleasure
Functional	This is what art does	Art is "deliberate communication through visual means"

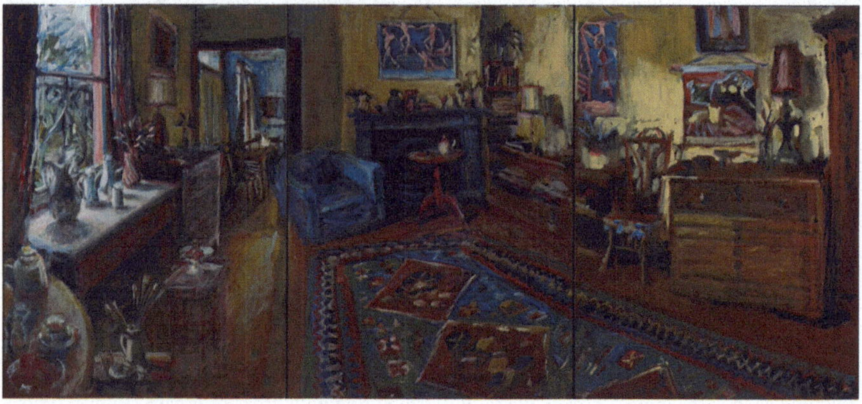

Fig. 3.6 *The Yellow Room* by Margaret Olley from the New England Regional Art Museum collections. (Permission from the Philip Bacon Gallery 20 December 2024)

conventions, but to demonstrate that would require that the patterns were repeated within the society of the makers, and that is not true for most of the earliest examples.

The exceptions are found at Blombos and Diepkloof (Fig. 3.28) in South Africa. At the first, all the attention for Blombos has been paid to the larger piece of ocher with a crosshatch pattern scratched on it (Fig. 3.27 Right) (e.g., Malafouris, 2013, 183–192). Just as with the Trinil shell, we tend to think that the pattern itself makes it significant, but at Blombos the cross-hatched pattern was matched by another piece which seems to repeat the actions, and thereby establish a convention of producing patterns involving crossed lines. I have discussed the argument in the Essay at the end of this Chapter, and in a paper with Culley and Davidson (2021), and will return to the importance of repetition in Chap. 7.

The dominant issue (Davidson, 2013, 22–23) is how human ancestors came to see that by making marks, which we might call the first objects, those same people could represent second objects (which I call subjects)—those other than the marks themselves—because the marks resembled (were a likeness of) some aspect of those subjects. Could they do so without the need for stories? Human ancestors unfamiliar with the capacity of marks to create a likeness of things that they knew in real life may have required that stories were told about the likeness to the second

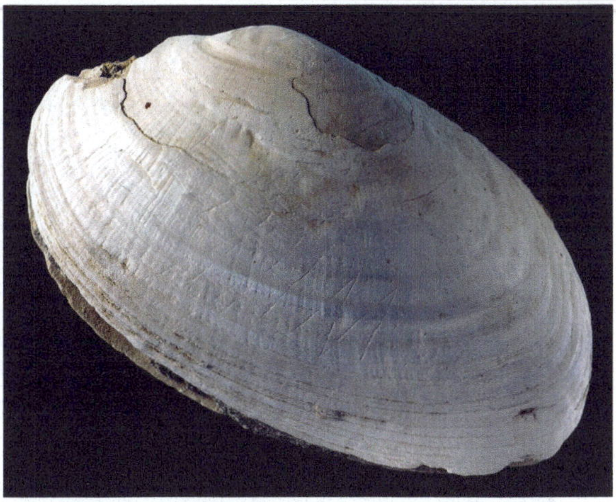

Fig. 3.7 Apparently patterned lines engraved on shell from Trinil, Java approximately 500, 000 years ago (Joordens et al., 2015). (Photo by Wim Lustenhouwer, (VU University Amsterdam). Registration number of the shell at the Naturalis Biodiversity Center RGM. DUB.1006.f)

objects in order for the seers to recognize the trick. That recognition was necessary for the possibility of creating marks with the intention of representing other things. And there is another key word in the creation of art—intention.

In the case of Olley's painting in Fig. 3.6, or any of the other images in this book the same question arises, just as it did that first time, when anyone wants to decide whether a work is or is not art: how do any marks come to represent something other than themselves? Once that convention was established, artists turned the question around to be: "When an artist wants to represent a particular subject, how do they make marks to achieve that?" Art schools and apprentices' masters devote themselves to teaching how to do it.

Operational definitions are altogether more difficult because they are like Davies's aim to produce a definition (Davies, 2016) and require us to be able to recognize the characteristics that make a thing that should be called "art." It could be said, in fact, that it is not a definition of art at all, but of whether the producer has achieved the requirements for us to call it art. It is very difficult using an operational definition to separate the art of Van Gogh from a photograph of the same image, or to say why wrapping paper or pictures hung on the walls of a motel are not art. My caption for the photograph of the three scarves on a couch (Fig. 3.8) is "But is it art?" or, indeed, has my photograph reached the requirements of the definition of art simply because it shows likenesses of images by Paul Klee, Georgia O'Keefe and Gustav Klimt?

Bob Layton gives a fourth definition of art that is popular among rock art specialists—art is "deliberate communication through visual means" (Layton, 1992, 1). This is, of course, not one of the three types of definition, but a fourth: we might call it a "functional" definition ("This is what art does"). It is problematic as a definition because the question of communication leaves open the possibility that it is not the "art" that communicates, so much as the associations people recall of the artist and their work and the art is incidental. In the same way as the operational definition, the

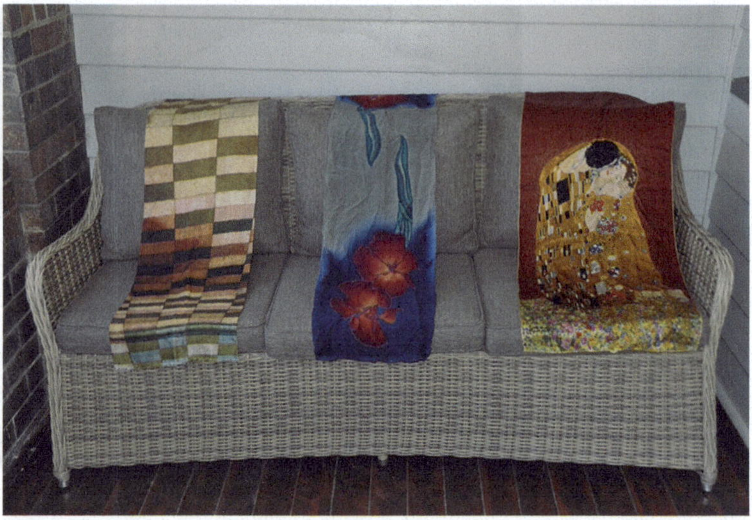

Fig. 3.8 "But is it art?" Three scarves on a couch, (from left to right) printed with likenesses of paintings by the well-known artists Paul Klee, Georgia O'Keefe and Gustav Klimt. (Photo Iain Davidson).

issue of communication is secondary to the question of whether the marks make up something that is art. The next chapter discusses the way art communicates.

Davies (2015) produced an operational definition with three parts:

> something is art (a) if it shows excellence of skill and achievement in realizing significant aesthetic goals, and either doing so is its primary, identifying function or doing so makes a vital contribution to the realization of its primary, identifying function, or (b) if it falls under an art genre or art form established and publicly recognized within an art tradition, or (c) if it is intended by its maker/presenter to be art and its maker/presenter does what is necessary and appropriate to realizing that intention.

As with so many attempts to define art, it might be simpler to emphasize instead the key words in this definition: skill and achievement; demonstrating aesthetics somehow (notice that Davies does not mention beauty, and aesthetics is undefined), or being said to be art by someone who may or may not be the producer of the art. The second point, of course, has the circularity that I have previously identified in Davies's attempt or Layton's—we need to know what art is to know whether a particular candidate meets the preconceived criteria. The trouble is that skill and achievement are demonstrated in many ways, such as basketry or building a factory to process Liquid Natural Gas. As for the others, both demonstrations and definitions are circular. First, the concept of what is aesthetically pleasing is not objectively identifiable and may, in fact, derive some of its nature from the qualities established by art. That is to say that we identify what is an aesthetic goal through living in a culture in which certain things are said to be art because they meet aesthetic criteria. And they are said to be art either by us or by the people who made them. This is a perfect formula for entrenching power in the eyes and minds of people who are interested in art. And, by choosing not to use the word "beauty"—the word in the Oxford definition—Davies appears to highlight the question of the relationship between the two concepts, beauty and aesthetics. That relationship, as I will show, is not unproblematic. I am attracted slightly by the gloss of this as "Art is art if it is art."

3.2.1 "Painting" by Numbers

There are several contexts in which archeologists use the idea of "art" and Robb (2017) distinguished distinct things, using different typefaces. As it is impossible (or, at least, very cumbersome and difficult) to talk about different categories of anything by reference to the font in which it is printed, I decided to elaborate upon these categories using numbers (Davidson, 2020)—Art, Art 1, Art 2 and Art 3—where the number rises as we move backwards in time (Table 3.2). Robb recognized that the word "art" can be used as a general concept or, by switching to a different font, as a term for "*a modern, Western set of concepts, institutions and practices,*" which I call Art 1. It can be used slightly differently by archaeologists "TO CONNOTE AN ARCHAEOLOGICAL CATEGORY OF MATERIAL, USUALLY REPRESENTATIONAL OBJECTS SUCH AS FIGURINES, PAINTINGS, ETC." which I call

Table 3.2 Comparisons between the use of different typefaces to refer to art (Robb, 2017) and the numbering system used here

Art	Art 1	Art 2	Art 3
	Modern Western concept	ARCHAEOLOGICAL STUFF	Before the emergence of writing
The Parthenon, Pompeii, Angkor Wat, The terracotta warriors of Xian, Duccio, Caravaggio, Turner, Hooper, or Roberts Rothko, Basquiat, Tracy Emin, Emily Kame Kngwarreye, or Banksy	Toledo Cathedral, *The Night Watch, Las Meninas*, Matisse's *Dance, Old Antequera, Guernica*	Site 3, La Bastida de les Alcusses, 'Ain Ghazal, Chauvet Cave, Altamira, Parpalló, Teotihuacan, Tenochtitlan, Chichen Itza, Main Cave, Greek Red Figure pottery, Nine Mile Canyon, Yarrowick	Bruniquel, Trinil, Berekhat Ram, Blombos, La Pasiega, Diepkloof, Leang Bulu' Sipong 4
	Robb, 2017 *Cambridge Archaeological Journal*	Robb, 2017 *Cambridge Archaeological Journal*	This book, Davidson, 2020 *JAMT*, Culley & Davidson, 2021

Art 2. I recognize a third category (Art 3) which includes all those examples of early archeological art before but leading to Art 2 and on to Art 1. One of the questions is whether all of Art 3 is really art. The proposition here is that the presence of early marking practices in Art 3 may tell us things we did not know about the evolution of the cognition of our ancestors that led to Art 2.

There is much content in this book about how human beings got to the position of creating art that meets a dictionary definition by historically passing through all the types of works which would not meet its terms. In our terms, then, Art 3 consists of stuff that might not be recognized as Art 1, but the people who produced it were, in some sense, the predecessors of those who made Art 1. Art historians, and Queensland graziers, might be less concerned about Art 3, and what I have to do is to show you what the differences were, and what the changing contexts of art were.

The aim here is to define art in such a way that we can recognize that it is not only the physical object such as the rock and paint of the Leang Bulu' Sipong 4 anoa (Fig. 2.12), though it entails it. Art is also contained in what the painter said and the reaction of the people who see the painting in that context, or knew the story passed down in their society of the hunt of the animal.

Art is involved in paintings by Duccio and other masters of church painting. The paintings arose from religious stories that Duccio did not need to tell because they were embedded in mediaeval culture and the contexts of the church both for him and for other people who saw them. On occasion, as in Duccio's painting, the producer felt it necessary to allude more explicitly to the story by including a specific biblical text. But some would have it that such fourteenth century masterpieces were not art (Shiner, 2001).

Art 1 consists of those things that are called Art in the modern world. They were made by artists such as Rembrandt, Velázquez, Matisse, and Leonard Lindsay. I would be happy also to include Duccio, Caravaggio, Turner, Hooper, or Roberts as well as Rothko, Basquiat, Tracy Emin, Emily Kame Kngwarreye, or Banksy. Add your own top five favorite artists. Art 2 is archeological art, such as the Moixente Warrior (Fig. 1.1) or Greek Red Figure Pottery (Fig. 4.8 as in the Essay at the end of Chap. 4), Iron Age mirrors, or the wall paintings of Pompeii (Fig. 1.8). Works about Art 2 are often about objects that have been transferred in one way or another to museums outside the country where they were produced—and transferability is an important issue since it is fundamental to the commerce in art in the last 300 years. For our purposes in this book, Art 3 is the earliest archeological art that some recognize as art, but some might not, such as that from Blombos (see Fig. 3.27 Right and discussion in the Essay). It is this group we want to understand and put into the context of all other categories of art.

I will show what sorts of works fall into each category, and this sort of discussion of meaning will enable me to address the question of rock art as art (see e.g., Porr, 2019). My initial conclusion is that the earliest art, Art 3 (Fig. 3.27 Right), differs quite fundamentally from the stuff everyone calls Art 1 (Fig. 3.6). I created a table to show the contrasts—which I call the seven types of ambiguity—and that will be the topic of Chap. 7. The conclusion of such an analysis was that during and through the period of Art 3, people made marks which persisted in the environment; the marks often had repeated patterns; once the capacity to make patterns was noticed through such repetition, people could attach meaning to patterns and sometimes told stories around that persistent patterning. I suggest that the stories and the marking had a profound influence on the cultural context of human minds and the cultural context that emerged came to be the context for all other forms of art. As Sutton and Anderson (1988, 4) wrote "art … connotes visible and intentional signs made by intelligent beings," but its existence also affected intelligence and hence what sorts of visible and intentional signs might be made by intelligent beings.

3.2.2 Art and Life

This sequence of behaviors can be understood if we also look at how the concept of art came into being during the evolutionary process—my ultimate aim in this book. We use the word art because that is what we are familiar with now wherever we live in the world. The big question becomes how, in the course of history, did stuff that was so widespread turn into the stuff that artists, critics, curators, philosophers or other interested people call art today? Are all the things and relationships that get called art really the same sort of thing, or should the word be confined to a short period of time in a small region of one continent with roots in one of the world's many religions? I am addressing central questions here. I am an archeologist, and, as we try to write what I call archeohistory, archeologists ask how, without supernatural intervention, things come in to being. It is all too easy to just say "and there was light" or "aliens built the pyramids" but if we thought for a moment we would realize that there are answers to these questions that do not involve the supernatural. My question is how making art became part of human behavior and became part of the history of relationships among people. In so doing, pictures also acquired respect equivalent to that which you might feel for another person. Gombrich (1995, 40) wrote:

> Suppose we take a picture of our favourite champion from today's paper—would we enjoy taking a needle and poking out the eyes? Would we feel as indifferent about it as if we poked a hole anywhere else in the paper? I do not think so.

Decades before, Franz Boas's book on *Primitive Art* raised awareness of the need to consider that the confines of the concept of art needed to be broadened beyond the limits imposed by many art historians of the time. It was originally published in 1927 and Helen Gardner's first edition of *Art through the Ages* was published the year before—in the decade before the advent of widespread color printing of books. My copy of *Primitive Art* is a 1955 facsimile of the original, published 5 years after the first edition of Gombrich, but 40 years before the 16th edition I have on my desk. Boas (1955, 3) (Fig. 3.9), a student in the 1870s in northern Germany, included a passage eerily echoed later by the Austrian Gombrich:

> I believe if a boy should observe someone spitting on his photograph and cutting it to pieces he would feel duely [*sic*] outraged. I know if this should have happened to me when I was a student, the result would have been a duel and I should have done my level best to do to my adversary *in natura* what he had done to me *in effigie* and I should have considered my success as a compensation for the harm done to me.

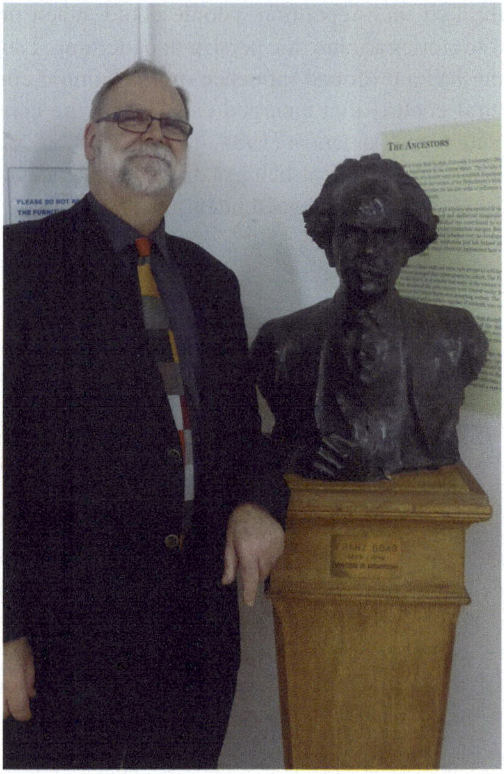

Fig. 3.9 The author (wearing a necktie influenced by the artist Paul Klee) standing next to a bust of Franz Boas, the founder of the Anthropology Department at Columbia University, and author of *Primitive Art*. I did not feel inclined to spit on the bust. (Photo in possession of the author)

Both Gombrich and Boas were writing about the mystique of the image, to the extent that anyone seeing it is reluctant to defile the paper on which the image is printed. In the process of making marks on paper turn into a likeness of an individual, the image they formed had become something more than marks. How did it happen that an image evokes the hesitancies and sensitivities of attitudes towards the subject it is a likeness of? That transformation is one of the mysteries of art. As a human being, I am interested in the art that surrounds all of us in the world and in the social divisions that are essential to the way we produce it and appreciate it. This book is about all these things.

Rock art is principally a concern of archeologists, yet even within archeology there has been a lot of discussion about whether art is the appropriate word to use when writing about it (see, for example, Bradley, 2009, Ch. 1), even if we call it Art 2. Some archeologists prefer to use the words, "pictographs" (for paintings—all images made by adding paint) and "petroglyphs" (for engravings—all images made by removing some of the rock), amongst other things, to avoid talking about art. Others insert a hyphen, so that readers will not be confused between rock-art and other art, to which it may not be related (Chippindale, 1985, 2001). I think they are related, and it is false to suggest they are not. Whether the hyphen reduces confusion is in the eye of the beholder.

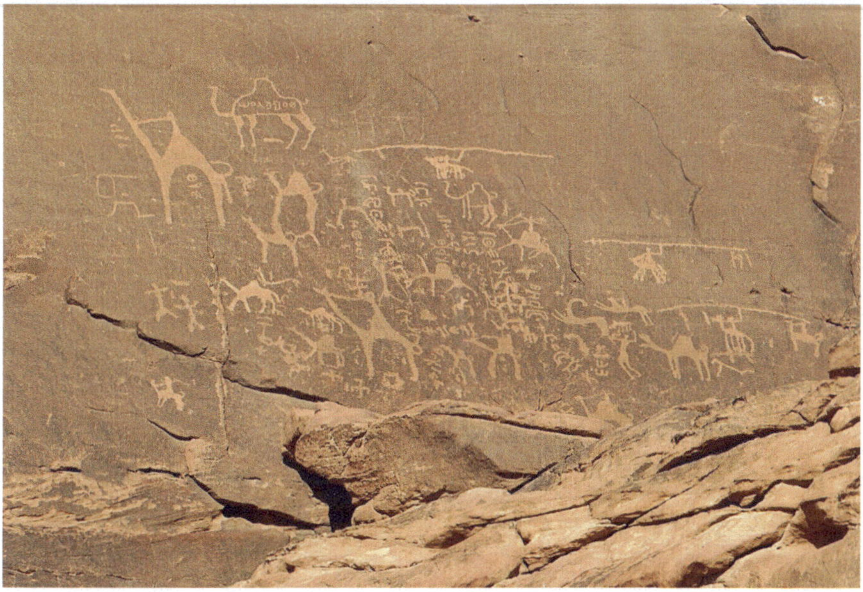

Fig. 3.10 Petroglyphs in Wadi Rum, Jordan. May 2014. (Photo Iain Davidson)

Among the general public, for whom rock art is not so salient, much thinking would align with the grazier Bill's contrast between art and scribbles. The important point is that rock art is composed of marks on rocks (Fig. 3.10). Much more important than my flippant statement would imply is that rock art persists on rocks in particular places.

Some would say that it seems to have little to do with the art in galleries or on the walls of houses or motels, but by the end of the book I will show that these different categories of art do have things in common.

Some rock art is self-evidently composed and like a work of art, such as the Great Hunt scene in Cottonwood Canyon in Utah (Matheny et al., 2004) (Fig. 3.11). You could put a frame around this image. I have discussed some of the indeterminacy of this image elsewhere, in particular whether it was all composed at the same time (Davidson, 2021). I argued that there seem to be two (or three) independent episodes to the construction of the panel, though there is no differential weathering which would allow us to identify that there was such a sequence. The images of the bighorn sheep seem to describe a natural winter grouping in which rams, ewes, lambs and others are together (Matheny et al., 1997) but the animals have little agency in the image—they are not doing anything. In addition, there are four, maybe five, (and a half) images of people who seem to be doing things with bows and arrows (but without impressing the sheep). Finally, there is a central figure that looks human, but has no arms, and it has a head that may be a bison headdress. I argued previously that it probably represented a ritual, interpreted by a narrative (Davidson, 2021).

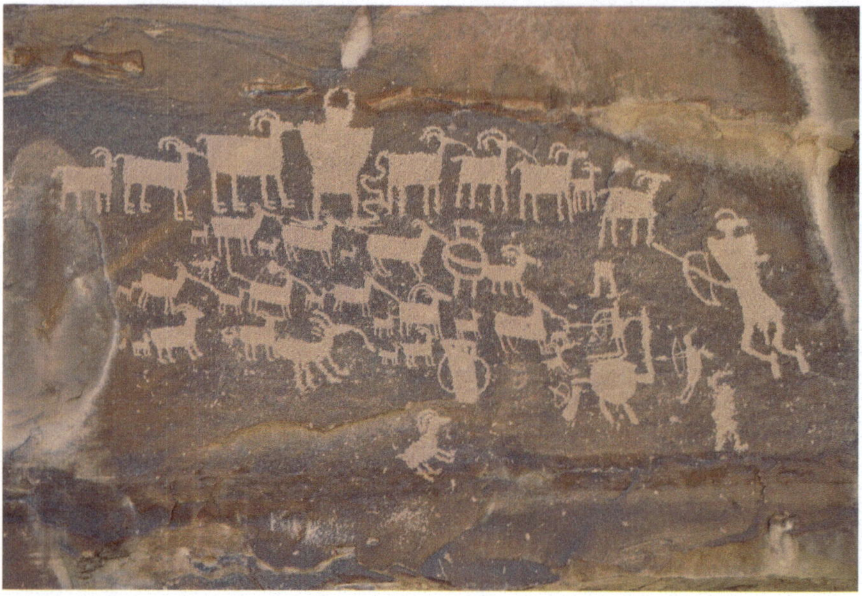

Fig. 3.11 The Great Hunt panel in Cottonwood Canyon, Nine Mile Canyon, Utah. (Photo Iain Davidson)

On the other hand, as we have seen, most art historians seem to mention rock and cave art just because it is there, with little attention to any nuances, and apparently little thought about how such phenomena are related to those that are their primary focus. Some (e.g., Davis, 1993), though, do engage in nuances that make the issue more complicated, especially about culturally determined aesthetics—we will consider what "aesthetics" is in a moment. Davis (2017) has returned to these questions with renewed vigor.

For some colleagues, art does not need defining and we muddle through without any definition. I once defined art as all the things we call art (Davidson & Noble, 1989, 136 fn. 6). That may be the best definition there is, as we shall see, but it may mean we are talking past each other. What I see as art is not what you see as art—and that might be one of its important characteristics. If it is true, then it seems really limiting to try to define art in such a way that we can all agree about which things fall into the category and which do not, especially as mark-making evolves through time from not being art to being art. I have dealt with the problem in detail elsewhere (Davidson, 2020; Davidson & Noble, 1989) and much of this book is a discussion of how early art, after passing through several stages in different parts of the world, turned into the art that people such as art historians, grazier Bill, or you and me, deal with. Is archeological art similar to other sorts of art, are there similarities between the things that I call art and those you do?

As an example of this problem, consider for a moment the First Nations Anaiwan rock art site at Bulagaranda, once called Yarrowick, on the Northern Tablelands of

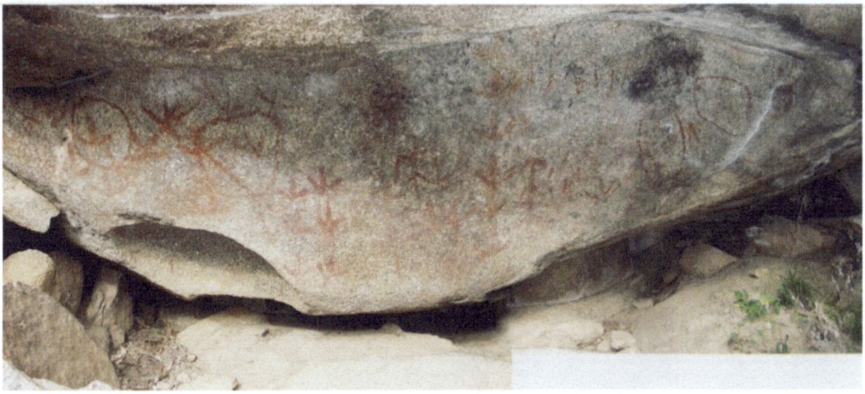

Fig. 3.12 Rock art at the Yarrowick art site, New South Wales, Australia (Photo Iain Davidson, 2007). (Permission to reproduce kindly given by Steve Ahoy, board member of the Mt. Yarrowick [*Bulagaranda*] committee)

New South Wales. This is a well-known rock art site that I have been visiting since 1974 and once heard an initiated Dhangatti Aboriginal elder, from an adjacent tribe, talk about that same year (Fig. 3.12).

I suspect that grazier Bill would suggest that this First Nations Art was not an example of art but of scribbles. And yet, there is in the Australian Parliament House a painting by the noted Australian artist Leah MacKinnon which she called *Landscape 3. Mount Yarrowick & Wilhelmshohe* (Australian Parliament House Construction Authority, 1989, 47) (Fig. 3.13). As MacKinnon said in an email (10 December 2024): "The series of diptych Landscape works were all depictions of comparisons between … The transplanting and overlaying of European/Anglo Saxon beliefs and cultural practices upon the indigenous land, beliefs and culture. The works themselves were conceptual comparisons of philosophies and cultural practices. 'Landscape 3' comparisons were set in the visual landscape of granite tors in the New England area of NSW". The work, therefore, represented what MacKinnon saw in the landscape combined with a story about cultural practices in that landscape, and showed how an artist can use art to emphasize, in different ways, the political situation.

The work was part of "an ambitious art program" which involved the formation of an "Arts Advisory Committee" and in which "artistic inspiration influenced the building's final form" (Australian Parliament House Construction Authority, 1989, 6–7). We can say, then, that some people consider the MacKinnon work as Art 1, yet it derives closely from the art at the site itself (which I used to regard as Art 2, though now I find there is no category for places of importance to Indigenous people except through this classification of the paintings). It could be argued that, in terms of the system of relationships I am arguing for here, the site in its setting is properly the art object (Fig. 3.14) either for the way any seers of the paintings must pass through the setting of the site on the rocks in the woodland, or from the way in which the painters chose a surface that nestles under the massiveness of the granite

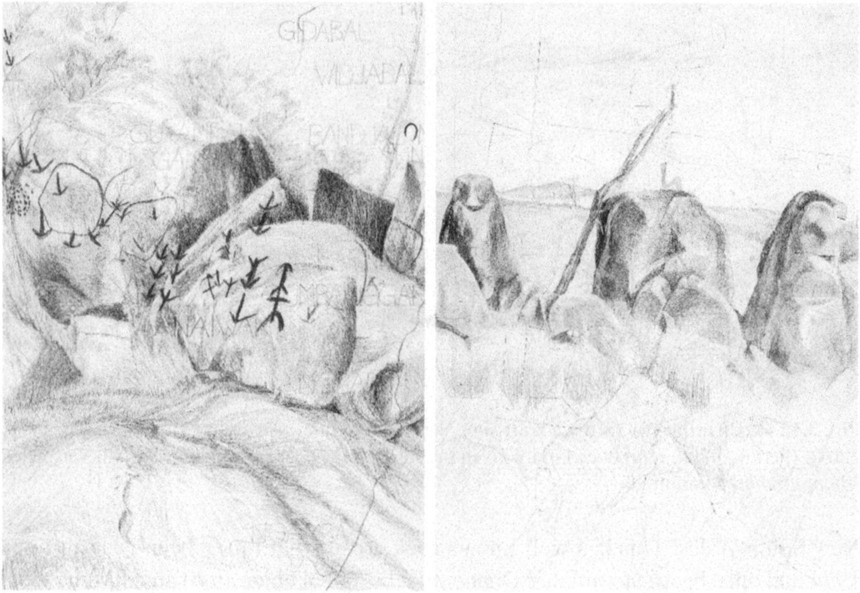

Fig. 3.13 *Landscape 3. Mt Yarrowick & Wilhelmshohe*. Diptych by Leah MacKinnon. (Permission granted by MacKinnon and by Parliament House)

tors. It seems more appropriate to consider that the whole formed an art-system consisting of the marks on the rock, the rocks in the surroundings as well as the people who made the marks, and the members of the society for whom the marks were important.

MacKinnon has told me that the art that was commissioned by the Parliament was a diptych which included the second image of the granite tors. The close relation between the image from MacKinnon and the image on the rocks suggests that grazier Bill's dichotomy between art and scribbles poses more questions than answers and highlights the social context of his judgment. I have no doubt that both the rock art and the Parliament House painting of *Landscape 3* should be called art, even though the original songs and stories may no longer be told.

3.3 Art and Beauty—Aesthetics

How do we address questions of skill, imagination, beauty, and emotion without seeing the production of the art and hearing the stories? How do those questions relate to the art-by-numbers I have just outlined? Is all art skillful, and is everything that is skillful inevitably to be considered as art? What are the processes necessary for someone to express their imagination? Does it always require skill? Does it always produce something of beauty, whatever the person's imagination, or is there art that is not beautiful by any criteria?

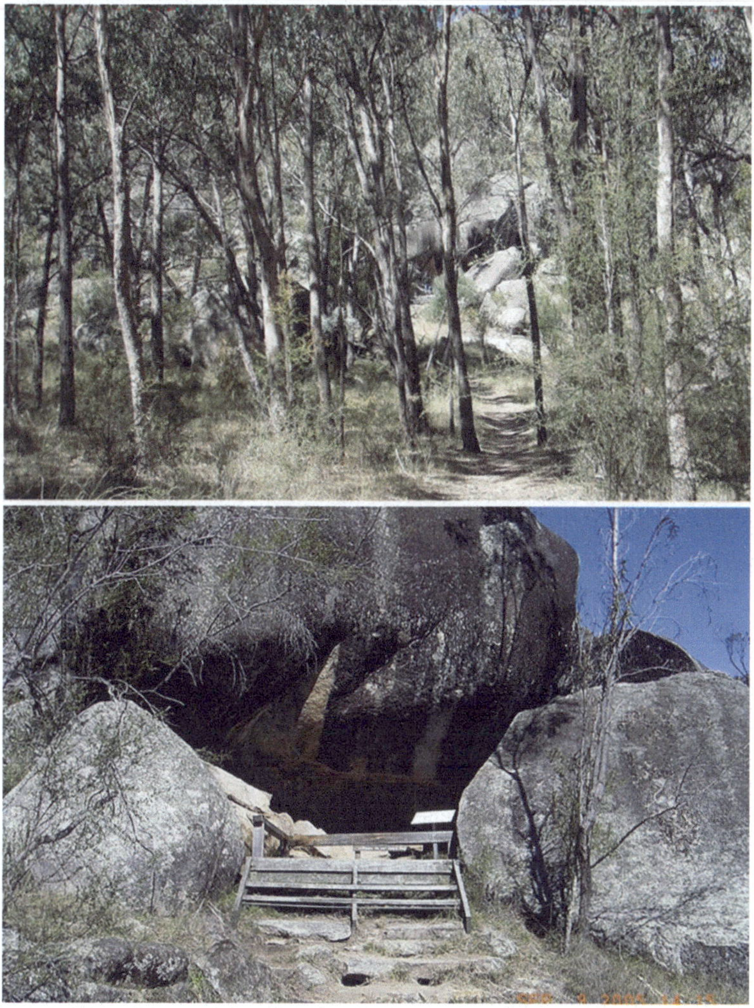

Fig. 3.14 Yarrowick Art site from the west. (Photos Iain Davidson, 2008)

These are all questions about the production of art, but there are other questions about the seers of the art which partly resemble the story about a tree falling in the forest (or its humorous equivalent about a man in the forest). Is beauty something that only humans are in a position to appreciate or is it an intrinsic quality of a thing? What is the relationship between skill and beauty? Is art about the interaction between us, with the eyes of beholders, and the thing of beauty? Is it true that things are beautiful, or do we think they are, but others do not? Is beauty something that things (including other people) "have" or is it about the relationship between us and those things? Is beauty different from one culture to another? Likewise, does any object stir emotion by itself—as I have heard people say they felt when confronted with a Rothko painting on the wall in front of them (I was unmoved)—or is it a property of the person, the painting and the whole context of knowledge of the painting? How is the emotion of the producer related to the emotion of the seer? How is the imagination of the producer related to the emotion of the seer? How is the emotion of the seer related to beauty (if any) produced by the artist?

Does it make a difference, for example, to know the history of a painting by Rothko that was exhibited in Harvard's Holyoake Building? Did Rothko's production of the painting require skill? What was he imagining when he painted it? Does anyone think it involved beauty at any stage? Why do people see work by Rothko and experience emotions? Is that emotion related in any way to judgments about skill, imagination or beauty? The painting was one of a series on display in the Holyoake Building in Harvard, and there the whole series deteriorated badly: the damaged images have been recovered from the original so that they can be exhibited again (Walsh, 2014). In the context of Rothko self-identifying as an artist, that might be enough to identify the image as art. For others, the whole process of production, donation, damage and restoration is part of the system that makes the experience of the image one of engagement with art. But is a damaged object still art or only the original undamaged work; or has some part of the process removed it from the category of art? Is Rembrandt's *The Night Watch* still a work of art even though for 300 years it has been exhibited with panels cut off it and will it be more or less of a work of Rembrandt's art if some artificial device "restores" those panels? Leonardo da Vinci's *Última Cena* (Fig. 7.22) was, after all, badly damaged within decades of its creation, and restored from a copy. Perhaps it is the story that makes such an image continue to be revered.

And then there is the question of the relation between art and aesthetics. Many authors write as if there is no difference. Boas, for example, in introducing the art of those we would call First Nations, states that "all human activities may assume forms that give them esthetic values" (Boas, 1955, 9), and then (p. 11) that "the will to produce an esthetic result is the essence of artistic work." For Pasztory (2005, 191) "[i]t is precisely the separation of the aesthetic aspect of works of art from their functional, social, and religious aspects that made it possible to see alien arts as nonthreatening and nonheretical." For James Clifford (1988, 222 fn. 4) some authors claim (as I do here) that art can only be understood in its original context, while others assert that aesthetic qualities lift other art into the category of "masterpieces" that have "universal or at least transcultural human sensibilities." Others have acknowledged the importance of aesthetics, even publishing one of their attempts at definition of art in the *Journal of Aesthetics and Art Criticism* (Davies, 2015).

Arnold (2020, 89) emphasized the importance of aesthetics, which, she says, entered epistemology in the mid-eighteenth century as a result of the recognition that sensory perceptions should have equal value with logic. Its original definition was "the science of sensible cognition" (Guyer, 2020), where "sensible" means something like "perceived through the senses." The Macquarie Dictionary (2017) has a philosophical meaning of aesthetics: "1. The science which deduces from nature and taste the rules and principles of art … 2. The study of the mind and emotions in relation to the sense of beauty." Shiner (2001, 140–146) addressed the appearance of this divided approach to what had once just been a question of taste. For him, the important distinction was between "crafts" which aimed at "sensual pleasure or bare utility" and "the fine arts" which were the subjects of "higher, contemplative pleasure." Yet it is possible to write of the "craft" activity of knitting in Shetland in the following terms (McHattie et al., 2018): "To speak of knitting is to speak of a mode of cultural production and describe it not simply as a reflection of some essentialist culture but as an arena for the constant process of negotiating the meanings that constitute culture and in so doing is intrinsic to the fabric of Shetland life." "[T]he constant process of negotiating the meanings that constitute culture" sounds very far from "sensual pleasure or bare utility" and much closer to contemplative pleasure, whether higher or not.

The conceptual distinction was designed to deal with the way the senses respond to art (Shimamura & Shimamura, 2012, 3) but seems to have become mired in the niceties of eighteenth century German philosophy.[1] Part of the problem was that the choice of arts which were to be "contemplated" was actually determined by the unquestioned beliefs in the elite status of the privileged people who selected them and who considered that the taste they took for granted—*their* taste—actually defined some objective qualities. Those unquestioned beliefs looked down, ignorantly, on the poor, on women and on people from outside Europe. Although Immanuel Kant did not ever visit Africa, and commented more than one hundred years before the British stole the Benin bronzes from Nigeria (Fig. 8.29) (Hicks, 2020; McDougall, 2008), Shiner (2001, 139) quotes him as saying that "the Negroes of Africa have by nature no feeling that rises above the trifling." On such a basis, the concept of the distinction between sensual pleasures and contemplative pleasure comes to us through Kant, and is, at least in part, a reflection of the cultural (and racist) values of his prejudice. The lack of aesthetic sensibility among Africans, according to Kant, is because of their "nature."

The distinction between sensual and contemplative emotions can be likened to that raised by Livio Dobrez (2022) between the sensual "I like it" and the contemplative "It is good." The first suggests that individuals only have subjective opinions about how they are affected by any work and the second that among a group of people (of the right class, gender and race, in Shiner's account) there can be objective criteria by which they can reach a single conclusion about the quality of art. The sensual is all about agency—who is doing the liking. The contemplative pretends to remove the agency of the seer(s)—except insofar as the agency is about gaining membership of a group qualified to make such judgments—because the pleasure is thought to derive from the qualities of the object being examined. Those properties are the ones that came to be called aesthetic.

[1] Others were disdainful—the Oxford English Dictionary quotes the architect Joseph Gwilt 1842 as saying "There has lately grown into use in the arts a silly pedantic term under the name of Æsthetics … it is however one of the metaphysical and useless additions to nomenclature in the arts in which the German writers abound,"—more Gwilt's attitudes, I suggest, than useful in understanding aesthetics.

There is a delightful definition of aesthetics (Palmer et al., 2013, 179): "we take aesthetics to be the study of those mental processes that underlie disinterested evaluative experiences that are anchored at the positive end by feelings that would accompany verbal expressions such as 'Oh wow! That's wonderful! I love it!' and at the negative end by 'Oh yuck! That's awful! I hate it!'" The implication would be that something is great because I love it, and that I hate some other thing because it is awful. The important conclusion, derived by Stephen Palmer and his colleagues from experiments to determine human attitudes to color and spatial composition, was that feelings were not as free of prior knowledge as the experimenters expected. In fact (Palmer et al., 2013, 86) "people like/dislike a specific color to the degree that they like/dislike all of the environmental objects that are associated with that color." Those unquestioned beliefs that Gombrich wrote about are actually important determinants of attitudes that get called aesthetic.

When people produce art, everything they make is material or has a form that can be perceived. Through such perceptions of the properties of the work, seers can generalize about such aesthetic properties and discern the nature of the art. An important element of that art is the attached stories (Green, 2014).

Let us suppose that all of that is true: the artist has a story, or at least something in mind, and uses their skill and imagination to produce an object which we call art, and the artist ensures that it is, perhaps, beautiful. I, a seer, look at the object and see that I think I like it, and many others have the same experience. This scenario suggests that there are two moments of judgment of the aesthetic: the judgment of the producer, and that of the seer. Both are judgments about the object, but they are slightly different concepts of the aesthetic and the idea that a communication has taken place between the producer and the seers is actually no more than an impression. Sometimes, when a whole lot of seers all have the same opinion about the object ("we all like it"), the aesthetic quality is said to be a property of that object ("it is good"). But it is, actually, a product of their shared cultural values.

So let me try to simplify this. I think that there are several things going on that are fundamental to what art is all about. First, people like certain things because they find them beautiful. Second, it is possible that beauty is a property of the things, and all people would find them beautiful. But different people have different responses to the same thing and find different things beautiful, so beauty may not be a property of things so much as that we only say they are beautiful if we think they are as a result of our experience in our community. Third, I want to think about why these first two statements are true. Is it because people are born with different innate properties of their visual systems or other aspects of their nervous or cognitive systems—things really have a quality called beauty—and people perceive that? Or is it because people are educated into particular attitudes to beauty and live in societies with common sets of attitudes to it—Gombrich's "unquestioned beliefs"—they like things that they call beautiful? Why did grazier Bill see scribbles where I and my companions saw art?

If I understand it correctly, one of the reasons for the invention of the concept of aesthetics was to try to bring some analysis to this dilemma. If it were possible to analyze the attitudes of many people to a single thing that they all like, it would be possible to conclude that the thing itself had a quality called "beauty." The people would have aesthetic attitudes to beautiful things, and beauty would be what some psychologists call "an affordance" of a thing they like (Gibson, 1966).

Now the situation I have just described is very simple: there is a thing (a subject) and an observer (a seer), and the seer makes a judgment on the basis of their emotional response to the thing. In the slightly more complex case of art, especially as I describe it in this book, there is a thing in the world or in the mind of a person (the artist—in Gombrich's view) such that the artist sees that subject and decides to produce a second thing (an object) that is in some way a representation of the first thing (the subject). Then there is another person—the seer—who sees the object and has an emotional response to it. There are, thus, two things that a seer might have an emotional response to: the subject the producer had in mind and the object (the work of art) they produced in response.

In this scenario, the artist sees the subject in the world or in their mind through the filter of the stories they tell about it or the cultural values within the society they share with the initial seer—Gombrich's unquestioned beliefs about what art is. Whether or not the subject or the object is called "beautiful" depends on those unquestioned beliefs, and, I think, is not a property, or affordance, of the subject or the object but a cultural product made by the artist in light of how they see the subject as a result of their cultural attitudes, and how, with those cultural attitudes, they are able to produce an object that represents that subject.

There are inevitably some further complications: does the seer have the same emotional response to the first thing, the subject (say, a Campbell's Soup Can) as they have to an artwork of the second thing, the object (say, the Andy Warhol image of an apparently identical soup can)? Alternatively, what is our response to an image of the same subject that was not produced by an artist (a photograph of a Campbell's soup can) or the recovery of an image of the same subject or object in an internet search on a mobile phone (Table 3.3)? Do the responses to the four questions depend on the order in which the objects are presented to the seers, or, if they can perceive that their judgments are being evaluated, do people respond differently when they see any of the four objects from their response?

Table 3.3 Possible categories of object generated from a single subject

Subject	Object 1	Object 2	Object 3	Object 4
Campbell's soup can	Warhol painting of one or more Campbell's soup cans	Photograph of Campbell's soup can	Photograph of Warhol's painting of Campbell's soup can	Image of Warhol painting on Internet search
The young women called Las Meninas	Whatever Velázquez is painting on the canvas in his picture called *Las Meninas*	The Velázquez painting called *Las Meninas*	Any of Picasso's paintings in his series of variations on Velázquez's *Las Meninas*	Either of the printed reproductions called Las Meninas in Fig. 1.10 in the Essay at the end of Chap. 1

In support of the claim I have just made, Palmer et al. (2013, 79–80) considered aesthetic judgments and decided that ostensive definitions do not work for aesthetic responses. Then they surveyed responses to both color and to spatial composition and concluded that "aesthetic response to both colors and spatial compositions is influenced by hidden knowledge about the observed colors and objects."

In other words, when the researchers tested the idea that aesthetic responses depended on the nature of the thing people were looking at—what some would call the affordances of the thing—they reached the more subtle variation that although the affordances of the thing were a factor, seers were strongly influenced by what they had seen before and what it meant to them.

3.3.1 Is There Any Sense of Beauty or the Aesthetic Among Non-human Animals Such as Apes

Charles Darwin (1871, 359) used the idea of aesthetics and pronounced that:

> On the whole, birds appear to be the most aesthetic of all animals, excepting of course man, and they have nearly the same taste for the beautiful as we have. This is shown by our enjoyment of the singing of birds, and by our women, both civilized and savage, decking their heads with borrowed plumes, and using gems which are hardly more brilliantly coloured than the naked skin and wattles of certain birds. In man, however, when cultivated, the sense of beauty is manifestly a far more complex feeling, and is associated with various intellectual ideas.

This confusing (as well as ethnocentric and sexist) passage seems to be unsure whether birds are on one hand beautiful to look at and to listen to and on the other have taste for the beautiful. It is not clear whether Darwin was saying that birds had aesthetic qualities because they are beautiful and, presumably, choose mates that are also beautiful, or because they produce beautiful songs. I find his comparison with women wearing bird feathers or brightly colored gems simply bizarre—as well as turning out ethnographically implausible—but it seems to be saying that adorning yourself with bird feathers makes you aesthetically pleasing in the manner of birds. In that interpretation, the report of a knotted skin worn as a necklace by a chimpanzee would be relevant (McGrew & Marchant, 1998), but it is a single instance of a single chimpanzee. Darwin offered no evidence for his opinion that the human sense of beauty is more complex than that of other animals and got away with it because we all think it is. In addition, as Palmer and colleagues showed, birds may be pretty and sing prettily, but we are more likely to say so if Darwin and I both learned within our cultures to like birds and their songs. And then there is the friar bird (Fig. 3.15).

The situation may be said to be slightly different in circumstances where humans are manipulating the actions of the non-human animals. I once bought a picture that was made by a dolphin at the theme park Seaworld on the Gold Coast in Queensland. It was actually created not by a dolphin but by the humans who put brushes with different colored paints successively into the mouth of the animal and moved the

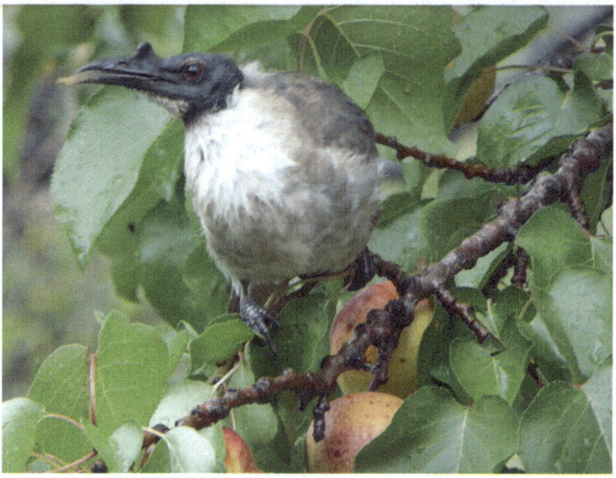

Fig. 3.15 Friar bird in the apricot tree. Even its mother wouldn't love it. (Photo Iain Davidson)

paper in front of the brushes. The keepers removed the paper when they judged the picture was complete enough.

The most famous of the examples where people manipulate an animal to produce commercial pictures is shown in videos of elephants painting pictures on surfaces mounted specially on an easel depicting (to our eyes) other elephants and plants.[2] There is, however, quite a lot of, mostly anecdotal, evidence of animals doing things that would be associated with art in some way: cockatoos keeping time with music,[3] Corvids undertaking all sorts of complex sequential tasks set up by humans.[4] Corvids may be the exception among these other animals because, for example, cameras operated in the forest without human presence have shown that the crows make and use tools in a manner that was previously unanticipated (Rutz et al., 2016; Troscianko & Rutz, 2015). But none of these animal examples seem to get close to creating art or appreciating it without the intervention of humans.

Although authors generally do not consider art when looking at the phylogeny of primates to determine how evolution generated humans on one hand and chimpanzees on the other, several million years ago human ancestors shared a common ancestor with apes, one of which evolved into humans and others into chimpanzees (Varki & Gagneux, 2017; Yaxley & Foley, 2019). Most humans produce art in some form or other and chimpanzees generally do not, so one evolutionary question concerns how art emerged in the human line from that last common ancestor. Various attempts have been made with captive chimpanzees in language programs, but all

[2] https://www.onegreenplanet.org/animalsandnature/why-making-an-elephant-paint-is-cruel-not-cute/

[3] https://duckduckgo.com/?q=cockatoo+music&t=opera&iax=videos&iai=https%3A%2F%2Fwww.youtube.com%2Fwatch%3Fv%3D5sSb1qF7m3w&ia=videos

[4] https://duckduckgo.com/?q=crow+problem+solving+youtube&t=opera&iax=videos&ia=videos&iai=https%3A%2F%2Fwww.youtube.com%2Fwatch%3Fv%3DcbSu2PXOTOc

seem to be overwhelmed by the various influences of the human participants (Noble & Davidson, 1996, 78–80).

In conversation with a distinguished primatologist, I once suggested that aesthetics would not be found among chimpanzees, and he was astonished that I would say so: against me he said that perhaps people had just not looked hard enough to find aesthetics. I thought I was saying that chimpanzees have no feelings about beautiful (or awe-inspiring) things. In practice, such observations become confused with others that can be interpreted in terms of symbolic or ritualized behavior. There are some published instances said to be relevant: leaf-clipping (Nishida, 1986); the grooming hand-clasp (the original as well as the more recent view of this in Goldsborough et al., 2023; McGrew, 1992, 68), and attitudes that have been interpreted as expressions of awe at the sight of waterfalls (Goodall, 1986, 67). One review of apes that I was involved in showed that the studies were biased against finding symbols among chimpanzees, but also that the existing evidence was very slight (Noble & Davidson, 1996, 50–52). Against this, James Harrod (2014) documented the instances he could find to fit with a definition of religion and concluded that there was evidence for religion, but not theology, among chimpanzees. Who knows where bias can be found?

It seems from all the evidence likely that chimpanzees do engage in behaviors to which humans can attribute meanings—but we cannot be sure whether the apes attribute them. These contexts and behaviors are of a type for which people would use or exclaim words and this gives some credence to claims for the cultural nature of their behavior (Byrne et al., 2004; Whiten et al., 1999). One such behavior is the practice of throwing stones into hollows in trees, reported from some study sites in West Africa (Kühl et al., 2016).[5] This seems to be ritualized, but it has been reported nowhere else among all the studies of chimpanzees. In almost all other instances, it seems to be the case that the emotional response of the animals is expressed in the presence of a natural phenomenon, and that the animals do not express their response by creating anything.

Some would argue that the exception is provided by chimpanzees in (at least) two experiments which appear to create novel works. In one experiment, chimpanzee Congo produced paintings with Desmond Morris (2013, Ch. 3) when he worked at London Zoo, though the chimpanzee did not continue that creativity when Morris left. The other, involving several individuals, was by John Matthews (2013) who first worked intensively with the production of pictures by children, then extended the experiment to work with several chimpanzees captive in Singapore Zoo.

Modern apes show little inclination to make anything like art, except when stimulated by humans who want to know if the apes can make art (Matthews, 2013; Morris, 2013, Ch. 3). Both Morris and Matthews have demonstrated by their careful analysis that there are systematic limits to the marking made by chimpanzees, limits that the animals seem to impose themselves. It seems to be possible to identify some rules about mark making, but not much more. Their marks seem to lack creativity,

[5] There were more authors on the report than the total number of stones thrown!

skill, imagination, beauty, and there is no sign that they have emotional power. Matthews documented how marks that we might interpret as scribbles which had nothing to do with art were actually an important part of the ontogenetic process for children (at least) learning to make marks that are meaningful to adults (Matthews, 2013). The scribbles are not lifelike, do not involve unquestioned beliefs, or involve acute observation. They are mostly ignored and do not persist for others to see, except on the refrigerators of devoted family. Paintings and engravings in rock art are very different—they were made deliberately and persist for others to see.

As a result, I am happy to argue that art appeared at some stage, along the human line but not the chimpanzee line, nor among the many other species that evolved during those millions of years. The earliest materials that look like art were the beginning of a process of emergence that fundamentally altered the way people saw the world, but they were localized and scarce (Culley & Davidson, 2021). At a later stage, art was made on rocks around the world, and it was accompanied by stories.

3.4 Art and Skill—Art and Craft

The question of the relations between art, skill, and craft is a little complicated. The word "art" has a history in Latin that relates to the Greek word "techne", and some writers think that discussion of art must take into account relationships to the ancient worlds of Greece and Rome. What I am trying to do in this book is to get away from such obsession with the ancient worlds that were once thought important in Western Europe. Sutton and Anderson (1988, 3–4) addressed the question of the existence of art in societies that do not appear to have a word for art. Adopting the point of view of linguistics they suggested that although Australian Aboriginal languages appear to have no word for art, they do have words for "sign, design, pattern, or meaningful mark." Although such marks also occur in nature, they are attributed to Dreamings when they do, and it is the action of Dreamings—relayed through story and song—that makes them into "intentionally meaningful forms, or signs." Such signs used not to be "part of an art market" "in a commercial sense" but could be "traded, bequeathed and …stolen for their high value" among Aboriginal people. If we are to free ourselves from the unquestioned beliefs that old ideas current at some times in Western Europe are somehow relevant to the whole concept of art that we are trying to understand, applicable to other times and places, we need to be able to show that something else is involved. In reality the whole question is qualified, as in the Aboriginal cases Sutton and Anderson were describing, by the elision between different sets of assumptions.

Robin Collingwood (1938, 9) attempted to argue his way out of the problem:

> we find [art's] *proper* [my italics—ID] meaning hedged about with well-established obsolete, analogical, and courtesy meanings. The only obsolete meaning of any importance is that which identifies art with craft. When this meaning gets tangled up with the *proper* one, the result is that special error which I call the technical theory of art: the theory that art is some kind of craft.

My purpose here is actually to reverse the "theory" at the end of this statement and argue that craft may be a form of art—but it is not always. Collingwood (admittedly writing a long time ago in a very privileged context with very specific cultural circumstances) envisaged a "proper" meaning of the concept of "art," but the issue, then, is: what sort of meaning is (or was) "proper"? Collingwood's answer was that proper art involved both expressing emotion and creation using imagination (Collingwood, 1938, 105–153). The important point I want you, the reader, to see here is not the rightness or wrongness of a solution to the relationship between art and craft that Collingwood proposed more than 80 years ago, but that there have been people who wanted to partition popular conceptions of "art" in order to give an authoritative ruling that one category was "proper", and others were not so. Such people still exist.

The idea of "craft" gets a bad press nowadays because it is presented to most people through badly made stuff sold at school fêtes and people's markets. In the context of this book, I refer to materials that were part of the apparatus of everyday life, such as baskets and textiles (Nettleton, 2010; Pöllänen, 2013) (like the *billums* of New Guinea we discussed earlier in this chapter) as well as attempts to educate people in the home-based crafts that were previously mainstays of life (Mason, 2005). As Ingold and his sources demonstrated, great skill was involved in the manufacture of the string bags on New Guinea, as it was in the traditional weaving of Maya Mexico (Greenfield et al., 2000). The reality is that wherever materials were made for a purpose and according to a traditional method, there was skill involved in making and decorating them. The presence of skill in craft does not preclude the conclusion that the form and decoration might be artistic.

In one revealing episode of contact between cultures involving the transfer of material culture from the newly "discovered" to the self-designated discoverers, Indigenous people from the Admiralty Islands traded obsidian-tipped artefacts, spears and daggers, with Europeans during the processes of interaction between the two groups of people (Torrence, 2000). Analysis of the obsidian artefacts collected showed much more agency by the people who produced them for transfer from their group to visitors. Adopting the perspective of membership in one group in the interactions might be compared with the designation of membership of "proper" groups. In other words, what gets designated as "art" may be a reflection of the positions of people within social structures. It would be open to members of particular groups to define the form and decoration of craft objects as "art", if they so choose, though it may also be that that designation would not be important to them.

The history of techniques in Renaissance painting involved the use of skills such as perspective and *chiaroscuro* to represent depth and hence to achieve more likeness or realistic representation (Haber, 1980); the use of oil paint (de Viguerie et al., 2009); the use of canvas which might persist beyond wood panels (Mecklenburg et al., 1998) or fresco on the walls of buildings (Brown, 1984, 273). Experimental reproduction of materials and techniques demonstrates the range of skills involved in many aspects of Renaissance arts (Smith, 2016), including those discussed in the fifteenth century by Cennino Cennini (1844). According to one guide to painting and drawing (Wilson, 2017): "In Cennino's world painting is a craft—a series of recipes and instructions that, if followed, will give a perfect result. I now realise that no book can really teach you how to make art."

I think that two things have muddied the waters about art and craft, particularly for archeologists. First is the motivation for defining "Fine Art" in the eighteenth century (Shiner, 2001), and second is the implications of the Arts and Crafts movement of the nineteenth century (Moro Abadía, 2006). Shiner (2001, Part II) described the history of the emergence of Fine Art, for which the adherents soon dropped the epithet "fine." Stiner adopted an approach "that links conceptual changes to specific social and economic factors, such as the rise of a market economy, the growth and status aspirations of the middle class, the increase in literacy, and the preservation of gender roles." As an example of the importance of the market, I point to the fact that almost any mention of an "artwork" in the press attaches a price to it. Sutton and Anderson (1988) emphasized that signs in Australian Aboriginal societies could be "traded, bequeathed and ...stolen for their high value." One of the outstanding examples of art in Australia is that produced in Central Australia since 1971 in and around Papunya west of Alice Springs (Bardon, 1991; Megaw, 1982; Myers, 2002). In 1971 Geoffrey Bardon, a white school-teacher, initially facilitated Aboriginal men in Papunya to produce paintings on boards with acrylic paints (Bardon, 1991). Although the history is more nuanced (Scholes, 2017) no one denies the important role of Bardon in the history. The men were so enthusiastic that in 1972 Bardon applied for a grant from the Australia Council for the Arts to continue the work on the grounds that the "great vitality and intelligence" of the "story paintings" "is a valuable contribution to the reputation of Aboriginal culture." For our purposes here, the point is that the application reported that the movement "had already earned sales of over $A3,000 from 170 works in just four months" (at an average of less than $20 per work—that was a bargain) (Myers, 2002, 125). It would appear that, even at that price, the market determined something of the value of the art.

Betty Churcher (2014, 96–100) described Arthur Streeton's *Golden Summer, Eaglemont, 1889*, one of the first paintings of the Heidelberg School of painting (Fig. 3.16), in almost purely monetary terms: sold on the first day of the Paris Salon of 1892, repurchased by Streeton in 1919 for UKP60, whereupon he altered it and resold it in 1924 for a record price of 1000 guineas, "so he understood well the market of the day" according to Churcher. The next sale was in 1985 for just over $1,000,000 and the purchaser William Hughes had the 1920s paint removed (because Streeton being dead by then could not argue about his artistic rights, and, moreover, it could be done safely). Finally, the painting was purchased by the National Gallery in 1995 for $3.55 million. As Churcher put it (p. 100): "It had broken records every time it went to market," as if that was more important than her judgments that this picture showed "An evanescent daytime moon" "a hillside of summer-bleached grass," "a magpie, so that the liquid notes of his call can accompany the shifting patterns of sun and shadows as the day winds to a close."

Take away the financial details, and the painting is still a representation of Australia as it was just after the country celebrated the first centenary of Europeans in Australia, and arguably the quality of the painting is more important than whether

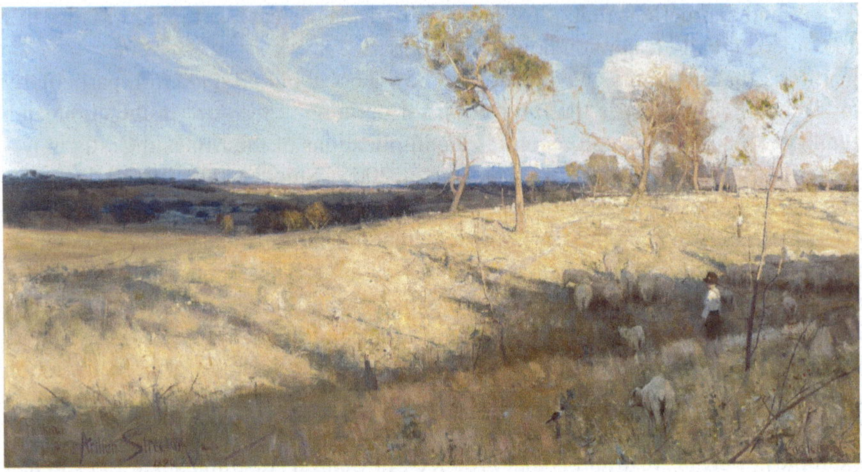

Fig. 3.16 Arthur Streeton's *Golden Summer, Eaglemont, 1889*. (National Gallery of Australia) (from Wikimedia Commons PD)

it met the market, appealed to the middle classes, the literate or men. It is art whether or not anyone has ever judged that it is "fine".

The consequence of creating a category of "fine" art, which was, of course, what Collingwood called "proper" art, was that the arbiters of the aesthetic from the eighteenth and nineteenth centuries had a further means of defining themselves as worthy of that sort of art and that those who did not meet their criteria were not. Thus, for example, Alice Gorman (2008) has shown that scarification and tattooing of the human body—surely not a craft by any definition—was used as an indicator of the relative "primitiveness" of the people who practiced it, as documented by Henry Roth (1900), and hence of the attitude colonial authorities might take to the people classified in that way. They were "primitive" but more-so if they practiced scarification than if they produced tattoos.

By the same token, and concurring with the judgment, the common attitude of colonizers in the nineteenth century (and earlier, no doubt) was that the peoples they were destined to colonize were "primitives." This, according to Oscar Moro Abadía (2006), was the colonial attitude—not one I am defending here—and that such people had craft but not art, at least in the judgment of the colonists. Since all surviving fisher-gatherer-hunters were subjects of colonization, it followed that nothing that they produced was worthy to be called art. As a result, although the art of the fisher-gatherer-hunters of the European Paleolithic had been known since the 1830s—all engraved on pieces of bone or ivory—in the opinion of the experts on the Upper Paleolithic of France, none of it was worthy of being called art. And that, according to Moro, was the main reason why it was so difficult for French archeologists to recognize that the paintings of the Iberian site of Altamira (Figs. 2.10 and 3.17) were indeed either old or art.

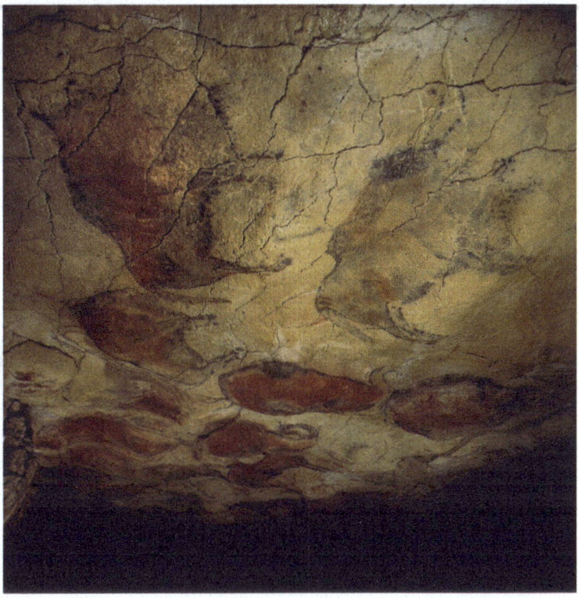

Fig. 3.17 The ceiling of Altamira with animals painted on bosses which created a 3 dimensional effect. (From Beltran & Saura Ramos, 1998, 63) (© Gobierno de Cantabria—MUPAC/Colección Pedro Saura)

The point is that the dogma of attitudes to the market and the social status of the people who said they were interested in art was enough to prevent an understanding of one of the great achievements of humanity—the cave art of the Ice Ages of Western Europe—in the face of multiple lines of evidence pointing the other way. At that stage people with a legacy of the definition of Fine Art and the dominance of the market should have abandoned the values of privilege and entitlement from which they sprang and recognized that there could have been art made by people other than the elitist groups to which they belonged. But that is not what happened.

In the eighteenth century the ramparts were defended by people with a clear idea of what should be included in the category "Fine" art, and they were mostly white males of a particular class and privilege. They were, in effect, making a statement about their own status using the medium of art as a resource of power. That, I suggest, is always how art has been used—as a resource by which one group of people established how important they were—either by owning it or aspiring to. We see the attitude continuing, not only through the art that is held in galleries and museums, but even on the streets where those in power insist on their right to remove painted murals that do not meet some unstated criteria (Figs. 3.18 and 3.19). The Newcastle example shows that over the years, street art was both destroyed by painting over by people with one attitude, modified (to great effect) by adding paint and slogan by people probably with another, and ultimately removed by the more powerful.

Skill was and is generally involved in the production of art, but also in things that many groups would not call art but might seek to disparage by calling craft. One way to look at the question of whether things should be called art or not is to ask questions about how people become members of groups who decide such things, and whether art is characteristic of one group of people while the products of others

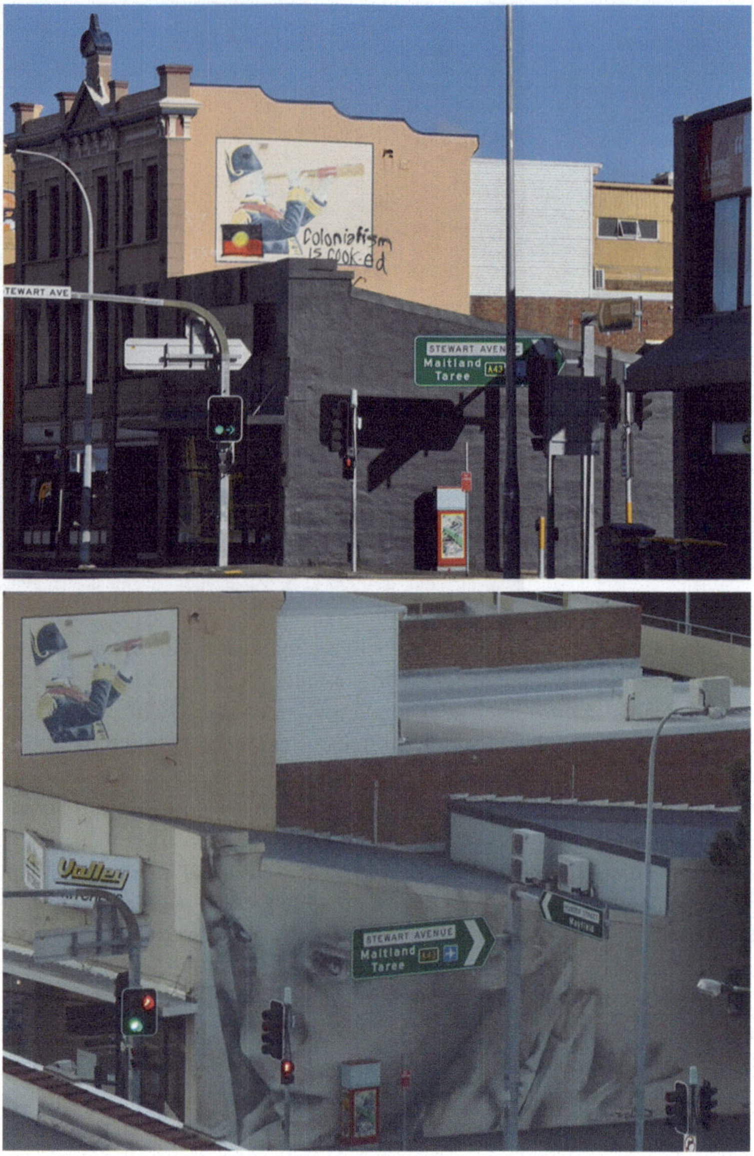

Fig. 3.18 Two views of the same part of Stewart Avenue, Newcastle, NSW. When I first saw them (March 2015), the lower image had a wall-painting by Guido van Helten at street level and above that on a separate building, there was an image of a mariner (presumably Captain Cook). In the upper image (October 2017), the van Helten painting had been painted over by Valley Kitchens and graffiti had been added to the Cook image: an Aboriginal flag, and the lettering "Colonialism is Cook-ed." All have now been removed and replaced by a commissioned painting by Adnate honouring the local Aboriginal Thirrilmun community (2018)

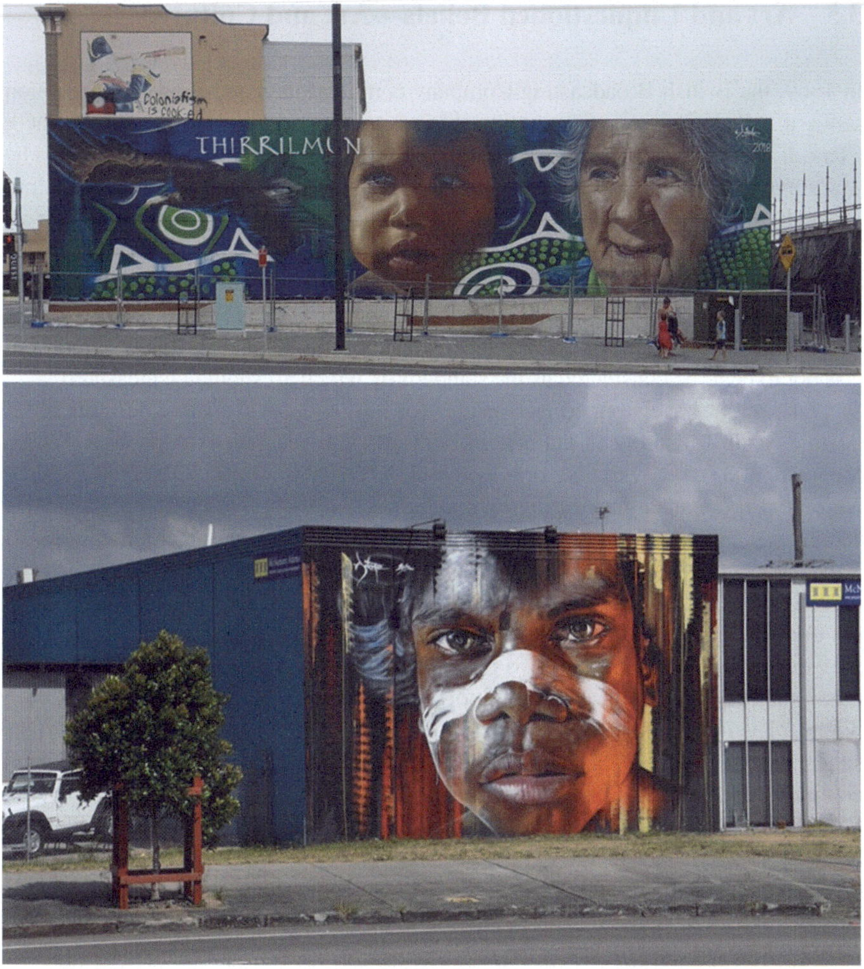

Fig. 3.19 Murals by Matt Adnate on Stewart Avenue, Newcastle. Bottom 2014, top 2018. The painting of the mariner has now been painted over (2024). A developer chose to destroy the art in the lower photograph. (Photographs Iain Davidson)

must be called craft, and then whether there is any useful or objective hierarchy by which one group is superior to the other. The issues are about the sociology of groups and membership and the rights to pronounce something art in the "proper" sense. Most people who write about the criteria for accepting something as "proper" art choose the criteria by which they define themselves as having such a right. Members of another group might choose other criteria, as is illustrated by my initial example of the grazier Bill who would see the paintings I see as rock art as "scribbles." I call them art, but he might have questioned my right to do so.

3.5 Art and Unquestioned Beliefs—Art and Culture

In 1969, the British Broadcasting Company commissioned and showed a television series it called *Civilisation: a personal view*, by the British art historian, Kenneth Clark. It was one of the landmark pieces of the early years of color television and he was some sort of authority on the paintings in its subject matter. When I began my research in Spain in 1971, my new friends pointed out that Clark omitted almost all mention of Spanish art. In fact, the book from the program proved to be even more offensively biased: "when one asks what Spain has done to enlarge the human mind and pull mankind a few steps up the hill, the answer is less clear" (Clark, 1969, 13–14). It is an excellent example of the way in which authorities define(d) what was and what was not to be to be considered as the standard by which art should be measured.

Clark might have stopped to consider for a moment the many contributions of Islamic Spain to European culture, in both arts and sciences (Dodds et al., 2008; Menocal, 2003), but with that airy dismissal, he chose to overlook the major contribution to European culture during several hundred years before the Renaissance. He sought redemption with little more than single remark and mere mentions of Velázquez's, *Las Meninas* (Fig. 1.10 Left) (p. 149 which, ironically, he labels "The greatest of all pictures"), Francisco Goya, Picasso, and Salvador Dalí. The point is not that Clark necessarily should have referred to Spanish artists, although he could have done, but that Spanish art was probably not part of Clark's unquestioned prejudices about any form of art. An author with a different perspective (Bass, 2008) wrote: "The lives of writers of the calibre of Lope de Vega and Góngora and artists such as El Greco and Velázquez did not just overlap in time and place; they shared common conceptions of their respective arts as well as social aspirations. ... On the literary side, the concept of sister arts informed what Frederick de Armas has described as 'writing for the eyes' ...." The point is that those "common conceptions" were exactly what I mean by selecting Gombrich's phrase "unquestioned beliefs." We all have them. And the literary side included Miguel de Cervantes and *Don Quixote de la Mancha* with its essential recursiveness, as Foucault (2002) insisted. It was not part of the cultural context of Clark's presentation. If his education had been different, he might have placed more emphasis on art in Spain, and the Islamic introduction of paper, rather than the shorter period of contribution from the artists of Italy and the Netherlands. One opinion expressed in the eighteenth century was that "Islamic Spain ... [was] the only redoubt of civilization in an otherwise barbarian Europe" (Díaz-Andreu, 1996, 71). The bias that Clark reflected is part of the way in which interest groups form opinion about what should be considered as art.

Other histories are possible. Most obviously, twentieth century art could not be discussed without Picasso, and Picasso cannot be understood without Spanish painters such as Velázquez. We get side-tracked, when talking about art, into a particular view of "the 'best' of what we see," just as some people recognize as "culture" only what they regard as the best among all the many things we learn of cultural behavior. Others make different choices about what is best, and consequently see things differently. Arabian, Chinese, African, Australian or Indian histories of art might spend less time on the achievements of parts of mediaeval Europe. Understanding what art "is" requires that we show some sympathy for arts produced outside a small part of Western Europe in the last 500 years.

The literature about art talks about a canon of famous examples and artists—what others call "art as we know it"—and puts them into a context of relationships among styles, techniques and traditions. By discussing such examples within their social context, the literature is removed from other examples that are not regarded as the greatest, within that context. This book is about art which might include marks that would not enter into such discussions—almost all archeological art (Art 2 and 3) would not. The purpose of this is to understand how, in the first place, we got to the position of calling some of it art and some not.

Pasztory (2005, 10) recounts how one art museum was offered a recent example of Indigenous art. They rejected it on the grounds that it was traditional and not art but recommended it be submitted to the "Primitive" art section of the same museum. When it was submitted there, it was also rejected—on the grounds that it was artistic and not traditional. The same story is told about the National Gallery of Australia at the beginning of the acrylic art movement (Megaw, 1982; Mitchell, 2016). Its collections are now one of the chief repositories of Australian art (Myers, 2002).

This does not augur well for the suggestion that any people might be able to identify art by pointing to some quality of an object itself. Instead, people who sit in judgment determine whether it is art, and how it should be regarded, based on their own unquestioned beliefs. Identifying art is something that is done by people. What gatekeepers require is not anything about the objects that might be called art, but about their education as gatekeepers—to become people who are interested in art (Myers, 1991). The issue becomes whether the recognition that something is art results from the culture and education of particular people, or whether the "work of art" "speaks" itself. This is a fundamental opposition that needs to be clarified.

In discussing the enigmatic and beautiful group of Aboriginal objects known as *toas*[6] (Fig. 3.20), from the desert country east of Lake Eyre in South Australia, Philip Jones (2007, 230) tellingly remarked that "The toas have been 'discovered' twice by the West during the past century—once as artefact and once as art. But they have rarely been scrutinised in their own right or measured against the historical events and personalities surrounding them." There is some suggestion that the nature of these objects was questioned at various times (Jones & Sutton, 1986, 54–62, 77). Some of these disagreements concerned interpretations by individuals who regarded themselves as experts: were they signposts which indicated locations to which people had travelled; did they include a systematic set of signs with an interpretable semiotic (Morphy, 1977); were they always in use or had they been produced for a market stimulated by the presence of the Lutheran missionary Georg Reuther (Jones & Sutton, 1986, 54–55)? Importantly, these objects were mostly vividly painted, and we find them beautiful now.

They were undoubtedly produced by the Aboriginal groups around the missionary station at Killalpaninna in South Australia, and the attitudes of those who described them have often been the principal determinant of the type of description.

[6] Peter Sutton tells me by email on 30 Jan 2023 that they should be called *thuwa*.

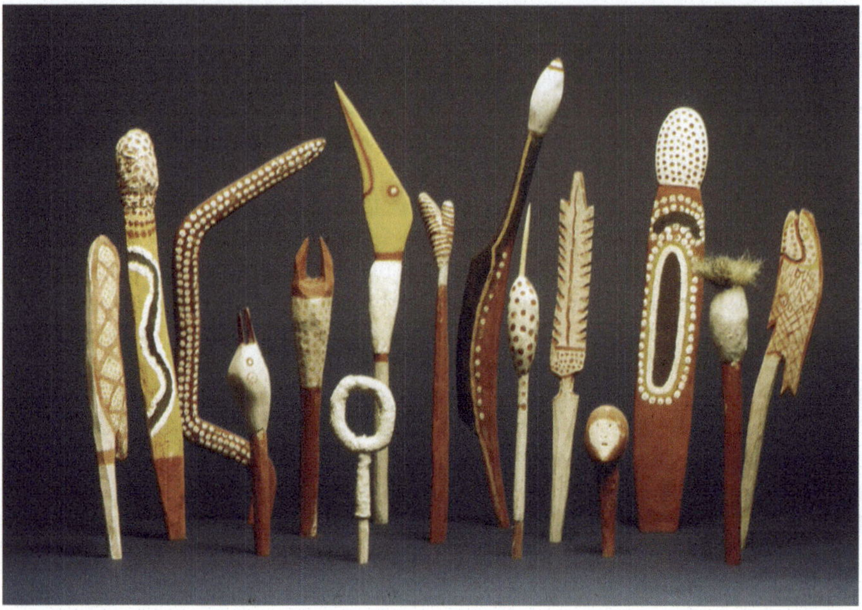

Fig. 3.20 Toas from the Lake Eyre region. Photo and permission by Philip Jones and the South Australian Museum (Cat 79–93). These were published in Sutton et al. (1989, Fig. 238) with the date 1904, and the attribution "Artists unknown"

Standing back from the controversy, these are great creations, imbued with the stories of the Diyeri people and their relationships to land, and whatever the minutiae of their history they were undoubtedly created by Aboriginal people. Stripped of the prejudices of the past—the beliefs non-Aboriginal people once held about Aboriginal people—toas deserve to be described as art (Morphy, 1998, 31), if only we could escape the attitude of some people to that designation.

These studies of Australian art show that storytelling accompanies all of its production, and all associated ideas must be learned. The same is likely true at all stages and places of artistic appreciation. A study of visitors to a museum of archeological art in Rome and another of contemporary art in Venice showed that the visitors to the Venice museum were of higher socio-economic status than the visitors to the Rome museum, and visit museums more often than them (Mastandrea et al., 2007). The visitors to the Venice museum were more motivated by the emotions generated by seeing the art. A sociological study of modern art in the United States (Dimaggio & Useem, 1978) concluded that there are four principal points determining the role of class in the consumption of modern art: (1) Arts appreciation is trained; (2) Arts appreciation is contextual; (3) Arts consumption enhances class cohesion; and (4) Arts consumption is a form of cultural capital.

These conclusions from modern surveys that take for granted what should be included in the arts, coincide to some extent with the definitions of culture that I used in exploring variation in stone tools, possibly the archeological objects least likely to be included in the arts (Davidson, 2016) (see Fig. 3.21). There I defined

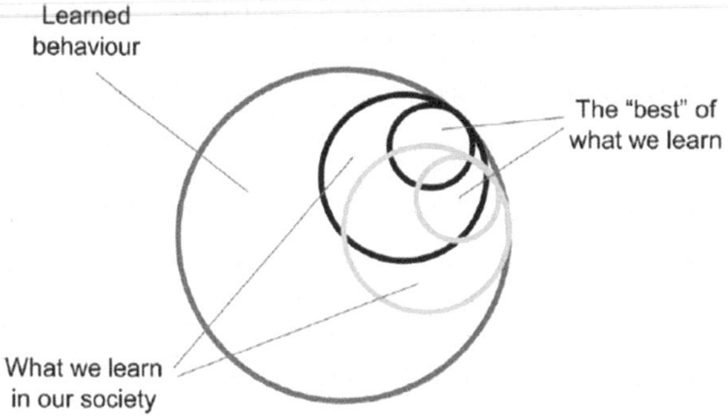

Fig. 3.21 Three dimensions of culture as learned behaviour. Cultural behavior generally is learned behavior; for each society there are particular bits of learned behavior and from one society to another those bits might overlap; then, within any society there is a bit of learned behavior which is particularly valued, again with some possibilities of overlap. (Model by Iain Davidson published in 2016, permission from Nick Conard)

culture as having several different meanings: (1) Learned behavior—characteristic of human and not in the vernacular understanding of other animals; (2) What we learn in our society—what makes us members of our society but not of the society of the neighbors; (3) The "best" of what we learn—either the opera or the Opera House in the Sydney Opera House sense (see, for example, Davidson, 2016, Fig. 10.1). As with other explorations of cognition in hominin and human evolution, what are the implications of those definitions of culture for the study of mark making? My purpose is to discuss the context of other models and to explore the ramifications for the study of the evolution and contextualization of art.

Learned behavior is the basis of cultural behavior among humans. Some non-human behavior, we now know, is also learned (Bugnyar, 2019; Tennie et al., 2020; Whiten et al., 2011), though this is not the place to elaborate about the implied relations between human and non-humans. Within the domain of human learned behavior, there are many ways it is revealed, but the aggregate of that for most people and those whose lives they share constitutes the culture of that particular group of people (the Culture of a Community). In almost all instances there is another group of people with slightly different shared and learned behavior and there is usually some overlap between groups who live nearby, and within any cultural group defined in that way, there are some things that the group accepts are the finest characteristics of stuff produced by the group (I previously called this Culture in the Sydney Opera House sense). As with the similar situation for other languages, in the case of the cultural example of the Spanish language that includes a *Real Academia Española,* founded in 1713, which lays down the rules about the correct way to speak—using the motto "Limpia, fija y da esplendor—it purifies, standardises and gives brilliance" though in some provinces there are spoken versions that are quite different. The Academy acts as a gatekeeper.

Gatekeepers confine some sorts of art within culture ("fine" art) but where the gatekeepers are absent, art partly lies outside the boundaries of "fine" art but it would still be part of the learned behavior of that particular group. Two cultures can overlap—they share some learned behavior but not all of it. "Culture" is not a fixed entity but one in which slight differences might mark the differences between different cultures. The criteria for art partly overlap from one culture to another. Culture is not a unitary thing: it varies from time to time and group to group; it does not determine what people do. What people learn to do defines the boundaries of their culture, and each work of art that gets to be seen shifts the culture slightly.

Thus, one group of people will consider Opera seen at the Sydney Opera House typical of art, while another would, instead, invest value in Country Music at the Grand Ole Opry House in Nashville, Tennessee (Davidson, 2016). The conclusion that "Arts consumption enhances class cohesion" is an inevitable consequence of class being a major element of the choices about who defines art. In fact, attitudes towards all sorts of cultural features, including art, are part of how people define and recognize their membership of groups, class, race, religious group, and maybe, in some circumstances, gender. At the same time, those attitudes are determined by the groups of which they are members (for a nuanced discussion of the semantics of culture see Goddard, 2005).

Pierre Bourdieu (1968) began his discussion of the sociology of art with a difficult passage:

> An act of deciphering *unrecognized as such*, immediate and adequate 'comprehension', is possible and effective only in the special case in which the cultural code which makes the act of deciphering possible is immediately and completely mastered by the observer (in the form of cultivated ability or inclination) and merges with the cultural code which has rendered the work perceived possible.

I take this to mean that you (in this case, the reader of this book—or, in the case of a work of visual art, the seer) and I can only understand the work if we are fully immersed in the culture ("What we learn in our society") of the producer who made marks (or in this case wrote sentences). It is immersion in the culture—complete mastery of it—that makes up the unquestioned beliefs Gombrich introduced into the criteria for art. It is the sense that your mastery is complete which leads to the beliefs being unquestioned. In addition, Bourdieu was recognizing that the work (of art) was a product of a series of conventions (cultural code) defined by (and defining) membership within a particular cultural group. In consequence, there is reciprocal feedback between culture and convention such that small changes in one can lead to larger changes in the other—in either direction. Neither art nor culture are fixed in their properties. Each person actually has a slightly different culture from every other person, because they learn different things, and is subject to slightly different conventions. In reality, mastery is never complete, and beliefs should always be questioned. Some homogenization is provided by people painting pictures so that others live(d) in a world of visual images, but importantly by imbuing those images with stories that were either already familiar (such as Jacob in Duccio's painting) or became familiar through the artist's telling and the persistence of the image.

There is a massive literature in philosophy which talks about the "indeterminacy of translation" (e.g., Quine, 2013) which is not the main issue here. I might flippantly suggest that that literature is itself an almost perfect example of that indeterminacy. The point here is that the indeterminacy which arises as a result of people turning marks in to art is a part of life's rich pattern. As we will see, it is what makes art open to different interpretations. We are not necessarily concerned with truth and one-to-one correspondence. We have moved on from the notion that unquestioned beliefs will always be right when they are those of our own society and culture, and wrong when they are someone else's. The purpose of this book is to make the reader question beliefs they might otherwise not be aware of. This level of indeterminacy is almost inevitable. Frances Morphy (2023) showed, as a linguist accustomed to decoding the subtleties of languages other than her own particularly involving art, that translation is almost impossible because, in the example of writing about art in Yolngu language, the words and the images are so imbued with unstated cultural values that in the English language "some things always remain beyond the scope of translation." Quine (2013, 71) described this indeterminacy as being a result of "a conflict of parts seen without the wholes."

Most obviously, Philippe Descola (2013) [2005], 2010), using examples illustrated by their works of art, discussed principles of the variation in the way people in different cultures conceive their relationships with the outside world (summarized in Davidson, 2012; Davidson, 2017). The details of the analysis are not relevant here, except the conclusion that it is possible to describe four different types of cultures, or ontologies, with different attitudes to the outside world and to their own ways of relating to it. In this scheme, people recognize their similarity or difference from the rest of the world both in their (external) physical nature and their (interior) mental nature. They may see themselves as physically part of or separate from other beings and as mentally similar to or different from other beings. The four patterns of belief about existence produced different sorts of art (for further discussion the reader is referred to Davidson, 2012, 2017). The categories Descola identified are not straightforward, as evidenced by a range of scholars, including Descola himself. The lesson we should draw from this, though, is that different cultures have different sets of unquestioned beliefs. They are likely to lead to identifications of art in different ways, and seeking to translate from one world-view to another we can fall into category mistakes.

We humans are the only creatures who think and behave as we do, but, at least since Darwin, we see our bodies as linked by the continuous process of evolution to the physical forms of other beings. Warner (1964 (1937), 409) argued that in Australian totemism "two aspects of society are being ritualized: the social relations of [people] to nature, and of [people] to [people]." It is arguable that that generalization underlies all four of the categories that Descola defined, though the particular social relations differ from one human group to another. For our purposes here, the unquestioned beliefs of societies differ from one to another, and are often manifested both through their rituals and in their art. I will return to totemism in Chaps. 6 and 9.

James Clifford (1988, 222–6) explored the interaction between art and culture when he was principally concerned with defining authenticity. He suggested that, much as Shiner (2001, xv–xvi) had puzzled about the transformation of ethnographic materials into art, so there was an identifiable process of materials moving from what Clifford called culture to his category of art. The argument seemed to explain some of how transformations take place, but failed because Clifford was using categories of both culture and art that are far too essentialist. He was implying that there is really only one definition for each of the entities and that materials can be classified as belonging to one or the other, either to culture or to art or their negative variations. In reality, as I have argued, the boundaries are much more flexible and overlapping.

Second, I think it can be argued that when Picasso, most famously, saw faces of Iberian and African art which revealed to him that he could represent faces in a way different from the realistic representations he was used to, he really did learn to see the world in a different way. Along with all the other ruptured conventions (such as his borrowing of the elongation of human forms and some of the framing from El Greco's *The Vision of St John* shown in Fig. 3.22) (see also Hockney & Gayford, 2020, 290–291) in his *Les Demoiselles d'Avignon*, the faces borrowed the shapes of Iberian or African masks but without the elements of ritual or other cultural context that the masked objects may have had before they were stolen from Africa to Paris.

While Picasso's original painting (of 1907) may have borrowed shapes, he incorporated the technique of seeing other things into his repertoire of representation, as he had done in his *Portrait of Gertrude Stein 1905* (Fig. 3.23). There is no hint, in the earlier portrait of Stein, that he was using shapes deprived of their ritual or other

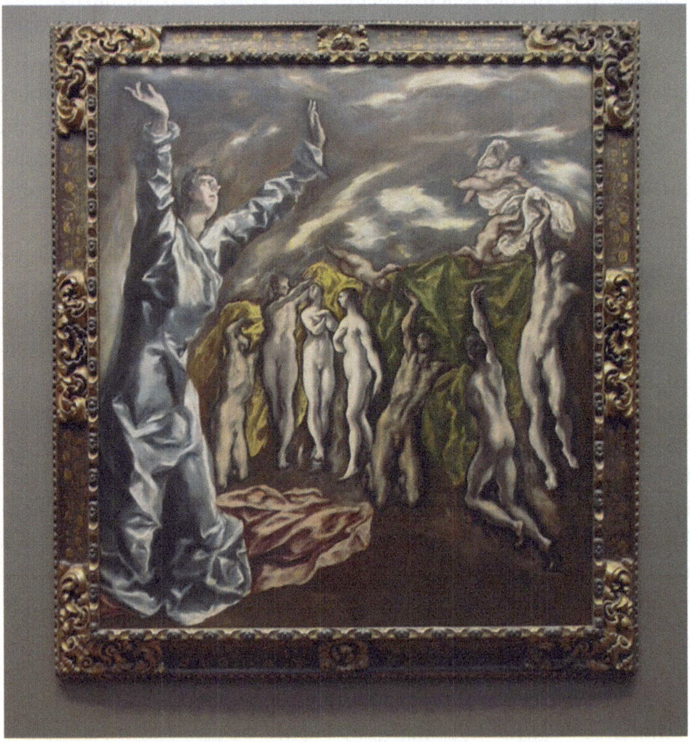

Fig. 3.22 The Vision of Saint John—painting by El Greco (Domenikos Theotokopoulos). Metropolitan Museum of Art, Rogers Fund 1956 MET_56.48_1. Copyright Wikimedia Commons, El Greco, CC0

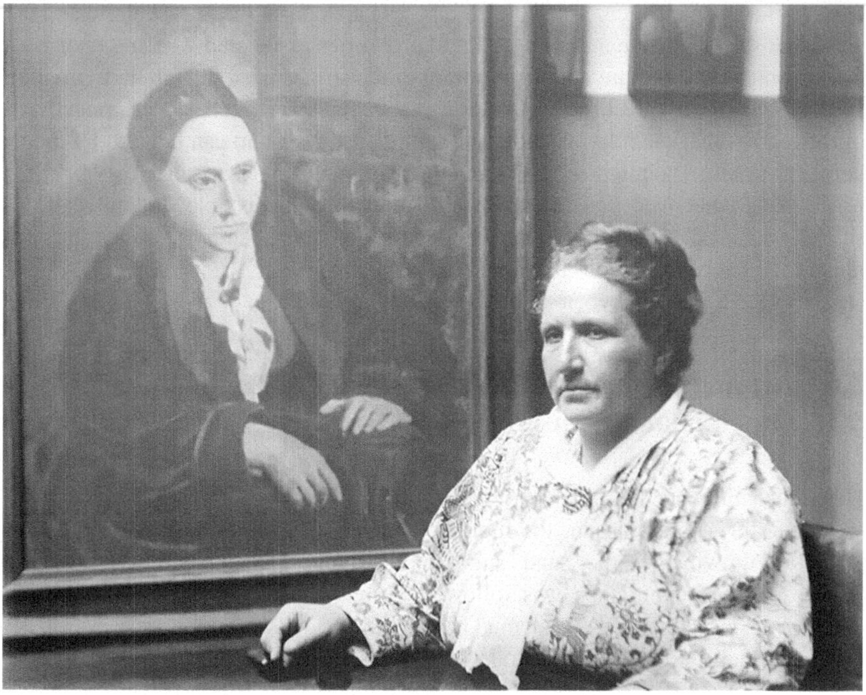

Fig. 3.23 1922 photo by Man Ray of Gertrude Stein seated in front of her portrait by Picasso of 1905. Public Domain. © *Man Ray Trust. ADAGP/Copyright Agency, 2025*

cultural context, just of the reality that the technique of using such stylized shapes could achieve dramatic results.

There are two elements of the relationships between people and art—which it is convenient to refer to as elements of the art-system. They are intertwined but can be considered separately, generally as culture and stories. Both elements, of course, are complex ideas and, as with art, there is a common understanding which makes it seem almost unnecessary to talk about the possible variations of meaning. Both also have quite specific roles in the argument we are unravelling here. We need to think of culture coming to have that name through the combined attitudes of people in societies (Davidson, 2016). In different cultures people produce works they are content to call art (though some cultures lack a word that can be translated as "art"), or some other name although we could recognize them as art using the sorts of discussion in this Chapter. Part of the way they reach such contentment is through the stories they tell others about the works they call art (Nowell, 2023) within an art-system.

The art-system is a part of the way in which people relate to each other within cultures and tell stories that are important to them about things they can see or think about. An art-system sets out (and reflects) the elements of the cultural values people share with others. In these relationships people produce art—the relationships between the producers and their material product are the art. Some of the material survives for people to see. Some of those people are part of the same culture and the art object stimulates the memory of the likely stories associated with it. Other people are outside the culture and are ignorant of the stories even though they can, like archeologists thousands of years later, recognize that an object has been created. All of this means that the relationships that define the art-system can be used by members of the society to relate to others and to the circumstances in which they live. We might expect that over time the art-system—the things, the people, and the relationships—was used as part of the way groups formed their identity and then the way history happened over thousands of years.

3.6 Art and Symbols—The Marks Stand for Something Other Than Themselves

I have been influenced by my experience in 1982 of seeing and hearing Wangkamadla woman, Mrs Annie Hansen, the mother of my friend Isabel Tarragó (mother of Avelina), drawing in the sand and singing songs (Davidson, 2013, 23–24} (Fig. 3.24). My experience was followed by reading the work of Nancy Munn among Warlpiri people in Yuendemu (1973), and the work of Jennifer Green (2014) among Arandic peoples of Central Australia. I have previously argued that the essential starting point in recognizing art is that marks were made (Davidson, 2020). It is not true that all of these marks in the sand were non-representational,

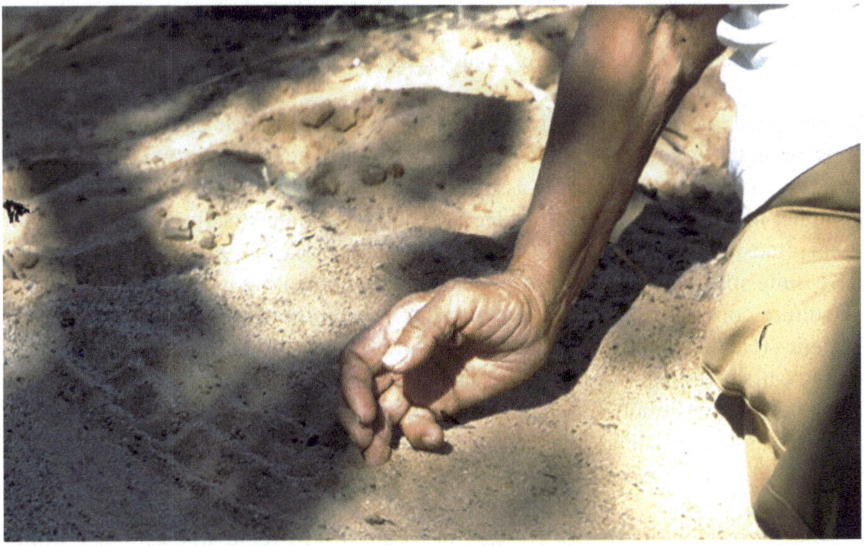

Fig. 3.24 Wangkamadla woman, Mrs Annie Hansen, drawing in the sand and singing traditional songs for her daughter, Isabel Tarragó, of the Mulligan River Aboriginal Corporation, 1982. (Photo Iain Davidson with permission of the Mulligan River Corporation)

non-lifelike, for many of them were simple images of the things the songs referred to. Rather, Munn described the art as "representational' (her word) but for many casual observers it looked "misleadingly" geometric. According to Munn, the signs people made in the sand as they told or sang stories could each have many meanings in the ritual realm and are actually "anchored in narrative communication". As I listened to Mrs Hansen and watched the smooth flow of the marks in the sand, it was clear that the songs have priority, and the sand drawings were explained by the song, and then erased after that moment had passed. I have discussed this previously (Davidson, 2020, 756):

> In a curious inversion of the usual order, the song stands alone, and the marks illustrate it. We, as students of art who have no knowledge of the song, think that the marks carry the meaning which they cannot do alone because in sand drawings the marks are continuously produced and then erased to be replaced by others. We archaeologists (and others) are observers in the absence of the producers and can only begin to approach coded meanings through recognizing the conventions or repetitions in the persistent marks. We cannot recover the songs, but we can be sure that they were necessary for the disambiguation.

Green (2014, 232) demonstrated with copious examples that: "The graphic symbols used in sand stories and in other Aboriginal art forms may well have a representational basis, but they are frequently multivalent, and cultural and contextual knowledge is needed to derive precise meanings." The implication is clear that, within the art-system, seers have to learn the cultural and contextual knowledge that will enable them to give meaning to ambiguous signs, and they do this through association between the signs and the stories/songs. The stories/songs are essential to the learning, initiation and enculturation, and the only material sign of that process is the (often incomplete) marks which we, and all others who have not learned the songs/stories, interpret as art.

This is a condition that more or less defines that the signs should be called symbols. Their capacity to be representational is determined by the particular conventions attached to them. Their relationship to the things they represent may have an important element of arbitrariness, and, following the logic of the more complete studies of Munn and Green, each sign can have many meanings which are clarified in the context of the song/story. I would argue that all non-representational marks have only ever had a meaning within the cultural system that ultimately derived from songs or stories which were absorbed through repetition of marks and their association with songs/stories. This is not specific only to modern Australian Aboriginal art.

Culley and I discussed the nature of symbols previously (2021, 25–27) emphasizing the three different categories of sign identified by Charles Peirce (1955[1931]), icon, index and symbol as I identified previously (Davidson, 2013, 21–22)—and note that the signs for horses have been discussed by others (Harnad, 1990).

> Iconic representations of a horse may become indexes of individuals, places, or even regions, such as "Horse Country" in the United States, simply through their repeated association. However, for an image of a horse to symbolize more abstract concepts, it must bring to mind *horse* through iconic reference, while yet guiding perception to an assigned meaning through conventions that specify the horse is not just a horse (Culley, 2016; Deacon, 2011)

In keeping with the argument I am expressing here, it is important to realize that, rather than use the simple word icon, it is better to think of iconic signs, so as to avoid confusion with the many journalistic stupidities surrounding the word icon. The practice would need to continue: with iconic signs, producers and seers see some similarity between the mark and the subject; with indexical signs, producers and seers see meaning through the repeated association of the sign with individuals, places or regions; and with symbolic signs, producers and seers see meaning through what we called "deliberate and purposeful mark-making with creative intent" (Culley & Davidson, 2021, 25). The way in which people get to use symbols is open to very complex discussion (Douglas, 2002; Eco, 1986; Harnad et al., 1976; Sperber, 1975) which is why the debate about the evolutionary emergence of language as a means of communication based on symbols has been heated and enduring decade after decade (Botha, 2020; Deacon, 2007; Dediu & Levinson, 2013; Harnad et al., 1976; Johansson, 2005; Lieberman, 1984; Mithen, 2024; Noble & Davidson, 1996; Zuberbühler & Bickel, 2022).

The arguments about the role of signs and symbols in the emergence of language are slightly peripheral to the argument here for the role of song and story in assigning meaning to symbols—though that role cannot exist until after a means of communication based on symbols had emerged in evolution. In all cases, it is essential that producers and seers have a role in making those marks into signs, and the purpose and intent is revealed through the song or story, in the first place, and subsequently, through the history of associations. Social learning is essential for signs to have shared meanings, with its associated impact on memory and cognition. By this means, some signs get to be called "proper" art, but in reality, it is just that social learning generates groups, but the hierarchy that is implicit in such group formation is not inevitable. The rank of a member of the group is given by the process of learning, not by the ranking of the materials that are learned about. Collingwood or Clark called their arts "proper" because they liked them, not because they were really art, and something else was not.

3.7 What, Then, Is Art—And Can We Tell it from Scribbles?

Many factors influence the perceptions, reception and consequences of the production of art and in turn define what it is. I began this chapter by listing some criteria for identifying art that I reduced to the following: art is lifelike, involves unquestioned beliefs, creativity and imagination, is beautiful, uses skill and acute observation, and creates emotion. Several of these characteristics are much more subjective than is convenient for a definition, or indeed for identification that something qualifies to be called art. I might, for example, be willing to sacrifice the criterion of "beauty" in light of, say, Goya's Black Paintings[7] from about 200 years ago (Fig. 3.25) or images such as the Picasso's sketches for *Guernica* from 110 years

[7] https://en.wikipedia.org/wiki/Black_Paintings

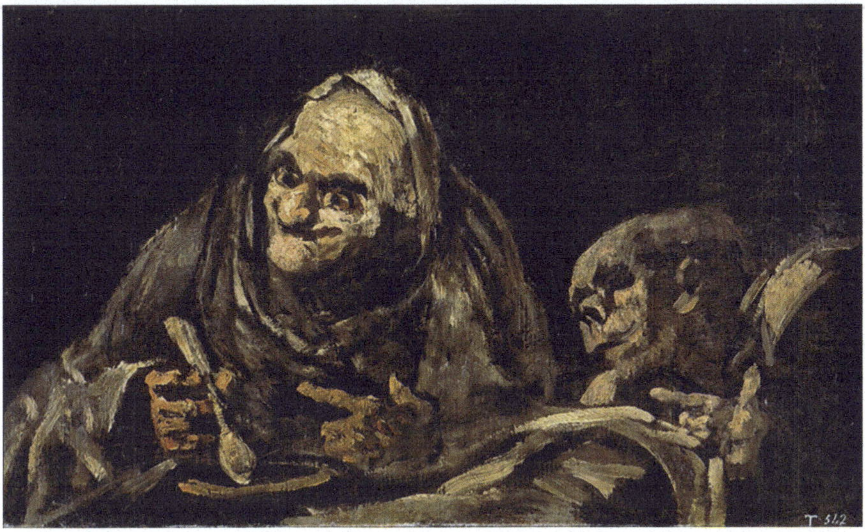

Fig. 3.25 Francisco Goya, about 1820. Old men eating soup. (Wikimedia Commons Public Domain)

later. It is arguable that Goya never intended for there to be any other seers of his Black Paintings, because he painted them on the walls of his house, so perhaps this should be excluded on those grounds. But then, perhaps, we would want to classify them as art because they were painted by an artist who was identifiable as one through other works. I fear that that criterion is full of awkward possibilities, and I will abandon it before I use it. I do not accept that I have to think that something is beautiful for us to call it art.

Ostensive definitions do not work because knowing that one thing is an example of art does not lead to the obvious identification of another. Lexical definitions tend to fail because they replace one lot of difficult words with another lot that are often defined in terms of the ones they are replacing. The search for an operational definition is particularly hopeless when individual judgment is involved. And the functional definition of art as a means of communication or something that creates emotion may recognize outcomes of the production of art but does not help anyone connect art now to the earlier steps that related to art over the course of human evolution. It is fundamental to such a definition that we understand how art comes to communicate or create emotion, but it is a secondary consideration to the question of whether it is art. I will discuss how the art-system manages to communicate in the next two chapters.

In two earlier treatments of art, I offered slightly different accounts of what we should be looking at, with an emphasis on the earliest examples in human evolution. First, the definition that Culley and I (2021, 3) used:

> We recognize that the earliest objects in the archaeological record which *may* constitute "art" include all intentionally made, non-functional, two- and three-dimensional marks on, and modifications to, portable and non-portable surfaces. These marks can occur on the

surface of utilitarian objects or on objects with no apparent function beyond decorative support (Davidson, 1989; Taçon & Chippindale, 1998).

This definition may be sufficient to recognize most examples of art from Art 1 and Art 2 and indeed much of Art 3.

The second definition recognizes that artists make and mark surfaces partly because they are stimulated by the sight or thought of something else. What they produce has an impact in the minds of the seers—an impact that we might call cognitive—partly because of what the producer says about their work, and partly because of the unquestioned beliefs within their culture. In recognition of this, I approximated a definition in a previous paper (Davidson, 2013, 42) which we will come back to near the end of the book: "Art is the making and marking of surfaces where the images have implicational meaning for the observer whether or not they know the full propositional meaning of the artist."

Two characteristics that I have discussed in this chapter are particularly important in the recognition of art:

Marks and art. I have previously pointed out that all art consists of marks made either by adding something, usually called pigment, to a surface, or by subtracting material from a surface (Davidson, 2020). In the evolution of art, those marks acquired meaning by the repeated associations between marks and subjects, sometimes including resemblance, and seers of the marks learned those meanings. The scribbles made by children (Matthews, 2013) are part of their process of learning to make meaningful marks guided by adults to the reality that marks can be meaningful.

Art and meaning. The production of marks constitutes the producer's way of showing to the world what is on their mind. It is a revelation of mind that does not seem to be available to other animals. The seer may be initiated into the producer's thoughts through the songs/stories that accompany the marks, but eventually those thoughts become part of the seer's ways of seeing—generally as unquestioned beliefs—as a result of conventions of mark making. In this way, marks reveal the intentions of the artist, and generate intentions of the seers. The intentions of these two different agents emerge whether they are the same or different, but the producer generates a response to the art they create. It is this set of interactions between the (at least) four elements of the production that I have chosen to call the art-system.

Berger (2008, 14), discussing two paintings by Frans Hals (Fig. 3.26) wrote about a version of this set of interactions:

> What is this 'seduction' he writes of? It is nothing less than the paintings working upon us. They work upon us because we accept the way [Frans] Hals saw his sitters. We do not accept this innocently. We accept it in so far as it corresponds to our own observations of people, gestures, faces, institutions. This is possible because we still live in a society of comparable social relations and moral values. And it is precisely this which gives the paintings their psychological and social urgency. It is this—not the painter's skill as a 'seducer'—which convinces us that we *can* know the people portrayed.

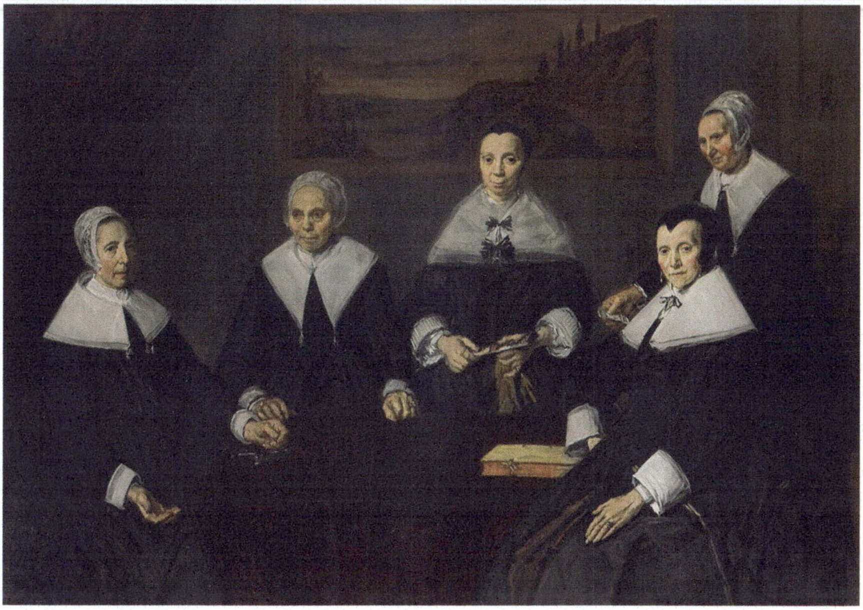

Fig. 3.26 Frans Hals 1664. *Regentesses of the Old Men's Alms House, Haarlem.* (Wikimedia Commons Public Domain)

In other words, in an art-system, the producer will have intentions, a story or a song but we, the seers, also have intentions of our own, which may be influenced by the particular story or song but they are principally possible (Berger's word) for us because we have, in Bourdieu's words, mastered the cultural code in which the image was created. We make things art through our critical eyes.

This art-system sets itself up for classifying groups of people who are part of it as producers or seers (Lamont et al., 2015). Generally speaking, membership of a particular group is defined by the relationship people may have with others who share knowledge of particular things, who have also mastered the cultural code. If they are both seers of the same product, and have shared the story with the other, then they will probably consider themselves members of a group with the other. I previously spelled out the steps by which humans and our ancestors acquired not just the relationships among members of interacting groups (Byrne et al., 2004), but as a result of novel cognitive perception of the meaning of the marks they exchanged with others (Davidson, 2023), also relationships between hominins and material things (Davidson & McGrew, 2005). Through them they formed relationships with places.

The context that the producer shares with more than one other person is very often a ritual context (Rappaport, 1999), such as an initiation (H. Morphy, 1991), and historically has been associated with religion (for the views of Durkheim, see the selection in Giddens, 1972, Ch. 11). These uses of art are, properly, aspects of the way people use art to communicate, as I will discuss in the following chapters.

Shiner (2001, Part II) described the history of the emergence of the practice and commerce in art in Europe in the eighteenth century, showing that intrinsic to it was a switch from the patronage which had supported religious art and court art to a commerce in pictures. Through the emergence of academies, as restrictive clubs of like-minded people with similar unquestioned beliefs, and through people grouping together on the basis of their "social relations and moral values" and their financial capacity to purchase art, attitudes towards what was worth purchasing were established. These involved judgments about the way in which paintings and sculptures were made, and their subject matter. The attitude that art was "invented" at that time is principally about the invention of a category of art that was deemed "fine" but is also about the restriction of what might be considered "proper" art to producers and seers in clubs and social groups who could afford to buy, and hence what the artist who did not have a patron could manage to sell.

Shiner's obsession, he says, began by wondering how art managed to migrate from the Field Museum, with a large collection of ethnographic items, to the Institute of Art in Chicago, and Pasztory has made a similar point. Picasso can be represented as adding images that look like Iron Age or ethnographic art to his painting of *Les Demoiselles d'Avignon*, but something else was going on because he was aware of the strength of the artistic vision involved in masks somewhat earlier. One of the revolutions of the twentieth century was for producers and seers of art to recognize that art had a role which was not just about some unquestioned rules of how to represent the world, nor about what was a suitable topic to be represented. Artists such as Picasso or Georges Braque, Marcel Duchamp or Andy Warhol could show that they saw the world in different ways and in doing so made seers see it differently too.

In Australia, cultural life has been transformed by the appearance and elaboration of Aboriginal art. The origins of this are various, principally because the art itself has always varied across the country. The best-known, perhaps, derives from the transference of artistic elements from sand paintings to persisting media of paint and support in Central Australia, as I documented earlier (for an introduction see the small book by Myers & Smith, 2024). But the history is longer, with rock art recorded on Groote Eyland in 1803 (Jones, 1989, Fig. 183) and on Depuch Island in 1840 (Stokes, 1846). Bark paintings were collected from Arnhem Land by 1877 (Sutton, 1989, Fig. 72) and *toas* were collected around Lake Eyre between 1888–1906 (H. Morphy, 1998, 84–88). It could be argued that what made a difference in the case of Central Australian acrylic paintings was that, facilitated by Bardon, the senior Pintupi men took the initiative in producing the first paintings in Papunya, that were sold in Alice Springs. The Pintupi painters thus achieved some sort of financial autonomy for the first time through their art.

One crucial point in the case of Aboriginal art in Australia has been that producers and seers approach the works in different ways. Although Morphy was able to document in detail the roles of bark paintings in the several levels of initiations of Yolngu people in eastern Arnhem Land (H. Morphy, 1991), the people found themselves in the position of needing to decide whether to sell paintings into the European (Australian) market, and then which paintings to sell, who should paint them, and how they should be marketed (H. Morphy & Layton, 1981, 58). As Howard Morphy and Bob Layton continued, in making decisions about this, people sought to preserve the power elders had over paintings while accepting that some images would leak out into the non-Yolngu world. In the process, people painted slightly different things, including more figurative images, but sought to maintain their agency in face of the pressures from outside. As we have seen in all instances in the art-system, the agency of the producers resulted in some changes in their art as a result of their decisions—and at the same time some changes to the balance of the social system. In turn, the seers through their own agency were able to see stories in the art, even though they were not the stories from which they had been denied access. And, as with the changes in art and aesthetics in eighteenth century Europe, overwhelming changes in the art-system resulted in the availability of a market for the product, and neither producers nor seers were wholly in control of the consequences.

Vincent Megaw (1982, 207) quoted Layton (1981 [1991], 11–12) to the effect that:

> the investigation of the artist's intent (as far as he can express it), and the response of fellow members of his community, are of primary importance if we are not simply to take objects from them and see whether such objects can be appropriated by *our* culture as works of art in *our* terms

Layton went on:

> ...it is impossible ever to investigate someone else's thoughts or intentions directly. They can be understood only *via* the means he uses to communicate them, and our understanding of such messages is based not on his experience, but on our own.

I fear that leaves us, now bolstered by Berger, Bourdieu and others, with the distinctly perilous situation of my early definition: art as all the things we call art (Davidson & Noble, 1989, 136 fn. 6) but now we can be certain that within an art-system, art is all the things we call art, but also all the things that producers of art call art. Our job becomes to see how art works, how it is used.

3.8 Essay 3 Art, Rock Art or Not Art (Scratch on Abalone Shells, Blombos and Collected Scratched Ocher Slabs from Blombos)

We all make marks in our environments, and so do almost all other animals that move on the surface of the ground (see, for example, Davidson, 2013). Humans, alone, seem to use visual signs (actually, indexical signs) of the passage of other animals as cues to pursue their prey. Other animals have not been shown to, though they follow scent successfully, and an ingenious experiment has shown that great apes can follow yogurt tracks to find hidden food (Völter & Call, 2014). So, there are marks that betray our actions, but generally we are not aware of the fact that we have left marks. What do we need if we want to use accidentally-made marks to produce art?

First, our ancestors needed to notice that conspecifics can make marks, then that such marks can be made deliberately. The last stage is for those ancestors to realize that those deliberate marks can carry meaning. As we now know from several examples, the difficulty we have is to work out when such marking was first noticed, then when it was done deliberately and finally how it came to carry meaning. These have proven difficult to establish because they are not trivial issues (d'Errico et al., 2015).

As archeologists, we can be sure that there was an object (we find it), and that there was a producer (the object is modified). The problem is to establish, first, that there were seers of the object (that people seeing the object would understand something by it), and second that there was a subject of that object (there was a thing or an idea the producer was trying to get across using the modified object). One of the elements of deciding that something was deliberately made is to establish that a pattern was repeated. I return to this in Chap. 7. But any marks such as tracks left in the ground would be repeated so repetition is not enough—such marks were often not deliberate, and generally did not carry meaning other than as indicators of the presence of the mark-makers. Stone tools had always involved repetition of motor actions and then of product, but other than indicating the presence of past knappers, it is not clear that meaning was attached (Davidson & McGrew, 2005).

Repetition was also a characteristic of the marks in two abalone shells found by Chris Henshilwood and Francesco d'Errico at Blombos, in South Africa, shown in Fig. 3.27 Left (Henshilwood et al., 2011). These were found scratched into the nacre in the bottom of the shells. As the abalone shells were next to each other and they had all the appearance of receptacles used for mixing ocher into wet paint, it was presumed that these marks were made in the process of mixing the paint. In all probability, no one saw the marks until the shells were excavated at the site in 2008 from levels dated to 100 thousand years ago. The paint must have been used to make marks that were deliberate, but it is less clear whether they carried meaning. What we lack, in the paint pots and associated marks at Blombos, is any indication of how the painting acquired meaning. We tend to assume it was there, but there is no evidence. Blombos is a cave site on the southern coast of South Africa from which people can look out to the Southern Ocean. In a sequence of ever more astonishing finds published since 1997, Henshilwood, d'Errico and their teams have published worked bone, shell beads (Henshilwood et al., 2004); engraved ochers (Henshilwood et al., 2009); bifacially flaked artefacts (Mourre et al., 2010); pressure flaking retouch

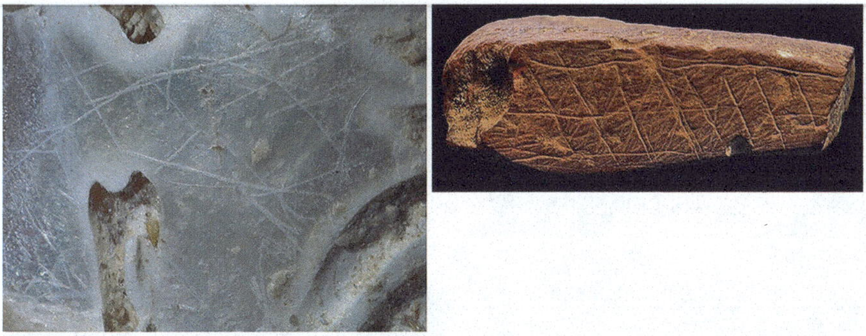

Fig. 3.27 Left: Scratches at bottom of an abalone shell used to mix paint 100,000 years ago at Blombos, South Africa (Henshilwood et al., 2011). Right: Engraved piece of ocher from Blombos dated between 100 and 75,000 years ago (Henshilwood et al., 2009). Permission from Chris Henshilwood

of stone tools artefacts (Henshilwood & Sealy, 1997); artefacts associated with the preparation of ocher as paint (Henshilwood et al., 2011); and heat-treatment of stone artefacts (Delagnes et al., 2016; Henshilwood, 2012). They changed the story.

At one stage I was preoccupied with the identification of early art (Davidson, 1992). This was because I was engaged in an argument about the earliest emergence of language, and art seemed to have something to do with it. It certainly does, but we probably did not have the answer in the 1980s. We did, however, define "modern human behavior" as involving "greater information flow, planning depth and conceptualization consequent upon the emergence of language" defined as the "symbolic use of communicative signs" (Noble & Davidson, 1991). Bill Noble and I wrote a book about our argument about language and evolution in 1995 (Noble & Davidson, 1996). Several discussions arising from the new finds at Blombos were framed in terms of their relevance to the emergence of language (d'Errico et al., 2003; Henshilwood et al., 2002; Henshilwood & Marean, 2003).

Among the early claims for art there is a peculiarity: they were all individual pieces that archeologists could not account for in some other way. I have dealt with some of these early claims jointly with Culley, and, by-and-large, my point is still valid: individual pieces are difficult to explain even though we might weave elaborate stories around them (Culley & Davidson, 2021; Davidson, 1990, 1991). These finds included the Makapansgat cobble (Bednarik, 1998) (Fig. 7.5), the scratched shell from Trinil in Java (Joordens et al., 2015) (Figs. 3.7 and 7.28 Left), the Berekhat Ram modified pebble from the Golan Heights (d'Errico & Nowell, 2000; Marshack, 1997) (Fig. 7.24); the Tata tooth from Hungary (Marshack, 1976) and the scratched bones from Bilzingsleben in Germany (Davidson, 1990; Mania & Mania, 1988) (Fig. 7.9). The task for the archeologist dealing with them, or other individual pieces, is to establish that whatever the "meaning" of the object to us, it was also meaningful to the society of the people who made it. I suggested that we could do that through identification of symbol-use through a convention (Davidson, 2003, 153): "We can infer a convention or code through which meaning might be recognized if we find depictive or non-depictive marks on objects in repeated patterns, restricted in time and distribution."

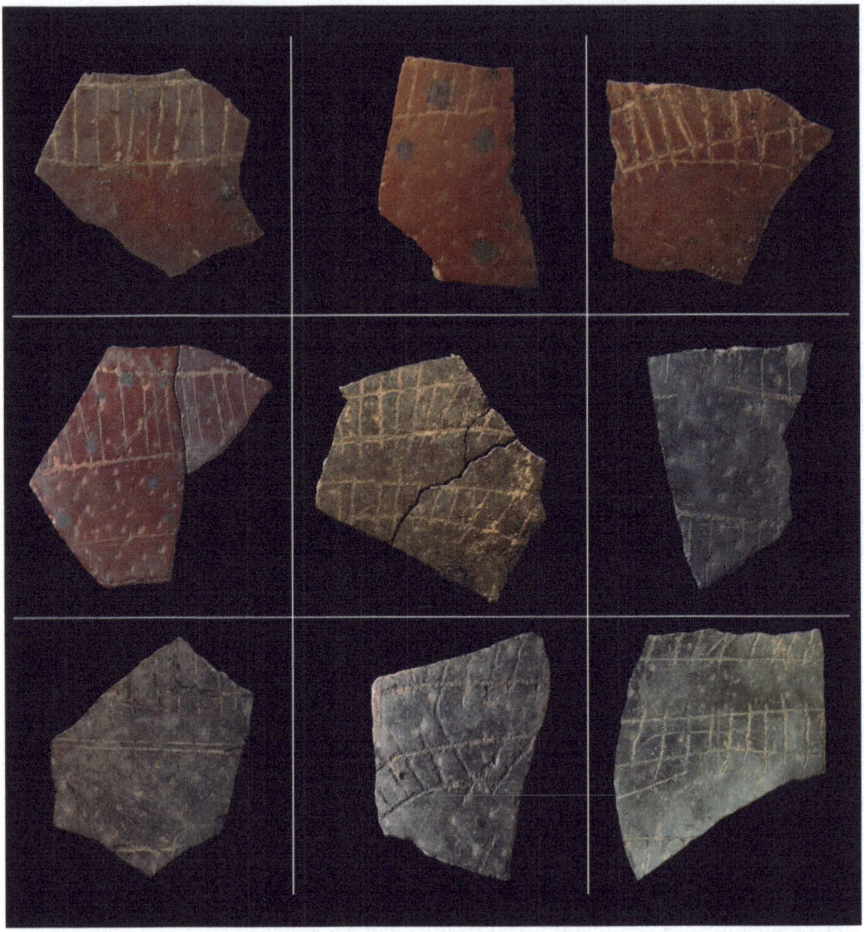

Fig. 3.28 Fragments of engraved ostrich egg shell from Diepkloof, South Africa, dated to between 100 thousand and 60 thousand years ago, similar to those from Klipdrift. (Permission from Pierre-Jean Texier 26/11/2024)

Once again, the finds from Blombos—all published after 1995—seem to suggest answers. First, the paint mixing at 100 thousand years ago was repeated—there were two examples. Second, the ocher with engraved patterns from 75 thousand years ago was also found with one elaborate example (Fig. 3.27 Right) and another one that had the same cross hatched design. So, repetition occurred unequivocally in patterned human activities both accidentally and deliberately and this was restricted in time and distribution. A similar repetition of patterns occurred with the marks made on ostrich egg shell at Diepkloof (Fig. 3.28) (Texier et al., 2013) and at Klipdrift (Henshilwood et al., 2014) at a period a similar to that for the ochre at Blombos.

But Blombos showed something more. There are several pieces of ocher from different layers at the site dating between 100 thousand years and 70 thousand (Fig. 3.29). Some scratching of ocher must have happened as part of the process of

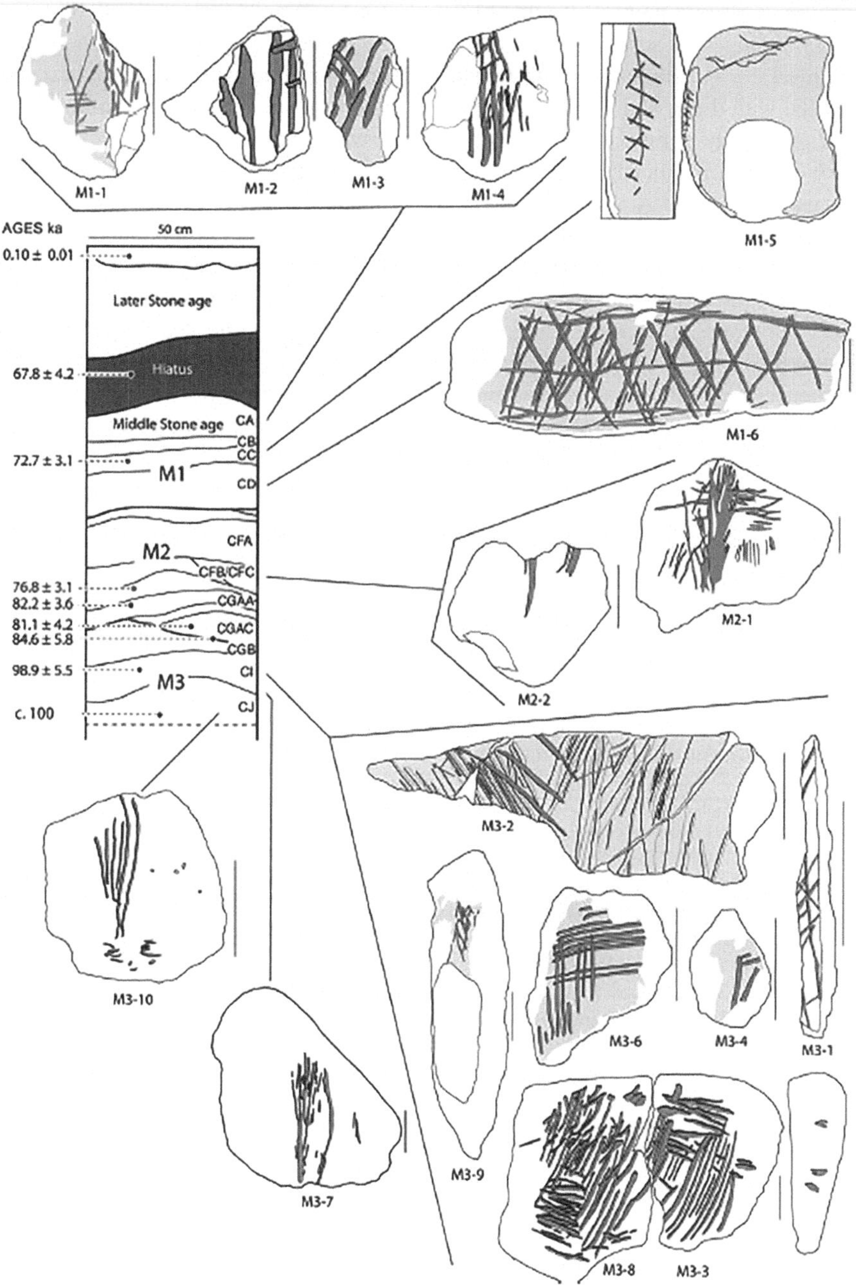

Fig. 3.29 Sequence of engraved ochre slabs from Blombos, South Africa, from about 100 thousand until about 70 thousand years ago. The two pieces with cross-hatching (M1-6 and M1-5) are from above and below the layer dated to 72.7 thousand years ago (from Henshilwood et al., 2009)

turning ocher into paint. But some of the markings are not consistent with that sort of scratching. The scratched ocher seems to be part of a process by which people left marks in a utilitarian context, they or other people noticed the marks and experimented with them, and finally people made patterns using the same technology.

I previously defined three categories of art which underlie this book. They are derived from a somewhat simpler categorization by Robb. Art 1 consists of those things that we call Art in the modern world made by "great artists" and Art 2 is archeological art, such as Greek Red Figure Pottery (such as that in the next Essay) or rock art. Art 3 is the earliest archeological art that some recognize as art, but some might not, including that from Blombos. It often involves repetition for archeologists to recognize from the pattern that it is an intentional production. There is a further question about how people turned one category of Art into any other. The sequence at Blombos may give us an insight into that, as the actions for one purpose (producing ocher powder) turned into those for another (producing patterns on ocher).

References

Andersen, B. (2015). Style and self-making: string bag production in the Papua New Guinea highlands. *Anthropology Today, 31*(5), 16–20.

Arnold, D. (2020). *Art history: A very short introduction*. Oxford University Press.

Australian Parliament House Construction Authority. (1989). *Expressing Australia: Art in parliament house*. Australian Parliament House Construction Authority.

Baguley, M., & Kerby, M. (2021). Beyond the metropolis: The [New] England Regional Art Museum (NERAM). *Australian Art Education, 42*(2), 107–123.

Bardon, G. (1991). *Papunya Tula. Art of the Western Desert*. McPhee Gribble.

Bass, L. R. (2008). The treasury of the language: Literary invention in Phillip III's reign. In S. Schroth & R. Baer (Eds.), *El Greco to Velázquez: Art during the reign of Philip III* (pp. 147–181). MFA Publications.

Bednarik, R. G. (1998). The 'Australopithecine' cobble from Makapansgat, South Africa. *South African Archaeological Bulletin, 53*, 4–8.

Beltran, A., & Saura Ramos, P. A. (Eds.). (1998). *The cave of Altamira*. Harry N. Abrams.

Berger, J. (2008). *Ways of seeing*. Penguin.

Boas, F. (1955). *Primitive art*. Dover Publications, Inc.

Botha, R. (2020). *Neanderthal language: Demystifying the linguistic powers of our extinct cousins*. Cambridge University Press.

Bourdieu, P. (1968). Outline of a sociological theory of art perception. *International Social Science Journal, 20*(4), 589–612.

Bradley, R. (2009). *Image and audience. Rethinking prehistoric art*. Oxford University Press.

Brown, P. F. (1984). Painting and history in renaissance Venice. *Art History, 7*, 263–294.

Bugnyar, T. (2019). Tool use: New Caledonian crows engage in mental planning. *Current Biology, 29*(6), R200–R202. https://doi.org/10.1016/j.cub.2019.01.059

Byrne, R. W., Barnard, P. J., Davidson, I., Janik, V. M., McGrew, W. C., Miklósi, Á., & Wiessner, P. (2004). Understanding culture across species. *Trends in Cognitive Sciences, 8*(8), 341–346.

Cennini, C. (1844). *A treatise on painting (written in 1437)*. Edward Lumley.

Chippindale, C. (1985). What is "prehistoric art"?: A definition and its consequences. *The Archaeological Review from Cambridge, 4*(2), 141–158.

Chippindale, C. (2001). What are the right words for rock-art in Australia? *Australian Archaeology, 53*, 12–15.

Churcher, B. (2014). *Australian notebooks*. The Miegunyah Press.

Clark, K. (1969). *Civilisation*. John Murray.

Clifford, J. (1988). *The predicament of culture*. Harvard University Press.

Collingwood, R. G. (1938). *The principles of art*. Oxford University Press.

Craig, B. (1967). The Houseboards of the Telefomin Sub-District, New Guinea. *Man, 2*(2), 260–273.

Culley, E. V. (2016). *A semiotic approach to the evolution of symboling capacities during the late Pleistocene with implications for claims of 'Modernity' in early human groups*. Arizona State University.

Culley, E. V., & Davidson, I. (2021). Art, sign, and representation. In N. Gontier, A. Lock, & C. Sinha (Eds.), *The Oxford handbook of human symbolic evolution* (2nd ed., pp. 331–368). Oxford University Press.

Danto, A. C. (2013). *What art is*. Yale University Press.

Darwin, C. R. (1871). *The descent of man and selection in relation to sex*. John Murray.

Davidson, I. (1989). Freedom of information: Aspects of art and society in western Europe during the last ice age. In: H. Morphy (Ed.), *Animals into art* (pp 440–456). Unwin Hyman, London.

Davidson, I. (1990). Bilzingsleben and early marking. *Rock Art Research, 7*, 52–56.

Davidson, I. (1991). The archaeology of language origins: A review. *Antiquity, 65*, 39–48.

Davidson, I. (1992). There's no art – To find the mind's construction – In offence. *Cambridge Archaeological Journal, 2*, 52–57.

Davidson, I. (2003). The archaeological evidence of language origins: States of art. In M. H. Christiansen & S. Kirby (Eds.), *Language evolution* (pp. 140–157). Oxford University Press.

Davidson, I. (2007). Tasmanian aborigines and the origins of language. In J. Mulvaney & H. Tyndale-Biscoe (Eds.), *Rediscovering Recherche Bay* (pp. 69–85). Academy of the Social Sciences. in Australia.

Davidson, I. (2012). Symbolism and becoming a hunter-gatherer. *L'art pléistocène dans le monde/Pleistocene art of the world/Arte pleistoceno en el mundo, Actes du Congrès IFRAO, Tarascon-sur-Ariège, septembre 2010, Symposium « Signes, symboles, mythes et idéologie... ». N° spécial de Préhistoire, Art et Sociétés, Bulletin de la Société Préhistorique Ariège-Pyrénées, LXV-LXVI*, Book: p. 292–293, CD: p. 1689–1705

Davidson, I. (2013). Origins of pictures: An argument for transformation of signs. In K. Sachs-Hombach & J. R. J. Schirra (Eds.), *Origins of pictures. Anthropological discourses in image science* (pp. 15–45). Halem.

Davidson, I. (2016). Stone tools: Evidence of something in between culture and cumulative culture? In M. N. Haidle, N. J. Conard, & M. Bolus (Eds.), *The nature of culture* (pp. 99–120). Springer.

Davidson, I. (2017). Images of animals in rock art: Not just "good to think". In B. David & I. J. McNiven (Eds.), *Oxford handbook of the archaeology and anthropology of rock art* (pp. 435–467). Oxford University Press.

Davidson, I. (2020). Marks, pictures and art: Their contribution to revolutions in communication. *Journal of Archaeological Method and Theory, 27*(3), 745–770. https://doi.org/10.1007/s10816-020-09472-9

Davidson, I. (2021). 1. Scenes and non-scenes in rock art. In I. Davidson & A. Nowell (Eds.), *Making scenes: Global perspectives on scenes in rock art* (pp. 16–31). Berghahn Books.

Davidson, I. (2023). Humans making history through continuities and discontinuities in art. *Cambridge Archaeological Journal, 33*, 637–654. https://doi.org/10.1017/S0959774323000057

Davidson, I., & McGrew, W. C. (2005). Stone tools and the uniqueness of human culture. *The Journal of the Royal Anthropological Institute,* 11(4), 793–817

Davidson, I., & Noble, W. (1989). The archaeology of perception: Traces of depiction and language. *Current Anthropology, 30*(2), 125–155.

Davies, S. (2015). Defining art and artworlds. *The Journal of Aesthetics and Art Criticism, 73*(4), 375–384.

Davies, S. (2016). *The philosophy of art* (2nd ed.). Wiley Blackwell.
Davis, W. (1993). Beginning the history of art. *The Journal of Aesthetics and Art Criticism, 51*(3), 327–350.
Davis, W. (2017). *Visuality and virtuality: Images and pictures from prehistory to perspective*. Princeton University Press.
Davis, W. (2022). *A general theory of visual culture*. Princeton University Press.
de Viguerie, L., Ducouret, G., Lequeux, F., Moutard-Martin, T., & Walter, P. (2009). Historical evolution of oil painting media: A rheological study. *Comptes Rendus Physique, 10*(7), 612–621.
Deacon, T. W. (2007). The evolution of language systems in the human brain. In J. H. Kaas (Ed.), *Evolution of nervous systems* (Vol. 5. The evolution of primate nervous systems,, pp. 529–547). Academic.
Deacon, T. W. (2011). The symbol concept. In K. R. Gibson & M. Tallerman (Eds.), *The Oxford handbook of language evolution* (pp. 393–405). Oxford University Press.
Dediu, D., & Levinson, S. C. (2013). On the antiquity of language: The reinterpretation of Neandertal linguistic capacities and its consequences. *Frontiers in Psychology, 4*. https://doi.org/10.3389/fpsyg.2013.00397
Delagnes, A., Schmidt, P., Douze, K., Wurz, S., Bellot-Gurlet, L., Conard, N. J., et al. (2016). Early evidence for the extensive heat treatment of silcrete in the Howiesons Poort at Klipdrift shelter (layer PBD, 65 ka), South Africa. *PLoS One, 11*(10), e0163874. https://doi.org/10.1371/journal.pone.0163874
d'Errico, F., & Nowell, A. (2000). A new look at the Berekhat Ram figurine: Implications for the origins of symbolism. *Cambridge Archaeological Journal, 10*(1), 123–167.
d'Errico, F., Henshilwood, C. S., Lawson, G., Vanhaeren, M., Tillier, A. M., Soressi, M., … Julienlo, M. (2003). Archaeological evidence for the emergence of language, symbolism, and music – An alternative multidisciplinary perspective. *Journal of World Prehistory, 17*(1), 1–70.
d'Errico, F., Vanhaeren, M., Van Niekerk, K., Henshilwood, C. S., & Erasmus, R. M. (2015). Assessing the accidental versus deliberate colour modification of Shell beads: A case study on perforated Nassarius kraussianus from Blombos cave middle stone age levels. *Archaeometry, 57*(1), 51–76. https://doi.org/10.1111/arcm.12072
Descola, P. (Ed.). (2010). *La Fabrique des images. Visions du monde et formes de la représentation*. Somogy et Musée du Quai Branly.
Descola, P. (2013 [2005]). *Beyond nature and culture (Par-delà nature et culture)* (J. Lloyd, Trans.). The University Press of Chicago
Díaz-Andreu, M. (1996). Islamic archaeology and the origin of the Spanish nation. In M. Díaz-Andreu & T. Champion (Eds.), *Nationalism and archaeology in Europe* (pp. 68–89). UCL Press.
Dimaggio, P., & Useem, M. (1978). Social class and arts consumption: The origins and consequences of class differences in exposure to the arts in America. *Theory and Society, 5*(2), 141–161. Retrieved from http://www.jstor.org/stable/656694
Dobrez, L. (2022). Aesthetic response, Kant, and rock art. *American Indian Rock Art, 48*, 155–166.
Dodds, J. D., Menocal, M. R., & Balbale, A. K. (2008). *The arts of intimacy. Christians, jews, and muslims in the making of Castilian culture*. Yale University Press.
Douglas, M. (2002). *Natural symbols*. Routledge.
Dutton, D. (2009). *The art instinct. Beauty, pleasure, and human evolution*. Bloomsbury Press.
Eco, U. (1986). *Semiotics and the philosophy of language*. Indiana University Press.
Foucault, M. (2002). *The order of things: An archaeology of the human sciences*. Routledge.
Gibson, J. J. (1966). *The senses considered as perceptual systems*. Houghton Mifflin.
Giddens, A. (Ed.). (1972). *Emile Durkheim: Selected writings*. Cambridge University Press.
Goddard, C. (2005). The lexical semantics of 'culture'. *Language Sciences, 27*(1), 51–73.
Goldsborough, Z., Schel, A. M., & van Leeuwen, E. J. C. (2023). Chimpanzees communicate to coordinate a cultural practice. *Proceedings of the Royal Society B: Biological Sciences, 290*(1991), 20221754.
Goldstein, C. (2021, June 23). Rembrandt's revered 'night watch' was cut up to fit through a door. With A.I., you can see it whole for the first time in 300 years. *Artnet News*.
Gombrich, E. H. (1995). *The story of art* (16th ed.). Phaidon.

Goodall, J. (1986). *The chimpanzees of Gombe. Patterns of behavior*. The Belknap Press of Harvard University Press.
Gorman, A. (2008). Chapter 5. The primitive body and colonial administration: Henry Ling Roth's approach to body modification. In R. McDougall & I. Davidson (Eds.), *The Roth family, anthropology and colonial administration* (pp. 93–103). Left Coast Press.
Green, J. (2014). *Drawn from the ground. Sound, sign and inscription in central Australian sand stories*. Cambridge University Press.
Greenfield, P. M., Maynard, A. E., & Childs, C. P. (2000). History, culture, learning, and development. *Cross-Cultural Research, 34*(4), 351–374. https://doi.org/10.1177/106939710003400404
Guyer, P. (2020). 18th century German aesthetics. *The Stanford Encyclopedia of Philosophy (Fall 2020 Edition)*. Retrieved from https://plato.stanford.edu/archives/fall2020/entries/aesthetics-18th-german/
Haber, R. N. (1980). How we perceive depth from flat pictures: The inherent dual reality in pictorial art enables us to perceive a scene as three dimensional at the same time we see that the painting or photograph is actually flat. *American Scientist, 68*(4), 370–380.
Harnad, S. R. (1990). The symbol grounding problem. *Physica D: Nonlinear Phenomena, 42*(1–3), 335–346.
Harnad, S. R., Steklis, H. D., & Lancaster, J. B. (1976). *Origins and evolution of language and speech* (vol. 280): Annals of the New York Academy of Sciences.
Harrod, J. B. (2014). The case for chimpanzee religion. *The Journal for the Study of Religion, Nature and Culture, 8*(1), 8–45.
Healy, S., Walsh, P., & Hansell, M. (2008). Nest building by birds. *Current Biology, 18*(7), R271–R273.
Henshilwood, C. S. (2012). Late Pleistocene techno-traditions in southern Africa: A review of the still bay and Howiesons Poort, c. 75–59 ka. *Journal of World Prehistory, 25*(3–4), 205–237. https://doi.org/10.1007/s10963-012-9060-3
Henshilwood, C. S., & Marean, C. W. (2003). The origin of modern human behavior: Critique of the models and their test implications. *Current Anthropology, 44*(5), 627–651.
Henshilwood, C. S., & Sealy, J. (1997). Bone artefacts from the middle stone age at Blombos Cave, Southern Cape, South Africa. *Current Anthropology, 38*(5), 890–895.
Henshilwood, C. S., d'Errico, F., Yates, R., Jacobs, Z., Tribolo, C., Duller, G. A. T., et al. (2002). Emergence of modern human behavior: middle stone age engravings from South Africa. *Science, 295*(5558), 1278–1280. https://doi.org/10.1126/science.1067575
Henshilwood, C. S., d'Errico, F., Vanhaeren, M., van Niekerk, K., & Jacobs, Z. (2004). Middle stone age shell beads from South Africa. *Science, 304*(5669), 404.
Henshilwood, C. S., d'Errico, F., & Watts, I. (2009). Engraved ochres from the middle stone age levels at Blombos Cave, South Africa. *Journal of Human Evolution, 57*(1), 27–47.
Henshilwood, C. S., d'Errico, F., Niekerk, K. L. V., Coquinot, Y., Jacobs, Z., Lauritzen, S.-E., et al. (2011). A 100,000-year-old ochre-processing workshop at Blombos cave, South Africa. *Science, 334*, 219–222.
Henshilwood, C. S., van Niekerk, K. L., Wurz, S., Delagnes, A., Armitage, S. J., Rifkin, R. F., et al. (2014). Klipdrift Shelter, southern Cape, South Africa: Preliminary report on the Howiesons Poort layers. *Journal of Archaeological Science, 45*(0), 284–303. https://doi.org/10.1016/j.jas.2014.01.033
Hicks, D. (2020). *The brutish museums: The Benin bronzes, colonial violence and cultural restitution*. Pluto Press.
Hockney, D., & Gayford, M. (2020). *A history of pictures: From the cave to the computer screen*. Thames & Hudson.
Holden, C. (2003). *Lionel Lindsay in Spain. An antipodean abroad*. Miegunyah Press.
Ingold, T. (1983). The architect and the bee: Reflections on the work of animals and men. *Man, 18*(1), 1–20.

Ingold, T. (2000a). Chapter 19. Of string bags and birds' nests. Skill and the construction of artefacts. In T. Ingold (Ed.), *The perception of environment. Essays in livelihood, dwelling and skill* (pp. 349–361). Routledge.

Ingold, T. (2000b). *Perception of the environment: Essays on livelihood, dwelling and skill*. Routledge.

Johansson, S. (2005). *Origins of language: Constraints on hypotheses* (Vol. vol. 5). John Benjamins Publishing.

Jones, P. (1989). Chapter V. Perceptions of aboriginal art: A history. In P. Sutton (Ed.), *Dreamings: The art of aboriginal Australia* (pp. 143–179). Viking.

Jones, P. (2007). *Ochre and rust. Artefacts and encounters on the Australian frontiers*. Wakefield Press.

Jones, P., & Sutton, P. (1986). *Art and land: Aboriginal sculptures of the Lake Eyre Basin*. South Australian Museum and Wakefield Press, Adelaide.

Joordens, J. C. A., d'Errico, F., Wesselingh, F. P., Munro, S., de Vos, J., Wallinga, J., et al. (2015). Homo erectus at Trinil on Java used shells for tool production and engraving. *Nature, 518*(7538), 228–231. https://doi.org/10.1038/nature13962

Kühl, H. S., Kalan, A. K., Arandjelovic, M., Aubert, F., D'Auvergne, L., Goedmakers, A., et al. (2016). Chimpanzee accumulative stone throwing. *Scientific Reports, 6*(1), 1–8.

Lamont, M., Pendergrass, S., & Pachucki, M. (2015). Symbolic boundaries. *International Encyclopedia of the Social & Behavioral Sciences, 2*, 850–855.

Layton, R. H. (1981 [1991]). *The anthropology of art* (2nd ed.). Cambridge University Press.

Layton, R. H. (1992). *Australian rock art. A new synthesis*. Cambridge University Press.

Lieberman, P. (1984). *The biology and evolution of language*. Harvard University Press.

MacKenzie, M. (2004). The Bilum is our mother. *Material Culture: Critical Concepts in the Social Sciences, 2*, 17.

Macquarie Dictionary. (2017). *Macquarie concise Dictionary* (7th ed.). Macmillan.

Madden, J. R. (2001). Sex, bowers and brains. *Proceedings of the Royal Society of London. Series B: Biological Sciences, 268*(1469), 833–838. https://doi.org/10.1098/rspb.2000.1425

Madden, J. R. (2008). Do bowerbirds exhibit cultures? *Animal Cognition, 11*(1), 1–12. https://doi.org/10.1007/s10071-007-0092-5

Mainwaring, M. C., Hartley, I. R., Lambrechts, M. M., & Deeming, D. C. (2014). The design and function of birds' nests. *Ecology and Evolution, 4*(20), 3909–3928.

Malafouris, L. (2013). *How things shape the mind. A theory of material engagement*. MIT Press.

Mania, D., & Mania, U. (1988). Deliberate engravings on bone artefacts of Homo erectus. *Rock Art Research, 5*, 91–109.

Marshack, A. (1976). Some implications of the Paleolithic symbolic evidence for the origin of language. *Current Anthropology, 17*(2), 274–282.

Marshack, A. (1997). The Berekhat Ram figurine–a late Acheulian carving from the Middle East. *Antiquity, 71*(272), 327–337.

Mason, R. (2005). The meaning and value of home-based craft. *International Journal of Art & Design Education, 24*(3), 261–268.

Mastandrea, S., Bartoli, G., & Bove, G. (2007). Learning through ancient art and experiencing emotions with contemporary art: Comparing visits in two different museums. *Empirical Studies of the Arts, 25*(2), 173–191.

Matheny, R. T., Smith, T. S., & Matheny, D. G. (1997). Animal ethology reflected in the rock art of Nine Mile Canyon, Utah. *Journal of California and Great Basin Anthropology, 19*(1), 70–103.

Matheny, R. T., Matheny, D. G., Miller, P. W., & Miller, B. (2004). Hunting strategies and winter economy of the Fremont as revealed in the rock art of nine mile canyon. In R. T. Matheny (Ed.), *New dimensions in rock art studies* (pp. 145–193). Brigham Young University.

Matthews, J. (2013). Seven spots and a squiggle: The prehistory of pictures. In K. Sachs-Hombach & J. R. J. Schirra (Eds.), *Origins of pictures. Anthropological discourses in image science* (pp. 311–352). Halem.

McDougall, R. (2008). Chapter 4. The making of *Great Benin*: Felix Norman Roth and Henry Ling Roth. In R. McDougall & I. Davidson (Eds.), *The Roth family, anthropology and colonial administration* (pp. 73–92). Left Coast Press.

McGrew, W. C. (1987). Tools to get food: The subsistants of the Tasmanian aborigines and Tanzanian chimpanzees compared. *Journal of Anthropological Research, 43*, 247–258.

McGrew, W. C. (1992). *Chimpanzee material culture. Implications for human evolution*. Cambridge University Press.

McGrew, W. C., & Marchant, L. F. (1998). Chimpanzee wears a knotted skin "necklace". *Pan Africa News, 5*(1), 8–9.

McHattie, L.-S., Champion, K., & Broadley, C. (2018). Craft, textiles, and cultural assets in the Northern Isles: Innovation from tradition in the Shetland Islands. *Island Studies Journal, 13*(2), 39–54.

Mecklenburg, M. F., Tumosa, C. S., & Erhardt, W. D. (1998). Structural response of painted wood surfaces to changes in ambient relative humidity. In V. Dorge & F. Carey Howlett (Eds.), *Painted wood: History and conservation* (pp. 464–483). The Getty Foundation.

Megaw, J. V. S. (1982). Western Desert acrylic painting—artefact or art? *Art History, 2*, 205–218.

Menocal, M. R. (2003). *The ornament of the world: How Muslims, Jews, and Christians created a culture of tolerance in medieval*. Back Bay Books.

Mitchell, S. (2016). "To the city for dancing": The women artists collection and the birth of commercial art practice at Yuendumu. In A. Simpson & G. Hammond (Eds.), *A cultural cacophony* (p. 86). Museums galleries Australia.

Mithen, S. (2024). *The language puzzle: Piecing together the six-million-year story of how words evolved*. Hachette.

Moro Abadía, O. (2006). Arts, crafts and Paleolithic art. *Journal of Social Archaeology, 6*(1), 119–141.

Morphy, H. (1977). Schematisation, meaning and communication in toas. In P. J. Ucko (Ed.), *Form in indigenous art: Schematisation in the art of aboriginal Australia and prehistoric Europe* (pp. 77–89). Australian Institute of Aboriginal Studies.

Morphy, H. (1991). *Ancestral connections: Art and an aboriginal system of knowledge*. University of Chicago Press.

Morphy, H. (1998). *Aboriginal art*. Phaidon Press.

Morphy, F. (2023, February 14). *When two worlds meet: "Translation" and its dilemmas*. Paper presented at the Madayin: Eight decades of Australian Aboriginal Bark paintings from Yirrkala, University of Virginia in partnership with the Buku-Larrŋgay Mulka Centre. American University Museum, Katzen Arts Center. (February 3–July 21, 2024 on display at The Fralin Museum of Art in Charlottesville)

Morphy, H., & Layton, R. H. (1981). Choosing among alternatives: Cultural transformation and social change in Aboriginal Australia and the French Jura. *Mankind, 13*(1), 56–73.

Morris, D. (2013). *The artistic ape: Three million years of art*. Red Lemon Press.

Mourre, V., Villa, P., & Henshilwood, C. S. (2010). Early use of pressure flaking on lithic artifacts at Blombos Cave, South Africa. *Sciences, 330*(6004), 659–662. https://doi.org/10.1126/science.1195550

Munn, N. (1973). *Walbiri iconography. Graphic representation and cultural symbolism in a central Australian society*. Cornell University Press.

Myers, F. (1991). Representing culture: The production of discourse(s) for aboriginal acrylic paintings. *Cultural Anthropology, 6*(1), 26–62.

Myers, F. (2002). *Painting culture: The making of an aboriginal high art*. Duke University Press.

Myers, F., & Smith, T. (2024). *Six paintings from Papunya: A conversation*. Duke University Press.

Nettleton, A. (2010). Life in a Zulu village: Craft and the art of modernity in South Africa. *The Journal of Modern Craft, 3*(1), 55–77.

Nishida, T. (1986). Local traditions and cultural transmission. In B. B. Smuts, D. L. Cheney, R. M. Seyfarth, R. W. Wrangham, & T. T. Struhsaker (Eds.), *Primate societies* (pp. 462–474). The University of Chicago Press.

Noble, W., & Davidson, I. (1991). The evolutionary emergence of modern human behaviour: Language and its archaeology. *Man: Journal of the Royal Anthropological Institute, 26*(2), 223–253.

Noble, W., & Davidson, I. (1996). *Human evolution, language and mind.* Cambridge University Press.

Nowell, A. (2006). From a paleolithic art to Pleistocene visual cultures (introduction to two special issues on 'Advances in the study of pleistocene imagery and symbol use'). *Journal of Archaeological Method and Theory, 13*(4), 239–249. https://doi.org/10.1007/s10816-006-9020-2

Nowell, A. (2017). Visual cultures in the Upper Palaeolithic. *Cambridge Archaeological Journal, 27*(4), 599–606. https://doi.org/10.1017/S0959774317000634

Nowell, A. (2023). Oral storytelling and knowledge transmission in upper paleolithic children and adolescents. *Journal of Archaeological Method and Theory, 30*(1), 9–31.

Palmer, S. E., Schloss, K. B., & Sammartino, J. (2013). Visual aesthetics and human preference. *Annual Review of Psychology, 64*, 77–107.

Pasztory, E. (2005). *Thinking with things: Toward a new vision of art.* University of Texas Press.

Peirce, C. S. (1955[1931]). *Logic as semiotic.* Dover

Pöllänen, S. (2013). 'The meaning of craft: Craft makers' descriptions of craft as an occupation. *Scandinavian Journal of Occupational Therapy, 20*(3), 217–227.

Porter, D., & Neuringer, A. (1984). Music discriminations by pigeons. *Journal of Experimental Psychology: Animal Behavior Processes, 10*(2), 138.

Quine, W. V. O. (2013). *Word and object.* MIT press.

Rappaport, R. A. (1999). *Ritual and religion in the making of humanity.* Cambridge University Press.

Robb, J. (2017). 'Art' in archaeology and anthropology: An overview of the concept. *Cambridge Archaeological Journal, 27*(4), 587–597.

Roth, H. L. (1900). On permanent artificial skin marks: A definition of terms. *The Journal of the Royal Anthropological Institute, 30*, 116–118.

Rutz, C., Sugasawa, S., van der Wal, J. E. M., Klump, B. C., & St Clair, J. (2016). Tool bending in New Caledonian crows. *Royal Society Open Science, 3*(8), 160439.

Scholes, L. (2017). Unmasking the myth: The emergence of Papunya painting. In L. Scholes (Ed.), *Tjungunutja: From having come together* (pp. 126–161). Museum and Art Gallery of the Northern Territory.

Shimamura, A. P., & Shimamura, I. (2012). Toward a science of aesthetics. In A. P. Shimamura & S. E. Palmer (Eds.), *Aesthetic science: Connecting minds, brains and experiences* (pp. 3–28). Oxford University Press.

Shiner, L. (2001). *The invention of art: A cultural history.* The University of Chicago Press.

Smith, P. H. (2016). Historians in the laboratory: Reconstruction of renaissance art and technology in the making and knowing project. *Art History, 39*(2), 210–233.

Soffer, O., & Conkey, M. W. (1997). Studying ancient visual cultures. In M. W. Conkey, O. Soffer, D. Stratmann, & N. G. Jablonski (Eds.), *Beyond art: Pleistocene image and symbol* (pp. 1–16). California Academy of the Sciences.

Solomon, S., Minnegal, M., & Dwyer, P. (1986). Bower birds, bones and archaeology. *Journal of Archaeological Science, 13*, 307–318.

Sperber, D., & Morton, L. A. (1975). *Rethinking symbolism.* Cambridge University Press.

Stokes, J. L. (1846). Discoveries in Australia; with an account of the coasts and rivers explred and surveyde during the voyage of the H.M. In *Beagle, in the years 1837-38-39-40-41-42-43, vol, 2*. T. and W. Boone.

Sutton, P. (Ed.). (1989). *Dreamings: The art of aboriginal Australia.* Viking.

Sutton, P., & Anderson, C. (1988). Introduction. In *Dreamings. The art of aboriginal Australia* (pp. 1–12). Viking. published in association with the Asia Society Galleries.

Sutton, P., Jones, P., & Memming, S. (1989). Survival, regeneration, and impact. In P. Sutton (Ed.), *Dreamings: The art of aboriginal Australia* (pp. 180–212). Viking.

Taçon, P. S. C., & Chippindale, C. (1998). An archaeology of rock-art through informed methods and formal methods. In: C. Chippindale, P. S. C., Taçon (Eds.), *The archaeology of rock-art* (pp. 1–10). Cambridge University Press, Cambridge,.

Tello-Ramos, M. C., Harper, L., Tortora-Brayda, I., Guillette, L. M., Capilla-Lasheras, P., Harrison, X. A., et al. (2024). Architectural traditions in the structures built by cooperative weaver birds. *Science, 385*(6712), 1004–1009. https://doi.org/10.1126/science.adn2573

Tennie, C., Bandini, E., van Schaik, C. P., & Hopper, L. M. (2020). The zone of latent solutions and its relevance to understanding ape cultures. *Biology and Philosophy, 35*(5), 55. https://doi.org/10.1007/s10539-020-09769-9

Texier, P.-J., Porraz, G., Parkington, J., Rigaud, J.-P., Poggenpoel, C., & Tribolo, C. (2013). The context, form and significance of the MSA engraved ostrich eggshell collection from Diepkloof rock shelter, Western cape, South Africa. *Journal of Archaeological Science, 40*(9), 3412–3431. https://doi.org/10.1016/j.jas.2013.02.021

Torrence, R. (2000). Ch. 5 just another trader? An archaeological perspective on European barter with admiralty islanders, Papua New Guinea. In R. Torrence & A. Clarke (Eds.), *The archaeology of difference: Negotiating cross-cultural engagements in Oceania* (pp. 104–141). Routledge.

Troscianko, J., & Rutz, C. (2015). Activity profiles and hook-tool use of New Caledonian crows recorded by bird-borne video cameras. *Biology Letters, 11*(12), 20150777.

Varki, A., & Gagneux, P. (2017). Chapter 9 – How different are humans and "great apes"? A matrix of comparative Anthropogeny. In M. Tibayrenc & F. J. Ayala (Eds.), *On human nature* (pp. 151–160). Academic.

Völter, C. J., & Call, J. (2014). Great apes (*Pan paniscus, Pan troglodytes, Gorilla gorilla, Pongo abelii*) follow visual trails to locate hidden food. *Journal of Comparative Psychology, 128*(2), 199–208. https://doi.org/10.1037/a0035434

Walsh, C. (2014). *A light touch for Rothko murals* (p. 2014). The Harvard Gazette.

Warner, W. L. (1964 (1937)). *A black civilization. A social study of an Australian tribe*. Harper Torchbooks

Watanabe, S., Sakamoto, J., & Wakita, M. (1995). Pigeons' discrimintation of paintings by Monet and Picasso. *Journal of the Experimental Analysis of Behavior, 63*, 165–174.

Whiten, A., Goodall, J., McGrew, W. C., Nishida, T., Reynolds, V., Sugiyama, Y., et al. (1999). Cultures in chimpanzees. *Nature, 399*(6737), 682–685.

Whiten, A., Hinde, R. A., Laland, K. N., & Stringer, C. B. (2011). Culture evolves. *Philosophical Transactions of the Royal Society B: Biological Sciences, 366*(1567), 938–948. https://doi.org/10.1098/rstb.2010.0372

Wilson, K. (2017). *Drawing and painting: Materials and techniques for contemporary artists*. Thames and Hudson.

Wu, W., Moreno, A. M., Tangen, J. M., & Reinhard, J. (2013). Honeybees can discriminate between Monet and Picasso paintings. *Journal of Comparative Physiology. A, 199*(1), 45–55.

Yaxley, K. J., & Foley, R. A. (2019). Reconstructing the ancestral phenotypes of great apes and humans (Homininae) using subspecies-level phylogenies. *Biological Journal of the Linnean Society, 128*(4), 1021–1038.

Zuberbühler, K., & Bickel, B. (2022). Transition to language: From agent perception to event representation. *WIREs Cognitive Science, 13*(6), e1594.

Chapter 4
Art and Communication

Abstract How does art communicate? A lot of literature is devoted to the idea that art communicates something by itself—that it has agency by itself—but in fact it is people who communicate, sometimes using the products of art. But what they communicate and the way they communicate is a product of the whole cultural context in which they live. These communications are embedded in what I define as six communications revolutions, which form the central core of the whole book. They involve what Marshall McLuhan (The medium is the message. In: *Media and cultural studies*, 2012 (1962), pp. 100–107) called "actual processes of thought."

I survey briefly the types of communications among non-human animals and point out that they do not generally result in conversations, but production and response depend very much on the context of communication. Generally speaking, non-humans do not respond to indexical signs of previous behavior of other animals. Most importantly, any suggestion that inanimate things actively send messages is confused, however appealing it might be, but inanimate things are often a third element of the interaction between two or more people (Schiffer, *The material life of human beings: artifacts, behavior, and communication*. Routledge, London, 1999).

Finally, I survey the work showing the similarity between the semantic roles experienced by humans, capuchin monkeys, and chimpanzees emphasizing their interactions with stones and with art. Human use of stones leaves a material product that persists, and our use of art not only persists but can only be shown to be art with the process of thought I mentioned before.

The Essay discusses how the creation of pictures necessarily involves interactions between producers and seers and the producers' concerns about something material (the ocher and its impacts), as well as concerns about both non-material things of the mind (the song/story) and material things related to the place where the interaction happened ("the place of the story of the whale").

I. Davidson, *Art or Scribbles? In the Eye of the Beholder: The Evolutionary Emergence of Visual Communication*, Interdisciplinary Evolution Research 11,
https://doi.org/10.1007/978-3-031-90458-5_4

4.1 Art Does Not Communicate by Itself

As I mentioned in the previous chapter, for some, the essence of art is communication (Layton, 1992), but that leaves the problem of how art communicates, and how it comes to communicate. The problems with the idea of art as communication became clear to me when I was listening to a conference paper by a scholar who was devoted to, indeed was an expert on, dance. She claimed, and few doubted her, that dance is a means of communication (for support of the view of dance as communications see Barnard, 2017). "Tell me," I asked (or at least my words were something like this), "when your episode of rehearsing or performing dance is over and you go to the pub with your fellow dancers, do you communicate in dance or by speaking?" The answer was obvious—in communication there are great advantages to speech, so how does communication work through any art?

One reason for writing this book is to explore how people use art, of any sort, to communicate, rather than to eliminate the role of communication as an aspect of art. Of course, I am using words. No one doubts that art is capable of communication, but it is not a definition of art to say that communication takes place through it. Does art enable people to communicate more or less than when they are only using the spoken or written word? Is the communication something that the art does or that the producer initiates, or does it depend on the mind and cultural background of the receiver of the communication? Does all art communicate equally, or does some—such as a scene in visual art—communicate in special ways? Leaving aside the annual "Dance you PhD competition" run by *Science*[1] (Bohannon, 2008), can you imagine reducing this (or any other) book to a few pictures with no words?

Just over 16 years after the encounter with the grazier Bill, I was asked to deliver the Annual Lecture of the Australian Academy of Humanities in Brisbane (Davidson, 2008). This lecture followed a symposium on "Humanities futures: new methods and technologies for humanities research" at which the Keynote Speaker, Mark Kornbluh, from the University of Kentucky (now at Wayne State), referred to the Internet as the second communications revolution. I knew he was wrong to call it the second revolution, because I had been publishing about the first revolution, the evolutionary emergence of language, and the digital revolution he was talking about came after a number of other revolutions in communication. For example, only a few years before the presentations by Kornbluh and me, the title of most influential person of the Millennium was awarded to Johannes Gutenberg (Füssel, 2001), who introduced to Europe the printing press using movable type. By that reckoning, the Internet must have been at least the third revolution after language and the printing press.

Before I delivered my presentation at the conference, I estimated that there had not been three, but actually six revolutions: (1) the emergence of language following a form of non-symbolic communication, which initially was only oral or gestural (Fay et al., 2022; Noble & Davidson, 1996) and Merlin Donald (1991) has called

[1] https://www.science.org/content/page/announcing-annual-dance-your-ph-d-contest

"mimetic;" (2) the production of visual, particularly representational, images (Culley & Davidson, 2021) though this is a matter of some recent discussion (Wilkinson, 2023); (3) the beginning of writing (Morin et al., 2020); (4) the invention of printing, especially with re-usable type, whether in East Asia (Pow-key, 1982) or in Europe (Eisenstein, 1979) (Fig. 4.1); (5) the invention of various means of communication at a distance, such as the semaphore, telegraph (Shepherd, 1996; Thomson, 1999), telephone (Gorman, 1997), and radio (Schiffer, 1991); and (6) the digital electronic revolution (Fig. 4.2) to which Kornbluh was referring. Note that, with the exception of the revolution represented by the emergence of language, all of the other revolutions involve some persistent, material product, often many of them. The information that was carried from signaler to recipient may be as ephemeral as the sounds of speech involved in the production of language, but art, writing, printing, communication at a distance, and the electronic revolution all depended on and left behind a material infrastructure.

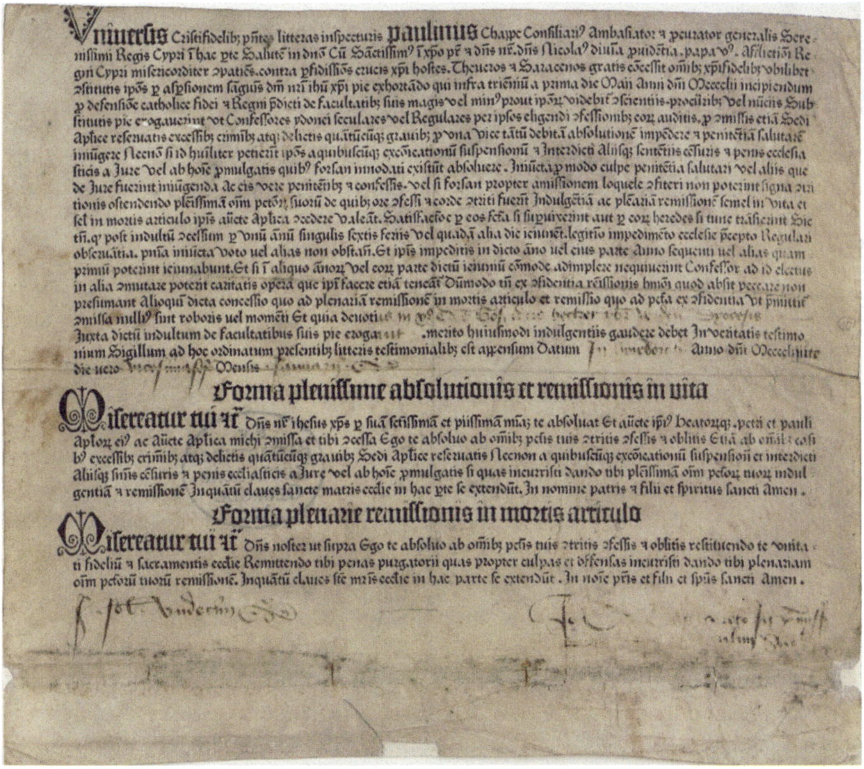
Universis Cristifidelibus presentes litteras inspecturis Paulinus Chappe Consiliarius ambasiator et procurator generalis ...

Forma plenissime absolutionis et remissionis in vita

Forma plenarie remissionis in mortis articulo

Fig. 4.1 1455 Indulgence of Pope Nicolas V printed by Johannes Gutenberg (https://dpul.princeton.edu/gutenberg/catalog/d217qp581). Supposedly the lead type was melted down to be used to make a different typeface after these indulgences were printed. Goettingen State and University Library. SUB Göttingen, shelfmark: 2 Bibl I, 5955:2 INC RARA, fol. 235r

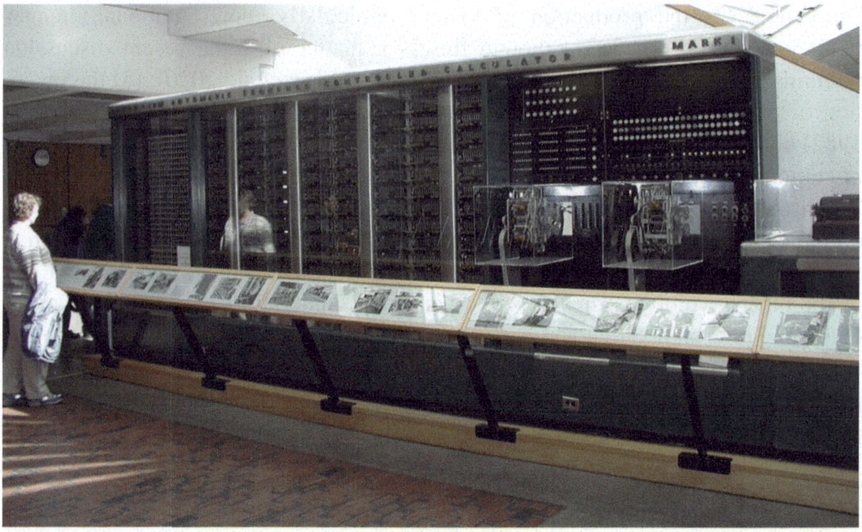

Fig. 4.2 First IBM Computer, in Harvard Science Building 2002. (Photo Iain Davidson—the face has been blurred to provide anonymity)

Each of these has been discussed in greater or lesser depth on previous occasions. Others have defined a list of communications revolutions that are slightly different. For example, Harold Innes and McLuhan (Carey, 1967) defined oral, written, print and electric means of communication (they wrote before electronic communication). As McLuhan (2012 [1962]) expressed it:

> The content of writing is speech, just as the written word is the content of print, and print is the content of the telegraph. If it is asked "What is the content of speech?" it is necessary to say, "It is an actual process of thought, which is in itself nonverbal."

I do not know whether he was influenced by it, but McLuhan's "actual process of thought" is very similar to the phrase used by Leonardo da Vinci that art is a *cosa mentale*. Robert Logan (2002) extended that earlier list to define five stages in the history of communications: the pre-verbal (adding a memetic phase, following Donald, 1991), the oral, the literate, the electric and the electronic ages (splitting the electric—because, of course, Innes and McLuhan worked before the invention of electronic media—and missing out print which McLuhan had included).

I thought, at the Academy conference, that the second of these revolutions—the initial production of art—did not seem to have substantial literature about how it was a revolution, and none at the length of a book. It may be that I was wrong about that, but presenting early art in the context of a discussion of communications is unusual. Donald (1991, 275–284) referred briefly to "visuographic symbolism" or "markings" but without significant analysis of their content, preferring to point to the significance of writing. He returned to art at greater depth in a later book (Donald, 2001) and a chapter (Donald, 2006) recognizing that "Art and artists are active players in the co-evolution of culture and cognition". To an extent this book is about that second revolution in human communication and how it fits with the others. It is about the production of visual, particularly representational, images, and how they constitute a major revolution in communication in human evolution. On this reckoning, digital communication, including the Internet, is "only" the sixth revolution in communication, though it has much in common with the others.

It seems plausible to put the issue the other way around and suggest that without words, an important part of the communication is lost. Research suggests that "effective understanding of scientific explanations requires a mapping between words and pictures" (Mayer & Anderson, 1991). Mike Schiffer (1999, 5–7) goes further and states "Virtually all communication and human behavior involve artifacts," which is an archeologist's version of the art historian's "As soon as there is form, there is message" (Pasztory, 2005, 10). Sometimes, a product of human activity does not have a definable form but Pasztory's statement it is still true because in a world where humans create form, lack of form is also a message, hence whatever we interact with includes the material. The consensus seems to be that art and artefacts add to or enrich communication by other means but in a curious conjunction, behavior that includes artefacts must also be involved with words.

Important in Schiffer's view is that communication results from inferences or "conclusion[s] about something arrived at through reasoning from evidence" which would almost always involve material artefacts (as a substantial part of the evidence) as well as the thought processes of the seer—conclusions and thought processes that largely involve words. I like to say that this is the nature of archeological arguments, and that hence archeology makes us human through the knowledge that we construct about others. We need, therefore, to understand something about human communication before we can understand the special place of art as an artefact in the interactive system of people—the art-system.

For archeologists, particularly, the inferences derive from materials that have persisted from times in the past to the present. The point of view I am adopting here is that such "persisting things" include art (with the exception of those examples where the artist seeks to destroy them deliberately before they can persist—as in Banksy's destruction of his own painting immediately following its auction[2]). The communication that involves art consists of inferences by the seer of the art made as a result of their reasoning—what I would call inferences of the type we make as archeologists. I suggest that all communication among people consists of the interaction between them together with the perceptions by the signaler of a material or mental thing and those of the receiver about the same thing as a result of that communication. All communication between two people includes a third thing.

[2] https://www.youtube.com/watch?v=iiO_1XRnMt4&t=59s

4.2 Where Does Art Fit into Communications Revolutions?

I began thinking about the way individuals communicate with each other when, with Bill Noble (Noble & Davidson, 1996), we were in the early stages of working out the evolutionary emergence of language. We began our work, I think, from a well-known diagram about communications by one of the founders of the linguistics of signs, Ferdinand de Saussure (Fig. 4.3) (Saussure, 1986). This showed information passing from the brain/mind of the signaler to the brain/mind of the receiver and back again. There is no hint in the image of how the message is generated in one nor the response generated in the other. Saussure seemed to suggest in this diagram that it just happens.

It becomes even more curious when we try to make such a system fit with the evolution of the communication system from a non-human to a human condition. And more difficult still to begin with such a model to search for archeological evidence that might indicate a step in the emergence of language based on symbolic communication. In discussing art and the way it communicates, I can demonstrate the problem by pointing out that there is no room for art or symbols in the image by Saussure. How does art fit into the communicative system with humans? I have suggested that we need to approach this question by considering how art is made up of marks made when people add something such as pigment to a surface or subtract something from a surface as in sculpting (Davidson, 2013).

Just to show that this is not a trivial issue, let me begin with an example of an animal leaving a mark. Leaving marks is, after all, one of the things that humans and other animals have in common. The paintings at Chauvet Cave were sometimes painted on cave walls that had been scratched by bears (Fig. 4.4). There is no confusion between the scratch marks of the bears and the intentional, subtle painted images of lions.

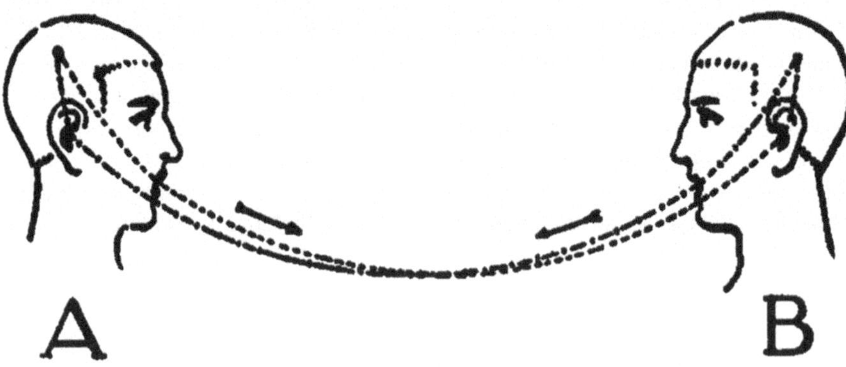

Fig. 4.3 Saussure's general model of speech (Daylight, 2017). I am arguing that this is quite inadequate in not specifying anything about the mental processes that lead to emission or reception of communications. What Leonardo called *cosa mentale*

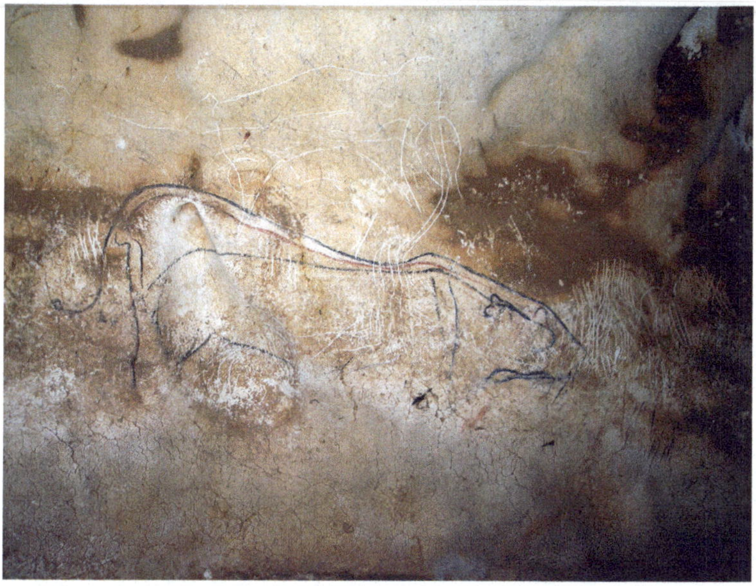

Fig. 4.4 Bear scratch marks on the cave wall in Chauvet Cave together with the images of a pair of lions while the female is in oestrus. (Photo Jean Clottes with permission)

Noble and I had to consider the significance of the differences between humans and our precursors back to the common ancestor we have with chimpanzees. We devoted a chapter to synthesize the evidence then known about primate communication pointing out the substantial differences between all non-human primates and humans, particularly in the state of the available evidence for vocal and non-vocal communication among non-humans (Noble & Davidson, 1996, Ch. 2).

The point, here, is not to get too bogged down in the details of animal communication, however fascinating that may be. Rather I want to emphasize that all land-based animals, whether or not they are communicating by making sounds, leave marks on the landscape where they move and so do their predators. Dorothy Cheney and Robert Seyfarth (1990, Ch. 9) devoted a whole chapter to the relevant observations, within a study best known for observation of the vocal calls of vervet monkeys. These monkeys and many others (Schlenker et al., 2016a, b), have a series of calls which they utter in the presence of predators (Cheney & Seyfarth, 1990). Many other animals such as chickens (Evans et al., 1993) also have such calls.

It is almost a cliché, now, that, when they perceive danger, vervet monkeys can be shown to give calls, quite but not entirely consistently. They give three different calls when they perceive danger depending on whether it is danger from the air, danger in the scrub, or danger on the ground (Fig. 4.5). Vervets who hear such calls generally respond in appropriate ways by taking action to avoid the danger. The different calls are commonly called "alarm calls" but there has been a lot of discussion about whether that is an appropriate name for them. Animals that perceive the threat give the calls and others respond to them and take appropriate action.

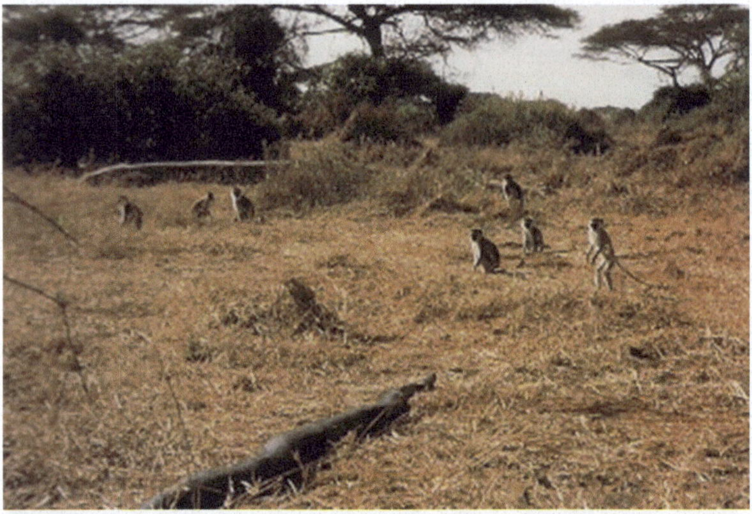

Fig. 4.5 Vervet monkeys in Amboseli carefully monitoring a python—danger on the ground. (With permission from the database Anthrophoto, the photographer Richard Wrangham and the researcher Robert Seyfarth)

Communication has taken place. This sequence is already two steps away from the simple diagram of Saussure: the signaler associates a visual stimulus with a threat and utters a vocalization; the recipient responds to the signaler's call, inferring a particular danger, and takes an action. Unlike the situation in human conversation, the response by the recipient of the call is not usually another vocalization.

Even the initial call, however, must be understood to be a response not just to the perception of the dangerous animal, but to memory of previous occasions when such a call was associated with similar danger. The recipient's response, likewise, depends on memory of association with previous threats of danger. Noble and I became very familiar with people who told us that language was a perfect communication system, but compared with this, it may well not be. The vervet calls and responses make up an almost perfect communication system because the signaler increases their own danger by making a vocalization, but probably benefits on other occasions when they are recipients; the receiver obtains information that enables them to avoid danger. There are three situations where things appear to go wrong, either because no animal makes the initial call, or because a recipient animal does not respond appropriately, or on rare occasions when an animal gives a false call (Cheney & Seyfarth, 1990, 213–215). Failure of the system increases the chance of death for both signaler and recipient. Although knowledge of the system is more nuanced now, one biologist (Slater, 1983, 15) put it: "The system is so precise that the signaler might almost be shouting the name of the particular predator." Natural selection has ensured that the system works perfectly almost every time—there is not much possibility of a better communication system than that.

In more recent work Cheney and Seyfarth (2018) have confirmed that although primate calls are very limited, they can be shown to vary according to the context in which they are given—particularly the context of interactions among the animals. It is consistent with what I am arguing here that the vocalizations communicate the danger in the context of the signaler perceiving the threat from another animal, and the recipient responds accordingly. This is not, therefore, a communication in the same mode as that illustrated by Saussure, but one which requires inputs from the context in which the communication takes place.

For our purposes, however, the important story is additional to this system. That exchange of calls from one monkey to another involves the perception of a predator and appropriate response (whether it is to the call or to the prompted perception of the predator). But, by and large, there is no material product which persists after the threat. It is only "by and large" because Cheney and Seyfarth witnessed vervet monkeys crossing the mark of the passage of a python (Cheney & Seyfarth, 1990, Ch. 9). The monkeys did not see the snake, but they could have seen the mark a python made moments before—as some would have it, the mark was indexical of the passage of the python, in the sense we discussed in Chap. 3. The mark might have alerted them to the danger, but in the absence of the python, danger was not immediate. The mark did not elicit the call which would have made the fellow animals run for safety. There appeared to be no identification of danger from the predator based only on the mark it left. In the same vein, the authors began Chap. 9 of their book with an anecdote about baboons coming across an animal freshly slaughtered by the predator lions which were resting nearby in the shade. Once the baboon saw the lions, they responded with their characteristic barks, but the indexical sign of the recent kill was not enough to prompt them.

Humans, seeing the marks of activity, respond differently. Remember that in Chap. 3, Boas and Gombrich both described an emotional response to inappropriate interactions with mere marks on paper. The vervets are a very long way from that. Charles Snowdon (1992) described the vervets' lack of response to marks as secondary indications as "puzzling" and Elisabetta Visalberghi (1992) sought to explain the lack as an absence of understanding of cause-and-effect relationships. Further careful experiments with the South American monkeys *Cebus apella* showed that they too recognized associations, but not cause-and-effect. Andy Whiten (1992) went further and suggested that chimpanzees, too, have a limited capacity to understand causality, and their calls in the wild seem to refer either to the callers themselves or to specific events (such as finding food). There is "no evidence of conveying anything about the causal structure of the perceived event" (Zuberbühler, 2022).

4.3 What Happens in Communication?

Among humans, the complexities of analyzing communication are similar to those among other primates, but even more so (see, for example, Noble & Davidson, 1996, Ch. 5). Communication cannot be conceived of as occurring between two isolated individuals as Saussure seemed to imply, but between two or more individuals embedded in their social and physical environments, as Bourdieu (1968) implied and we discussed in Chap. 3. First, a signaler utters a communication based on their social and cultural context and perceptions of the thing they want to communicate about. That utterance may be about something that they and others can perceive or something that exists only in the mind of the signaler by virtue of the way they think about it. Then, the recipient receives the communication and in order to decode it, they make an inference based on their shared experience with the signaler.

All human communication is of this nature. The signaler and the recipient are animate and communicate about their individual contexts including the things and people around them in one of two ways. Either they utter something as they perceive the things around them, or an utterance reflects the knowledge of such things they hold in their minds by what McLuhan referred to as "an actual process of thought". In the human case, of course, some of the "things around them" are items they have themselves created, and Donald (1991) emphasized that the shift from "biological memory" to external storage of information—using things that I have referred to here as "persisting"—was one of the most revolutionary in human history. Human knowledge of their act of creation is something they hold in their minds, or they know by their inferences from the persisting material products of similar acts by other people.

Non-human animals certainly manipulate stone material and even get some nutritional benefit from it (Izar et al., 2022; Ottoni & Izar, 2008). When they manipulate stones, they leave battered and broken stones as an indication of their activity (Mercader et al., 2007; Proffitt et al., 2016) but they may not pay attention to those material products, since the immediate benefit of the nut is all they are interested in—they do not hold the stones in their minds. It may be difficult to define any patterning of the flakes (what we would recognize as "form" to use Pasztory's word) that have been produced because they were unintentional (Proffitt et al., 2023) and they do not show that the broken stones must have been left by a particular species. Likewise, chimpanzees seem to show no interest in the flakes accidentally removed from stone anvils (Davidson & McGrew, 2005), although Christophe and Hedwige Boesch showed that chimpanzees in the Taï forest of the Côte d'Ivoire, in their pursuit of food from nuts, do move nutting stones around the forest systematically from one season to the next (Boesch & Boesch, 1984). They seemed to hold the hammers and the nutting trees in their minds, but they did not shape the hammers. Otherwise much of chimpanzee interest in material things is in plant parts, is transitory and does not survive in any recognizable way from one visit to a termite mound (as it were) to another (Davidson & McGrew, 2005). In addition, all monkey and ape artefacts seem to lack form.

Humans and their ancestors—whom I will call hominins—seem to be the only communicators who repeatedly interact with persistent material culture, beginning with the carrying of stone raw material in early hominin sites from 3.4 Million years ago. One of the earliest signs of hominin activity has been said to be the cut marks on bones at the site of Dikika, where the marks seem to have been made by sharp flakes that were removed from the site after use (Domínguez-Rodrigo et al., 2012; McPherron et al., 2010; Thompson et al., 2015). Because the sharp flakes were removed, we do not know whether hominins made them by striking them off another stone (which would then be a core) using a third stone (which would be a hammer). In consequence, if there was production of flakes, there must have been a persisting place with hammer and core, as well as the transport of flakes and the removal of edible portions of meat (steps of this argument are outlined in Davidson, 2019, 87–92).

Some early archeological sites allow archaeologists to make the inference (despite the lack of communication between people) that tool production involved hammerstones, flakes and the cores from which hominins struck flakes (and Rob Foley pointed out that one hominin would have been able to detect the presence of another from the sounds of knapping). Some sites consist of the remains of knapping left by hominins consistently producing and using such stone tools but, significantly, the remains are found at sites where hominins first produced then removed tools, and left behind other broken stones (Davidson & McGrew, 2005; Delagnes & Roche, 2005).

At some level, all communication is really simple. You speak, I hear. That is communication. There is, of course, some communication among plants, where, for example, chemistry of one plant can affect the growth of another (Levitt & Lovett, 1985) in a process said to be "analogous to one more usually associated with animals", or the scent of the flower can attract the pollinating insect using "signals that already have a role in the animals' own communication systems" (Schiestl, 2010). Generalizing a little about all communication, there is a signaler that emits a communication, and there is a receiver that accepts it. But there are a few other things going on. First, almost all animals communicate, and many would argue that plants do too. Many years ago, I wrote that: "The animate world is replete with communication. Almost all of the senses of one creature or another guide them, their predators or their prey to determine their friendly and unfriendly interactions" (Davidson, 1991, 44). I should have emphasized that those communications also guide individuals in their interactions with other individuals of the same species. The issue Noble and I were dealing with at that time was about the way in which such communication turned into language, and how the hominins communicating in this way came to recognize that their utterances could mean anything. It turned out that even the most experienced primatologists studying the utterances of monkeys (Cheney & Seyfarth, 2018) concluded that "When compared with humans, nonhuman primates have small vocal repertoires that show little acoustic modification during development. These limitations pose a dilemma for those interested in the evolution of language." Cheney and Seyfarth emphasize that while the nature of calls is limited, there is substantial manipulation of the contexts in which the calls are given so as to convey honest indications of the signaler's intentions and "give rise to an open-ended, highly modifiable, and cognitively rich set of meanings" (Cheney & Seyfarth, 2018, 1978).

Unlike some, I will maintain fiercely that only a living agent can utter a message—can be the signaler—that communicates to an agent (patient) which is its receiver. Any suggestion that inanimate things actively send messages is confused, however appealing it might be. First, if the inanimate thing seems to send a message, it is only sent because the person recognizes that the inanimate thing can be used in signaling, and in using it, they—the person—will become the signaler. Second, the receiver has an independent interaction with the inanimate thing and recognizes that the signaler is using the inanimate thing in their communication. There are no actions by the inanimate thing, independent of the agency of people, either in sending a signal or in receiving a communication. It is simply a case that human agents communicate about (or in ways that involve) the inanimate thing. Conkey (1989) once wrote, incorrectly: "the meaningfully constituted material record is not an 'expression' or 'reflection' nor even a 'record,' but an active, constructing, constituting agency, which does not express meaning, but produces it." My response was that "the material record, being an object and not a person, cannot ever be an agent in the process of producing meaning" (Davidson, 1997).

Conkey (1987) also once wrote, this time correctly: "We cannot expect to capture 'meaning' as a single, inclusive, empirical entity or category of our inquiry," because both the producer and the seer are human and meaning starts as something the producer understands about a subject and ends as something the seer understands about the object. The meanings are a product of (at least) two people and (at least) one material thing: the meanings are not themselves material.

The issue is complex, as Cheney and Seyfarth showed, because neither the signaler nor the receiver exists in a vacuum, so that vervet calls are not just from one individual to another, but from one individual to other animals with a common history during which the animals developed a social relationship among them. But once that is recognized, among humans, at least, many aspects of their personhood enter into communicative interactions between people. Schiffer (1999, Ch. 3) outlined ways in which humans involve all aspects and movements of their bodies in their communications. It would be very easy for me to say that these features communicate a message from one individual to another, but I refrain from saying that because in almost every case in which a person infers something about another's behavior using these bodily features, the signaler has some control over those features, and the receiver has a history of experience of such features. The interaction involves the active behavior of both parties and is not a product simply of the state of the feature. People communicate with other people through the states of their bodies.

Then, of course, in addition to interaction between people and their body states, they also interact with material things, such as the stone flakes they have used as cutting tools ever since their ancestors first knapped. Bill McGrew and I argued that the act of knapping left unused remains in places and that these constituted a further opportunity for interaction with material but tied to a specific place (Davidson & McGrew, 2005). In the simplest communication between people, just as in that between non-humans, the receiver has immediate access to the information from the signaler, and none afterwards because vocal information fades. But the number of people who have access to information at the same time can vary according to the location. If the communication is tied to a particular location, then the number of people who can receive a audible communication varies according to the acoustics of the place. That is the appeal of research into the acoustics of art sites (Díaz-Andreu & García Benito, 2012).

4.4 How Do People Communicate Using Art?

These questions which relate human and non-human behavior have become enmeshed in questions about the commonalities and the differences across species, particularly those that followed the emergence of language among humans (Barnard et al., 2007). I emphasize here that both humans and chimpanzees could be shown to engage in quite similar sets of semantic roles (Byrne et al., 2004, Table 4.1) using

Table 4.1 Semantic roles as defined by Fillmore

Semantic role	Role as defined by Fillmore—roles of chimpanzees and humans	Capuchin monkey stone tool use	Hominin stone tool making	Picture-making
Agent	*Instigator* of an event	Pat fetched the nut	Kim fetched the stone	Fred fetched the ocher
Counter-agent	Force or resistance *against which* the action is carried out	Pat hit the nut with a rock	Kim hit the stone with a rock	Sandy rubbed the ocher on a rock
Object	Entity that *moves* or *changes* or whose position or existence is *under consideration*	The nut broke	The stone broke	The ocher marked the rock
Result	Entity that comes into *existence* as a result of the action	Pat ate the nut	Kim made a flake	Sandy made a mark
Instrument	Inanimate *stimulus* or immediate *physical cause* of an event		Kim used the flake to cut meat brought by Kelly	The mark made me think of a whale
Dative	Animate being *affected* by the action named by the verb		Kelly gave the meat to Kim	Sandy told a story about a whale to me
Experiencer	Animate being having a given *experience* or *mental state*	Pat felt well-fed	Kim felt well-fed	Sandy and I knew some things about whales
Locative	*Location* or *spatial orientation* of the state or action named by a verb		Kim walked away from the flake scatter	Sandy remembered this as the place of the story of the picture of the whale

Taken from Byrne et al. (2004, Table 1). Semantic roles in picture-making and stone-tool making (after Davidson, 2013, Table 1). Semantic roles as inferred from video of *Cebus* stone tool use (Izar et al., 2022)

a definition of "semantic roles" from the work of the linguist Charles Fillmore (1968). Although I was involved with this analysis, my initial view was that the roles are not semantic when they are applied to chimpanzee behavior, but that they are when they are aspects of human behavior. As it happens, they have similar characteristics in both species, and it is just that humans engage with persisting materials and other animals do not. We, people, have a word for the roles, and other animals do not.

In the fullness of time human communication takes advantage of that persistence of the material in quite specific ways which I can illustrate using the same roles we highlighted in 2004. When hominins started to make tools by striking flakes off the margins of larger pieces of stone—that is to say, knapping it—they created semantic roles for their use of knapped stone that were unremarkable in the context of those that humans share with chimpanzees (Table 4.1). The same set of roles applies among humans and their ancestors who made stone tools but with a significant difference: the persistence of intentionally knapped stone carries six of the roles through time often at a particular place (Davidson & McGrew, 2005). The hominins left scatters of knapped stone from which they had removed used artefacts.

These material products are the things that archeologists use to make inferences about the remote past, and any primate that recognizes the relationship between cause-and-effect could come to do so. The two differences are in the Dative and Experiencer roles (shaded in Table 4.1). These are about relationships between hominins or people and do not necessarily involve a persistent product. Persistence seems to be an important feature in the evolutionary process that led to humans, because, in evolutionary terms, it may be what allowed the selection of the inference of cause-and-effect relationships. From the point of view of communications, the Dative role involves communication from one hominin to another, but no evidence of that would persist after the event. The Experiencer role, as in the case of the monkeys, involves emotions but does not necessarily involve communication.

I then applied the same analysis to the production of pictures (Davidson, 2013, Table 1), with results that seem, to me, to belie the ability to interpret visual products, such as art, independently from the stories the producer had in mind. In particular, as with the production of stone tools, some roles involved a physical product—the ocher or the mark. The production of pictures was distinguished in the Dative and Experiencer roles, as well as adding distinctions in the Instrumental and Locative roles (all shaded in Table 4.1). In all four of these, the roles involved conceptual rather than physical products—the meaning of the mark, the story about the meaning, consequential knowledge about the subject of the mark, and the impact that had on the relationships among the elements of the art-system and the place.

The simple point is that the sequence of semantic roles involving art (Table 4.1) shows that the creation of pictures necessarily involves three features: communication as a result of interactions between agents (in this case, people—the producer and the seer); their concerns about something material (the ocher and its impacts); and the association between the place, the persisting product and storytelling at the place. People could use the persistence of the marks at particular places to tell stories to others and that allowed continuing communication among the people. The marks themselves did not communicate unless the seer knew the stories already.

The creation of pictures necessarily involves both non-material things of the mind (the song/story) and material things related to the place where the interaction happened ("the place of the story of the whale"). Unlike the examples of simple comparison with the behavior of chimpanzees (Byrne et al., 2004), rock art has persistence in the environment in common with the semantic roles identified for making of stone tools. People, who were necessarily seers of the art or knowers of the stories told at the time of creation of the art, interact with places after the art has been produced. They are part of the art-system. That persistence affects the way they interact. Unlike the example of stone tools, what persisted with the rock art was not only the material evidence of people being in the place but intangible mental aspects of the art. The communication takes place because of the "process of thought" by the mark-maker when telling the story associated with marks at that place, and the related process by the seer of the marks with or without the presence of the producer.

From the beginning of the making and use of stone tools to slice things (see, for example, Davidson, 2019), particularly useful things like meat, six of the associated roles include a material product which is left behind, but two of the roles leave no persisting product. Those two roles, however, involve feelings of the animals concerned, emotional reactions over which they have little control. One of those two roles, I presume, is also felt by monkeys (exemplified by the feeling of being well-fed), the other is a role that monkeys do not exhibit. Finally, in the case of art, four of the roles can only be described because they involve a mental construction—an "actual process of thought" in McLuhan's phrase, or a *cosa mentale* in Leonardo's. In the examples we gave when discussing humans and chimpanzees in 2004, those four semantic roles involved affective (emotional) behavior but did not involve the same intentional processes of thought (Byrne et al., 2004).

The simple way to approach the observations in Table 4.1 is this. People and chimpanzees both exhibit a number of (at least eight) well-known semantic roles. Monkeys appear to exhibit only five of those roles. So all humans, apes and monkeys seem to share these roles: they can be the instigator of an event; they carry out actions against a force or resistance; the consequence of their actions is that something moves or changes; there is a result of their actions; that result is usually a desired one. At the same time, apes and humans behave as if they recognize the inanimate, hence material, stimulus or cause of events, but other primates do not. Among humans, but not among apes so far as we can tell, such an event can be a mental one. Among humans and apes, but not other non-human primates, that can be a social exchange of something that results from the actions, and in the case of humans, alone, that could be a mental thing, such as a story. Finally, among humans and apes, but not other non-human primates, these actions create a memory of the location where the events took place. Again, this may involve material aspects of the place, but only among humans is there an association between the story and the place.

Now there is a way to deal with this range of variation exhibited by a range of animals belonging to different parts of the same clade, related but not in ancestral relationship to each other. That way involves an attractive story about evolution from animals that exhibit one combination of semantic roles to another with a larger number of roles. Such stories are common among people, including some scholars, who are not familiar with either primatology, archeology or comparative studies of communication or behavior. That popular story is misleading. We did not descend from living species of apes or monkeys, but we all have a common ancestor with each other, such that the Last Common Ancestor we shared with chimpanzees evolved separately both into humans and into chimpanzees. The shorthand way of saying what Table 4.1 implies about relationships is to suggest that evolution took place from the simpler state (say, chimpanzees) to the more complex state (say, humans). In reality some part of the variation among the animals in the simpler state evolved to become apes, and another part of that same variation evolved into humans. As I have argued elsewhere, we tend to find it tempting to fill in the gaps between the different states I have described among animals in the modern world as a means to tell a story about how the relationships worked out in the past (Davidson, 2023). In reality we have hardly any evidence for such a process. What the table does, however, is to highlight the importance among human ancestors of persisting products of their actions, on the one hand, and of mental constructs—actual processes of thought—on the other.

The question then becomes how such necessary mental aspects became an intrinsic part of art. Let me consider three options (for a summary of early cases see Culley & Davidson, 2021). First, the "art" may consist of scribbles through which people do no more than scratch the surface on which they leave the marks, like many of the marks on the surface of the 500,000-year-old shell from Trinil (Joordens et al., 2015) (Figs. 3.7 and 7.28 Left). Second, such marks may have form, such as geometric patterns, like the grids from Blombos 73,000 years ago which occur after a long sequence of what looks like formless scratching of ocher rocks to extract powder (Henshilwood et al., 2009) (Fig. 3.29) or the hashtag from Gibraltar more than 39,000 years ago (Rodríguez-Vidal et al., 2014), or they might be non-representational in other ways, not as scribbles but with form (in Pasztory's sense), as in the 60 thousand year old paintings in several caves in Spain including La Pasiega (Hoffmann et al., 2018) (Fig. 4.6).

Third, the marks made included some that were representational—were, in fact, iconic representations of the objects they represent (Fig. 4.7).

The issue is, therefore, how marks in each of these categories come to have a meaning that can be shared with others. I will address this in detail in Chap. 6. For now, let us just recognize that any of the three categories I have outlined—scribbles, marks with form, which could be geometric, and representational marks, which may or may not be complete outlines of the things they represent—requires an initial storytelling about them for seers to understand the meaning or intention a producer may have in making them. In reverse order, a picture in the form of a hand stencil from Sulawesi must be accompanied by a story about why hand stencils were made. The common statement is that people "just" put an image of their hands on a cave wall to indicate their presence or some such. But it is not a simple matter.

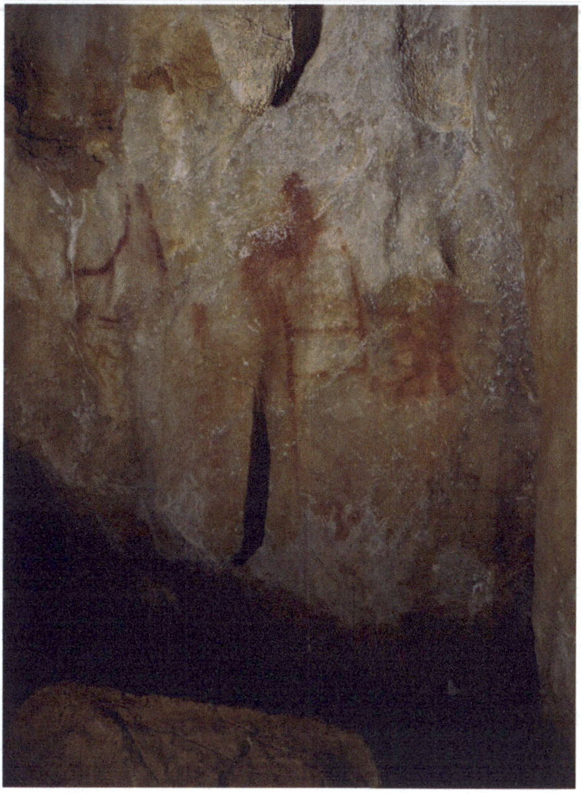

Fig. 4.6 Panel 78 in La Pasiega, (Cantabria, Spain). The scalariform (ladder shape) composed of red horizontal and vertical lines dates to older than 64,000 years and was made by Neanderthals. (Hoffmann et al., 2018). (Credit: C.D Standish, A.W.G. Pike and D.L. Hoffmann) (Courtesy of and with permission of A.W.G. Pike.)

Other animals do not do it. Not all people do it when they see a rock wall. The practice is intrinsically cultural and supported within the community by the telling of stories about the significance of hand stencils. By the same token, a painting of an anoa or a warty pig could be "just because they were there" but that is not enough. The whole process of selecting and making the paint and its survival (Huntley et al., 2021) implies forethought that is distinctively human, and the selection of particular species was not just obvious but accompanied by stories. That is intrinsic to the roles described in Table 4.1. Without these relationships, the marks cannot be considered as art.

Going back through the list, patterned or geometric designs seem obvious to us today because we are used to making or seeing patterns. But to make them in the absence of prior cultural experience of such things is highly unlikely. Yet it would be uncontroversial to suggest that such patterns had some significance or meaning. We will consider such things in Chap. 7. Finally, on this list, the presence of scribbles is problematic. We have seen that young children draw scribbles before they make marks that have meaning (Matthews, 2013), but they do so in a process of learning to make marks that do have meaning, sometimes where adults were making more structured art (Wisher et al., 2025). There seems to be little sign that other animals make marks to which they assign meaning. In other words, scribbles are much more significant than the grazier Bill could have imagined. In the world of

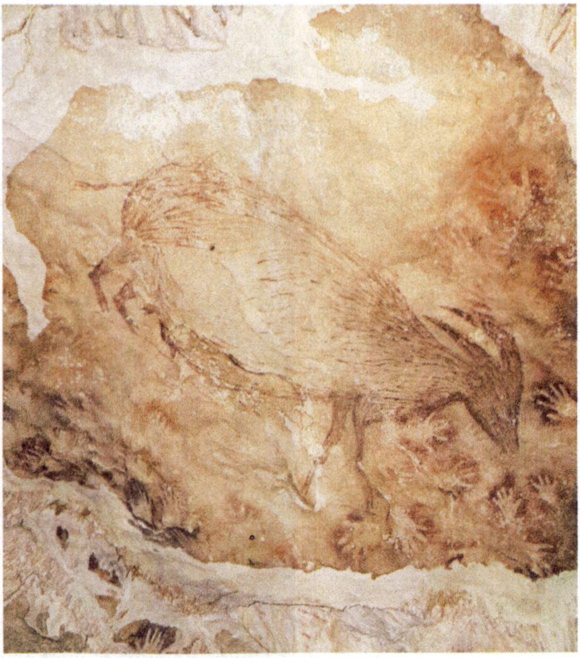

Fig. 4.7 Painting of an anoa surrounded by negative handprints on a cave wall at Gua Uhallie, Sulawesi. (With permission of Adam Brumm, Maxime Aubert, and Adhi Agus Oktaviana)

people who are used to making marks, scribbles might be incidental but more often they have a role in an art-system. Humans may make such marks to have the meaning "I am making marks that do not have form and therefore do not have a meaning in the sense of the other marks I make." Pasztory would be proud.

In the ordinary course of events, people communicate face-to-face, using words or gestures and these do not last more than a few milliseconds. All that remains is the memory of them in one or other party to the communication. Access to the information is immediate and short-lived and generally does not include many people. Once images in any of the three categories are produced on a persisting medium such as a rock wall, the situation is quite different. For rock walls that are visible to all at any time (give or take some cultural restrictions) some part of the communication is available to a much larger group of people at the one time who can be told about the story associated with the art, or rather some story whether or not it was the original.[3] Moreover, they and others can revisit the place and recall and retell elements of the story. If the rock walls, the persisting medium, were not visible to all, as in deep caves, then the restricted context limited the numbers who could share in the information except under special circumstances.

[3]There is a literature about the role of story telling or narrative in the origins of language (e.g., Ferretti, 2021). This must be in error, on the view that language is a form of communication using symbols (Noble & Davidson, 1996), and there are no narratives without symbols. In the narrative I am constructing here, art and its persistence played a role in the emergence of narratives within languages.

4.5 Essay 4 Art and Communication (Erotic Miniature and Odysseus Passing the Sirens Red-Figure Vases)

The creation of pictures necessarily involves interactions between producers and seers and the producers' concerns about something material (the ocher and its impacts), as well as concerns about both non-material things of the mind (the song/story) and material things related to the place where the interaction happened ("the place of the story of the whale"). These provide the subjects of the art as well as its context. Rock art persists in the environment so that it affects the way people interact with places after the art has been produced. What persisted with the rock art was not only the material evidence but intangible mental aspects of the art, especially if they were derived, in part, from the fact that the marking entered into a ritual that the survivors remember. These features are the object created by the producer, who made meaning by engaging in a series of conventions that enabled people to identify with what was going on. Once the conventions of that intangible oral tradition were broken, observers' connection to the ritual was gone, storytelling in rock art might have been connected to the original storytelling only if it was created in scenes. If they persisted, the story could be forgotten, and a new seer could interpret the images in the scene as telling that or another story (Davidson, 2021).

Both rock art and images painted on pottery would be included in archeological art, which would be Art 2 in my scheme. In the left-hand picture in Fig. 4.8, a scene from a red-figure vase of the 2.5 thousand years ago (fifth century BC) shows a young man and a young woman looking as if they will soon be joined together in the fact of the lovemaking.[4] In 1819, nearly ten years before the Berlin Museum acquired the jug bearing the image, the English poet, John Keats, wrote in his *Ode on a Grecian Urn* (Keats, 1956, 209–10):

> Bold lover, never, never canst thou kiss,
> Though winning near the goal – yet, do not grieve;
> She cannot fade, though thou hast not thy bliss,
> For ever wilt thou love, and she be fair!

The point here is that the actions in the scene in the painting are plain. There is a woman paying attention to a young man who appears to find it stimulating. But as Keats understood (whether or not he had seen THIS jug[5]), and as we can predict from our understanding of the cognition of humans, there is more than one possible outcome—however unlikely that seems to be. We will tend to assume that the most obvious outcome was the consummation that the male appears devoutly to wish. That is what we think the scene communicates to us. Those with knowledge of the circumstances of the classical Greek world can add details about the status of the actors in the scene. In fact, there is a more interesting narrative where the male is

[4] https://artsandculture.google.com/asset/true-love-unknown/DgEftaMymJN8Vw?ms=%7B"x"%3A0.5%2C"y"%3A0.5%2C"z"%3A8.844263622087718%2C"size"%3A%7B"width"%3A3.237353871038893%2C"height"%3A1.2374999999999996%7D%7D

[5] And it is an intriguing possibility that he could have.

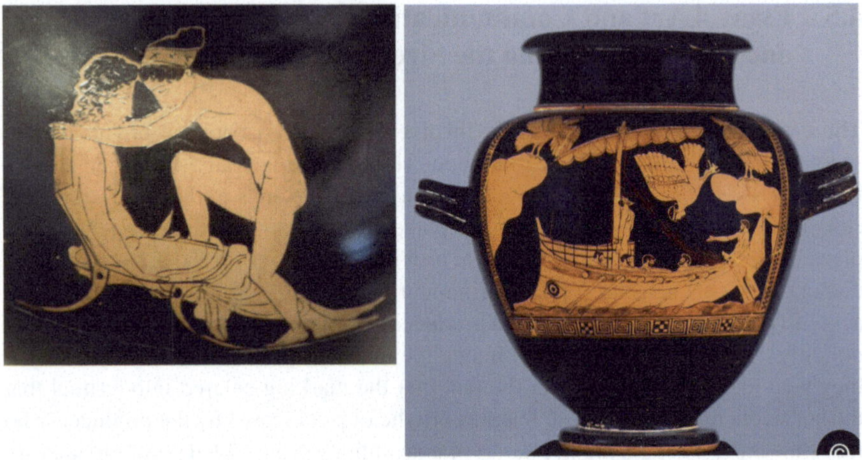

Fig. 4.8 Left Greek Urn kiss Shuvalov Painter, Oinochoe, erotic scene with young man and hetaira; Antikensammlung (F 2414) Berlin CC BY 3.0 <https://creativecommons.org/licenses/by/3.0>, via Wikimedia Commons. Right The ship of Odysseus passing the Sirens; British Museum 01613406857 The Siren Vase, stamnos, Attic, Attica, Vulci

frustrated as the woman rejects him, and another where the female is rejected despite the man's biological urge. We, the seers, can or cannot choose between them just on the basis of the figures in the scene. The story we tell, the choice we make, depends on the relationship we have with the painting, whether we are knowledgeable about the context. It is the implicational meaning of the scene—derived from what we see, together with our knowledge of such interactions. Delightfully, the obvious outcome would be a propositional meaning—it depends only on the information in the painting as we see it, and not at all on our mental construct of the context of such paintings.

The second image from a Greek vase (Fig. 4.8 Right) also depicts a scene. The so-called "Stamnos Vase" is currently in the British Museum[6] This image is about the story of Odysseus sailing past the Sirens and being tied to the mast so he could hear them sing (while his rowers had their ears blocked with wax ["stopped against the song"]) (Homer., 1805, Book XII, 194–239). If we did not know the story, this would be a nice painting of a man chained to a mast while others rowed the boat and some creatures with wings and human heads appeared to plunge about them in the sky.

In an argument unrelated to art, philosophers argue about what they call the "Ship of Theseus paradox." The paradox is about the possibility of a boat made from the spare parts Theseus carried, and whether a boat made of those spare parts would be a second ship or the same one. I argued that this paradox is partly resolved

[6] "Stamnos Vase" British Museum number 1843,1103.31 https://www.britishmuseum.org/collection/object/G_1843-1103-31 (accessed 20 November 2021), purchased from Alexandrine Bonaparte, through James Millingen in 1843.

through the material evidence which constitutes the stuff and substance of cultural heritage as it is generally appreciated (Davidson, 2015), and partly through the narratives on which the heritage depends:

> Material culture can be restored, but the intangible songs and stories, of the people who are custodians of the heritage, may be more difficult to recover unless they survive in material form. It is the interplay of material cultural heritage and the intangible heritage associated with it that is strongest. The changing ship of Theseus can be interpreted best when there is, in the same frame, a song of Homer, however much that too changes.

In the argument of this book, it is the cultural context that gives the clue. In societies who know the *Odyssey*, it is obvious what is going on in the scene of the ship. But if you do not know the *Odyssey*, then the pot is a pretty scene, but the interpretation is thin.

References

Barnard, P. J. (2017). What do we mean by the meanings of music? *Empirical Musicology Review.*

Barnard, P. J., Duke, D. J., Byrne, R. W., & Davidson, I. (2007). Differentiation in cognitive and emotional meanings: An evolutionary analysis. *Cognition & Emotion, 21*(6), 1155–1183. https://doi.org/10.1080/02699930701437477

Boesch, C., & Boesch, H. (1984). Mental map in wild chimpanzees: an analysis of hammer transports for nut cracking. *Primates, 25*, 160–170.

Bohannon, J. (2008). Can scientists dance? *Science, 319*(5865), 905–905.

Bourdieu, P. (1968). Outline of a sociological theory of art perception. *International Social Science Journal, 20*(4), 589–612.

Byrne, R. W., Barnard, P. J., Davidson, I., Janik, V. M., McGrew, W. C., Miklósi, Á., & Wiessner, P. (2004). Understanding culture across species. *Trends in Cognitive Sciences, 8*(8), 341–346.

Carey, J. W. (1967). Harold Adams Innis and Marshall McLuhan. *The Antioch Review, 27*(1), 5–39.

Cheney, D. L., & Seyfarth, R. M. (1990). *How monkeys see the world.* The University of Chicago Press.

Cheney, D. L., & Seyfarth, R. M. (2018). Flexible usage and social function in primate vocalizations. *Proceedings of the National Academy of Sciences, 115*(9), 1974–1979. https://doi.org/10.1073/pnas.1717572115

Conkey, M. W. (1987). New approaches in the search for meaning? A review of research in "Paleolithic AArt". *Journal of Field Archaeology, 14*, 413–430.

Conkey, M. W. (1989). The structural analysis of Paleolithic art. In C. C. Lamberg-Karlovsky (Ed.), *Archaeological thought in America* (pp. 135–154). Cambridge University Press.

Culley, E. V., & Davidson, I. (2021). Art, sign, and representation. In N. Gontier, A. Lock, & C. Sinha (Eds.), *The Oxford handbook of human symbolic evolution* (2nd ed., pp. 331–368). Oxford University Press.

Davidson, I. (1991). The archaeology of language origins: a review. *Antiquity, 65*, 39–48.

Davidson, I. (1997). The power of pictures. In M. Conkey, O. Soffer, D. Stratmann, & N. G. Jablonski (Eds.), *Beyond art: Pleistocene image and symbol* (Vol. 23, pp. 128–158). The California Academy of Sciences.

Davidson, I. (2008). Scales and balance: archaeology, cultural heritage and sustainability. The academy lecture 2007. *Proceedings of the Australian Academy of the Humanities, 31*, 121–150.

Davidson, I. (2013). Origins of pictures: an argument for transformation of signs. In K. Sachs-Hombach & J. R. J. Schirra (Eds.), *Origins of pictures. Anthropological discourses in image science* (pp. 15–45). Halem.

Davidson, I. (2015). *Heritage, the Ship of Theseus and the Song of Homer*.
Davidson, I. (2019). Evolution of cognitive archaeology through evolving cognitive systems. In K. A. Overmann & F. Coolidge (Eds.), *Squeezing minds from stones* (pp. 79–101). Oxford University Press.
Davidson, I. (2021). 1. Scenes and non-scenes in rock art. In I. Davidson & A. Nowell (Eds.), *Making scenes: global perspectives on scenes in rock art* (pp. 16–31). Berghahn Books.
Davidson, I. (2023). Humans making history through continuities and discontinuities in art. *Cambridge Archaeological Journal, 33*, 637–654. https://doi.org/10.1017/S0959774323000057
Davidson, I., & McGrew, W. C. (2005). Stone tools and the uniqueness of human culture. *Journal of the Royal Anthropological Institute, 11*(4), 793–817.
Daylight, R. (2017). Saussure and the model of communication. *Semiotica, 2017*(217), 173–194. https://doi.org/10.1515/sem-2016-0038
Delagnes, A., & Roche, H. (2005). Late Pliocene hominid knapping skills: The case of Lokalalei 2C, West Turkana, Kenya. *Journal of Human Evolution, 48*(5), 435–472.
Díaz-Andreu, M., & García Benito, C. (2012). Acoustics and Levantine rock art: auditory perceptions in La Valltorta Gorge (Spain). *Journal of Archaeological Science, 39*(12), 3591–3599. https://doi.org/10.1016/j.jas.2012.06.034
Domínguez-Rodrigo, M., Pickering, T. R., & Bunn, H. T. (2012). Experimental study of cut marks made with rocks unmodified by human flaking and its bearing on claims of~ 3.4-million-year-old butchery evidence from Dikika, Ethiopia. *Journal of Archaeological Science, 39*(2), 205–214.
Donald, M. (1991). *Origins of the modern mind. Three stages in the evolution of culture and cognition*. Harvard University Press.
Donald, M. (2001). *A mind so rare*. MIT Press.
Donald, M. (2006). Art and cognitive evolution. In M. Turner (Ed.), *The artful mind: cognitive science and the riddle of human creativity* (pp. 3–20). Oxford University Press.
Eisenstein, E. L. (1979). *The printing press as an agent of change: communications and cultural transformations in early modern Europe*. Cambridge University Press.
Evans, C. S., Evans, L., & Marler, P. (1993). On the meaning of alarm calls: functional reference in an avian vocal system. *Animal Behaviour, 46*, 23–38.
Fay, N., Walker, B., Ellison, T. M., Blundell, Z., De Kleine, N., Garde, M., et al. (2022). Gesture is the primary modality for language creation. *Proceedings of the Royal Society B: Biological Sciences, 289*(1970), 20220066. https://doi.org/10.1098/rspb.2022.0066
Ferretti, F. (2021). The narrative origins of language. In N. Gontier, A. Lock, & C. Sinha (Eds.), *The Oxford handbook of human symbolic evolution* (2nd ed., pp. 679–700). Oxford University Press.
Fillmore, C. J. (1968). The case for case. In E. Bach & R. T. Harms (Eds.), *Universals in linguistics theory* (pp. 1–88). Holt, Rinehart and Winston.
Füssel, S. (2001). Gutenberg and today's media change. *Publishing research quarterly, 16*(4), 3–10.
Gorman, M. E. (1997). Mind in the world: Cognition and practice in the invention of the telephone. *Social Studies of Science, 27*(4), 583–624.
Henshilwood, C. S., d'Errico, F., & Watts, I. (2009). Engraved ochres from the Middle Stone Age levels at Blombos Cave, South Africa. *Journal of Human Evolution, 57*(1), 27–47.
Hoffmann, D. L., Standish, C. D., García-Diez, M., Pettitt, P. B., Milton, J. A., Zilhão, J., et al. (2018). U-Th dating of carbonate crusts reveals Neandertal origin of Iberian cave art. *Science, 359*(6378), 912–915. https://doi.org/10.1126/science.aap7778
Homer. (1805). *The Odyssey of Homer translation by A. Pope Esq*. W. Sullaby.
Huntley, J., Aubert, M., Oktaviana, A. A., Lebe, R., Hakim, B., Burhan, B., et al. (2021). The effects of climate change on the Pleistocene rock art of Sulawesi. *Scientific Reports, 11*(1), 9833. https://doi.org/10.1038/s41598-021-87923-3
Izar, P., Peternelli-dos-Santos, L., Rothman, J. M., Raubenheimer, D., Presotto, A., Gort, G., et al. (2022). Stone tools improve diet quality in wild monkeys. *Current Biology, 32*(18), 4088–4092. e4083.

Joordens, J. C. A., d'Errico, F., Wesselingh, F. P., Munro, S., de Vos, J., Wallinga, J., et al. (2015). Homo erectus at Trinil on Java used shells for tool production and engraving. *Nature, 518*(7538), 228–231. https://doi.org/10.1038/nature13962

Keats, J. (1956). *The poetical works of John Keats*. Oxford University Press.

Layton, R. H. (1992). *Australian rock art. A new synthesis*. Cambridge University Press.

Levitt, J., & Lovett, J. V. (1985). Alkaloids, antagonisms and allelopathy. *Biological Agriculture & Horticulture, 2*(4), 289–301.

Logan, R. K. (2002). The five ages of communication. *Explorations in Media Ecology, 1*(1), 13–20.

Matthews, J. (2013). Seven spots and a squiggle: the prehistory of pictures. In K. Sachs-Hombach & J. R. J. Schirra (Eds.), *Origins of pictures. Anthropological discourses in image science* (pp. 311–352). Halem.

Mayer, R. E., & Anderson, R. B. (1991). Animations need narrations: An experimental test of a dual-coding hypothesis. *Journal of educational psychology, 83*(4), 484–490.

McLuhan, M. (2012 [1962]). *The medium is the message. Media and cultural studies: keyworks* (pp. 100–107).

McPherron, S. P., Alemseged, Z., Marean, C. W., Wynn, J. G., Reed, D., Geraads, D., et al. (2010). Evidence for stone-tool-assisted consumption of animal tissues before 3.39 million years ago at Dikika, Ethiopia. *Nature, 466*(7308), 857–860.

Mercader, J., Barton, H., Gillespie, J., Harris, J., Kuhn, S. L., Tyler, R., & Boesch, C. (2007). 4,300-year-old chimpanzee sites and the origins of percussive stone technology. *Proceedings of the National Academy of Sciences, 104*(9), 3043–3048. https://doi.org/10.1073/ProceedingsoftheNationalAcademyofSciencesoftheUnitedStatesofAmerica.0607909104

Morin, O., Kelly, P., & Winters, J. (2020). Writing, graphic codes, and asynchronous communication. *Topics in Cognitive Science, 12*(2), 727–743.

Noble, W., & Davidson, I. (1996). *Human evolution, language and mind*. Cambridge University Press.

Ottoni, E. B., & Izar, P. (2008). Capuchin monkey tool use: overview and implications. *Evolutionary Anthropology: Issues, News, and Reviews: Issues, News, and Reviews, 17*(4), 171–178.

Pasztory, E. (2005). *Thinking with things: toward a new vision of art*. University of Texas Press.

Pow-key, S. (1982). *Early Korean typography* (pp. 120–133). Po_chin_caiCo. Ltd.

Proffitt, T., Luncz, L. V., Falótico, T., Ottoni, E. B., de la Torre, I., & Haslam, M. (2016). Wild monkeys flake stone tools. *Nature, 539*(7627), 85–88. https://doi.org/10.1038/nature20112

Proffitt, T., Reeves, J. S., Falótico, T., Arroyo, A., de la Torre, I., Ottoni, E. B., & Luncz, L. V. (2023). Identifying intentional flake production at the dawn of technology: A technological and 3D geometric morphometric study. *Journal of Archaeological Science, 152*, 105740.

Rodríguez-Vidal, J., d'Errico, F., Pacheco, F. G., Blasco, R., Rosell, J., Jennings, R. P., et al. (2014). A rock engraving made by Neanderthals in Gibraltar. *Proceedings of the National Academy of Sciences, 111*(37), 13301–13306. https://doi.org/10.1073/pnas.1411529111

Saussure, F. D. (1986). The linguistic sign. In R. E. Innis (Ed.), *Semiotics. An introductory anthology* (pp. 28–46). Hutchinson.

Schiestl, F. P. (2010). The evolution of floral scent and insect chemical communication. *Ecol Lett, 13*(5), 643–656.

Schiffer, M. B. (1991). *The portable radio in American life*. University of Arizona Press.

Schiffer, M. B. (1999). *The material life of human beings: artifacts, behavior, and communication*. Routledge.

Schlenker, P., Chemla, E., Schel, A. M., Fuller, J., Gautier, J. P., Kuhn, J., et al. (2016a). Formal monkey linguistics. *Theoretical Linguistics, 42*(1–2), 1–90.

Schlenker, P., Chemla, E., & Zuberbühler, K. (2016b). What do monkey calls mean? *Trends in Cognitive Sciences, 20*(12), 894–904.

Shepherd, S. (1996). The significance of the overland telegraph line, 1872/1901. *Journal of Northern Territory History, 7*, 41–46.

Slater, P. J. B. (1983). The study of communication. *Animal Behaviour, 2*, 9–42.

Snowdon, C. T. (1992). The sounds of silence. *Behavioral and Brain Sciences, 15*(1), 167–168.

Thompson, J. C., McPherron, S. P., Bobe, R., Reed, D., Barr, W. A., Wynn, J. G., et al. (2015). Taphonomy of fossils from the hominin-bearing deposits at Dikika, Ethiopia. *Journal of Human Evolution, 86*, 112–135.

Thomson, A. (1999). *The singing line. Tracking the Australian adventures of my intrepid Victorian ancestors*. Anchor Books.

Visalberghi, E. (1992). Is lack of understanding of cause-effect relationships a suitable basis for interpreting monkeys' failures in attribution? *Behavioral and Brain Sciences, 15*(1), 169–170.

Whiten, A. (1992). Mind reading, pretence and imitation in monkeys and apes. *Behavioral and Brain Sciences, 15*(1), 170–171.

Wilkinson, D. (2023). On the ontological significance of naturalistic art. In C. Watts & C. Knappett (Eds.), *Ancient art revisted. Global perspectives from archaeology and art history* (pp. 47–66). Routledge.

Wisher, I., Riede, F., Matthews, J., Pagnotta, M., & Tylén, K. (2025). Children as playful artists. *Hunter Gatherer Research,* 1–39. https://doi.org/10.3828/hgr.2024.38

Zuberbühler, K. (2022). Event parsing and the origins of grammar. *WIREs Interdisciplinary Reviews: Cognitive Science, 13*(3), e1587. https://doi.org/10.1002/wcs.1587

Chapter 5
Graphic Elements in Communication

Abstract Art enables people to talk about the past and the future as well as the present—to reflect on them. Art is a system of meanings communicated to members of society such that anyone seeing it knows something about that society's values, even if they do not know all of them. I call this "dual expression" to refer to the thing itself and the idea associated with it. Stories make marks art. The people who make the object can think about it, engage in an "actual process of thought," and tell stories about it. This process of thought requires a much more complex cognition than any other animal has been shown to engage in. We take many things for granted about communication. Many of them derive, ultimately, from the discovery of the dual expression that allows people to use things and their literal surfaces to communicate visually, even in ways that are not worth a thousand words for every viewer. In the variation of dual expression, the literal surface can remain the same but the deeper cognitive meanings can change through the associations of ritual. In early revolutions, art combined with ritual to fix both story and ritual in a particular place and generate the reflections we now take for granted. Ritual had an unexpressed function to provide validation for the associations between the stories and the material expression of them through the art.

Writing uses visual means to make an oral communication permanent. Once writing existed in those places where it emerged, the message of the story could be written down and understood by another person who had learned to read it. And it did not matter when this reading happened—it was part of what has been called "asynchronous communication"—such that the communication or story that accompanied the visual message was less ambiguous and could be read at any time. Writing could transmit the message across great distances because it was not tied to time or place.

Printing changed the game in Europe, despite the initial lack of color. Although it presented problems of validation, printing made it possible for two people (or more) to be reading the same book in different places at the same time and others to be reading the same book at different times with an unchanged story. Communication

I. Davidson, *Art or Scribbles? In the Eye of the Beholder: The Evolutionary Emergence of Visual Communication*, Interdisciplinary Evolution Research 11,
https://doi.org/10.1007/978-3-031-90458-5_5

at a distance was implied by reading the same book at different places. Communicating directly at a distance happened initially through visual signals conveyed by semaphore but later by electric telegraph. Ultimately the conquest of delay was added to the other triumphs represented by the use of the visual (art) to tell stories, overcoming the immediacy of stories. The problem was not so much the tyranny of distance as the tyranny of time. The electric telegraph more or less required the development of Morse code which was a near-digital means of reducing the finite elements of written language to something that could be on or off. Digital code was the breakthrough that led to electronic communication, ultimately including the transmission of color images of pictures and other things.

5.1 The Implication of the Art-System and the Graphic World

In the last chapter I examined a definition of art that emphasizes its role in communication. In order to do this, I had to establish that, whether they are closely related to humans or not, no other animals do anything like what humans do. They do not communicate in the same way—vervets make calls about eagles but they do not sit around and tell stories about the eagle they saw this morning or the one they might see tomorrow; other animals do not recognize material culture in the same way—each time they fish for termites, chimpanzees make and discard an old termiting wand and then they make a new one; and other animals do not make marks to which they can attribute meaning—indeed whether or not they have any concept of meaning, they leave no behavioral or material sign of it. Other animals do not tell stories or live in societies defined by such stories. They do not reflect on them, and, as a gloss on more elaborate description of the mental systems (Barnard et al., 2007, 2016), I will refer to the reflective mind. As I argued previously (Davidson, 1999, 264), whatever the remarkable interactions between bonobos, gorillas, vervets, or other non-human primates, they do not undertake the studies of our species that we have of theirs and not just because of differential power. "There are no ape or monkey humanologists, archaeologists, psychologists, or linguists. Only humans play word games; only humans pun." Only people produce paintings and engravings unaided by other animals.

I think we can be sure that art, as I have defined it, is a distinctively human achievement. Art enables people to talk about the past and the future as well as the present—to reflect on them. Art is a material sign both of your own past activity and a visible sign to others that someone has engaged in such activities—it is an indexical sign of that. Art is a system of meanings communicated to members of society such that anyone seeing it knows something about that society's values, even if they do not know all of them. I call this "dual expression" to refer to the thing itself and the idea associated with it. Almost certainly, that art will only reflect the values of one part of any society (for example, Procter, 2020; Spies-Gans, 2022)—but even awareness of such divisions of sexism, racism and class allows us to know about the values of the whole society. Dual expression is a property of the reflective mind.

Myers (following John Kean in Myers & Smith, 2024, 51) described discussions among painters during the early stages of the production of art at Papunya about whether they were painting a ceremony or a place where the ceremony happened, concluding "the ceremony represents the ancestral action at the place that left the features of the place as one sees them now." This was an example of the dual expression I am defining here.

The art-system carries with it a couple of important differences between what humans do and what all other animals do. We alone define the places we have been by marking them and telling stories about them. We alone tell such stories in terms of what McLuhan called "actual processes of thought". Those processes are both those of the artist and of ourselves as seers. As a result of those processes of thought, and the combination of meanings and understandings that the system produces, art shows that humans are cognitively different from other primates—but this does not mean that the moment the first art was produced in SE Asia, Africa or even Europe then all humans suddenly had their cognition transformed. Humans alone use art in their communication but that does not mean that all humans are artists, though we live in societies pervaded by art. The art itself does not communicate. Art and its derivatives pervade and dominate almost all our communications, but that process begins with the location and its implications so that communication through art is connected with a fixed place, at least at the beginning, together with the stories about the place, and reflections upon it and the art. Once upon a time that place was a cave or rock shelter; now it is more usually a studio, museum, gallery or a wall outside in public view (such as a mural) or inside in private domestic space.

One idea underpins everything else: material evidence of art shows that its producers have engaged in "actual processes of thought", and this combination makes us cognitively different from other primates. (It may be that other primates engage in "actual processes of thought," but they leave no material sign of it.) That cognitive difference, however, derives from the production of art. The first humans who made art because they engaged in actual processes of thought were cognitively different as a result. In addition, by teaching others that the marks they made carried meaning because of those processes, those others also became cognitively different.

It is easy enough to argue that the cave art of Western Europe involved such processes. Whatever interpretations have been sought over the years, and I do not intend to spell them out when others have done so (see, for example, David, 2017; Guthrie, 2005; Lawson, 2012), all of them require that the paintings (or other objects) executed regularly in caves needed some storytelling by the painters either about the images themselves or about why they were doing them. We can gauge the special nature of those acts from the fact that, for anyone other than the painters to see the work, they had to make a special purpose visit into the cave. Even the fact that art was also created in the open-air, as in the Côa Valley in Portugal (Fig. 5.1), implied knowledge in the community of the different contexts in which art was created.

We know from the presence of footprints of children and others in the caves (Pastoors et al., 2015, 2017), from the varied range of hand-stencils and other marks

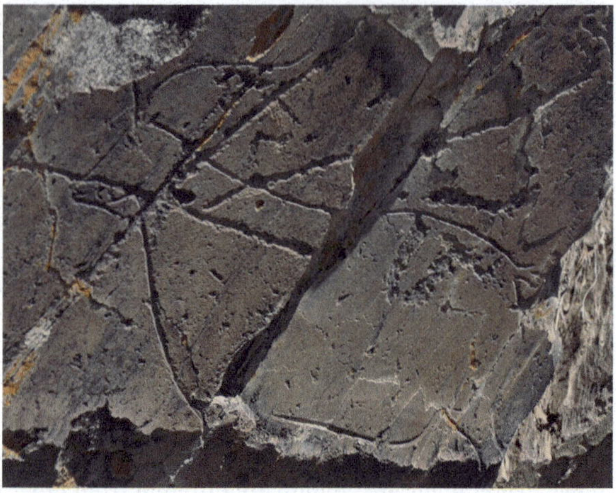

Fig. 5.1 Engraved Ibex facing to the left (on the right of the main slab of rock) at the open-air site of Penascosa, Portugal which is part of the group of sites known as Foz Côa (see, for example, Batarda Fernandes et al., 2017). (Photo Iain Davidson)

(Nowell, 2021, Ch. 3 and 118), from the production of finger fluting by children and adults, males and females (see, for example, van Gelder & Nowell, 2021) that, if they were, indeed, not just marks with no meaning other than "I (someone) was here" or they were "scribbles" they required some conversation between those making the marks and those seeing them. It would be possible to imagine that a python leaving its track, or a dog walking on wet concrete, could be showing that it was there but in the absence of all the other necessary conditions, it is implausible. Stories make marks art.

It sounds like the cliché "Monkey see, monkey do," to say that producers see things and use marks to make art in response to what they see. Nowadays we would dress up such a cliché in the language of brain scientists such as being a product of the presence of mirror neurons in the brain, which fire when one animal sees another carrying out a task (e.g., Arbib, 2003). Whether or not mirror neurons are involved in the production of art, that creation is more pragmatic. A producer makes a mark—produces an object—when they see a subject, but it is not a simple matter of a stimulus producing a response. In addition, I am arguing, the people who produce the object can think about it, engage in an "actual process of thought," and tell stories about it (as spelled out by Barnard et al., 2007). This process of thought requires a much more complex cognition than any other animal has been shown to engage in.

It would, in principle, be possible to respond pragmatically without the mental processes of reflection and thought. These two conditions are described in the model of mental evolution, primarily due to Phil Barnard, that complements the emphasis on the semantic roles of humans in different actions (Table 4.1). In that model, the "actual processes of thought" belong only to modern humans, while the simpler mental condition with little more than pragmatic responses was a condition of earlier cognition (Barnard, 2010, 2019; Barnard et al., 2007, 2016). I would argue that perhaps such a simpler pragmatic response was present among Neandertals with their simpler, less repetitive, occasional productions of stone-arrangements (Jaubert et al., 2016) (Fig. 5.2), occasional use of feathers (Blasco & Peresani, 2016;

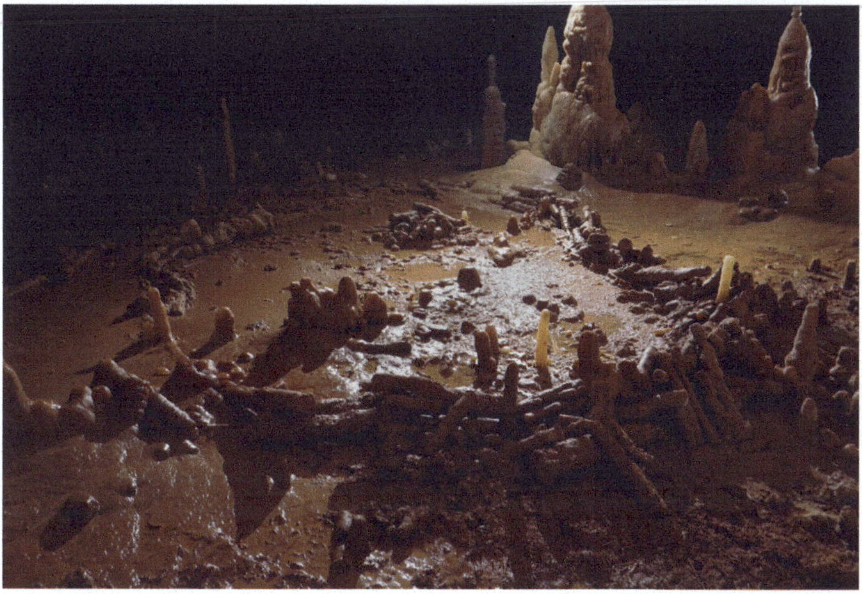

Fig. 5.2 Bruniquel Cave, stalagmite structures dating back 175,500 years (Jaubert et al., 2016). (Photo E. Fabre for the Bruniquel team, with permission)

Finlayson et al., 2012) or pendants (Radovčić et al., 2015) (Fig. 7.26), or markings inside caves (Hoffmann et al., 2018; Rodríguez-Vidal et al., 2014) (see Fig. 4.6).

Once people had established, in the past, that they could make things that we now consider as art, and that they could tell stories about it in the place where it was created, several things followed. Most importantly, people now lived, as we now live, in a world where visual signs have meaning, and mark the landscape (as shown in the modern painted murals in Newcastle in Figs. 3.18 and 3.19). Our culture is imbued with the sense that images and ideas are all around us. John Berger (2008, Ch. 7) ended his revolutionary television series about *Ways of Seeing* with examples which sought to show the relations between publicity images and the ways of seeing other art, pointing out that there is always more about the interactions with such images than the thing in the picture itself. In 2023, the Italian Tourist bureau produced posters using an image that included a human figure in modern dress with the face of Sandro Botticelli's Venus from *The Birth of Venus*, in modern pose viewing famous Italian sites (Fig. 5.3).

A heritage professional (whom I choose not to name) criticized the campaign because it "trivializes our heritage in the most vulgar way" (Lawson-Tancred, 2023). It seems that some people think that art can only be presented in one way, whereas these are, just, visual signs of the sort that surround everything we do, and such as we can see all over the landscape. Following Chap. 3 I would suggest this is a clear example of the boundaries of art being patrolled by those who are "interested in art." The idea of such an image is not, after all, very original, as the painting on the

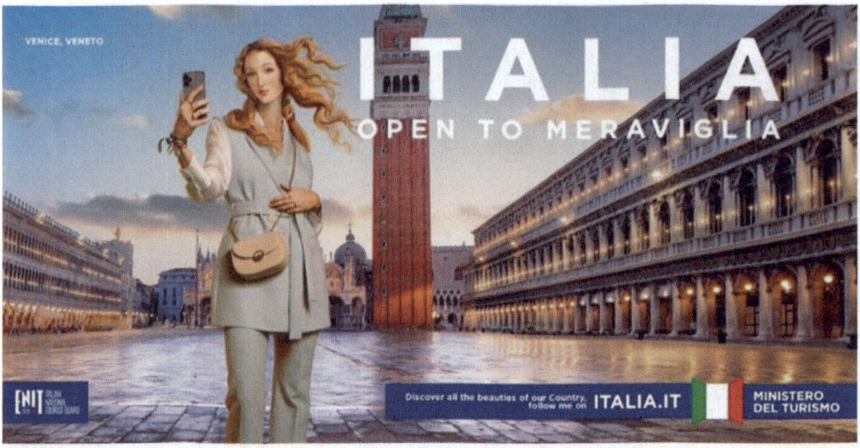

Fig. 5.3 Botticelli Venus, in modern dress, taking a selfie on her mobile phone in Saint Mark's square, Venice. Publicity poster for Italian Ministry of Tourism. Aspects of the square pre-date the sixteenth century Botticelli painting

Brisbane traffic signal box (https://www.brisbane.qld.gov.au/things-to-see-and-do/experiences-and-culture/public-art-in-brisbane/street-art/art-on-traffic-signal-boxes-artforce-brisbane) shows (Fig. 5.4), and Berger (2008, 25) began his Television series by cutting out a remarkably similar face of Venus by Botticelli from the work *Venus and Mars*, only revealing later that he had cut up a reproduction. This act led him to the later conclusion that "Reproduction isolates a detail of a painting from the whole. …An allegorical figure becomes a portrait of a girl." One of the questions we might ask is about the importance of the ability to fit art into a frame.

Visual signs pervade our culture, even though not all of them are or were ever called art (though whether they should be is another question). The Italian Tourism posters were showing the world that Italian culture manifests itself in many different ways through art (older than the eighteenth century, in fact), architecture, landscape, fashion and food (and they even have phones) which all have historical context.

We have become accustomed to the fact of images in our lives, without concerning ourselves that once upon a time there were no such images nor that all images were produced by others with some meaning in mind. Their meanings were not necessarily the same as the meanings that others, including us, have understood over the years—that is the nature of art, a product of the spaces between the elements of the art-system. Botticelli painted his Venus which is now seen with a mobile phone long before there were such things, and under the circumstances the painted subject, Venus, never went to Venice nor, in all likelihood, did the model. It is this pervasiveness, together with the inevitability of a dual expression of literal surface and deeper cognitive meanings, that makes the initial production of visual images so revolutionary. In the example of the Italian tourist poster, the dual expression derives from the initial fact that people take "selfies" in St Mark's Square in Venice, the duality derives exactly from the use of the 500-year-old image thus invoking a famous

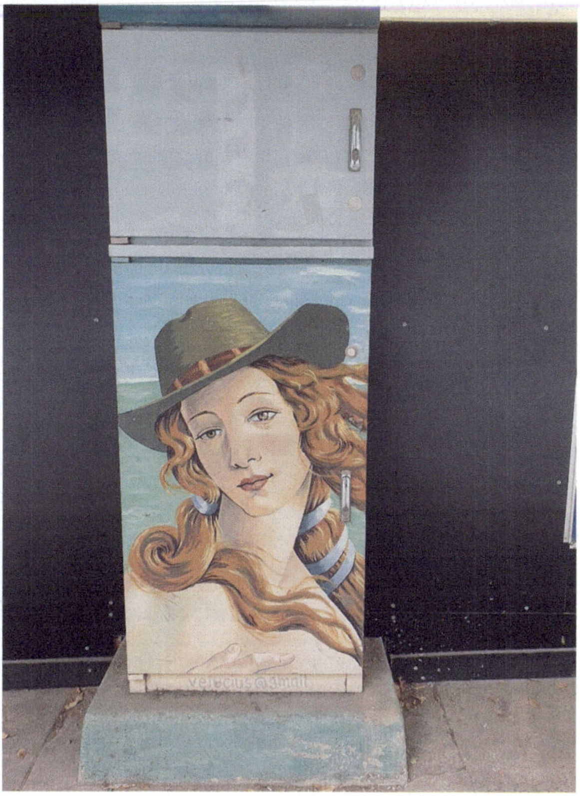

Fig. 5.4 Head of Venus copied from Botticelli's *The Birth of Venus*, wearing an Australian Akubra hat, painted by Velucius on a Brisbane traffic signal box (https://www.brisbane.qld.gov.au/things-to-see-and-do/experiences-and-culture/public-art-in-brisbane/street-art/art-on-traffic-signal-boxes-artforce-brisbane). (Photo Iain Davidson)

Italian painting with the famous, older, Italian landscape. We are tricked (it is not much of a trick) into recognizing that both painting and architecture are part of Italian heritage, but both painting and buildings have survived, so that they are modern enough for people to be taking photographs on their phone cameras, in the same place where in the eighteenth century Caneletto (Giovanni Antonio Canal) painted several images (Fig. 5.5).

This duality is now all pervasive in art, but was not necessarily present earlier than art. As an example of that duality in modern art, Munn (1973) demonstrated that Warlpiri art in Central Australia consists of what some might see, "misleadingly", as geometric, whereas it is really "representational" (her word). The signs of the graphic art carry multivalent meanings in the ritual realm even though they are actually "anchored in narrative communication." Warlpiri art is not unusual in this respect. In a similar fashion, Myers (2016, 40) described how (and why) "Doublings" were a characteristic of the art of the adjacent Pintupi people whose images might also be seen misleadingly as geometric, but which have meanings which were often best understood in a secretive and sacred sense. In this way, he said, they represent "one thing also being simultaneously something else." Indeed, in many respects the art of Central Australia, as in Papunya, always wove two categories of meaning (Myers & Smith, 2024, 51).

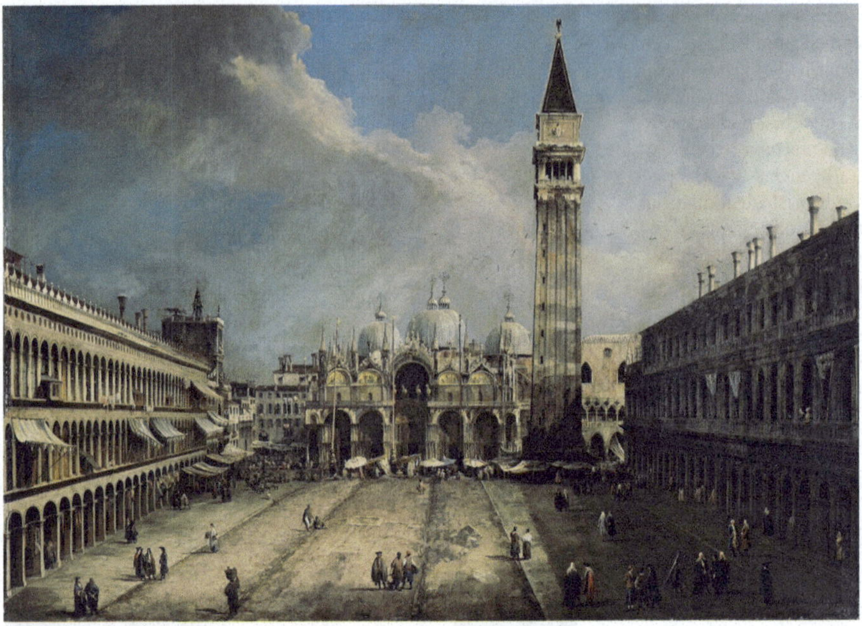

Fig. 5.5 Canaletto (Giovanni Antonio Canal) The Piazza San Marco in Venice ca. 1723–1724 Museo Nacional Thyssen-Bornemisza, Madrid Inv. no. 75 (1956.1)

After the second of the revolutions I defined in Chap. 4—the revolution of "the production of visual, particularly representational, images"—there was an expectation among those who knew about art and architecture that images, such as the tourist poster, had meanings that were not just what you saw in the original image. All subsequent revolutions in communication had an important graphic element as well as social consequences derived from that dual expression. One way or another, visual signs pervade our lives in almost everything we do, but especially our means of communication.

There were more revolutions to come: Writing, Printing, Communication at a distance, and Digital communication. In this chapter I will discuss each of these briefly, with more discussion of writing than of the other revolutions. I will discuss the more recent revolutions primarily to put their visual components as art into modern context and consequences. We will find, in Chap. 9, that these three revolutions need to be seen broadly when we consider the art of Aboriginal Australia. This is not a study of the emergence of each of these technologies but intended to illustrate the visual component of them and point to the consequences in terms of society and communication within it. Many of the things we take for granted about communication derive, ultimately, from the discovery of the dual expression that allows people to use things and their literal surfaces—the woman with a phone in a square with a tower—to communicate visually, even in ways in which the intention of the artist, such as the cognition about Venus, St Mark, phone cameras and image making, could not control the intentions of the viewer. The aggrieved heritage specialist could not get over the fact that the heritage entrusted to her had been used in a way she, and, presumably, Botticelli, had not intended. I think they miss the point about art, though maybe Botticelli would not have been so narrow-minded.

5.2 Ritual and the Word of Mouth

Communication is rarely preserved straightforwardly because it is not material, but an outcome of the relation between the subject of the art and the producer's story about making the object—an aspect of the dual expression. The marks were accompanied by a story from the producer, and then became a reminder that there was a message. That is an important shift. David Olson (1994, 70) suggested that "visual graphic systems [picture writing] served as mnemonics by means of which a suitably trained expert could recover important cultural information." The reminder existed because the message from the producer became part of the culture because the seers were told about it. People knew that the art had been made by someone with something to say. The question is: "How did (what was the process by which) individual acts of production of things with associated storytelling move to become such an integral part of widespread cultural beliefs?" The answer, I suggest, lies in ritual, which also engendered trust in what people knew. As Piers Kelly put it[1] "the oral channel is doing almost all of the heavy lifting in terms of the information load," but it is an oral channel with very particular characteristics. The oral transmission of the story needed to be accompanied by repeated acts that have material products and the repetition, if not necessarily the details of the story that was transmitted, needed to be generally known in the community. The stories were transmitted in ritual by both verbal and visual means and became embedded in the culture of the community through repetition. Olson used the word mnemonic—we need do no more than recognize that the power of communication in art derives in part from sharing individual memory from one person to another.

As a result of his lifetime of research, primarily among the people of New Guinea, Roy Rappaport (1999, 24) defined ritual as denoting "the performance of more or less invariant sequences of formal acts and utterances not entirely encoded by the performers." June Ross (2003) explored this conceptualization and determined that it could be extended to apply to studies of rock art (Ross & Davidson, 2006). The niceties of this reach a long way. As Françoise Dussert (2000, 5), writing about modern Warlpiri[2] people of Central Australia, wrote: "Warlpiri life *is* ritual

[1] https://bravenewwords.info/2023/01/08/thoughts-on-proto-writing-in-the-upper-palaeolithic/ accessed 2 April 2023.

[2] Different authors use various spellings of this language name. In this section, Dussert uses "Warlpiri" and Nancy Munn uses "Walbiri". They are, essentially, the same named people, though "studied" by different people with their own different backgrounds and at different times. No doubt they also had differences in the way they heard the language of others.

insofar as the expression of a spiritual connection to the ancestors extends beyond the formal ceremonial sphere. The links can be made manifest in momentary asides during hunting trips or songs sung during sudden pangs of grief over missing loved ones (which can include pieces of land as much as kin)." In this way, through asides or through spontaneous expressions, the fact that people have participated in ritual acts impacts the whole way of life of people who might otherwise restrict access to ceremonies. Rituals constitute their culture. In Zambia, Victor Turner (1974, 15) found "in an *Ndemba* ritual context, almost every article used, every gesture employed, every song or prayer, every unit of space and time, by convention stands for something other than itself." In fact, I would say "in addition to itself" because the thing is still the thing, but by having symbolic value it entails something more. I suggest that this pervasiveness of symbolic meaning of everything is widespread in non-literate societies. It derives in part from the association of the oral channel—through memory—with the visual and material world.

People acquire much important knowledge of their culture through informal interactions with adults as they are growing up, both as children among adults, and in exchanges between adults (Nowell, 2021). Low-risk experience can be acquired through horizontal transmission between peers that may involve pragmatic experiences in the environment. The foraging practices required for survival as an adult might be taught by other adults to make the risk as low as possible (summarized in Sterelny, 2022).

Much more information beyond foraging practice is essential to the sort of cultural life Dussert described. It is imparted more formally, by vertical transmission from adult to young person or initiated to uninitiated, in contexts that vary throughout a person's life. Howard Morphy (1991) documented this successive revelation and showed how senior people teach many of the meanings through initiation ceremonies (that is to say, rituals). Those meanings are learned by the people who then know them and teach them to others; they are not immutable properties of the material things, but they have some properties that are intrinsic and others that are assigned "by convention," as Turner would put it. Elders use material things such as art in telling stories about the relations between things and the unfathomable worlds they are related to (for a brief introduction to these ideas see Davidson, 1995). Kim Sterelny (2022) called the secret knowledge in such telling "arcane and the arbitrary," pointing out that much of the information that was shared was socially constructed: "Adventurous-enough individuals might discover for themselves the routes linking reliable waterholes in the Western Desert [of Australia], but they cannot discover the totemic dreamtime figures linked to those sites, nor the rights, obligations, practices, and prohibitions that derive from those links. These are not objective facts to discover." That set of relationships, about any resources, social, geological, subsistence, or spiritual, needs to be learned by members of a social group from others more powerful who know the group's Dreaming—the complex of cosmological beliefs that defined the group's relation to the land. The Yolngu art-system Morphy described combined both geometric art such as he illustrated himself and figurative art, which might consist of more iconic signs (Morphy, 1999). Morphy tested the iconicity of figurative art by isolating an iconic painting of an animal from its context and asked the artist himself to name the species of animal he had painted, suggested that even iconic signs require context to be understood (Davidson, 1995, 891).

Otherwise, figurative art consists of more or less iconic signs—the images resemble aspects of what they represent—and is interpretable by people at low or no level of initiation and could be said to have "outside" meanings accessible without initiation (Morphy, 1991). As an example of this look at Fig. 1.2, where the horses' heads have some similarity to real horses, but they are not part of representations of whole animals, and certainly they are only representations. They are iconic signs of horses, not horses, and the person looking at them in the photograph has absolutely no level of initiation, at least into the society of the painters, only some into the society of people who study such things! Because of the location in a special place, it is difficult to argue that they were not painted to enhance a story. The combination made them part of an "inside" initiation, even if they have an "outside" meaning to us.

In Australian documented art-systems, almost all of the art revealed in secret initiation was geometric (and would be called a part of "inside" meanings by Morphy, 1991) and needed exegesis to link meaning to the marking (Morphy, 1980), as Nancy Munn (1973) and subsequently Jennifer Green (2014) have pointed out with the geometric art of Central Australia. Geometric designs, such as those at the South African archeological sites of Blombos (in Fig. 3.27, Henshilwood et al., 2009) or Diepkloof (Texier et al., 2010) (Fig. 3.28), would always have required meanings of non-iconic signs to be encoded, possibly through initiation, within the culture for them to have meaning. Ironically, if we think back to the grazier Bill whom I mentioned in the Introduction, the simpler images, which he might have recognized as art because they were iconic signs, belong to more public and outside contexts (Fig. 2 Left). It is the less easily interpreted imagery that has to be explained to initiates when they are admitted to the inside of the society (Fig. 2 Right)—and these grazier Bill might have looked down upon as "scribbles."

Ritual aspects of art and society were important in the evolution of cognition. That importance means that it becomes much more relevant to find convincing evidence for ritual in the archaeological record. Ritual is not just a label of last resort for archaeologists to describe things they cannot understand, what April Nowell (2023) alludes to when saying "There is a joke among archaeologists that we label an artefact as 'ceremonial' [or ritual, ID] if its purpose is not readily discernible."

I previously argued that we can get at the concept through the strange distribution of sites, from around 20 thousand years ago in Western Europe, with art done on stone slabs as at Parpalló in País Valencià, as such evidence (Davidson, 1989). In that instance, I argued, such art was produced on stone slabs in very large numbers at only a small set of sites across Western Europe between 27 and 13 thousand years ago, and that these probably constituted central places where people taught information of importance, enhancing the value of the knowledge, and their own power, by restricting access to that knowledge to places where large numbers of plaquettes were deposited (Davidson, 2012) (See Figs. 2.2 and 2.3 in this book). It will become important in my argument later in the book that, on the basis of my analysis of the representation of animals in the art and in the bones, during the course of the 14 thousand years of use of such a ritual there seems to have been a cognitive change within the sequence at Parpalló. This change corresponded with changes in the material culture of art, stone, bone and antler tools, none of which were a product of climate change despite the end of the period of maximum low sea-level and cold (Davidson, 2005). That cognitive change opened new opportunities for people to expand north to places which had previously had colder climates. On that evidence, art and ritual were, thus, part of the cultural context of Western Europe—rather as it had been assumed but not demonstrated previously.

Such rituals would only have worked if they were part of oral cultures which transmitted canonical messages that were fundamental to the continuation of the society, and, I suggested, the capacity to transmit messages derived from the power associated with restriction of access to the ritual (Davidson, 1989). The history of communication is one of changes in the control of power that can be transmitted orally, through writing, printing, communication at a distance or electronically. In all cases, the power involved a visual message—the surface—which was incomplete without the accompanying story, which could become internalized cognition. Leah Minc (1986) pointed out that, in the Arctic, oral traditions work at two scales: folktales, songs, and histories served to mitigate the impact of seasonal shortages or those that operated over a short-term such as from year to year; sanctified rituals, on the other hand, tended to be relevant to resource stress which showed itself over several generations. Perhaps the past tense is not appropriate there: retelling of stories about past *tsunamis* in Simuele Island, west of Aceh Province in Indonesia, seems to have been responsible for the negligible death toll on that island arising from the 2004 *tsunami* (Sutton et al., 2020). The epicentre of the earthquake was between Simuele and Sumatra, and the *tsunami* killed hundreds of thousands in several different countries, including nearby mainland Aceh province in Sumatra. But not in Simuele because an oral tradition about surviving previous *tsunamis* saved them.

Walter Ong (1982, Ch. 2) showed how much the use of ritual routines penetrates all storytelling by describing how Milman Parry analyzed the most famous[3] of all oral epic songs, Homer's *The Iliad* and *The Odyssey* (the image of Odysseus in Fig. 4.8 Right illustrates a story from *The Odyssey*). Parry showed that the oral poems included many formulaic phrases, often invoking visual images that tie one object to another metaphorically (e.g., "wine-dark sea", "rosy-fingered dawn"), which were the common-place of everyday speech and appropriate to the metrical rhythm of the poems. The redundant nature of such formulae, or "fixed patterns of thought" (Parry, 1930, 77) made the lengthy works more memorable and embedded in the culture of ancient Greeks, part of their cognitive understanding, as an outcome of such ritual utterances. Indeed, they were embedded in Greek culture for much longer as a result of their use in the teaching of Greek into the third century AD,

[3] I write as one who was brought up with these stories. Of course, in other cultures there are other oral traditions, in some cases known to far more people. I discuss some of these in Chap. 8, including *The Mahabarata* and Confucius.

1800 years ago (Horsley, 2000, 62) about 1000 years after the Homeric poems were written down.

Ong went further and showed that similar traditions of combinations of invariant sequences of formal utterances were present in oral traditions wherever they have been looked at. Jack Goody (1998) pointed out, however, that "Oral performance in literate societies is undoubtedly influenced to different degrees by the presence of writing and should not be identified with the products of purely oral cultures." The use of standard phrases seems to be widespread in oral traditions around the world, and in some instances was accompanied by drawings (McLaughlin, 2013), but it is difficult to quantify the impact of literacy on the modern form of oral performance since most of us need to identify it through literature that is already written down.

The point of setting out the relations between marks and meanings is that the dual expression that is characteristic of all art consists of the marks we can all see as well as the story behind their production, which we can't. Without that storytelling, I would say, all we have is the literal surface and not the dual expression. The world-wide pattern for oral traditions is that the oral channel has characteristics that make its words easier to learn, the utterances are "not entirely encoded by the performers," and through repetition are part of ritual as defined by Rappaport. In addition, that "story behind the production" gives the marks the fundamental characteristics of symbols: the art may be arbitrarily related to the subject it represents, but by conventions shared within the society its form and style is often learned in ritual. If writing, or any other technology, creates opportunities for any form of access to the story other than the shared conventions of the oral story, then there would have been options for change in that set of relationships because stories could be accessed without knowledge of the conventional context of their telling. Arbitrariness allows people to make subtle changes to the symbols and their meaning as appropriate, particularly through "actual processes of thought." In the dual expression, the literal surface can remain the same, but the deeper cognitive meanings can change through the associations of the ritual.

Fixing the image in place determines the number of people who have access to information immediately and it does not alter the numbers who have access to information at the same time, because that is limited by the numbers who can witness the production of the art object. By making the material object of the art persistent, on rock for example, it was possible for several people to have access to some of the same information at different times, but there was still a limit on the numbers who could see it at any one time, it was tied to a place, and the relationship between the visual sign and the story was not fixed. Some work on archeoacoustics has shown that some rock art sites—where the art persisted on the rock wall to be seen after the ritual performance—were chosen to ensure that the spoken message could reach a large audience (e.g., Mattioli et al., 2017).

Travelling has been common for one reason or another among all people, including seers who knew about the art and its message. They might travel while foraging or farming, when visiting others perhaps while arranging partners, including during ceremony, when engaged in a spiritual round or on pilgrimage, for trade or exchange, or for other reasons. Sometimes people travel over large distances and can speak of information they have learned, including that obtained through mark-making. In the absence of the marks themselves, however, there would have been little confirmation of their oral account. Those who heard about it from others needed to have a means of trusting the bearer of the message, an intrinsic part of the process of information validation (Moore, 1983). I will discuss the question of corroboration in the following sections.

5.3 Writing: "to give colour and body to thoughts"

In the system of dual expression, writing has a special place, but the history is complex (Morin et al., 2020). In one view "All full writing systems are in fact representations of language rather than representation of ideas" (Olson, 2008) yet these two types of representation are and were clearly related to each other. Writing allows the sharing of the deeper cognitive meanings of the literal surface of an image through the shared coding of the words themselves, but it fixes those meanings. This fixing makes it possible to know the meanings without knowing the visual, literal surface. We often ignore the fact that writing is, itself, visual such that its literal surface can be seen by anyone who has the opportunity to read the words. The learned ability to extract the meaning by reading the words led on to the deeper cognitive meaning. The Luxor obelisk (Fig. 5.6) that now stands in the Place de la Concorde in Paris is covered with hieroglyphs that include the appropriately French sentiment: "I give thee perfect health, I give thee life, stability and perfect happiness" (Gorringe, 1882, Ch. III). At the time the French obtained the obelisk, the Egyptian Museum described it as a "specimen of antique art."

Writing is generally said to have been invented independently in Mesopotamia about 5500 years ago (Nair, 2021), in China about 3500 years ago, and in Mesoamerica around 2900 years ago (Schmandt-Besserat & Erard, 2009). As Denise Schmandt-Besserat and Michael Erard state it: "Alphabets (such as the Roman alphabet), syllabaries (such as Cherokee or Ethiopic, which represent consonant + vowel syllables), abjads (such as Arabic, which represent only consonants), and logosyllabaries (such as Chinese, which represent words) are all types of writing systems." Those authors go on to argue that in Mesopotamia writing emerged from an earlier token system used for economic purposes, and that art and writing were quite different forms of communication with different roles in society. A more recent synthesis traced the origins of writing to scenes on seals marked with iconic images, including signs that resemble early cuneiform, rather than to such tokens (Kelley et al., 2024).

Art of the sort we are considering was visual and accompanied by stories. Art and writing may be different forms of communication, but art began with stories that gave it meaning, and writing not only fixes such stories, but also has a significant visual component. Writing may have begun as a way to record economic transactions and then became a means by which people could write stories and other narratives, but it inevitably remained visual. We tend to ignore the artistic qualities

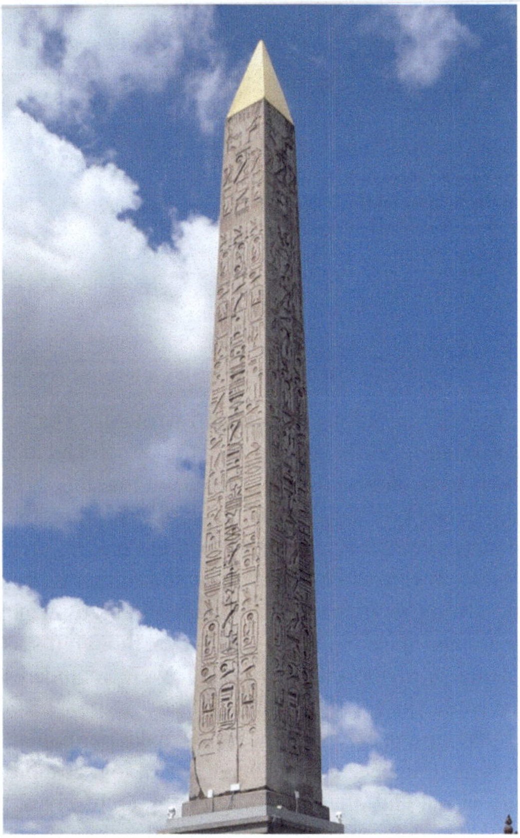

Fig. 5.6 3200 year old Egyptian Obelisk (possibly the source of the name of the cartoon character Obelix) in the Place de la Concorde, Paris, showing carefully carved hieroglyphs. (Photo Iain Davidson). An account of the removal to Paris, and translation of the text, is detailed in Gorringe (1882, Ch. III). The text includes "I give thee perfect health, I give thee life, stability and perfect happiness." That message cannot be obtained merely by looking at the obelisk

of the literal surface of writing partly because a deeper cognitive meaning was identifiable by decoding it.

By the time writing was put on public display 4000 years ago, script itself had become an object of display such as the Mesopotamian legal Code of Hammurabi (now, for some reason, in the Louvre, Paris) carved carefully on a freestanding basalt stela along with the sculpture that topped the column (Fig. 5.7 shows a detail of the writing and the sculpture at the top of the stela). As Olivier Morin et al. (2020) argue, though, the journey from shared marks to communication that could be read consistently at different times and places was not straightforward. Kelly[4] has defined writing in the following way: "Writing is an unembodied code that permits a context-independent reconstruction of linguistic information." That is a linguist's definition about writing, not about the emergence of it, and does not mention the effort required for people to learn to write and read. One of the earliest attempts to define writing that Kelly could find was that in the 1755 edition of Denis Diderot's encyclopedia or *Dictionnaire raisonné des sciences, des arts et des métiers* (de Joucourt,

[4] https://www.youtube.com/watch?v=q6xLAKAL7Ck

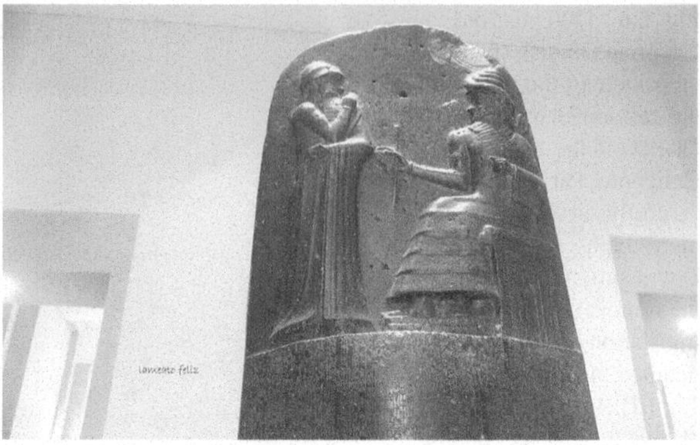

Fig. 5.7 4000 year old Code of Hammurabi carved on basalt stela found at Sippur in Mesopotamia. In addition to the sculpted figures, the photo shows the fine nature of the writing of the legal code. (Photo by lameato feliz on Flickr)

1755). There, Louis de Joucourt wrote of "That ingenius art of painting the word & speaking to the eyes, and by various strokes of drawn figures, to give colour and body to thoughts." Kelly pointed out that this definition emphasized the use of writing to embody thoughts rather than to represent sounds. I am not going to enter that debate because it would take us too far away from the chief concerns of this book. Rather I want to emphasize something that Kelly did not which is the visual aspect of writing, as in the Code of Hammurabi, by which people engaged in "painting the word and speaking to the eyes" even though it was not through pictograms. The two elements of the dual expression were included together without the need for an invisible story.

Writing uses visual means to make permanent an oral communication. The letters of alphabets and the characters of logographic systems are made up of signs that are visual and often meaningless by themselves, whatever their initial history. The process of learning to write is still arduous, and always has been, though some maintain that learning to read it is easier (McKitterick, 1992, 3). Creating and spreading it for the first time must have been extremely difficult, particularly given the painstaking efforts all of us have all gone through as we learned to read and write—efforts which have been in evidence since the first archeological examples (Robson, 2001).

Writing can fix a message because it records the words of speech in a visual form. When it is produced on a portable medium, it can be carried to places other than those where it was produced. Other visual forms of art have been made on portable objects, as at Apollo XI cave in Namibia (Rifkin et al., 2015; Vogelsang et al., 2010) (Fig. 5.8) and at Parpalló in Valencia (Davidson, 2012; Villaverde Bonilla, 1994) (Fig. 2.3), though we have little information about them being carried around and no evidence that the message was conveyed faithfully. Moreover, there is no evidence that the visual component entailed codes that had anything to do with writing, rather an oral message would have been sufficient.

Once the message is removed from the place where it was originally made, and from the context of the rituals that may have accompanied it, those who heard from others about the stories of the art needed to have a means of trusting them. In Australia early European colonists sometimes sent written messages secured in a

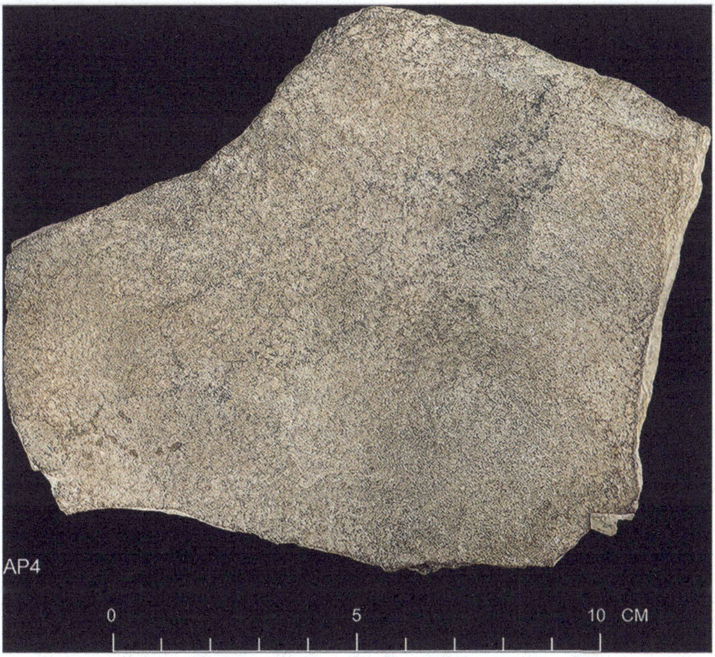

Fig. 5.8 30-thousand-year-old painted image, probably of *Rhinoceros*, on a plaquette from Apollo XI Cave, Namibia. (Rifkin et al., 2015, with permission)

"cleft stick" using Aboriginal runners—we will see the problems about verifying the message in Chapter Nine (Foster, 1998). Writing involved the invention of signs that could represent the sounds of language with sufficient consistency that some people could learn the code—could learn to write—others could learn to decode them—could learn to read—and extract a consistent message. We can only judge this through the demonstration of consistency through repetition.

There were several outcomes of writing and reading. The written words allowed the persistence of the message, and not just of the images that accompanied it, independent of the seer's knowledge of the spoken story that went with a painting. They separated the story from the context in which it was told, and at the same time, told a story that could be read anywhere by anyone. The medium on which the message was written could be moved to another location, yet still be read without the place or the producer. As Jack Goody (1998, 77) said, "[writing] provides a fixed, society-wide code enabling us to establish more or less perfect linguistic communication over distance in time and space…" The reading of that message did not necessarily require the ritual context in which it was previously transferred from producer to seer, and any reader could read it many times, and on occasions separated in time. People could read writing in different times and places, and they could also read the message when it was separated in place. But because of the lack of ritual context, and because of the lack of the independent visual message, there could be opportunities for "cheating" and some form of validation of the message was needed.

Mesopotamian codes and cuneiform developed a system of validation of written message, for example, by wrapping a message inside an envelope closed with the impression of a cylinder seal, or with marks confirming the contents of the envelope (Gándara Vázquez, 2007, 256–57; Wright & Johnson, 1975, 272). This validation technique enabled receivers of messages to have some trust in them without needing the context of the original production of the dual expression when the message (the writing) was read without its producer or the context.

For our purposes here, when people wrote words on a persisting medium and separated the message from a particular place, they emphasized the message itself and the words it was written with. They brought into existence the elements and structures of language in the form of material things. In many cultures, such writing was committed to the medium with the same beauty and style that was otherwise accorded to other forms of art (Fig. 5.9) and in some circumstances could be included in religious sites. When people built the *Mezquita de Cordoba* around 961 (1062 years ago) and surrounded the *mihrab* with Kufic[5] Qur'anic texts they were creating deliberate echoes of the Mosque in Damascus (Dodds et al., 2008, 24), and they further confirmed that by establishing Cordoba's huge library. The dual

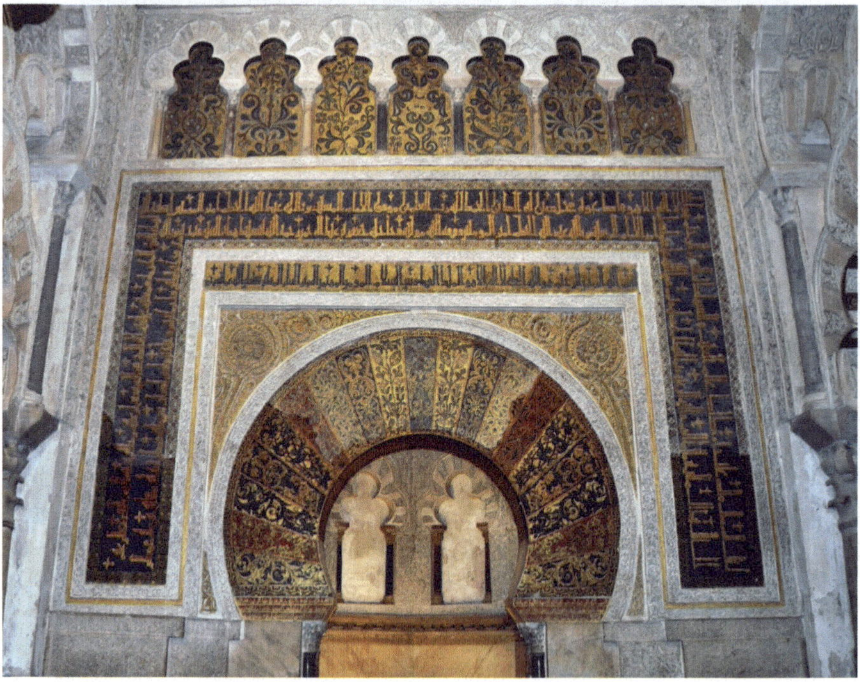

Fig. 5.9 Qur'anic texts in 1000 year old Kufic Arabic writing in linear panels around the Mihrab in the Mezquita de Cordoba, Spain. (Photo Iain Davidson 22 Oct 2013)

[5] I am grateful to Rebekah Roberts for discussion of this text

expression came from the existence of the *mihrab* indicating (approximately) the direction of Mecca, the associated story of the reason for pointing to Mecca, together with the Qur'anic texts confirming the story, for anyone who could read them. The art and its context carried with it memory of other circumstances. In general, ritual sites inside caves did not include writing (Kelly, 2023) and, although some marks in cave art have been interpreted as writing (Bacon et al., 2023), even if the claims are to be believed (and for a cogent argument against it see Nowell et al., 2024), the marks were only used in the context of special access. The lack of use of the code in any other context would have meant that it could not be learned without special visits and would serve no purpose outside the cave where it was produced.

The emergence of writing changed the relationships within the art-system. The words themselves became objects of thought and separated the words from the things they referred to; the writing itself could be composed of art objects entailing the same four-part set of relationships as in other forms of art, with the addition that the things themselves could be interpreted by all who had learned to read them. To turn around a well-known Christian text (see Fig. 5.15), this was not so much "In the beginning was the word" as "When people finally began writing words, they became aware that the word itself was important." Words themselves had probably existed as parts of formulaic expressions before they were analyzed into separate components (Wray, 2002).

The recognition that written words were composed of signs for the sounds that made them up enabled people to write down the ideas that speech conveyed, given that there was a code to represent them and a medium to record the word on. Ingold (2000, 399–400) reminded us that the invention of writing has actually distorted the nature of language because it became possible to analyze its composition into words and grammatical structures. Olson (1994, 68) suggested that "Awareness of linguistic structure is a product of a writing system not a precondition for its development." Just as Philip Barnard and others previously argued that in human evolution, combinations of parts to create composite artefacts required the prior conceptualization of "parts" (Barnard et al., 2016, 61; Davidson, 2019, 87–91), so, through writing, "**words** (ID emphasis) are now distinguished from both things and from names of things; words as linguistic entities come into consciousness" (Olson, 1994, 75).

Once all of these things were in place, there was a material product which had a physical form, just like any other "work of art" but, unlike images on rock or cave walls, these works could be transferred across space and through time. The written words can be read by those trained in the script; they can be deciphered by others not so-trained (Chadwick, 1961); and the medium to which the message was entrusted could be carried to a new location, as it were "in a cleft stick"—a practice described among Aboriginal messengers, where it could be read in a completely different context but given some authority by the "cleft stick" (Foster, 1998). Somewhat more than a cleft stick was required to relocate the Egyptian obelisks now in Paris, London and New York.

For the transmission of reliable information over large distances, writing supplied the technology of confirmation. The stories and even the form of marks that accompanied them could be committed to a portable medium. We are familiar with

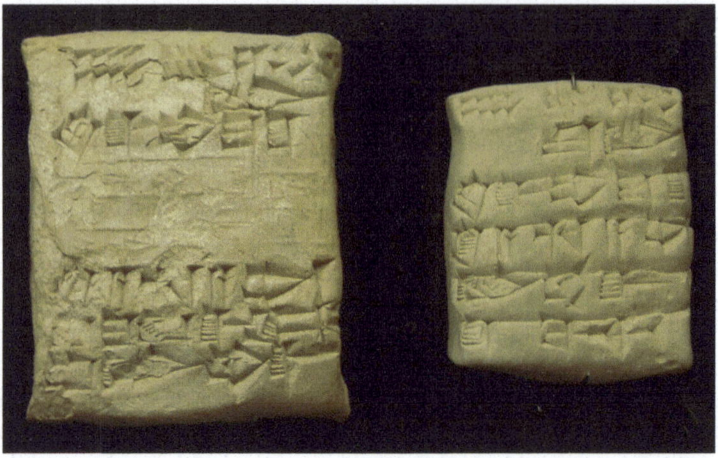

Fig. 5.10 Clay tablets impressed with cuneiform. Held in the Museum of Antiquities, University of New England. The left hand tablet, dated 4450 years ago, contains 9 lines of text bearing the seal of the scribe Ur-gar and a presentation scene dated to the ninth year of Shu Sin. The right hand tablet contains 7 lines of text recording payment of 48 gur or barley for work involving earth and bitumen by Lugal-Anns-tum (information from the Museum of Antiquities catalogue). (Photo David Elkins)

paper, but that was a late invention between 1800 and 2100 years ago (Tsuen-Hsuin, 1985, 2). In Mesopotamia, the new medium was clay marked by a stylus with a wedge-shaped end (Fig. 5.10). Provided there was a common understanding of the way marks were put down in the medium, it could be "read' in the recipient community wherever a medium was marked in that way and confirmed by the seals and envelopes we have discussed. The marks of accounting could be used for many purposes and ultimately carry a narrative. Of course, that sounds simple to us—you, for example, are reading something that has been printed (or has standardized form as a digital file which is akin to print) but once would have begun life as writing—but the challenges were formidable. Everyone has to learn both to write and to read—no one does either naturally. Telling the story of those challenges would occupy a book by itself. But the implications were unexpectedly wider.

The story of the emergence of writing (Kelly et al., 2021) and its impacts on societies previously dominated by oral histories and traditions (Goody, 1977; Olson, 1986; Ong, 1982; Vansina, 1985) are generally well-documented—even if they are controversial (Cole & Cole, 2006; Halverson, 1992). The controversy turns, in part, on the distinction between the meaning of the utterance and the definition of a word within such an utterance (Olson, 2008), and, in turn, on the authors' perceptions of how cognition changes. One person's cognition changing because of their ability to write and read does not mean that all members of the society have a different cognitive ability. But because, as the example from Dussert showed, some of the knowledge gets shared between those who know and those who don't, the conditions exist for broader change. This is not a process where a cognitive feature became general once a substantial number of individuals had it, but a much more subtle process. It is not the case that once one person's cognition is changed (or that of 100 people) then everyone's cognition is. This more than anything demonstrates that cognition is not just subject to a biological switch turning on in the brain but is part of the process of enculturation.

Jan Vansina argued that oral history is not necessarily accurate in all details of the history, partly because its content was negotiated. The process of negotiation was shown in the documentary movie about a burial ceremony in Arnhem Land, *Waiting for Harry* (Clunies Ross, 1989; McKenzie, 1980).[6] Yet all history begins with such orality (Atkinson, 2002, Introduction). Goody (1998) pointed out that there are significant differences between the oral cultures that do not know writing and oral traditions that do. In his view, Parry's conclusions arose because both Homeric poems, at the time they were written down, were products of literate cultures (inevitable for oral texts that come down to us in the written form that Parry had to analyze). Goody's (1977, 26–29) experience with oral transmission in the absence of literacy was that oral recitations were composed without slavish adherence to an earlier form of words. Writing fixes the narrative whenever the message gets written down to the extent that "one cannot imagine a novel or a symphony in a society without writing" (Goody, 1977, 26). This resembles the argument that William Shakespeare's plays were continually worked and re-worked to suit patrons and audiences until finally they were collected together in the printed version (Shiner, 2001, 48–50).

So what is the link between the art and writing? Peter Daniels' (1996, 3) definition of writing is: "a system of more or less permanent marks used to represent an utterance in such a way that it can be recovered more or less exactly without the intervention of the utterer." The best-known link between art and writing would seem to be through what are called pictographs—images of ideas. Daniels excludes these from his definition of writing on the grounds that languages contain many elements which cannot be represented pictorially through pictograms. The best-known battleground for this opposition is the 2400 year old Rosetta Stone (Fig. 5.11), the tri-lingual text removed by French troops, and then taken from them and, subsequently from Egypt, by British troops. The Stone, which is currently in the British Museum, contains three identical texts: the one in Greek could be read, but the other two—Egyptian hieroglyphics and Egyptian demotic—could not. It is said that the Englishman Thomas Young recognized that some of its hieroglyphs represented the sounds of language, and that the Frenchman Jean-François Champollion expanded the insight importantly, leading to the decipherment of Egyptian hieroglyphics and demotic script (Griffith, 1951). There seems to be more than a little chauvinism and conflict surrounding groups with different interests and different power in this story, as in so many others.

---

[6] https://www.roninfilms.com.au/feature/10159/waiting-for-harry-from-aiatsis-collection.html (downloaded 27 November 2024)

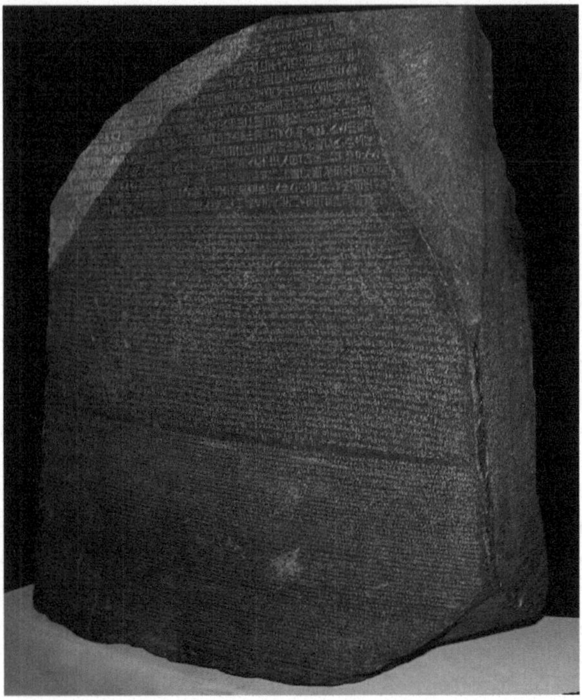

Fig. 5.11 The Rosetta Stone, of 2400 years ago, removed by British troops from Egypt. (This file is licensed under the Creative Commons Attribution-Share Alike 4.0 International license. Attribution: © Hans Hillewaert. From Wikimedia Commons)

Some scripts took marks which had resemblance to something in the form of iconic signs and transformed them into forms of writing in which the individual signs had their own relation to sounds, and their own rules about combination. Some time after the use of such coded signs as writing, they became objects of public display. Thus, the Rosetta Stone is undoubtedly an imposing material artefact. Its original creation and display can only have demonstrated that visual signs pervaded society in Egypt, and the stone itself would have called to mind the numerous public and concealed carvings of hieroglyphs in many places in Egypt.

Just as visual imagery had some resemblance to the thing it represented, so too it had (and has) a relationship with the knowledge of such things in the cultural contexts of the society that produced it. Writing was initially composed of iconic signs and their stylization led to the emergence of the letters and characters of writing (see for example Overmann, 2022). Writing is obviously visual but as it comes to represent words, the visual has a quality of being more than the marks on paper (or other medium) and comes to make up signs of the meaning of the words. In this, it resembles the cognitive importance of the art-system in general—the image stands for much more than its surface appearance. As Overmann (2022, 11) suggests, "later forms [of writing] no longer resembled or depicted; reading and writing them involved and depended upon a set of neurological and behavioral reorganizations acquired through training and practice." (We might question whether neurological reorganization was essential.) The visual impact of writing, then, became suitable for public display through massive objects such as the Rosetta Stone, and through public display of hieroglyphic texts in Egypt.

Writing achieved several things but evolved with its own peculiar problems. First, it fixed the message. I remember one of my students studying the work of an ethnographer from more than 50 years earlier who recorded geographic information about people he was studying. The student was also talking to modern people who descended from those the ethnographer had first described, and he was convinced that problems arose when anyone wrote down oral history. Sometimes people from the ethnographer's culture wrote it down incorrectly, but whether they did or not, the writing fixed the message. At one level it does not matter whether it really was recorded incorrectly—the oral knowledge of the people could have changed, or perhaps it did not—but Western society has tended to give more credence to the written word than to the oral. That, after all, is the basis of much of the nineteenth to twenty-first century fights for land rights in countries where Indigenous people had been colonized and much of their culture—people and social facts—destroyed. Colonizers on the ground, in their determination to acquire land, did not accept oral evidence of rights (Burke et al., 2021; Davidson et al., 2020)—as grazier Bill's great grandfather probably argued about the Aboriginal people he was in conflict with near Bowen—but First People and colonial authorities recognized that rights clearly had existed then, and continue to exist now (Reynolds & Marks, 1987). Attitudes of the literate to writing, therefore, might be said to alter views about the rights of people to their own land. The literate thought that codification of rights in writing was fundamental; those whose systems were not based on writing thought no such thing. What that opposition really did was to alter the nature of conflicts over rights to that land—it changed the balance of power in favor of the literate—at least until the recognition of Native Title in Australia, which is recognized as a common law right (French & Lane, 2002).

In addition, and as a part of that process of altering the relation between how people see themselves and how others might see them, writing permitted the stabilization—the fixing—of all sorts of stories, including rights, but also those that were intrinsic to the process of art production. Such written stories ended up as legal documents such as the Code of Hammurabi about 4000 years ago (Fig. 5.7); as great works of literature, such as *The Epic of Gilgamesh* about 4300 years ago, one of the earliest cuneiform documents, written on clay tablets, initially found at Nineveh (George, 1999); or later oral poems that were written down about 3000 years ago, *The Iliad* and *The Odyssey*.

Whatever the medium, there was both an immediate material product, and a product that could be accumulated in libraries or other stores. Some texts were inscribed with exquisite care onto rock walls, as in the Behistun Inscription (Fig. 5.12 see detail in Fig. 8.25) in Iran carved about 2700 years ago. The public display of the historical accounts of Darius's succession to the Persian throne were accompanied by images of his power, exemplifying the characteristics of other rock art—the visual images accompanied by a story, the classic dual expression. The text was written in three languages (rather like the later Rosetta Stone) and contributed to the decipherment of cuneiform script, and unlike the stories that accompany rock

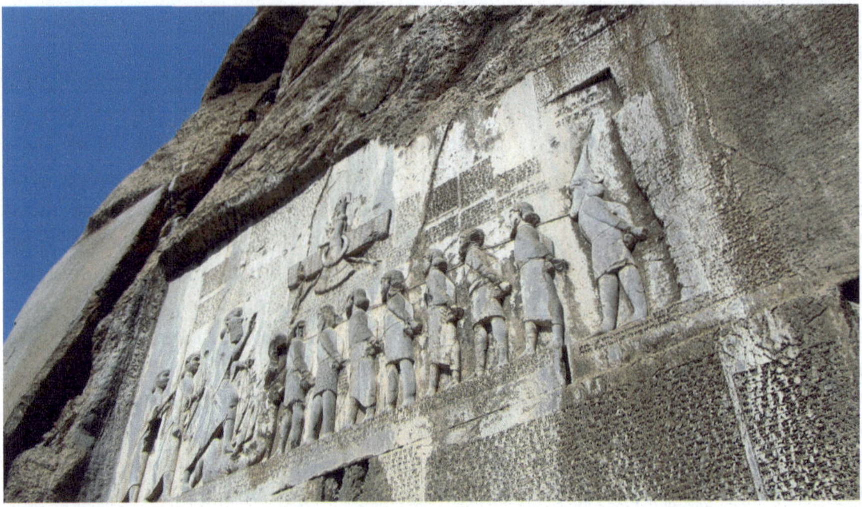

Fig. 5.12 Behistun Inscription carved around 2700 years ago in Western Iran. (By PersianDutchNetwork—Own work, CC BY-SA 4.0, https://commons.wikimedia.org/w/index.php?curid=50210803 (Downloaded 2 April 2023))

art in non-writing societies, the story was written, displayed, and could be read by some.

And power is a very important part of the story. It is said that Ashurbanipal acquired power in Nineveh, in Mesopotamia, by the accumulation into his library of texts written on clay tablets (Puchner, 2017, 28–45). Such writing could be used to produce texts that were sent from one person to another sometimes of trivial importance, other times with information that allowed the receiver to control the knowledge and acquire power.

Writing needed to be learned both by the writer and by a potential reader. Of course, in early art, there was also a requirement for people to be taught about the art, probably in ritual and probably with stories, but the stories could not be transferred directly. You had to be there. Once writing existed, the message of the story could be written down and understood by another person who had learned to read it. And it did not matter when this reading happened—it was part of what has been called "asynchronous communication" (Morin et al., 2020).

It turns out that almost all societies that have a form of writing also have a distinctive type of formal education both because that is necessary for students to learn to read and because written texts provide a material means of transmission of knowledge. Writing provided an essential part of the way society reproduces itself; interactions between the artist (writer) and a subject produced an object (a text) that a student can learn from. The visible form of writing played a fundamental role in the presentation of writing both in its public display and in its private teaching.

Through all of this, it emerges that writing is itself visual and combines within any written text both elements of the dual expression that is fundamental to art: the literal surface visuality and the deeper cognitive meanings. Yet it separated the parts of the dual expression—the literal surface that could be seen in cave art did not need to accompany the storytelling that led to deeper cognitive meanings. The power of the place where the art was performed was reduced. Yet there was still the opportunity to use the art—the writing—to maintain power through restriction of access to the objects that were written on, as Ashurbanipal showed by accumulating cuneiform tablets in a library at Nineveh. Likewise, the public display of images and text tied the images to particular people, such as Darius at Behistun.

At the same time, writing came to emphasize the text of the message, together with the medium on which it was inscribed. In cases such as the tablets of Mesopotamia, and later papyrus, parchment and paper, the message could be transported, liberating it from necessary connection to one place, but allowing many people access to the message at any time and in many places. Moreover, writing enabled texts to be reproduced, albeit not rapidly, such that a message could be received at the same time in many different places—all those worthy clerics in scriptoria, and, of course, the secretaries (Moss, 2023). Those places could be widely spread showing the capacity of writing to disseminate a message over great distances. In this way, writing made possible changes in communication beyond the local transmission of information.

5.4 Printing

There is a story, both well documented (Eisenstein, 1983) and accessible (Man, 2010; Puchner, 2017, Ch. 7; Winchester, 2023, Ch. 3), that Gutenberg introduced printing using re-usable type into Europe. He needed great skills and experience to adapt a press from other purposes to make one suitable for printing, to make type, and ink; he was also lucky in the fact of the diffusion of papermaking from its origins in China and passage through the Islamic world of Spain. The success of Gutenberg's printing required these and the transfer of the technology of the press to the printing of papers, pamphlets, products of priests, or the financing of the Church's fight against Islam. What I need to do in this chapter is not to spell out all of these details (many details are in Eisenstein, 1983). Rather, my story is that Gutenberg's achievements created a technology for printing, which would eventually be used for apparently conflicting purposes—such as the printing of the papers for the corrupt practice of indulgences (Fig. 4.1), on one hand, and of Martin Luther's *Theses* denouncing indulgences (Fig. 5.13), on the other. That liberalization had artistic impact through the destruction of some religious institutions and other collectivities of scribes thus changing the working conditions of artists.

One outcome was the proliferation of printed works that became visible all over rather than confined to churches and Universities, creating new expectations of a world of people who could read. The relatively newly invented reading spectacles were of little use if there was nothing to read (Fig. 5.14). But where books were once confined to religious or scholarly buildings, now the world was awash with

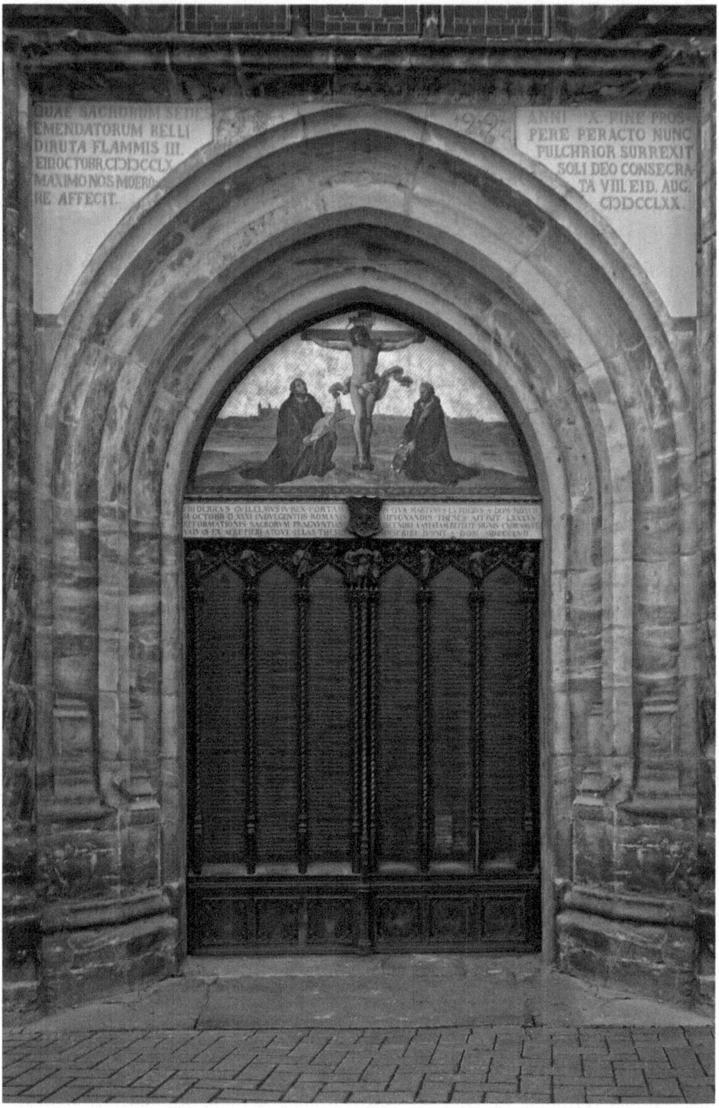

Fig. 5.13 Doors of All Saints Church in the Castle of Wittenberg which has been cast in bronze with the 95 Theses against the use of indulgences Luther is said to have nailed to the original door. The spread of his objections to indulgences is due to the printed version of the theses that was circulated in the early sixteenth century (Photo by Chris Roth from Flickr permission on FlickrMail 20 Feb 2025)

books and pamphlets, many of them identical and the visual expectations of the world were different.

The other important outcome of Gutenberg's introduction of printing and his activities around it was the impact on the economy. Gutenberg's own activities were very much directed, as a matter of urgency, at the commercial outcomes, because he

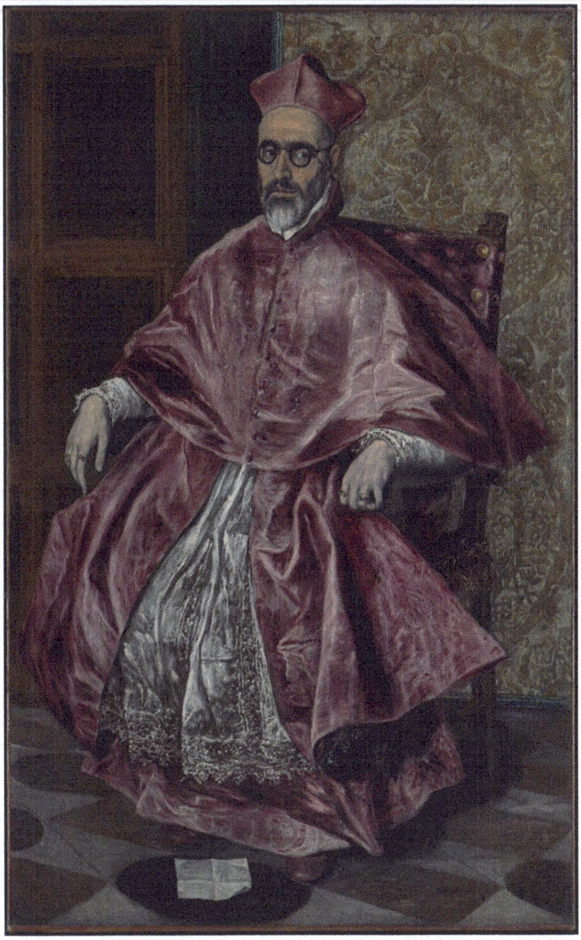

Fig. 5.14 Portrait of Cardinal Fernando Niño de Guevara (1541–1609), wearing spectacles. El Greco (Domenikos Theotokopoulos) (Greek, Iráklion (Candia) 1541–1614 Toledo). Metropolitan Museum, New York. (Public Domain. H. O. Havemeyer Collection, Bequest of Mrs H. O. Havemeyer, 1929)

could not have afforded to develop the system otherwise. These measures included the printing of Catholic indulgences, and the choice of the Vulgate Bible as a first big project. Gutenberg was keenly aware of the importance of the visual appearance of the products of his presses (Fig. 5.15), and ensured quality control of that first book by making it look better than text written by monks in Abbey scriptoria.

The economics of printing spread further, as Jeremiah Dittmar (2011, 1145–1146) has shown that cities that adopted printing relatively early in the fifteenth century grew more rapidly in the two hundred years following than those that did not. That was impacted in some instances by the deaths in religious conflict between Protestants (who came into existence partly as a result of printed documents) and Catholics (who stood to suffer from the impact of printing because they had no control over the use of the press to print work that opposed indulgences, for example).

The stark contrast needed to understand the impact of European printing in the last 500 years is shown by a simple statistic from a much shorter period. It is estimated that the Library of Cordoba, around 1000 years ago held in excess of 400,000 volumes (Wasserstein, 2017), the same size as one estimate for the library of Alexandria which was destroyed, perhaps, 800 years earlier (Phillips, 2010). Of course, none of the books in either Cordoba or Alexandria was printed (Dodds et al., 2008, 24) and at the time of the Cordoba Library, no other library in Europe held more than 1000 manuscript books. In the twelfth century, Reading Abbey Library listed 196 volumes, of which there were 57 Romanesque books; between the late 1130s and late 1140s there were about 24 books; and between 1150s to the 1170s about 15 (Gullick, 2016). In the 50 years after Gutenberg's introduction of the printing press using moveable type, and its ancillary inventions, in excess of nine million volumes were printed. The numbers for Reading Abbey, one of the major libraries of medieval Europe, may be inaccurate in the details, but the differences in orders of magnitude would survive any correction.

Gutenberg's printing press had in common with the alphabet and language itself that the same small number of elements could be combined to produce radically different messages. The wide spread of the printing of Luther's theses was very much responsible for the fact that the spread of Protestantism and the Reformation was strongly associated with printing (Dittmar, 2011, 1161–1162). The spread, of course, was not of the abstract concept, so much as the material product that carried it.

Popular though they may have been there could have been no validation or corroboration of the effectiveness of indulgences, though paying money to rid yourself of something you have never had could be validation enough. The validation of the bible was encased in ritual and the power of the Church, so Gutenberg chose the Latin version from 200 years earlier in the thirteenth century, the Vulgate, avoiding the perceptions of parochialism that could have arisen from use of a German text. Then he ensured that the printed version looked somewhat like that produced in monastic scriptoria, but enhanced it by right justifying the text, and yet still leaving room for the visual illumination of initial letters that people were familiar with (Fig. 5.15).

Finally, validation was provided by investment in book production particularly through the loans Gutenberg got from his wealthy sponsors. The reader should assume that, although there are exceptions (as argued by Johns, 2009), a publisher would not invest in a book in which falsified or distorted words can be attributed to an author incorrectly and without their approval.

The most famous example of the need for validation is Albrecht Dürer's woodcut print of a Rhinoceros from 1515 (Rookmaaker, 1978) (Fig. 5.16). It was a validation that did not happen in Dürer's lifetime. Dürer cut the woodblock after reading a description, originally given orally then written down, of an Indian rhinoceros (Hoare, 2022, 31–37). He was able to make many copies because of the relatively recently invented printing press, and he distributed the prints, some as gifts, some for sale. Several of the images that found their way into subsequent publications are subtly different from the original, but the print fixed the animal—the species—in

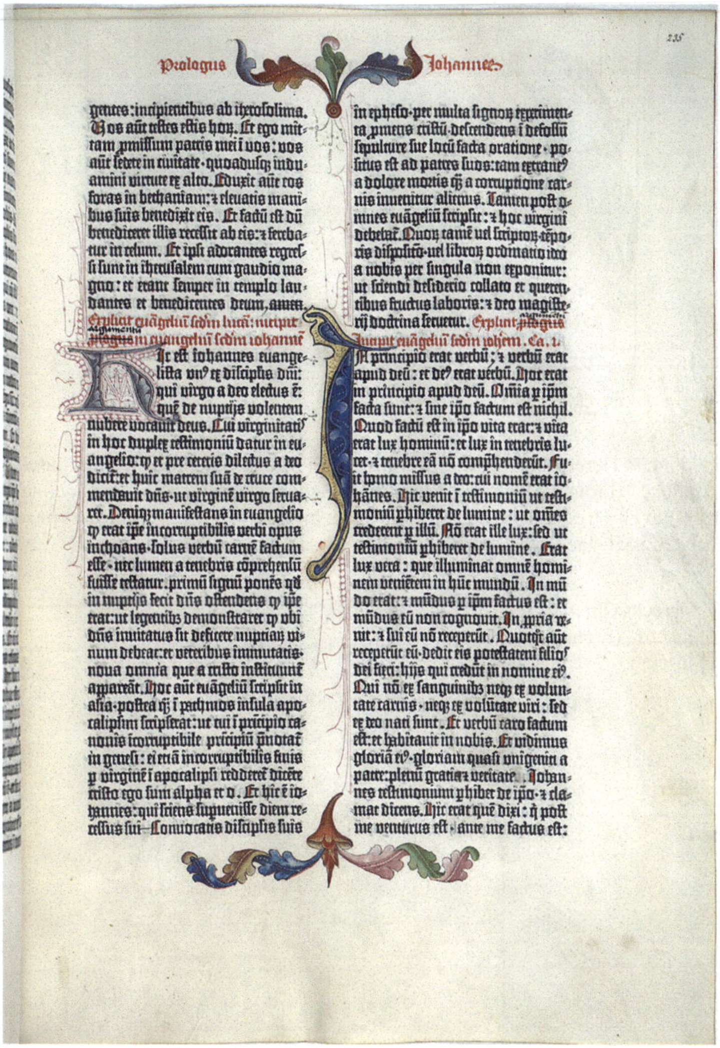

Fig. 5.15 First page of John's gospel from the Göttingen Gutenberg Bible. The start is in the middle of the right hand column with the words "In principio erat verbi." Goettingen State and University Library, SUB Göttingen, shelfmark: 2 Bibl I, 5955:2 INC RARA, fol. 235r

the public imagination, even though neither Dürer nor the subsequent viewers had seen a rhinoceros, and were not likely to do so (Leitch, 2017). We can only speculate that Dürer's 1502 painting of a *Hare* (Fig. 5.17) could have influenced some people, who had seen that animal. Just as he depicted the hare with apparently accurate detail, so his picture of the rhinoceros could be believed. You, the reader, and I know about the hare because of the wide diffusion of the image in print, but how many people knew about it in 1515?

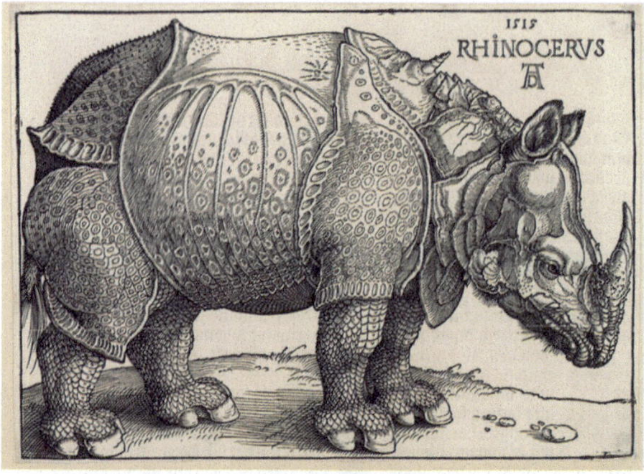

Fig. 5.16 Albrecht Dürer Rhinocerus 1515. Print from woodblock which was cut based on a verbal description. (From Albertina Museum, Vienna. https://sammlungenonline.albertina.at/?query=search=/record/objectnumbersearch=%5b3073%5d&showtype=record#/query/93bcbc2f-50d1-4ae2-a01a-e162fa9c3425)

Fig. 5.17 Albrecht Dürer Painting of a *Hare*. (From Albertina Museum, Vienna)

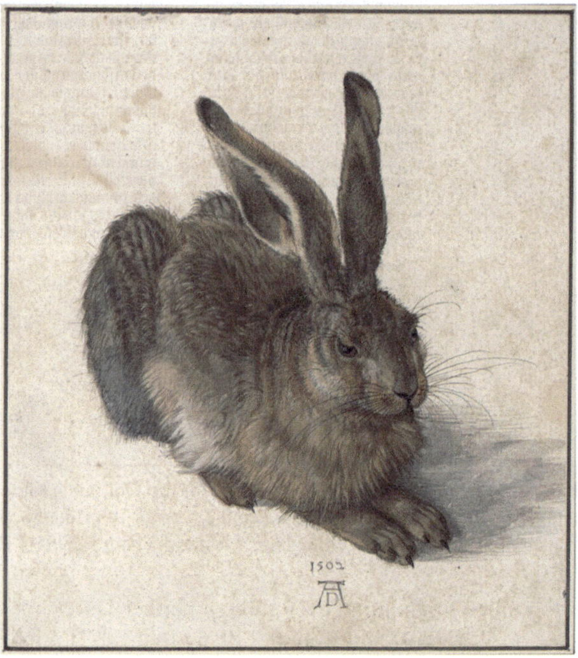

We can now see such works and, for example, *Las Meninas* (Fig. 1.10) whether by Velázquez or by Picasso, reproduced in photographs and printed in books, but before color film, color photographs, and color printing, relatively few people could have known the details of the pictures without travelling to see them, their colors, context, and community.

In the nineteenth century, copies of paintings were accessioned by the English National Gallery (McEvansoneya, 2008) and elsewhere (Macartney, 2010) in the form of prints, initially, then copies and most recently photographic reproductions. In 1972, it was a source of amazement to me as a student to see so many people copying famous paintings in the Museo del Prado of Franco's Madrid—but of course color prints of such paintings were not readily available then (so monochrome photographs mostly accompanied the elegant storytelling about Spain in Morris, 1964, 32–34). The first editions of Gardner's (1926) *Art through the Ages* were copiously illustrated, but almost entirely in monochrome photographs. Before the advent of the photograph, access to knowledge about art was even more generally restricted, with a heavy class bias to allow for those who could afford to travel, perhaps on the Grand Tour (Chard, 1997; Prown, 1997). One of the functions of the Grand Tour, particularly as it led to the establishment of the discipline of archeology, was to obtain specimens of old art from famous places in order to carry the loot back to the growing private cabinets of curiosity of rich people and museums of London, Paris, Berlin (Orrells, 2011) and New York (Sorabella, 2003). Before the color photograph it was really difficult to see the colors of, say, the Sistine Chapel without having visited it. The confinement of the message that accompanied rock art, and its persistence on the rocks, would have restricted the distance over which a reliable oral message could be transmitted. In that limitation it was not completely unlike any other type of art before the advent of writing, printing and especially color printing. You had to be there.

Printing changed the game, despite the initial lack of color. It made it possible for two people (or more) to be reading the same book in different places at the same time and others to be reading the same book at different times with an unchanged story. No longer were written words and the stories they relayed confined to the work of monks beavering away at bibles in scriptoria, nor of copyists around universities providing texts, often other religious texts, for students. Printing provided, instead, a democratization and a visible presence throughout the world of printing presses. Above all, printed works released the word from the power of the church and cognoscenti and expanded the world of knowledge and its visual presence. Eventually that process of democratization led to newspapers and the common spread of knowledge, at least until unscrupulous proprietors destroyed their credibility.

5.5 Communication at a Distance

When I first conceived of the set of revolutions in communication that we are dealing with in this chapter, I was thinking of the importance of the telegraph in connecting the World with Australia and other places (e.g., Thomson, 1999). In earlier revolutions, art combined with ritual to fix both story and ritual in a particular place and generate the reflections we now take for granted. Although the situation was not exactly person-to-person communication, a small number of people probably told the important stories to small numbers of other people who became seers through the interaction. The outcome was that stories and images became fundamental elements of the culture of the whole society even if the full set of knowledge was only possessed by an in-group of insiders.

Unlike the knowledge that is or was an intrinsic association of rock art, written texts involved knowledge that one person would write, and send a set of words to one or more others who had learned to read or get someone to read the words to them. In fact, the "story" included in the written text could be seen and read by several others, so that this was getting close to one-to-many communication, and with it, characteristic asynchronous communication which released the storytelling from some of its tethering to time. The power of the insiders shifted to include those who controlled access to writing and reading, such as teachers in schools or their equivalents. Candida Moss (2023) has documented how early Christian literature depended for its diffusion on secretaries, who were often enslaved people. In addition, writing allowed some to accumulate texts in libraries—of which Ashurbanipal's library at Nineveh is the best known (Kilgour, 1998, 20–21). It may have had as many as 30 thousand tablets, including the *Epic of Gilgamesh*. The insiders then included librarians, controllers of archives and collections, often in religious contexts. Printing enabled those in control of that means of production to produce multiple copies easily which could then be dispersed. That released the accurate and ritualized message from its link to a place or a small number of places, and from centralized insiders in religious and educational organizations. In addition, it enabled communication that was not tied to any particular time. The accumulation of such printed works in libraries in different places released the communication from the place of origin, but it also released the message from time—it could be referred to many times without reference to the original acts of communication.

The next release from constraint was the communication of a message (fairly) rapidly from one place to another. Geoffrey Blainey, famously, referred to the history of Australia as a story of overcoming the "tyranny of distance" (Blainey, 1966). That distance could be covered, by sailing ship, steam ship, or ultimately by aircraft of one sort or another. The question was the time it took and that was what the telegraph was for: for some the electric telegraph meant the connection of the places of commerce of the British Empire (Shepherd, 1996). Graeme Davison (2016), in discussing Blainey's phrase, highlighted the reduction of the impact of distance through the introduction of direct flights into Australia: a journey that once took him more than a month by ship could be accomplished by plane in a day and a half. The distance was the same, but the time had reduced significantly—it was a tyranny of time not distance. Likewise, a written or printed message could have been carried on the ship or by a messenger on foot or horseback, but the introduction of the telegraph allowed brief oral messages to be sent more quickly. This was the motivation for the earlier construction of the Chappe telegraph (actually, semaphore) stations across Western Europe so that Napoleon Bonaparte could send instructions to troops in France and Spain.

Semaphore is a word that applies not only to modern (recent) railway signals, where the code must be agreed beforehand or the trains will crash, as they did in

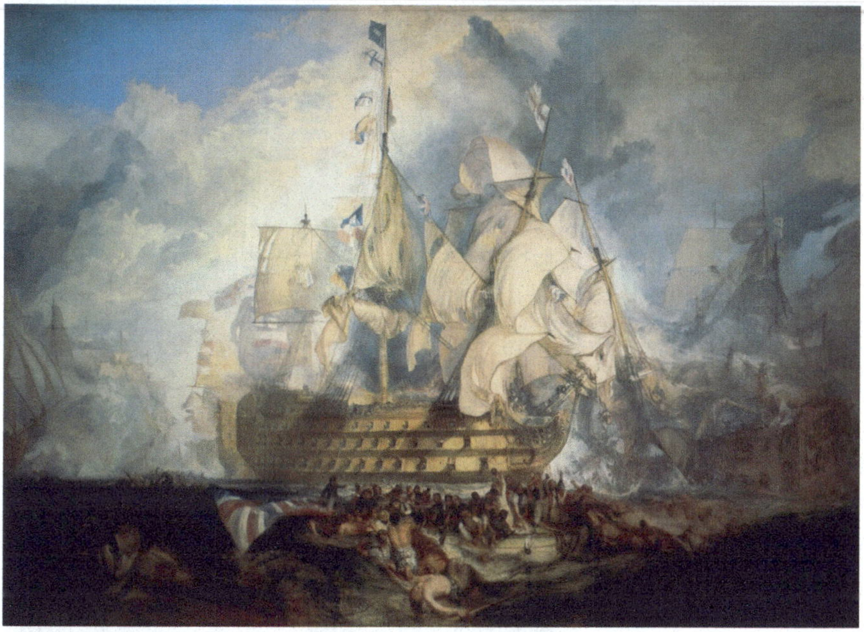

Fig. 5.18 The Battle of Trafalgar, 21 October 1805. Oil on canvas (1822–1824), 2615 × 3685 mm. National Maritime Museum, Greenwich, London by J.M.W. Turner. (Public domain from https://commons.wikimedia.org/wiki/File:Turner,_The_Battle_of_Trafalgar_(1822).jpg). The painting shows the flags for the last three letters of the famous semaphore signal "England expects every man to do his duty"

India as I was writing this: the word remains in some languages (Spanish and Portuguese, for example) as a word for traffic lights. It also applies to more historic use of flags by ships (Fig. 5.18) and others to send messages at a distance without anyone needing to travel over that distance. Semaphore has in common with smoke signals and fire beacons, that there must be a prior existence of a shared oral convention of the meaning of particular patterns.

There were problems with the semaphore system because it worked by line-of-sight signals, so it could not be used at night or in bad weather, and a large number of signal stations needed to be set up that were close together. In fact, for our purposes here, the semaphore is prior and has that quality of visual expression that is important to my argument (Schiffer, 2008, 41). Curiously, as Schiffer points out, in light of the later development of electric communication, the invention was put in place first, despite the infrastructure required, because there was not a means of efficiently implementing an electric system. Other transmissions of signals could be quite complex, but, prior to the transmission of electric signals by wire they always involved a visual element (Tully, 2009) (Fig. 5.19).

Electric signals improved on line-of-sight though repeater stations were still needed (the loss of signal problems with various attempts at a telegraph based on electricity are detailed by Schiffer, 2008). In Australia the telegraph stations were

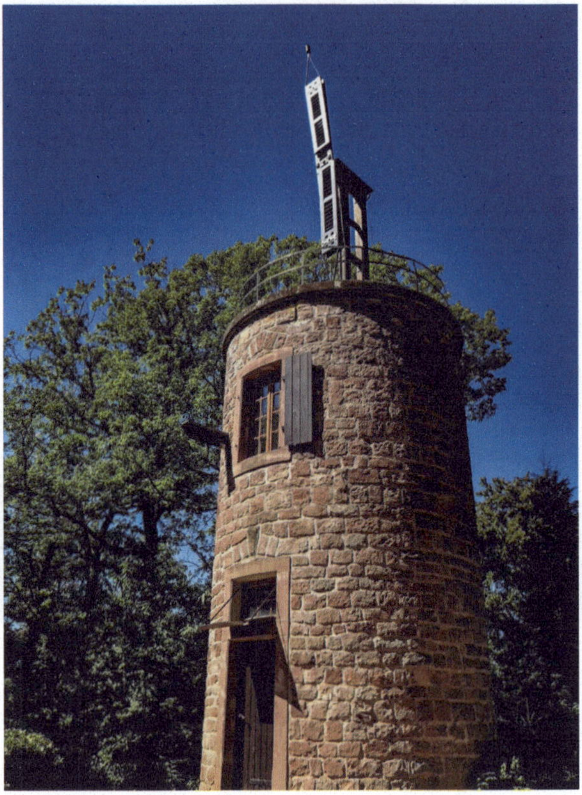

Fig. 5.19 Semaphore station in Alsace. (Photo by Jürgen Lübeck from Flickr with permission on FlickrMail 24 May 2024)

extremely isolated outposts of Europeans on Aboriginal land and were famously vulnerable to attack by Aboriginal landowners (Jones, 2023) (Fig. 5.20).

Once electric telegraphs were in use, the most famous examples were the telegrams used during major wars which announced the death of a soldier to (usually) the parents. In these cases, it was mostly the symbolism of the physical object and its deliverer which conveyed all that the parents did not want to know.

By using visual signals of this sort, people added the conquest of delay to the other triumphs represented by the use of the visual (art) to tell stories, enhancing the immediacy of stories. The fundamental initiative was still the expectation of meaning, to which was added writing and printing, which conquered space, time and size of audience. By moving from the visual communication of semaphore to electric communication, the tyranny of time was reduced, but not overcome completely, but you could say that what was needed for that was the development of digital communications derived in part from the dots and dashes of Morse code used on the telegraph.

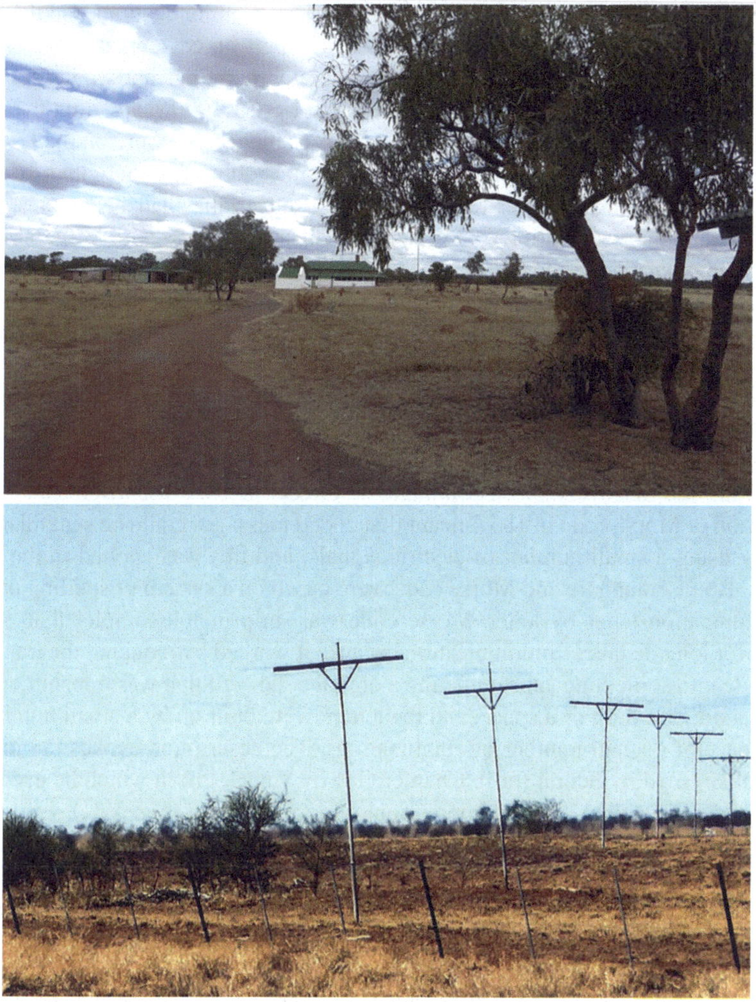

Fig. 5.20 Top: The isolation of the Telegraph Station at Tennant Creek, Northern Territory. Bottom: the visible sign of the telegraph posts across the landscape. Insulators were often used by Aboriginal people as raw material for their tools and symbolic objects. (Photos Iain Davidson)

5.6 Digital Revolution

Jean Marie Chauvet and his colleagues described the discovery of paintings in Chauvet Cave like this: "In the evening of 18 December 1994 we set foot, the first for perhaps twenty thousand years, in one of the most beautiful decorated caves in the world: Chauvet Cave" (Chauvet et al., 1995). The art is now known to be more than 28 thousand years old (Quiles et al., 2016). As remarkable as the discovery and paintings in the cave were, the thing I am interested in here is what happened next.

Only 11 days later on 29 December, through various communications and the complications of other timetables, the pre-eminent expert on cave art in France, Jean Clottes, was contacted and able to travel to see the paintings and vouch for their authenticity and quality. Only 28 days (one day for every 1000 years) after the original discovery by French cavers the photographs of the images were on the Internet, and people like me could see them on my electronic equipment, more than 16 thousand kilometers away in Australia. The same images that had been transmitted across continents were printed in color in local newspapers. Unlike the struggle against other tyrannies, the one against time and distance was reduced to 4 weeks—not quite an instantaneous end then to the tyranny of time, but 30 years later I can speak in real time in seminars about storytelling in the United States or talk about images with family members in England who can see the same as me.

The eighteenth century dream of rapid, long-distance electric communication failed and confronted the reality of the simpler mechanical semaphore system that worked (Schiffer, 2008). Eventually semaphore for major messages was replaced by exploiting the discovery of electromagnetic effects in 1820 (Schiffer, 2008). The invention of Morse code in 1837 meant that coded messages could be sent relatively simply using a small number of electric signals, and this was applied in the early 1840s. Both semaphore and Morse code were based on a system of spelling out the communication letter by letter. Morse code was, in principle simpler than semaphore for long distance communication because it worked by reducing the transmitted code into something approximating a digital code—all that was transmitted was either a dot or a dash or a space, and the letters were built up by a small number of sequences of a small number of combinations. The description evokes the idea of pointillism used by Seurat (or in Chapter One by Van Gogh) in which he used dots of color juxtaposed to give the impression of more vibrancy (Fig. 5.21).

The digital revolution applied to the telegraph overcame the length of time taken to deliver a brief message from one place to another, but the conversion of the written message to a code made up of combinations of two elements, dots or dashes in the classic rendition, foreshadowed binary code that was fundamental to the electronic revolution. It is that which has given us digital computers, and the transmission of images from the cave wall, to the digital camera, to the airwaves all over the world which could be printed in newspapers as far away as it was possible to imagine.

The final stages (for now) of the revolutions in communication led from the telegraph, which more or less introduced digital electric signals delivered from point to point, to the wired phone which did the same only it converted the digital electric signal to voice using a microphone. The process initially involved an individual sending electric signals using Morse code to one other person who decoded them, transcribed them and delivered the written message to the person receiving. The next innovation in transmission was the broadcast radio which sent signals generated by electricity out from a transmitter so that they could be converted by a suitable receiver into sounds which could be heard simultaneously by many. Once that

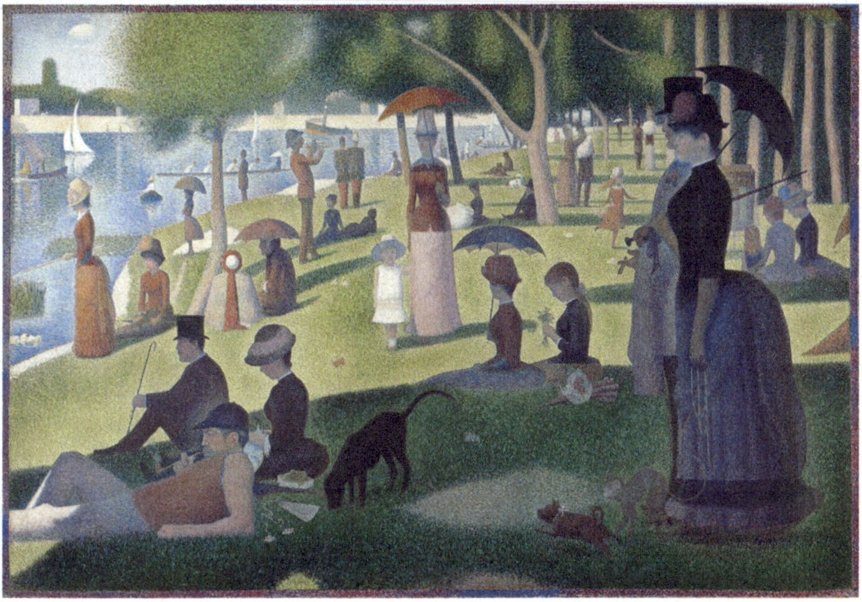

Fig. 5.21 A Sunday on La Grande Jatte, Georges Seurat, 1884. The Art Institute, Chicago. (Public domain via Wikimedia from https://www.artic.edu/artworks/27992/a-sunday-on-la-grande-jatte-1884)

was expanded to include the transmission of images through television, the stage was set for broadcast storytelling of the type exemplified by those television programs *Civilisation* by Clark, or *Ways of Seeing* by Berger. The medium expanded enormously the range of seers informed about the visual appearance of art with colors and forms—all introduced by an individual with their own stories about the images. That could be followed by means of restriction of access to such storytelling which enabled individuals or corporations to control story and content, and the financial value of them. At the present time, we can make still or moving images of ourselves or others and transmit them simultaneously to another person using a similar device, just like the Venus woman in St Mark's Square in the Italian advertisements.

5.7 Essay 5 Graphic Elements in Communication (Lloyd Rees Dredged Through Europe and Tipping's "Watermark" and Multiple Meanings)

I previously defined 6 revolutions in communication: the emergence of language; the production of visual, particularly representational or iconic, images; the beginning of writing; the invention of printing, especially with re-usable type, whether in Europe or in East Asia; the invention of various means of communication at a distance, such as the telegraph, telephone and radio; and the digital electronic revolution. After the emergence of language, whenever that was, the other communication revolutions all involved visible arts. Language tends to fade. In general, once it has been spoken it cannot be recovered. It disappears. Some people have remarkable memories of spoken communications, but, even when they commit speech to memory, it is unusual that it can be corroborated. Two people may have the same memory of a spoken event, but a third person cannot know, just from hearing the repetition of those memories that what they are hearing is what was originally said. Even if a hearer commits speech to memory, its transmission to another person lacks corroboration. But once speech is made visible through the material culture of mark-making, everything changes.

In the image on the left, I have reproduced a print of a picture by Lloyd Rees that my wife and I purchased at the New England Regional Art Museum. In Chap. 1, I cited the Australian painter Brett Whiteley, writing to his Australian inspiration, Lloyd Rees, as he lay dying (extract from letter previously published in Davidson, 2008) rejoicing that "They contained nature AND ideas, they contained naturalism but seemed also very invented".

The Rees print is best understood first as a simple landscape of Mt. Warning in northern New South Wales (seen from somewhere around Byron Bay), but then through its relationship to the history of landscape representation in European art, as exemplified by the Cezanne painting at the top of Fig. 5.22 Left. At some stage, as in most rock art, landscape was not represented at all. At another stage it was principally an imagined backdrop against which the more important human and material events were set. But Dürer and others depicted landscapes where the marks on the page showed some resemblance to what could be seen of the ground and set the human and material events against that. As it happens, this Rees print has slight similarity with one of those landscapes. No such landscape can be seen on rock art.

The image on the right is of a sculpture from 2000 by Richard Tipping called *Watermark*. It is located at the Powerhouse, an entertainment venue in Brisbane. As the producer writes on his web site:[7]

> This sculpture is meant to evoke the power of the Brisbane River sweeping around this curve at New Farm, triggering memories of the massive floods of 1893 and 1974, and anxieties about the next. The word "flood" appears in thick solid letters, styled in a *sans serif*

[7] http://www.richardtipping.com/artworks/public-art/watermark-brisbane.html accessed 19 November 2021

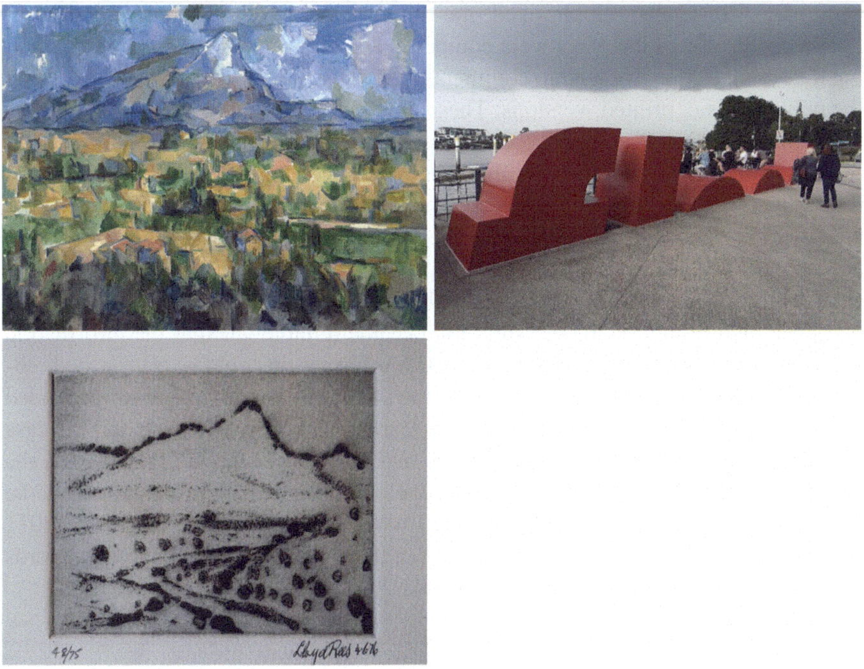

Fig. 5.22 Left: Top, Cezanne painting of Mont Sainte-Victoire 1904 (E1936-1-1 Philadelphia Museum of Art) (Public Domain); Bottom Lloyd Rees print (in possession of the author), which we might say, with Brett Whiteley, had been dredged through the art of Europe. (Photo Matt Cawood. © Alan and Jancis Rees/Copyright Agency, 2025) Right: Richard Tipping Watermark, 2000, a steel sculpture 15 m long which references the history of floods on the Brisbane River (New Farm, Brisbane). (Photo Iain Davidson)

> font, with only the top third showing, like a headline literally going under. The title *Watermark* means both a mark showing the height to which water has risen, and a design impressed into paper which is visible when held to the light, guaranteeing authenticity. Giving weight to words, and seeing letters as sculptural forms, mirrors the ways we make thought concrete. Our words become shapes which reflect the energies we speak, and the ideas and passions which they contain.

Then:

> When I was commissioned to make *Watermark* in 2000, I was told that there would never be another flood in Brisbane, thanks to the Wivenhoe Dam. Unfortunately history repeated in January 2011, and remarkably the sculpture became the literal watermark showing the final height of the flood a couple of centimeters above its ground level.

This dramatic artwork exemplifies the point. Written text makes thoughts concrete and gives them an air of permanence that the spoken word does not have. Writing is intrinsically both material and visual. Unlike speech, the written word can survive through time, can be stored in libraries, can be passed from person to person. The concept of permanence of the word led to printing and the other communications revolutions.

Shiner (2001, 50) emphasized the importance of the printed version of Shakespeare. Where the plays were composed and altered by several people including the acting troupe, we tend to comment on Shakespeare's artistic genius based on the published, i.e. printed, versions. These have settled the spoken word around the seen word and given the seen word more power.

In addition to their visual qualities, the revolutions in communication affected the number of people who have access to information in a particular place; the numbers who have access to information at the same time but in different places; or who have access to the same information at different times and places. Some of these changes in communication also affected the distance over which information could be transmitted reliably—the transmission became reliable whether the information was accurate or not—or the speed with which information could be transmitted reliably from one place to another. Some of the changes in communication depended on the ways in which the new means were learned by communicators, and the ease of access to the means of communication, and power over that access. Such is the power of the visual especially when the medium is persisting. I first saw the *Watermark* work of art in 2019. As such I thought it was a reminder of the 2011 floods, but in reality it "reflects the energy" of the earlier floods and only afterwards those of 2011, still less those of 2022 (Cook, 2023).

References

Arbib, M. A. (2003). The evolving mirror system: A neural basis for language readiness. In M. H. Christiansen & S. Kirby (Eds.), *Language evolution* (pp. 182–200). Oxford University Press.

Atkinson, A. (2002). *The commonwealth of speech*. Australian Scholarly Publishing.

Bacon, B., Khatiri, A., Palmer, J., Freeth, T., Pettitt, P., & Kentridge, R. (2023). An upper Palaeolithic proto-writing system and phenological calendar. *Cambridge Archaeological Journal*, 1–19. https://doi.org/10.1017/S0959774322000415

Barnard, P. J. (2010). From executive mechanisms underlying perception and action to the parallel processing of meaning. *Current Anthropology, 51*(s1), S39–S54. https://doi.org/10.1086/650695

Barnard, P. J. (2019). Sticks, stones, and the origin of sapience. In K. A. Overmann & F. Coolidge (Eds.), *Squeezing minds from stones* (pp. 102–127). Oxford University Press.

Barnard, P. J., Duke, D. J., Byrne, R. W., & Davidson, I. (2007). Differentiation in cognitive and emotional meanings: An evolutionary analysis. *Cognition & Emotion, 21*(6), 1155–1183. https://doi.org/10.1080/02699930701437477

Barnard, P. J., Davidson, I., & Byrne, R. W. (2016). Toward a richer theoretical scaffolding for interpreting archaeological evidence concerning cognitive evolution. In T. Wynn & F. Coolidge (Eds.), *Cognitive models in Palaeolithic archaeology* (pp. 45–67). Oxford University Press.

Batarda Fernandes, A., Reis, M., Escudero Remirez, C., & Vázquez Marcos, C. (2017). Integration of natural stone features and conservation of the Upper Palaeolithic Côa Valley and Siega Verde open-air rock-art. *Time and Mind, 10*(3), 293–319.

Berger, J. (2008). *Ways of seeing*. Penguin UK, London.

Blainey, G. (1966). *The tyranny of distance: How distance shaped Australia's history*

Blasco, R., & Peresani, M. (2016). Humankind and the avian world: Zooarchaeological evidence for inferring behavioural evolutionary signatures. *Quaternary International, 421*, 1–5. https://doi.org/10.1016/j.quaint.2016.09.048

Burke, H., Wallis, L. A., Davidson, I., Cole, N., Barker, B., & Hatte, E. (2021). The shape of absence: Community archaeology and the heritage of the Queensland Native Mounted Police, Australia. *Journal of Community Archaeology & Heritage*, 1–14. https://doi.org/10.1080/20518196.2021.1996155

Chadwick, J. (1961). *The decipherment of linear B*. Pelican Books.

Chard, C. (1997). Grand and ghostly tours: The topography of memory. *Eighteenth-Century Studies, 31*(1), 101–108.

Chauvet, J.-M., Deschamps, E. B., & Hillaire, C. (1995). *La Grotte Chauvet à Vallon-Pont-d'Arc*. Seuil.

Clunies Ross, M. (1989). The aesthetics and politics of an Arnhem Land ritual. *The Drama Review, 33*(4), 107–127.

Cole, M., & Cole, J. (2006). Rethinking the Goody myth. In: D. R., Olson, M. Cole (Eds.), *Technology, literacy, and the evolution of society: Implications of the work of Jack Goody*. (pp. 305–324). Lawrence Erlbaum Associates, London.

Cook, M. (2023). *A river with a city problem: A history of Brisbane floods*. Univ. of Queensland Press, St Lucia, Queensland.

Daniels, P. T. (1996). The study of writing systems. In: P. T. Daniels, W. Bright (Eds.), *The world's writing systems* (pp 3–17). Oxford University Press, Oxford.

David, B. (2017). *Cave art*. Thames & Hudson.

Davidson, I. (1989). Freedom of information: Aspects of art and society in western Europe during the last ice age. In H. Morphy (Ed.), *Animals into art* (pp. 440–456). Unwin Hyman.

Davidson, I. (1995). Paintings, power, and the past: Can there ever be an ethnoarchaeology of art? (Review of *Ancestral Connections: art and an Aboriginal system of knowledge*. [H. Morphy]). *Current Anthropology, 36*(5), 889–892. Retrieved from http://links.jstor.org/sici?sici=0011-3204%28199512%2936%3A5%3C889%3APPATPC%3E2.0.CO%3B2-2

Davidson, I. (1999). The game of the name: Continuity and discontinuity in language origins. In B. J. King (Ed.), *The origins of language. What nonhuman primates can tell us* (pp. 229–268). School of American Research.

Davidson, I. (2005). The painting and the tree: Symbolism in the Upper Palaeolithic. A tribute to a great Basque scholar. *Munibe (Antropologia-Arkeologia), 57*, 197–205.

Davidson, I. (2008). Scales and balance: Archaeology, cultural heritage and sustainability. The academy lecture 2007. *Proceedings of the Australian Academy of the Humanities, 31*, 121–150.

Davidson I (2012) What a carry on? Portable art and changes of symbolic meaning. *L'art pléistocène dans le monde / Pleistocene art of the world / Arte pleistoceno en el mundo, Actes du Congrès IFRAO, Tarascon-sur-Ariège, septembre 2010, N° spécial de Préhistoire, Art et Sociétés, Bulletin de la Société Préhistorique Ariège-Pyrénées, LXV-LXVI*, Book pp 268–269, CD: pp 1559–1570

Davidson, I. (2019). Evolution of cognitive archaeology through evolving cognitive systems. In K. A. Overmann & F. Coolidge (Eds.), *Squeezing minds from stones* (pp. 79–101). Oxford University Press.

Davidson, I., Burke, H., Sullivan, L., Wallis, L. A., Artym, U., & Barker, B. (2020). Cultural conflict in text and materiality: The impact of words and lead on the northwest Queensland colonial frontier, Australia. *World Archaeology, 51*(5), 724–740. https://doi.org/10.1080/00438243.2020.1711153

Davison, G. (2016). Distance and destiny. *Inside story*. Retrieved from https://insidestory.org.au/distance-and-destiny/

de Joucourt, L. (1755). Écriture. In J. l. R. d'Alembert & D. Diderot (Eds.), *Encyclopédie, ou, Dictionnaire raisonné des sciences, des arts et des métiers* (Vol. 5, pp. 358–360). Briasson, David, Le Breton, Durnad.

Dittmar, J. E. (2011). Information technology and economic change: The impact of the printing press. *Quarterly Journal of Economics, 126*(3), 1133–1172.

Dodds, J. D., Menocal, M. R., & Balbale, A. K. (2008). *The arts of intimacy. Christians, jews, and muslims in the making of Castilian culture*. Yale University Press.

Dussart, F. (2000). *The politics of ritual in an aboriginal settlement: Kinship, gender, and the currency of knowledge / Francoise Dussart*. Smithsonian Institution Press.

Eisenstein, E. L. (1983). *The printing revolution in early modern Europe*. Cambridge University Press.

Finlayson, C., Brown, K., Blasco, R., Rosell, J., Negro, J. J., Bortolotti, G. R., et al. (2012). Birds of a feather: Neanderthal exploitation of raptors and corvids. *PLoS One, 7*(9), e45927. https://doi.org/10.1371/journal.pone.0045927

Foster, R. (1998). Paper yabber: The messenger and the message. *Aboriginal History, 22*, 105–115.

French, J. R., & Lane, P. (2002). The common law of native title in Australia. *Oxford University Commonwealth Law Journal, 2*(1), 15–45.

Gándara Vázquez, M. (2007). *El análisis teórico en ciencias sociales: Aplicación a una teoría del origen del estado en Mesoamérica*. (PhD). Escuela Nacional de Antroppolia e Historia, Mexico

Gardner, H. (1926). *Art through the ages: An introduction to its history and significance* (Vol. 1). Harcourt, Brace.

George, A. (1999). *The epic of Gilgamesh. A new translation*. Allen Lane the Penguin Press.

Goody, J. (1977). *The domestication of the savage mind*. Cambridge University Press.

Goody, J. (1998). Memory in oral tradition. In C. Darwin, K. Patterson, & P. Fara (Eds.), *Memory* (pp. 73–94). Cambridge University Press.

Gorringe, H. H. (1882). *Egyptian Obelsiks*. The Author.

Green, J. (2014). *Drawn from the ground. Sound, sign and inscription in Central Australian sand stories*. Cambridge University Press.

Griffith, F. L. (1951). The decipherment of the hieroglyphs. *Journal of Egyptian Archaeology*, 38–46.

Gullick, M. (2016). Reflections and observations on Romanesque manuscripts and charters from Reading Abbey. *Reading Medieval Studies, 42*, 1–24.

Guthrie, R. D. (2005). *The nature of paleolithic art*. The University of Chicago Press.

Halverson, J. (1992). Goody and the implosion of the literacy thesis. *Man*, 301–317

Heather Phillips.. *Library Philosophy and Practice*. Retrieved from https://web.archive.org/web/20120418191647/http://unllib.unl.edu/LPP/phillips.htm#_ftnref8

Henshilwood, C. S., d'Errico, F., & Watts, I. (2009). Engraved ochres from the Middle Stone Age levels at Blombos Cave, South Africa. *Journal of Human Evolution, 57*(1), 27–47.

Hoare, P. (2022). *Albert & the whale. Albrecht Dürer and how art imagines our world*. Pegasus Books.

Hoffmann, D. L., Standish, C. D., García-Diez, M., Pettitt, P. B., Milton, J. A., Zilhão, J., et al. (2018). U-Th dating of carbonate crusts reveals Neandertal origin of Iberian cave art. *Science, 359*(6378), 912–915. https://doi.org/10.1126/science.aap7778

Horsley, G. (2000). Homer in Pisidia: Aspects of the history of Greek education in a remote Roman province. *Antichthon, 34*, 46–81.

Ingold, T. (2000). Chapter 22. Speech, writing and the modern origins of "language origins". In T. Ingold (Ed.), *The perception of environment. Essays in livelihood, dwelling and skill* (pp. 392–405). Routledge.

Jaubert, J., Verheyden, S., Genty, D., Soulier, M., Cheng, H., Blamart, D., et al. (2016). Early Neanderthal constructions deep in Bruniquel cave in southwestern France. *Nature, 534*(7605), 111–114. https://doi.org/10.1038/nature18291. http://www.nature.com/nature/journal/vaop/ncurrent/abs/nature18291.html#supplementary-information

Johns, A. (2009). *The nature of the book: Print and knowledge in the making*. The University of Chicago Press, Chicago.

Jones, P. (2023). Aboriginal interactions with the overland telegraph line, 1870-1880. *Journal of Telecommunications and the Digital Economy, 11*(1), 1–17.

Kelley, K., Cartolano, M., & Ferrara, S. (2024). Seals and signs: Tracing the origins of writing in ancient South-west Asia. *Antiquity, 1-19*. https://doi.org/10.15184/aqy.2024.165

Kelly, P. (2023). *Thoughts on 'proto-writing' in the Upper Palaeolithic*. Retrieved from https://bravenewwords.info/2023/01/08/thoughts-on-proto-writing-in-the-upper-palaeolithic/?utm_source=pocket_saves

Kelly, P., Winters, J., Miton, H., & Morin, O. (2021). The predictable evolution of letter shapes: An emergent script of West Africa recapitulates historical change in writing systems. *Current Anthropology, 62*(6), 669–691. https://doi.org/10.1086/717779

Kilgour, F. G. (1998). *The evolution of the book.* Oxford University Press.

Lawson, A. J. (2012). *Painted caves. Palaeolithic rock art in Western Europe.* Oxford University Press.

Lawson-Tankred, J. (2023). *Art Critics and Government Officials Slam Italy's 'Humiliating' Tourism Campaign Turning Botticelli's Venus Into an InfluencerArtnet News 26 April 2023*

Leitch, S. (2017). Dürer's rhinoceros underway: The epistemology of the copy in the early modern print. In *The primacy of the image in Northern European Art, 1400–1700* (pp. 239–255). Brill.

Macartney, H. (2010). The reproduction of Spanish art. In N. Glendenning & H. Macartney (Eds.), *Spanish art in Britain and Ireland, 1750–1920* (pp. 103–129). Tamesis.

Man, J. (2010). *The Gutenberg revolution.* Random House, London.

Mattioli, T., Farina, A., Armelloni, E., Hameau, P., & Díaz-Andreu, M. (2017). Echoing landscapes: Echolocation and the placement of rock art in the Central Mediterranean. *Journal of Archaeological Science, 83*, 12–25. https://doi.org/10.1016/j.jas.2017.04.008

McEvansoneya, P. (2008). John Savile Lumley and the copies after Velázquez in the National Gallery, London. *Hispanic Research Journal, 9*(5), 437–457.

McKenzie, K. (Writer) (1980). *Waiting for Harry.* Australian Institute of Aboriginal Studies.

McKitterick, R. (1992). *The uses of literacy in early mediaeval Europe.* Cambridge University Press.

McLaughlin, C. (2013). *Lakota war book from the Little Bighorn: The pictographic "Autobiograpy of Half Moon".* Peabody Museum Press.

Minc, L. D. (1986). Scarcity and survival: The role of oral tradition in mediating subsistence crises. *Journal of Anthropological Archaeology, 5*, 39–113.

Moore, J. A. (1983). The trouble with know-it-alls: Information as a social and ecological resource. In J. A. Moore & A. S. Keene (Eds.), *Archaeological hammers and theories* (pp. 173–191). Academic.

Morin, O., Kelly, P., & Winters, J. (2020). Writing, graphic codes, and asynchronous communication. *Topics in Cognitive Science, 12*(2), 727–743.

Morphy, H. (1980). What circles look like. *Canberra Anthropology, 3*(1), 17–36.

Morphy, H. (1991). *Ancestral connections: Art and an Aboriginal system of knowledge.* University of Chicago Press, Chicago.

Morphy, H. (1999). Encoding the dreaming – a theoretical framework for the analysis of representational processes in Australian Aboriginal art. *Australian Archaeology, 49*(1), 13–22. https://doi.org/10.1080/03122417.1999.11681648

Morris, J. (1964). *The presence of Spain.* Faber and Faber.

Moss, C. R. (2023). The secretary: Enslaved workers, stenography, and the production of early Christian literature. *Journal of Theological Studies, 74*(1), 20–56.

Munn, N. (1973). *Walbiri iconography. Graphic representation and cultural symbolism in a Central Australian society.* Cornell University Press.

Myers, F. (2016). Doublings. In S. Gilchrist (Ed.), *Everywhen: The eternal present in Indigenous Art from Australia* (pp. 52–59). Yale University Press.

Myers, F., & Smith, T. (2024). *Six paintings from Papunya: A conversation.* Duke University Press.

Nair, R. B. (2021). Archewriting: The symbolic evolution of script and narrative. In N. Gontier, A. Lock, & C. Sinha (Eds.), *The Oxford handbook of human symbolic evolution* (Second ed.). Oxford University Press.

Nowell, A. (2021). *Growing up in the ice age: Fossil and archaeological evidence of the lived lives of Plio-Pleistocene children.* Oxbow Books.

Nowell, A. (2023). Children of the Ice Age. *Aeon.* Retrieved from https://aeon.co/essays/what-was-it-like-to-grow-up-in-the-last-ice-age

Nowell, A., Bahn, P., & Le Quellec, J.-L. (2024). Evaluating the evidence for lunar calendars in upper Palaeolithic parietal art. *Cambridge Archaeological Journal*, 1–18. https://doi.org/10.1017/S0959774324000155

Olson, D. R. (1986). The cognitive consequences of literacy. *Canadian Psychologist, 27*, 109–121.
Olson, D. R. (1994). *The world on paper*. Cambridge University Press.
Olson, D. R. (2008). Footnotes to Goody: On Goody and his critics. In *Ecritures: sur les traces de Jack Goody*. Presses de l'ENSSIB.
Ong, W. J. (1982). *Orality and literacy: The technologizing of the word*. Methuen.
Orrells, D. (2011). Burying and excavating Winckelmann's history of art. *Classical Receptions Journal, 3*(2), 166–188. https://doi.org/10.1093/crj/clr005
Overmann, K. A. (2022). Early writing: A cognitive archaeological perspective on literacy and numeracy. *Visible Language, 56*(1), 8–44.
Parry, M. (1930). Studies in the epic technique of oral verse-making. I. Homer and Homeric style. *Harvard Studies in Classical Philology, 41*, 73–147.
Pastoors, A., Lenssen-Erz, T., Ciqae, T., Kxunta, U., Thao, T., Bégouën, R., et al. (2015). Tracking in caves: Experience based reading of Pleistocene human footprints in French caves. *Cambridge Archaeological Journal, 25*(3), 551–564. https://doi.org/10.1017/S0959774315000050
Pastoors, A., Lenssen-Erz, T., Breuckmann, B., Ciqae, T., Kxunta, U., Rieke-Zapp, D., & Thao, T. (2017). Experience based reading of Pleistocene human footprints in Pech-Merle. *Quaternary International, 430*, 155–162.
Phillips, H. (2010). *The Great Library of Alexandria? Library Philosophy and Practice* (Published August 2010). https://nbn-resolving.org/urn:nbn:de:0168-ssoar-190807
Procter, A. (2020). *The Whole Picture: The colonial story of the art in our museums & why we need to talk about it*. Cassell.
Prown, J. D. (1997). A course of antiquities at Rome, 1764. *Eighteenth-Century Studies, 31*(1), 90–100.
Puchner, M. (2017). *The written world: How literature shaped history*. Granta Books, London.
Quiles, A., Valladas, H., Bocherens, H., Delqué-Količ, E., Kaltnecker, E., van der Plicht, J., et al. (2016). A high-precision chronological model for the decorated Upper Paleolithic cave of Chauvet-Pont d'Arc, Ardèche, France. *Proceedings of the National Academy of Sciences, 113*(17), 4670–4675. https://doi.org/10.1073/pnas.1523158113
Radovčić, D., Sršen, A. O., Radovčić, J., & Frayer, D. W. (2015). Evidence for neandertal jewelry: Modified white-tailed eagle claws at Krapina. *PLoS One, 10*(3), e0119802. https://doi.org/10.1371/journal.pone.0119802
Rappaport, R. A. (1999). *Ritual and religion in the making of humanity*. Cambridge University Press.
Reynolds, H., & Marks, M. (1987). *The law of the land*. Penguin Books Ringwood, Ringwood, Victoria.
Rifkin, R. F., Henshilwood, C. S., & Haaland, M. M. (2015). Pleistocene figurative *art mobilier* from Apollo 11 Cave, Karas region, southern Namibia. *South African Archaeological Bulletin, 70*(201), 113–125.
Robson, E. (2001). The tablet house: A scribal school in Old Babylonian Nippur. *Revue d'assyriologie et d'archéologie orientale, 93*(1), 39–66.
Rodríguez-Vidal, J., d'Errico, F., Pacheco, F. G., Blasco, R., Rosell, J., Jennings, R. P., et al. (2014). A rock engraving made by Neanderthals in Gibraltar. *Proceedings of the National Academy of Sciences, 111*(37), 13301–13306. https://doi.org/10.1073/pnas.1411529111
Rookmaaker, L. C. (1978). Two collections of rhinoceros plates compiled by James Douglas and James Parsons in the eighteenth century. *Journal of the Society for the Bibliography of Natural History, 9*(1), 17–38.
Ross, J. (2003). *Rock Art, Ritual and Relationships. An archaeological analysis of rock art from the central Australian arid zone*. (PhD). University of New England, Armidale
Ross, J., & Davidson, I. (2006). Rock art and ritual: An archaeological analysis of rock art in arid Central Australia. *Journal of Archaeological Method and Theory, 13*(4), 305–341. https://doi.org/10.1007/s10816-006-9021-1
Schiffer, M. B. (2008). Chapter 5: Telegraphic visions. In *Power struggles* (pp. 41–48). The MIT Press, Cambridge, MA.

Schmandt-Besserat, D., & Erard, M. (2009). Origins and forms of writing. In C. Bazerman (Ed.), *Handbook of research on writing* (pp. 7–26). Routledge.
Shepherd, S. (1996). The significance of the Overland Telegraph Line, 1872/1901. *Journal of Northern Territory History*, (7), 41–46.
Shiner, L. (2001). *The invention of art: A cultural history*. The University of Chicago Press.
Sorabella, J. (2003). *Grand tour*. Heilbrunn Timeline of Art History. Retrieved from https://www.metmuseum.org/toah/hd/grtr/hd_grtr.htm
Spies-Gans, P. A. (2022). *A revolution on canvas: The rise of women artists in London and Paris, 1760–1830*. Paul Mellon Centre for Studies in British Art.
Sterelny, K. (2022). Ethnography, archaeology, and the late Pleistocene. *Philosophy in Science, 89*(3), 415–433.
Sutton, S. A., Paton, D., Buergelt, P., Sagala, S., & Meilianda, E. (2020). Sustaining a transformative disaster risk reduction strategy: Grandmothers' telling and singing tsunami stories for over 100 years saving lives on Simeulue Island. *International Journal of Environmental Research and Public Health, 17*(21), 7764. Retrieved from https://www.mdpi.com/1660-4601/17/21/7764
Texier, P.-J., Porraz, G., Parkington, J., Rigaud, J.-P., Poggenpoel, C., Miller, C., et al. (2010). A Howiesons Poort tradition of engraving ostrich eggshell containers dated to 60,000 years ago at Diepkloof Rock Shelter, South Africa. *Proceedings of the National Academy of Sciences, 107*(14), 6180–6185.
Thomson, A. (1999). *The singing line. Tracking the Australian adventures of my intrepid Victorian ancestors*. Anchor Books.
Tsuen-Hsuin, T. (1985). Science and civilisation in China. In J. Needham (Ed.), *Volume 5. Chemistry and chemical technology, part I: Paper and printing*. Cambridge University Press.
Tully, J. (2009). A Victorian ecological disaster: Imperialism, the telegraph, and gutta-percha. *Journal of World History*, 559–579.
Turner, V. (1974). *The ritual process: Structure and anti-structure*. Penguin Books.
van Gelder, L., & Nowell, A. (2021). Chapter 13: Scene makers. Finger Fluters in Rouffignac Cave (France). In I. Davidson & A. Nowell (Eds.), *Making scenes: Global perspectives on scenes in rock art* (pp. 195–206). Berghahn Books.
Vansina, J. (1985). *Oral tradition as history*. James Currey Publishers, London.
Villaverde Bonilla, V. (1994). *Arte Paleolítico de la Cova del Parpalló*. Servei d'Investigacio Prehistorica, Diputacio de Valencia.
Vogelsang, R., Richter, J., Jacobs, Z., Eichhorn, B., Linseele, V., & Roberts, R. G. (2010). New excavations of middle stone age deposits at Apollo 11 Rockshelter, Namibia: Stratigraphy, archaeology, chronology and past environments. *Journal of African Archaeology, 8*(2), 185–218. Retrieved from http://www.jstor.org/stable/43135517
Wasserstein, D. (2017). *The Library of al-Ḥakam II al-Mustanṣir and the culture of Islamic Spain*. Routledge, Abingdon, Oxon.
Winchester, S. (2023). *Knowing what we know. The transmission of knowledge from ancient wisdom to modern magic*. William Collins.
Wray, A. (2002). Dual processing in protolanguage. Performance without competence. In A. Wray (Ed.), *The transition to language* (pp. 113–137). Oxford University Press.
Wright, H. T., & Johnson, G. A. (1975). Population, exchange, and early state formation in southwestern Iran. *American Anthropologist, 77*(2), 267–289.

Chapter 6
Meaning, Expectations, Agents and Scenes

Abstract Earlier chapters have enabled us to put the long history of art into a context of both dual expression and the struggle against the tyrannies of time and place. This chapter begins with a summary of prior arguments about this with discussions of ritual and the importance of place. Then I refresh our memories about the graphic elements in aspects of communication usually better known for other features. Writing was important both in being relatively persistent and in fixing stories, but also freeing stories from attachment to particular places. Printing then enabled people to free the stories from particular structures of control, such as the church, and it became possible for the stories to be spread far and wide. There was still a problem of time—the stories could not be disseminated rapidly across distances, until the widespread use of the semaphore and then electric and electronic transmission. This would be an attractive story if it were one of the gradual evolution of ways of permitting people to see art, but there was no straightforward sequence in any one place. Rather I describe here the elements that make up the present means of knowing about art, and in doing so, I want to draw attention to the futility of placing emphasis on the art of the last 300 years in Europe and America (primarily) when we need to consider how that sits within the whole range of art around the world and at different times.

I then move to discuss the concept of "meaning" in art and emphasize that the result of the early history of mark-making was to ensure that when people made deliberate marks the seers of those marks would expect that there would be an associated meaning. Yet the specific meaning could be lost without knowledge of the stories that accompanied the mark-making. Thus, two concepts go together: the expectation of "meaning" and the indeterminacy of "meaning." I then discuss ways in which scenes and the stories that can be read from them and the narratives that accompany them reduce the extent of the indeterminacy, and in appropriate cultural contexts can remove it altogether.

I. Davidson, *Art or Scribbles? In the Eye of the Beholder: The Evolutionary Emergence of Visual Communication*, Interdisciplinary Evolution Research 11,
https://doi.org/10.1007/978-3-031-90458-5_6

6.1 Tracing the Long History of Art—Dual Expression and Overcoming Tyrannies of Time and Place

We have reached the stage where we can draw together some of the threads of the argument in order to allow us to move on to a narrative about the role of art in the interaction between art and scribbles and the evolutionary emergence of variation in the art of the Sahulian First Nations.

6.1.1 Ritual and Dual Expression—The Importance of Place

The earliest art for which we have good documentation seems to have involved the four elements of the *art-system* I have been describing. A *producer* (artist) had a thought as a result of which they produced an *object* ("a work of art")—it could have been a representation of a thing, a *subject*, in the environment, or a *coded* version of the thought expressed geometrically or through other non-representational marks—while telling a *story* to the *seer* (you and me, perhaps) of the thing. The story was accompanied by *ritual*, because it seems to have involved some repetition, was told in a special place away from everyday activities, and at a special time because people needed to know when to go to the special place. The story was related to the subject necessarily but even that did not allow people to recover the story unaided if they did not already know it. The special circumstances of the ritual—particularly repetition at a special time and place—ensured that the seer would remember more elements of the relationship between the art and the story. As producers (artists) engaged in an "*actual process of thought*" which Leonardo da Vinci referred to as a *cosa mentale* (see, for example Da Vinci, 2012), when making marks deliberately, they created the cognitive expectation in seers that they were not just marks but that there was a meaning beyond them. This involved the seers knowing more (in their minds) than the visible signs themselves which they could see had been produced at a particular place. The marks persisted in a place, but the seers could carry a meaning in their minds even when they had left that place.

We can think of this arrangement as "*dual expression*," in which, as I showed in the previous chapter, the producer can refer to a thought or subject—the literal surface—but the deeper cognitive meanings in the minds of both producer and seer might be different at different times, enhanced through the strength of the ritual and the cultural context. It is important to remember here the distinction I made in Chapter One that rituals are not all religious (though that is the point of the expression "canonical meaning" in Rappaport's [1999] scheme) but might include, for example, the ritual of museum visitation (Duncan, 1995). It includes Philip Larkin's (2012) "I take off/ My cycle-clips in awkward reverence" in "*Church Going*" https://allpoetry.com/Church-Going. The "reverence" is a secular act, though the church services need not be. Larkin's reverence results from being a person embedded in the mental, social, and cultural context of church services, whether or not he was a church-goer at the time.

The cognitive influence on the minds of participants in the art-system allowed the *reflectivity* that is one characteristic of humans (Barron et al., 2023; Davidson & Noble, 1989): we—you and I—can think about what we see, not just look at it. This is an outcome of our interactions within the system—society and culture—we live in, and not necessarily just a product of the biology of our brains. One of the important points here is that these meanings are conventional because the meanings are defined by—that is to say, limited by—culture in consequence of the ritual, and yet the actual relationship between the object and the subject can involve arbitrary differences. In this way all such objects are and were symbolic. Early art, then, whether in Sulawesi or Swabia, involved communications between one person and a small number of others using symbols which were associated with *particular places*, often fixed in the environment.

6.1.2 Writing and Fixing the Story—Freeing the Art from Time and Place

At some time later and in a small number of regions, writing emerged that could *fix a story*, rather than the place, at one telling of it. The emphasis was transferred definitively to the story rather than the place, the ritual, the people and the story. Writing enabled more people to see, read or hear the same message at different times and in different places, at least in a world where a text was taken from place to place to be read. Successive changes in communication changed the relations between people who produced communications and other people who received them, and between people, place, time and distance. We can summarize the revolutions in communication in Table 6.1.

The story could also be copied, and people could validate what they read by comparing the copy to the original document. This allowed *asynchronous communication*, though not without some difficulties (Morin et al., 2020). People could communicate free of their need to restrict the message to particular places and particular times. Given that until recently there were few people who had the ability to read (and perhaps fewer to write), some could control the story, and those who controlled it could accumulate power to themselves. The written text was both the story and the visual accompaniment of it—and eventually the writing itself became a thing to be produced carefully so that it too became the work of art (Fig. 5.9). Nowadays Chinese calligraphic script, for example, is produced as visual art, though the lack of durability of the text might violate a traditional definition of writing (Piers Kelly personal communication 30 Nov 2023) (Fig. 6.1).

Table 6.1 Characteristics of changing communication

	People	Place	Time	Distance
Language	Transmit and receive	Fixed	Fading	Short
Art	Producers and seers	Usually tied, but options for transport	Persists	Fixity means only the stories travel
Writing	Anyone can learn to read	The written message can travel	The written message can be read at different times	The message can be carried in a cleft stick or from "Ghent to Aix"
Printing	Central producers, many receivers	Works can be spread to many different places	Identical works persist in many places	Identical works can be spread very widely
Semaphore	Central producers, many seers	Message meant to be seen at a distance	Fading	Means of transmitting one message over distance
Electronic communication	Central producers, many seers. Interactive	Message can be seen at a distance	Message can be stored for seeing later	Message can be interactive at any place and any time anywhere

6.1.3 Printing—Releasing Control of the Message

As with writing, the advent of printing also changed the nature of interactions involved in communication. The really significant change in the structures of power arising from use of the printing press was that the same press could be used (as it was) to print works that had quite contradictory content. In Europe, it was used to print bibles which meant that the church and its functionaries were not the only source for such books. The press could be used, for example, to print bibles in the vernacular, so that ordinary people could understand them and were no longer controlled only by the power of those who spoke Latin. And the same press could be used to print secular books, making the literature independent of the scriptoria associated with churches and abbeys. In the case of early printing, the press was partly paid for through printing the indulgences of the church (Fig. 4.1) at the same time as it was used to print Martin Luther's theses that pointed out their vacuity (Fig. 5.13). Printing was capable of producing multiple copies of works, as exemplified by the Rhinoceros of Dürer (Fig. 5.16)—a work of art, of course—which was seen by so many that his vision of an animal he had never seen became among Europeans the expectation of what a rhinoceros was like for hundreds of years, until they finally saw one. This, if nothing else, exemplified the problem of validation of the communication but to some extent that problem was partly solved through the knowledge of the costs and risks of publishing large numbers of copies of a work, together with the credentials (and integrity) of the publisher. Until recently, cost was a gatekeeper for the truth.

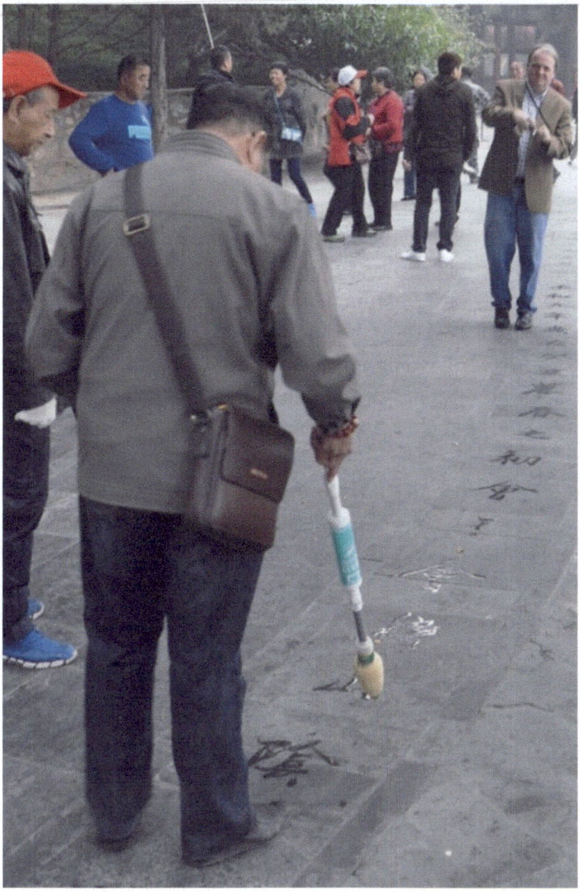

Fig. 6.1 Writing Chinese calligraphy as art for tourists (who mostly can neither read nor write it) in the grounds of the Summer Palace, near Beijing, October 2014. (Photo Iain Davidson)

6.1.4 Getting the Good News from Ghent to Aix Without a Horse—Semaphore, Telegraph and the Tyranny of Time

Printed works reduced many of the difficulties of asynchronous communication, at least in part because identical printed works could be disseminated and seen in different places and at different times. People could get the story no matter when or where it was originally told. Getting a single communication to be known at a distance presented different problems; urgency when people began to live at a large distance from those with whom they wanted to communicate; a problem to be solved when that separation hindered military or commercial activity. At first, communication at a distance involved the physical translocation of a written or printed message by foot, beast of burden, or ship—as in the feverish horse ride (and death of the horse) in Robert Browning's, 1845 poem "*How they Brought the Good News from Ghent to Aix*" (Browning, 1845). The need for speeding up the transmission was exemplified by Napoleon establishing an infrastructure of semaphore stations by which he and other military could communicate over large distances relatively quickly (Fig. 5.19) without the fear that the horse would die too soon. Some of the semaphore stations survive as a visible sign of the urgency of communication.

Semaphore could not, of course, communicate (very far) across the sea (Fig. 5.18) (any more than galloping horses could), so the need for other rapid communication is further demonstrated by the fact that in the War of 1812 the last hostility between the British and the Americans—the Battle of New Orleans, in the Americas—began and ended on 8 January 1815 two weeks *after* the agreement to end the war was signed, in Europe where Britain was at home, in the Treaty of Ghent on 24 December 1814 (Carr, 2007). Ratification of the Treaty by the US Congress, in Washington DC about 1700 km away, could not happen until more than a month later on 15 February 1815 primarily because of the difficulty of communicating the Treaty across the Atlantic in winter, necessarily by boat.

Experiments with electric transmission, by which the installation of wire-based telegraphs began to overcome *the tyranny of time* for colonies such as Australia, led to a form of communication, the Morse code, which anticipated the simplicity of binary/digital communication. Conversion of binary code was essential for the telephone which converted speech at one end into pulses of electricity in a microphone and then converted those into speech at the other end. Thus was born first the radio, which broadcast that speech, then electronic communication, which in the fullness of time permitted computers and mobile phones capable of conveying color images at a distance—either broadcast deliberately, or accidentally when people hold loud conversations on their mobile phones.

6.1.5 A Note on Variation in Art and Communication

This rapid summary of the ways in which the visual and the artistic contributed to changes in communication at different times and in different places sounds like a macro story of progress where one thing led to another, each time better than the one before. It was much more complex. There was no logical progression whereby the combination of story and image exchanged between small numbers of people led to the invention of writing which would allow asynchronous communication between larger numbers of people. The very early art of the east Mediterranean region such as the paintings of Çatalhöyük (Davidson, 2016), or the stone carvings of Göbekli Tepe (Peters & Schmidt, 2004; Schmidt, 2010) (Fig. 6.2) in Turkey appear to have no precedents, and there must be hundreds of similar cases of independent invention without cumulative improvements in technique.

Despite the contrary assumption implicit in the arguments of some (e.g., Van Huyssteen, 2010), the Turkish images are still without any relationship to the cave art of Western Europe and the whole assemblage of early art in the east Mediterranean bears no comparison to that in France and Spain (Bar-Yosef, 1997; Davidson, 2012a). On the other hand, Lewis-Williams and Pearce (2005, 114) have written (imaginatively—but fancifully) of Western European cave art representing "spirit

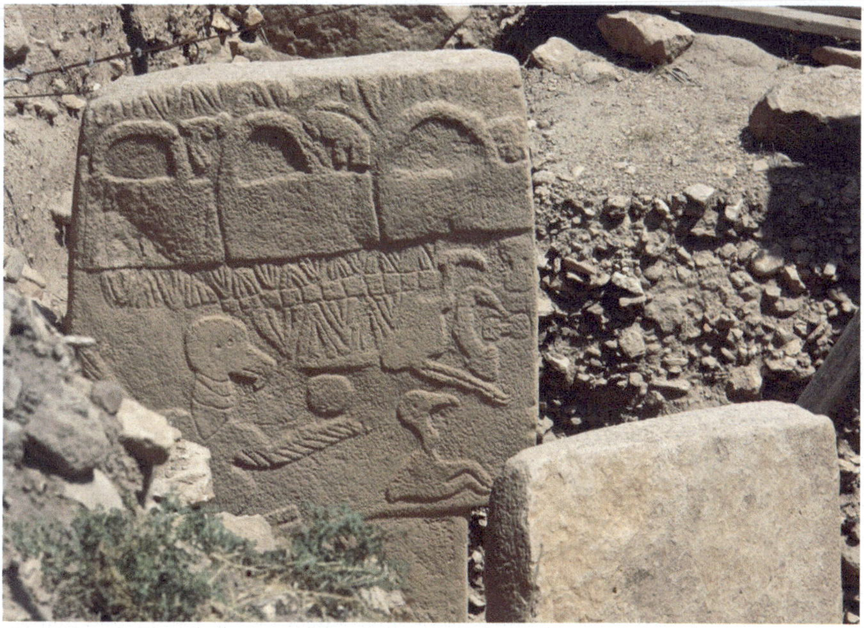

Fig. 6.2 Vulture Stone, Gobekli Tepe, Sanliurfa, South-east Anatolia, Turkey. It depicts a human head in the wing of a vulture and a headless human body under the stela. There are various figures like cranes and scorpions around this figure. https://commons.wikimedia.org/wiki/File:Vulture_Stone,_Gobekli_Tepe,_Sanliurfa,_South-east_Anatolia,_Turkey.jpg. Sue Fleckney, CC BY-SA 2.0 https://creativecommons.org/licenses/by-sa/2.0, via Wikimedia Commons

animals [that] came out of the walls of the caves … thus, in a sense, *giving birth* to spirit animals." They continued: "it seems probably that some aspects of this more focused notion were present, … , at Çatalhöyük." My point is that we can overlook such implausible comparisons because there is no reason to believe that there has ever been any connection, whatever we think of the fantastic notion of spirit animals.

Most obviously, there was no continuity between the earliest art and the later invention of writing. In Southwest Asia and Northeast Africa, writing was invented in societies that were closer to each other in time and space, but it was nevertheless very different in the two places. Moreover, there was much variation within the sorts of art that routinely get lumped together, say, as Paleolithic art (Davidson, 1997, 2012b).

My macro argument is not one about continuity but about the sorts of principles that can be extracted by analyzing examples from different places and different times. In the end, this is not an argument about a continuous sequence of events so much as a demonstration that from the very beginning of the production of art, there has been variation both within and between regions, and from time to time. I once attempted a synthesis of the art of the Western European Upper Paleolithic only to find that there were distinctive stages in the way that art was used some of which were shared in other places, such as Australia, not through particular images but in the way the art-system seems to have functioned (Davidson, 1997):

> …in two very separate parts of the world humans invented means of marking individuals. In the European case, the sequence of events that followed seems to suggest that identification by this means played a role in identifying segments of society when open social networks presumably made mobility between groups more frequent. Subsequently there was an emergence and spread of more closed social networks, which, from the manner in which we have identified them, seem to have involved the establishment of associations between people and some form of center of their world. … Following on from this, the marking of places through symbols painted and engraved on cave walls might be involved with the identification of the manner of corporate ownership of territories.

The details of knowledge of this set of variations may have changed over the last 30 years, though some aspects (e.g., marking individuals) are still unchallenged. The important point here is that the art over a longish period in one region showed substantial variation, and that that variation always involved the engagement between people and the art that they and their fellows created. How it did so, I contend, means that we must contemplate, at least, storytelling about the art at the time it was made.

Isolating one segment of that variation—the much later art of Western Europe produced in the last three hundred years—and giving the name "Fine Art" and hence "art" to that, and only that, is to misunderstand the nature of the interactions between producers, their subjects and objects, their stories and the seers who appreciate the art. The very nature of art as a human product has always been tied up with that set of relationships. Art is so much more than the defenders of the importance of "fine" art would claim. It is too important to be left to them.

6.2 The Expectation of Meaning and the Indeterminacy of Meaning

My last vignette or chance remark is about meaning. In 1989, a few months after our encounter on Maggie's Creek with grazier Bill, I visited the Museum of Modern Art (MOMA) in New York to see the exhibition of paintings produced by Picasso and Braque between 1907 and 1914. It was a very striking exhibition because the curators assembled some of the most powerful paintings by the two artists from the beginnings of the art movement called Cubism, marked by *Les Demoiselles d'Avignon* of Picasso. Wherever possible, the curators of the exhibition had paired the works in such a way that many observers (like me) had difficulty knowing who had painted which. In my hearing, one old person said to another that she liked the paintings, but she wished she knew what they meant—expressing, I think, a universal sentiment that is quite independent of class, age, gender, or cultural background—though members of some such groups might vary in their willingness to express it. It seemed to me that this got to the essence of what art is about: we expect that it should have "meaning," but if we are honest, we probably have to admit, at least presented with nothing except the object produced by an artist, that we do not know what the meaning is—or that my meaning might not be the same as yours. I could go further and suggest that although we expect meaning we all actually have very little idea what "meaning" might be.

It was only afterwards that I was able to read the catalogue of the exhibition where Braque was quoted as saying (in Rubin, 1989): "Picasso and I [Braque] said things to one another that will never be said again … that no one will ever be able to understand … things that would be incomprehensible, but that gave us great joy … ." The important point here is that art is not just about the picture on the canvas that ends up on the walls of MOMA. It is not just about producing marks by addition or subtraction, by making music or writing books, but involves the storytelling between the producer and the audience. There are three options: one where the "meaning" stands alone and is self-evident from the art produced—as where there are scenes which involve at least one agent in the image doing something; the second option requires a story from what the producer says about the work; in the third option, the observer can bring their own meaning to the work, either because they were part of the audience who heard what the producer had to say, or because they are living, breathing, thinking human beings who have been brought up with cultural conventions that generate reactions to the stimuli produced by an artist. They expect meaning, and their prior knowledge and experience reduces any indeterminacy in the marks themselves.

In addition to the deep relationship and understanding and the shared discussion of art that he had with Braque, Picasso had a similar relationship with Matisse. Both men are said to have commented that: "When one of us dies, there will be some things the other will never be able to talk of with anyone else" (in Cowling, 2002, 74). These two remarks, both involving Picasso, make the point clearly that one of the things that turns something into art is the fact that not only does the artist produce it, but artists talk about it, and people learn what the artist has said about it. Then what you and I (or the two women in the Museum of Modern Art) see in the art is not necessarily the same as any of those artists or seers saw or said. The meaning may be different, yet it is not wrong. This uncertainty, this imprecision of meaning, this ability to reflect on the object the artist produced and hence contributed to our own cultural knowledge, is the very essence of art.

Betty Churcher (2014, 12) described this aspect of works of art for the seer as follows:

> … it is impossible to convey their full power. For this you must stand in front of them. If you give it enough time you'll find that an exchange takes place between you and the work. The French artist Marcel Duchamp claimed that this exchange is what we call the 'work of art'. It's your imagination and thoughts meeting the work of art halfway. Of course, if you know a little about the artist's motivation it helps, but according to Duchamp, no exchange, no work of art.

At issue is the question of agency, and the cultural context of the way we learn what to imagine and think. For some, the art itself has agency—they claim the exchange (whether for Churcher or Duchamp) is prompted by the work of art ("if you know a little"). That would be what William Mitchell (1996) called "a dubious personification of inanimate objects." As I wrote in Chap. 4, and it is so important I shamelessly repeat it here, for Conkey (1989) the material record is "an active, constructing, constituting agency, which does not express meaning, but produces it." I argued against this that Conkey may have had some argument for such a position among the participants in the performance of production of the art ("your imagination and thoughts"), but not for non-participants. "The material record, being an object and not a person, cannot ever be an agent in the process of producing meaning. …" (Davidson, 1997, 128). In the same place, I continued: "meaning is not a property of the 'art' or any other material, but of the interaction between the human agents and the material." In short, it is what Churcher refers to as "your imagination and thoughts meeting the work of art"—your knowledge derived from your position in your own culture, your own art-system. This is currently not a fashionable view, but it is inevitable if agency is tied to animacy, as it is in most dictionary definitions. To attribute agency to non-living things is to use a metaphor or to confound a system which has more than one element (a human and an object) with the object as the sole agent. Laura Ahearn (1999) discussed the linguistics of the concept as an anthropologist to show how agency is not unconstrained, but limited by the social, cultural and linguistic conditions of the actors. She does not consider that the agent could be non-animate. As a contrast, Lambros Malafouris (2008, 2013) has followed Gregory Bateson (2000) in considering the "Blind Man's Stick." Does the mind of the blind person extend beyond their body, because they are able to navigate so well using a stick? There are slightly similar examples of this sort of phenomenon such as the common experience of "phantom limbs" (Hill, 1999) or philosophical discussion of the number of people represented by a pregnant woman (Meincke, 2022). In phantom limb pain, an amputee experiences feelings which are present only in the lost body part but in phantom form: "Phantom pain is pain sensation to a limb, organ or other tissue after amputation and/or nerve injury" (Kaur & Guan, 2018). In speculating about pregnancy, there are several phenomena, most obviously the fact that many women feel that their identification with their child continues even after birth (though perhaps the child does not feel that identification for long) (Kingma, 2019). While these examples are difficult to relate to art perhaps, they indicate that the agency of the individual may not be coterminous with the existing bodies of conscious agents, but they do warn that agency will be difficult to identify in things completely unrelated to those bodies. Material objects, such as art, do not seem to me to be in any of the same categories as other agents.

Malafouris (2013, 164–7) drew attention to the experimental results of Iriki and Sakura (2008) who showed that training monkeys to use tools altered their brains. As Iriki and Sakura noted, extension of this process in relation to stone tool use in the natural world, would not only change the hominin mind during evolution but have fundamental impact on the animal's perception of themselves as a "self" separate from the tool they were using. Importantly, those changes and that evolution are/were not just a result of biological changes but a product of the interaction of biology, materials and behavior of the animals (hominins) concerned. Agency, then, is not in the material but in the interactions with the material of humans/hominins with particular brain states.

If it is really true that things have agency, then we should note that there could be a question whether the mind of the stick includes the person. I think that proposition is absurd. Things cannot wonder whether people have agency. Agency is strictly a matter of human thought and behavior, and they interact with material things. There are consequences, of course, since we tend to follow up the idea that things have agency by seeking the intentions of agents that do not exist by proposing the existence of gods and spirits. That is a related issue, but (mercifully) a different one.

The anthropologist, Alfred Gell (1998), was concerned with the agency of art, and I have railed against it already several times. Let us unpack a passage from one of the people who has taken the claim seriously. Robb (2015, 636) suggested that "In Gell's view, art exerts an agency on people by affecting them in particular ways." As I have indicated, this is a major source of error for both Gell and Robb. People react to art, but it cannot have any agency at all because it is inanimate (there is a similar sentiment expressed by Morphy, 2009). Robb goes on: "It is material culture designed to do relational tasks. It can make remote spiritual or political beings present (as on a crucifix or a coin)." It is a simple argument in the context of the four-part account of art I am arguing here: it is humans who design the material culture for a purpose (or, of course, in the knowledge that it will serve many purposes); the art does not make any beings remote, nor make them salient in human perception. The producer may seek to be obscurantist about the way they are turning their vision of the subject into an object, or else the seers do. In either case the art is a result of (in Robb's words it is "designed to do") a whole set of culturally determined stories and conventions about representations which either producer or seers, or both—humans all of them—bring to the sight of the material. Robb (2015) goes on "reassure the worried (as with the paintings in a doctor's office)" or calm those in unfamiliar territory such as a motel room, again principally as a result of a set of expectations produced through convention; "awe the viewers (as in Gell's example of the Trobriand kula canoe prow-board)" though by now you the reader will be familiar with the idea that the prow-board does no such thing, but rather the conventional reaction of seers to the sight of an elaborately carved prow is to have respect and awe for the people in the boat which carries it (Gell, 1998, 69–72); "transform a person (as with a ritual mask), assert class, taste, or cultivation (as with images intended to be objects of connoisseurship), or assert the authority of the state (as with elite propaganda)", whereas those accepting the conventions about the use of masks are people who are embedded in cultures (for Torres Strait masks see Lui-Chivizhe, 2022). People in those cultures recognize that individuals can transform themselves using masks with cultural consequences about their relationships with other individuals or organizations, as seen in other societies, such as those of Torres Strait (Lui-Chivizhe, 2022).

Gell's description of Trobriand prow-boards mentions the "powerful magical associations they have." Yet, when he turned to the explanation of the inequality produced by the magic, he admitted that the impact is due to the "disparity of powers between artists and spectators"—not to the agency of the art of magic. Chris Gosden (2001) tackled this question by looking at the effects on people: "An object with new or subversive sensory qualities will send social relations off down a new path, not through any intention on the part of the object, but through its effects on the sets of social relations attached to various forms of sensory activity." It was people who created or pointed out such an object; people identified its sensory qualities; people engaged with others in relation to the object and its qualities. The object never initiates anything, and always requires a human agent in order to appear to have agency.

Even (especially?) in the modern world, art is embedded within a culture in which people talk about, and tell stories about, the works. The acceptance of Duchamp's *The Fountain*, and others of his "ready-mades" such as the first of them,

The Bicycle Wheel (Fig. 6.3), results from an expectation within our culture that the observer should be able to appreciate the work because it is presented as art, and seers think more about it than they can see on the surface.

The urinal or the bicycle wheel do not have the agency to make themselves into art, Duchamp and the seers did that. Neither of those examples of "ready-mades" was considered art until Duchamp said they were. In the end, people make the work itself famous for being famous whatever the seers know or see in it. So, the stories about the work become as important as the work, and seem to suggest, erroneously, that the work had a role in that. It is equally an error to personify "culture" so that we think that "a culture talks about and tells stories about a work". Culture as a "thing with agency" does not exist. Cultures are entities that are constantly being constructed and their boundaries defined by what people who belong in that culture do and talk about. Those processes of doing and talking constantly change the boundaries of culture. That is how cultures change.

It is possible to represent a single category of subject in a million different ways. People easily interpret some of them from their iconic resemblance to their subject;

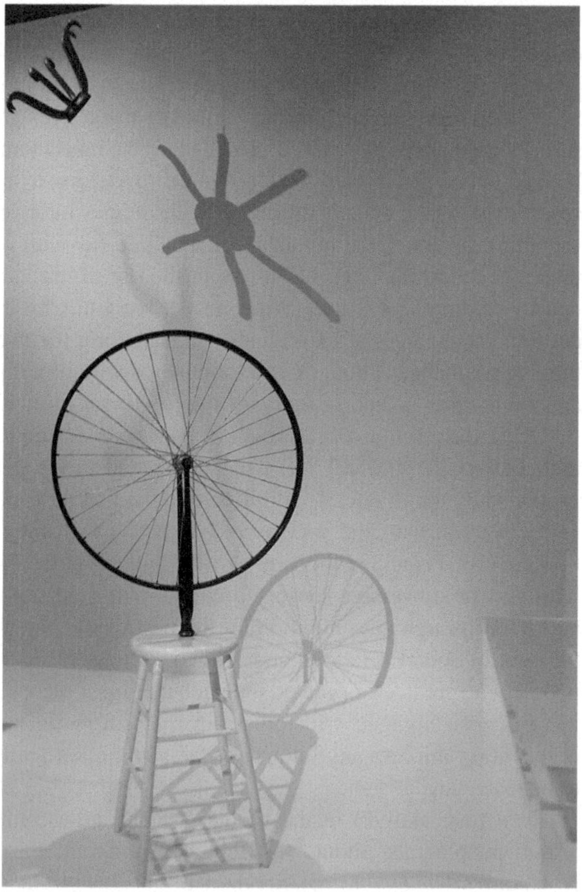

Fig. 6.3 Duchamp's *Bicycle Wheel,* and his *Hatrack* at the Art Gallery of NSW 2019. (Photo Iain Davidson)

for others the seers must know a story to identify more than the literal surface of the object—the dual expression. All can carry a story through the embedding of the object and the subject in the lives, cultures, and histories of the observers, and most of them do carry a story that people know, and we change our cultures as they do so. In the exhibition of his works, the Art Gallery of NSW in August 2019 quoted Duchamp as saying: "It's the viewer that makes the work", which accords well with what I am arguing here.

6.2.1 The Meaning of "Meaning"

The question of "meaning" is, nevertheless, complicated. Any Internet search for "Art and Meaning" will uncover a large number of different approaches at different lengths by people from different disciplines with an interest in art. I need to address it, as I have done before. In a previous paper (Davidson, 1997), I wrote a section called "The Meaning of 'Meaning'", in recognition that many of the advances in the study of the paintings, engravings, drawings and stencils of the Stone Age had come from agreeing with Conkey's (1987) observation that "We can never expect to capture 'meaning' as a single, inclusive, empirical entity or category of inquiry." To illustrate what I mean by agreeing with her, I give examples of the ways in which I used the idea of meaning in a later survey paper about a careful selection from the bibliography of Pleistocene art (Davidson, 2017b). My purpose was to outline the way in which the authorities had used the concept of "meaning" rather than to express my view of that use. Some of these uses I would be happy to argue against (elsewhere).

Where to look for meaning:

1. The history of art begins with the capacity to turn mark-making into the making of meaningful marks (Davis, 1986)
2. It is useful to look at the origins of pictures, rather than art, in order to come to terms with the difficulty of defining meaning as part of the process of the emergence of art (Davidson, 2013; Sachs-Hombach & Schirra, 2013)
3. Bone tools in early sites in Germany and adjacent countries were marked in patterned ways such that the patterning could have carried meaning (Hahn, 1972)
4. Meaning is symbolic and a product of language "Fully syntactic language[1] is arguably an essential requisite to share and transmit the symbolic meaning of beadworks and abstract engravings such as those from Blombos Cave" (Henshilwood et al., 2004, 404)
5. The meaning of Paleolithic art is part of how we define what it was to become human (Tattersall, 1998).

[1] It is worth noting that Botha (2009) argued cogently that the Blombos team had NOT established a case for "fully syntactic language."

Why meaning is important:

6. Style can be one marker of symbolic meanings of image (Pigeaud, 2007)
7. Ritual is/was one way in which people give meaning to signs whether they are indexical (a constant product of an activity), iconic (having some resemblance to the subject) or symbolic signs (often arbitrarily related to the subject but having meanings defined by convention) (Davidson, 2013). I discuss the original meanings of these words in Chap. 7: "Painting by Numbers. Seven Types of Ambiguity in Art".
8. Meaning of individual images is not as important as the meaning of the whole set of images in context (Lewis-Williams, 1981)
9. Meaning is something that pertains to the whole of Western European Cave Art and that includes shamanism (Clottes & Lewis-Williams, 1998)
10. Meaning in cave art is fundamental to the way people told narratives using images (Fritz & Tosello, 2007).

What sorts of things are associated with meaning:

11. The nature of the images, including the geometric signs within a corpus, may remain the same despite changes from one time to another in the meanings associated with them (Davidson, 2012c; Ross, 2005; Taçon et al., 1996)
12. Unpatterned lines may nevertheless have meaning but we do not know how to get at that meaning from the past (Layton, 1991). We can only use our own ways of seeing things.

Meaning and money:

13. In the society of modern Australian Aboriginal people, the notion of "art" had little meaning prior to the creation of a commercial art market (O'Connor & Fankhauser, 2001, 297).

This multiplicity of associations for the word "meaning" in one set of art studies is a product of three features. The first feature is the simple expectation that when we see marks that we call art, we expect that there was something more than the marks themselves. That *expectation of meaning*, displayed by the women at the Cubism exhibition, is the most important aspect of art, and maybe its emergence is one of the most important events in the evolution of human behavior. This is the dual expression that we have discussed before. The object is not everything—what makes up the art-system for the seer is the producer, the subject, the object and the story. As those four elements change, so will the views of the people about the meaning of the art. We must avoid the simplification that the object is an agent when the object is only one part of an art-system determined by people. It is our (thoughtless) expectations that create the illusion of the agency of things.

The second feature is the imprecision or *indeterminacy* of the very concept of meaning, such that the search for meaning sets up something like an infinite regress—if we find out what one aspect means in terms of something else, then we need to ask what that something else means, and so on. In the end, such a regress can only lead to an ultimate answer for which we do not know the meaning. As a result, we are tempted to say the ultimate answer might be something we invent such as the supernatural or some other unfathomable agent. It is the inability to discover any agency in things that leads us constantly to seek an intentional agent such as a god or spirit that is always beyond our comprehension. All definitional systems, as I implied in chapter 3: "Art, Rock Art or Not Art", depend on a small number of irreducible terms that cannot be reduced to any further degree. If we seek a meaning of those irreducible things, we will end up with some form of mysticism.

The third feature is relatively trivial (perhaps). The bibliography I analyze here is one that I chose, and I wrote the annotations, so you are free to argue that the meanings I present here represent nothing more than *my* view about meaning and art studies at that time. I like to think that is no bad thing, but you may disagree. My choices are not of course indications that the art is an agent in the creation of meaning. I made the choices.

In this set of examples, there is a suggestion that we might be concerned about the production of marks at different scales—marks, motifs, panel, site, art and site archeology, groups of sites, relations between groups of sites (some of which we explored in Davidson et al., 2005). Christopher Chippindale (2004) characterized these differences of scale running from the thousandths of meters to thousands of meters, from the individual mark to the figure, to the placement of figures on a surface, and finally to the place of similarly marked surfaces in the landscape. At the base, is the individual brush stroke which is part of the construction of the work of art but is not artistic to the seer. This can be illustrated by annotations with which Paloma Alarcó explained the painting by Braque of *Park at Carrières-Saint-Denis* (Fig. 6.4) for the Museo Thyssen-Bornemisza (https://www.museothyssen.org/en/collection/artists/braque-georges/parc-carrieres-saint-denis):

> In these proto-Cubist landscapes Braque continued with the simplification of form and exploration of a new sense of perspective begun at L'Estaque the previous year, borrowing from Cézanne's painting a device that is essential to the new language: the vibrations, called *passages* by Pierre Daix, that caused forms to be merged. As Tomàs Llorens has pointed out, Cézanne achieved this effect by means of "short brushstrokes, generally in cold, dark tones, with which the master of Aix [Cezanne, one of whose paintings is shown in Fig. 5.22 Left Top—ID] marks, as if with a shadow, some of the edges of the main volumes in his paintings," in such a way that although the image becomes less legible, the painting "gains unity, rhythm and stylistic coherence." It is precisely this commitment to Cézanne's two-dimensional treatment of space that Christopher Green highlights in Braque's work when discussing the latter's major role in the development of early Cubism.

The same would be true of the individual marks made by pounding one rock against another to create a petroglyph. As we move through the scales, we encounter the interpretation of individual figures—"is that a painting of an eland?" (Vinnicombe, 1976), for example (Figs. 2.4 and 2.5). But then we might only be concerned with them in the context of other aspects of the art and culture—were the eland associated with other animals and people (see how this was analyzed by Lenssen-Erz et al., 2021, 100, Fig. 6.4)? Once we have considered that, were eland an element of the belief system associated with trance dances at the time of painting, as they were in some regions one hundred years ago, or were they painted because they were good to eat? (Lewis-Williams, 1987). Next, we might want to ask what is the meaning of that aspect of the art within the culture—what was the role, if any (Solomon, 2013), of trance dances, tricksters and shamans in the culture? (Clottes &

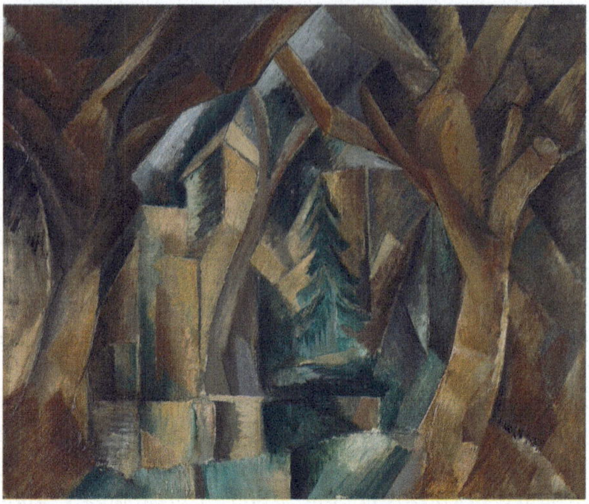

Fig. 6.4 Georges Braque *The Parc at Carrières-Saint-Denis 1909* Museo Nacional Thyssen-Bornemisza, Madrid. Inv. no. 479 (1978.63). One of Braque's first paintings after his military service (Rubin, 1989, 364)

Lewis-Williams, 1998; Guenther, 2002). The next stage asks about the meaning of beliefs about relationships between people and other animals (Descola, 2013 [2005]), but, as Ingold (2018, 169) pointed out, there is not just one relationship which we can define as a single meaning, but an ongoing process: "It is, in this sense, a never-ending quest for an answer to the problem of what being human, or what living in this world, actually means."

"Meaning", then, is a complex entity which can be understood in many ways and at different scales such that there is an *indeterminacy of meaning*—and all of those aspects that matter are the ones people, producers or seers, bring to the art in question—the expectation of meaning. In the study of religion in modern societies, the anthropologist Clifford Geertz (1973, 125) suggested that there needed to be two stages: "first, an analysis of the system of meanings embodied in the symbols which make up the religion proper, and second, the relating of these systems to social-structural and psychological processes." I think that something similar happens with art. Why can some systems of marks be considered as symbols that have meaning? What I would suggest is that it is the indeterminacy in the system of meanings that allows the interpreter (the seer) to complete the meaning by inserting some agent responsible for the symbolism that the society might choose to call a spirit. Once that aspect has been determined by a ritual then seers are embedded in a social system in which an inanimate agent comes to have a role in the social and psychological processes. It is the indeterminate nature of each person's point of view which defines that art is the material we use to understand our own world. Sometimes, then, in order to fill in the gaps left by the indeterminacy of meaning, people invent spirits as agents or invent other worlds that they claim to know about, but others do not. There is nothing that requires art to do this, but it arises from the expectation of meaning and the indeterminacy with which it shows it. In general, some members of society are able to manipulate such beliefs in spirits, often through ritual, to achieve their own relations of power in their society. Spirits and power are thus consequences of the indeterminacy of meaning.

Element	*Range of meanings*	*Category description*
13. 	Nest Hole Water hole Fruits and yams Tree Hill Prepared food Fire Upright fighting stick Painting material Billy can Egg Dog, when curled up in camp Circling (as, e.g., dancing around), or any encircling object	Closed, roundish items, or encircling, nondirectional movement

Fig. 6.5 Range of possible meanings for a circular item, from research by Nancy Munn (1973)

There are clear examples of the combinations of images with songs and stories (Green, 2014; Munn, 1996) (see also the comments in relation to Fig. 3.24). A simple sign such as a circle can have multiple meanings (Fig. 6.5) (see also Morphy, 1980). At one level, the meaning is in the performance. The image, the meaning and the story are irrevocably intertwined.

The same would be true of many works of art all over the world, but the overwhelming importance of this process is that it leads to the position of the observers of the Braque and Picasso exhibition—the conviction that there is an unknown meaning to be reflected on and discovered, somehow beyond what they themselves see before their eyes and know in their minds. In reality, what seers are seeking is the meaning that the producers told stories about. There is no single meaning of the marks every time such marks are produced. The important point is that as a result of all the mark-making, storytelling, and viewing of art objects we the seers expect that there is (now as there was in the past) meaning to the marks and we ignore the fact that that meaning may not be wholly determined by the art object. Meaning is a consequence of social relationships.

6.2.2 Cave Art and Meaning

European cave art, consisting of multiple well-executed paintings found in European caves, is one of the prime examples of what we defined as Art 2 in Chap. 3. The early years of interpretation of it were based on the confusion caused between 150 and 100 years ago because the caves were occupied by people regarded as fisher-gatherer-hunters and Europeans were unwilling to accept that art could be made in such societies (David, 2017; Moro Abadía, 2006; Moro Abadía & González Morales, 2004, 2005; Palacio-Pérez, 2010). The search for understanding of this phenomenon led to a search for "meaning" of that art because people had little more than stereotypical and pejorative attitudes to fisher-gatherer-hunters. There was an expectation of meaning. Various early attempts to understand why there could be something so artistic among such early people included "art for art's sake", sympathetic magic, fertility magic, or ritual (for one recent summary of such search for "meaning", see Solomon & Bahn, 2023), ironically, often by appeal to the literature on people who lived as fisher-gatherer-hunters. The multiplicity of proposals suggests that there was an indeterminacy of "meaning" of the images, although, in keeping with the view that there are different scales of meaning, the images themselves were unequivocally of bison, horses, ibex, deer or pigs.

At one stage that search for "meaning" in cave art was mixed up with understanding of the relationship between totemism and art, amongst other things. The confusions in the to-and-fro of the emergence of beliefs about totemism in Australia and its associations is described by Lester Hiatt (1996). A recent discussion of this literature can be found in Deborah Bird Rose (2024). The early discussions involved considerable confusion about what totemism is, and the confused mingling of stereotypes and evidence. Identifying prehistoric beliefs in totemism by looking at prehistoric art, particularly when so separated in time, culture, and space was made more complicated because 130 years ago the anthropological literature on totemism was only just emerging (indeed, anthropology itself was just emerging) through studies such as those of Australians. The idea that totemism (as it was 100 years ago or is today) might be a relevant concept in discussing European cave art of many thousands of years ago depended on assumptions already being made. These were quite independent of knowledge about European cave art, or the beliefs associated with Australian social organization, about the relationship between those people from the very remote past (say, in Western Europe) and modern people (say, in Australia).

Claude Levi-Strauss (1964), an anthropologist of Indigenous South Americans as well as a theoretician, was much occupied with the topic but he did not demonstrate a strong connection between the European art and totemic beliefs except that the art often showed animals and, among modern people, totemic beliefs sometimes also involved animals, but, he said, totemism was good to think with. It has reappeared with scholars such as Philippe Descola (2006), another anthropologist of Indigenous South Americans and theoretician, who now occupies Levi-Strauss's Chair. Descola attempted to systematize the structural relationships among the variations in material and mental natures of belief (Chap. 3), but the connection between the anthropology of modern people in one continent and the evidence from the past in another is still elusive, as I have discussed elsewhere (Davidson, 2017a).

To illustrate the historical context of these over-generalizations, we can consider some of the evidence of modern Australians. There were two main early studies of modern Australian First Nations at about the same time: the first, in Northwest Central Queensland by Walter Roth, did not emphasize the totemic beliefs of the Pitta Pitta, Yulluna and others (Roth, 1897); the other became a major source of information about the totemic beliefs of the Arrernte of Central Australia, mostly through the collaboration of Frank Gillen, the local telegraph operator who knew the Arrernte well, and Baldwin Spencer, the Professor of Biology at Melbourne University, who became Gillen's friend (Spencer & Gillen, 1899). The publication of the first book by Spencer and Gillen contributed to the literature about the "meaning" of European cave art (Ucko & Rosenfeld, 1967), though the earlier book by Roth did not. Wrongly or rightly, Gillen displayed contempt for Roth (Mulvaney et al., 1997) at least in part as a result of his own preconceptions and prejudices (see also Mulvaney, 2008, 219–20). It is hardly surprising that there was some confusion about totemism when two studies so close to each other in time and space could show such different attitudes towards it.

In addition, Spencer and Gillen ended up in dispute with Carl Strehlow (Austin-Broos, 2010), the Lutheran pastor who worked with Arrernte elsewhere as a missionary, about his belief in the possibility that the Western Arrernte believed in a High God. At the same time as this ethnographic work, the rediscovery of a significant quantity of art in caves in Europe led interested people to begin to speculate about what it all meant in relation to their understanding of the behavior of fisher-gatherer-hunters surviving around the world. In one third of the original volume (Cartailhac & Breuil, 1906) about the first painted cave to be reported in Europe—Altamira, in northern Spain—Émile Cartailhac and the Abbé Henri Breuil made comparisons between the art of Altamira and the rock art of fisher-gatherer-hunters around the world, including a section on the engraved rock art reported more than 60 years earlier in 1842 from Depuch Island in Western Australia (Wickham, 1842) (Fig. 6.6). It may be that the beliefs by archeologists about totemism among the people who produced cave art in Europe were an attempt to identify whether the European painters also had High Gods. There was subsequently a very muddled literature about the relationship between art and beliefs (as outlined, for example, by Ucko & Rosenfeld, 1967, Ch. 3).

One view was that totemism was a word anthropologists used to describe a set of social concepts that had varied expressions. Without more definition and discrimination in every case, any discussion could be misleading. I have previously cited Warner's study of the Murngin to the effect that in totemism "two aspects of society are being ritualized: the social relations of [people] to nature, and of [people] to [people]" (in Davidson, 2017a, 446). Warner's study (Warner, 1964 [1937]) of the group of people now generally known as Yolngu (Morphy, 1991) was particularly informed by his relationship with Mahkarolla from that society (Sutton, 2009, 167–77). The importance of Indigenous individuals making cultural choices about how to communicate with anthropologists who are not from their society is probably generalizable beyond Australia. The Brazilian anthropologist Eduardo Viveiros de Castro (2013, 478) mused about Indigenous people in South America: "It is precisely because the anthropologist very easily takes the native to be an other *subject* that he cannot see him as an *other* subject, as an Other figure that, more than subject or object, is the expression of a possible world." Our knowledge of other cultures derived from studies of them by people from our society is filtered through the knowledge our culture already knows. Who knows whether both parties, anywhere, understood the same things by the words used for any concepts they were discussing about any modern societies and any of their archeological counterparts at the turn of the nineteenth century? Notably, there is no mention of such an individual providing information to Walter Roth with a facility to communicate by selecting and translating concepts from their culture to Roth's. It is conceivable that the knowledge anthropologists reported 120 years ago reflected, in part, the beliefs

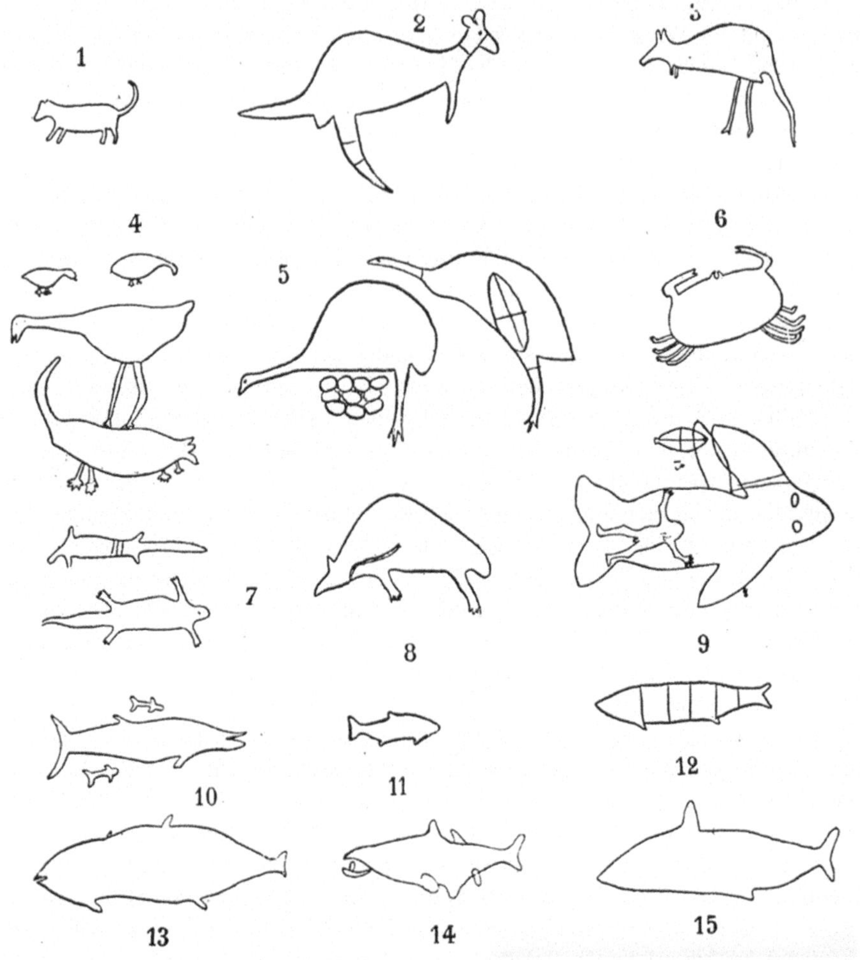

Fig. 6.6 Australian rock engravings from Fig. 151 in Cartailhac & Breuil, 1906. The caption indicates that images 1, 3, 4, 6, and 10 are from Depuch Island, Western Australia described in 1842. The others are from New South Wales

they wanted to discover because of the astute inter-cultural communication—we might almost call it code-switching—by their informants. The applications of such compromised knowledge to the study of art were fraught with difficulty.

Those difficulties spill over into many other aspects of how we know things. Philosophers have speculated about the very possibility of translation because of the indeterminacy of meaning (e.g., Quine, 1970). Sometimes the application of a word such as totemism (or shamanism), in studies of societies other than those of the scholars using the term, reflects the limited vision of those scholars, and may not be an accurate translation of the meaning of the concept at all, unless there has been an attempt to specify the relationships underlying that concept. Indeterminacy is everywhere in cross-cultural communication—and if we accept that each one of us becomes slightly different culturally from anyone else as we make marks that generate the expectation of meaning, then artist and seer communicate imperfectly in different possible worlds.

Totemism, then, is a word that encompasses the different systems of beliefs of a range of various societies, so it has limited utility as a descriptor of the beliefs of societies of the distant past, without further assumptions about the relevance of modern evidence from people with different histories and social organizations. Just because both the modern and the ancient societies seem to have identifiable relationships with animals does not mean that those relationships are or were of the same sort. It is better to avoid giving a name to the concept (such as, totemism or High Gods) and concentrate on those three elements that Warner identified: (1) ritualized relationships; (2) social relations between people and nature; (3) social relations between one group of people and another.

When I say that art is too important to be left to those who are interested in art, I am not just referring to the literature on totemism. Rather, analysis demonstrates some truths—the three identified by Lloyd Warner—about the sort of art that was produced as an intrinsic part of social life: ritual, social relations between the object and the outside world and social relations between people. It is open to question whether the truths may apply to art produced for the market or for museums, or galleries (or doctor's offices and motels), though it seems likely that these concepts, without giving a restricting name such as totemism, define important aspects of all art. The concepts have always been contested, but they are fundamental to all our understanding of art. Art is made with ritual—not just a specific ritual with a defined set of beliefs—and ritual defines the ways we relate to its subject matter and to other people. Meaning is generated through ritual and relationships. That is enough to account for the expectation of meaning. We all witness mark-making as a product of ritual and relationships, and the indeterminacy of meaning—the experience each one of us brings to the ritual and relationships—generates sets of expectations that differ from individual to individual.

6.3 Scenes—Reducing the Indeterminacy of Expected Meaning

When seers of art construct a story from the material products they see in front of them the result is mind altering because it begins a process of reducing the indeterminacy of meaning of art objects. It is, of course, a feature of the distinctly human behavior that is called archeology to make inferences from material evidence alone (I have discussed this creative process in Davidson, 2023). One way in which all people, not just archeologists, make such inferences is through the identification of scenes depicted in the art (Davidson & Nowell, 2021b). When people look at scenes they do two things: first, they follow the story more easily, hearers become seers and the actors in the scene are agents in the story; and second, they are able to identify a narrative in the absence of a teller, such that it is clear that there are actions even if we do not know all of the roles. Scenes do not always remove all indeterminacy, but within a set of cultural expectations they certainly reduce it.

The discovery at Leang Bulu' Sipong 4 in Sulawesi (more than 12 thousand km away from the European art caves) shows that the painters in Wallacea—the islands that always lie between mainland South East Asia and the mainland of Australia, even at the normal periods of time when sea-level is much lower—depicted a scene of confrontation between large animals (wild pigs, *Babirusa*, and dwarf bovids, *Anoa*) and several figures that seem to be both human in stance and, at the same time, non-human "therianthropes" because masks or imagination shaped their heads (Aubert et al., 2019) (Fig. 2.12 and 2.13). It is not a simple scene because seers still need to evaluate the human-like figures, but it was painted in a way that allowed painters to tell a story about the relationships between humans and animals. There is earlier art, for example at Leang Tedongnge which dates 45.5 thousand years ago (Brumm et al., 2021), but this scene is among the earliest representational rock art currently known in the world at more than 43.9 thousand years ago.

It has been dogma for much longer than it should have been that the art of the European caves did not show scenes. Indeed Breuil (1930), then the almost unchallenged authority, contrasted European art with that in South Africa: "There is thus a capital difference between European Upper Palaeolithic and South African art. The first is, 'par excellence,' the art of isolated figures; the second is that of figures grouped into scenes." A couple of figures of rhinoceros from Chauvet Cave, in the Ardèche Region of France, consist of a small number of lines (Fig. 7.2) or a single line (Fig. 6.7) which must be completed by the mind of the seer to make up a figure of the animal (Clottes, 2001).

In both cases, the images are examples of what Breuil would have called "isolated figures". The single line image in Fig. 6.7 can be seen as a figure of the back of a rhinoceros produced vertically, from bottom to horn on a panel which also includes a figure of a *Megaceros* and the earlier scrapings of the cave wall. Elsewhere in Chauvet Cave such scrapings contributed to preparing surfaces for painting, here they have been said to be the representation of an erupting volcano 38 thousand years ago (Nomade et al., 2016, esp. Fig. 3C). These three classes of imagery—the single line rhinoceros, the "complete outline" of a *Megaceros*, and the possible volcano—do not seem to be related to each other on the panel and were produced independently (Baffier & Feruglio, 2001, Fig. 114) (Fig. 6.7). Their juxtaposition does not make them a scene.

The widely publicized claims for the lack of scenes in cave art have not stopped people, among whom cartoonists are prominent in forming perceptions, from repeating the error of suggesting that cave art consists of scenes of people hunting animals (often with bow and arrow which has its own problems of anachronism) (see, for example, the Banksy image of a person blasting off a wall-painting, reprinted in Davidson, 2020). That, of course, is an example of the application of a stereotype of scenes conventional in public expectations of the art of fisher-gatherer-hunter people.

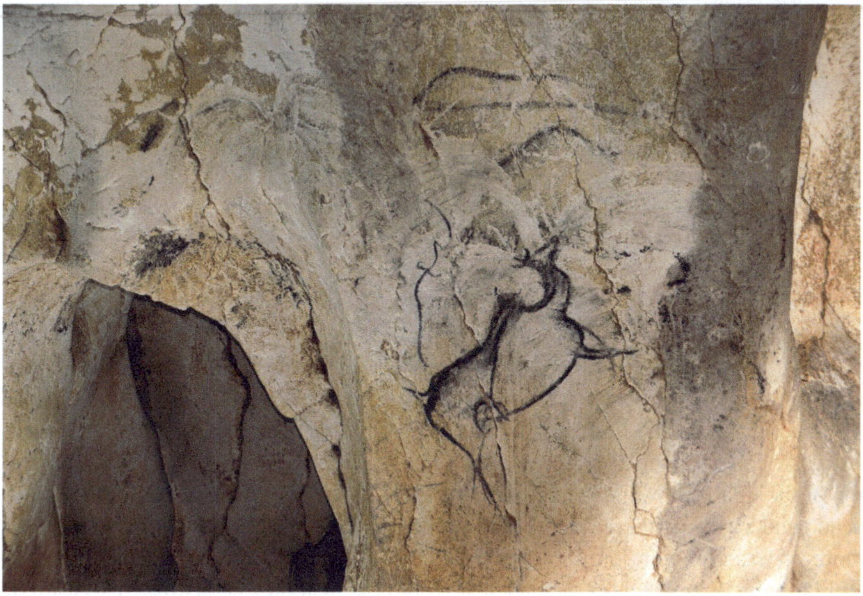

Fig. 6.7 Chauvet Cave panel with the *Rhinoceros* back drawn with a single line so that the idea of the whole animal is implied by the line of the outline of the back of the animal (placed vertically on the cave wall), and its ears (from Clottes, 2001 Fig. 114. Photo Jean Clottes). To the right, Megaceros and other paintings and scrapings. The image includes the scraped figure that has been interpreted as an image of a volcanic eruption—the convergence of the erupting spray (Nomade et al., 2016) is where the head of the *Megaceros* was drawn. (Photo and copyright Valérie Feruglio & Camille Bourdier/French Ministry of Culture, with permission)

Faced with professional claims of lack of scenes in European cave art and a public perception that cave art included them, I considered the different consequences of panels with single or multiple animals of single or multiple species (Davidson, 2021, Fig. 1.1a–g on p. 20) and then moved on to consider what was implied by those combinations. As a result of this, I suggested that a suitable definition of a scene might be "a scene can be identified from a set of images in spatial proximity to each other from which, without any knowledge other than the images themselves, an observer can infer actions taking place among the actors represented in the images" though Nowell and I went on to assemble the different definitions used by a range of authors (Davidson & Nowell, 2021a), including the possibility that seers might infer that an isolated figure engaged in a "perceived event" constitutes a scene (Dobrez, 2021). It is that process of inference—the observation of interactions among the depicted actors or their implied context—that reduces the indeterminacy, but the process is usually constrained by cultural conventions about what can be considered a scene. In fact, recent studies demonstrate that there are several different ways in which art is formed up into scenes (Davidson & Nowell, 2021b) and the seer's experience of such scenes influences subsequent interpretations.

One response has been to look in the caves to find painted images that people seem to have painted with the elements of scene-painting similar to those seen in frames on the walls of the dwellings of the rich and the galleries of the "great." A study of another of the Sulawesi sites, Leang Timpuseng, more than 35 thousand years old, showed that a line was drawn below an image of the wild pig, *Babirusa* (Aubert et al., 2014, Fig. 2b), which, using conventional criteria, might identify it as a scene of an animal standing on the ground (Fig. 6.8).

A later study from France showed that there was a small number of caves, quite close together, with images in which there also seems to be a line indicating the ground level on which the animals stood (Fritz et al., 2016). This seems to be a simple confirmation of the representation of the contexts by which those who are interested in art might identify scenes, though it stops short of landscape art.

In fact, there is an argument that scenes were quite common, but not always easy to recognize by such conventional criteria. You could not put a frame around them. Scholars have been trying hard to find evidence that the same conventions about representation of scenes were indeed present, such as the images of rhinoceros from a panel in La Salle du Fond in Chauvet Cave (Fig. 6.9). The painting is not shown with obvious boundaries like a framed picture. Here, it is sufficient to state that the top of the figure shows multiple (certainly seven) horns of rhinoceros, and multiple heads and backs of the animals. Below there is an apparently isolated figure of a more-or-less complete outline of a rhinoceros, which is one of a whole group of similar animals painted on the panel. It would be difficult to construe this group of paintings as anything other than a scene. We could infer actions taking place among the animals in a herd of rhinoceros.

Indeed, Marc Azéma (2015) has argued that, in this sector of the cave, some of the paintings on the panels around these seem to follow a sequence of hunting by

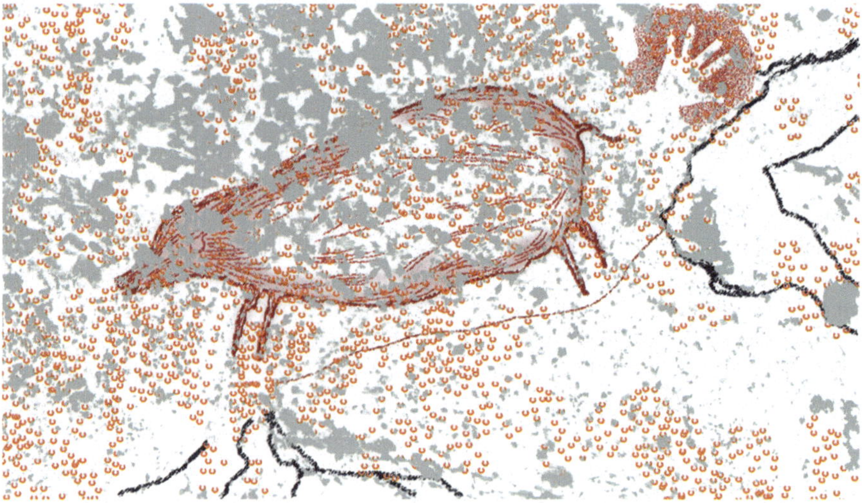

Fig. 6.8 Drawing of *Babirusa* appearing to stand on the drawn ground-surface at Leang Timpuseng. The figure is dated to more than 35.4 thousand years ago (Aubert et al., 2014). The ground line is visible on the unaltered photographs of the painting but is difficult to enhance enough to be reproduced here. (Drawing with permission of Adam Brumm and Maxime Aubert)

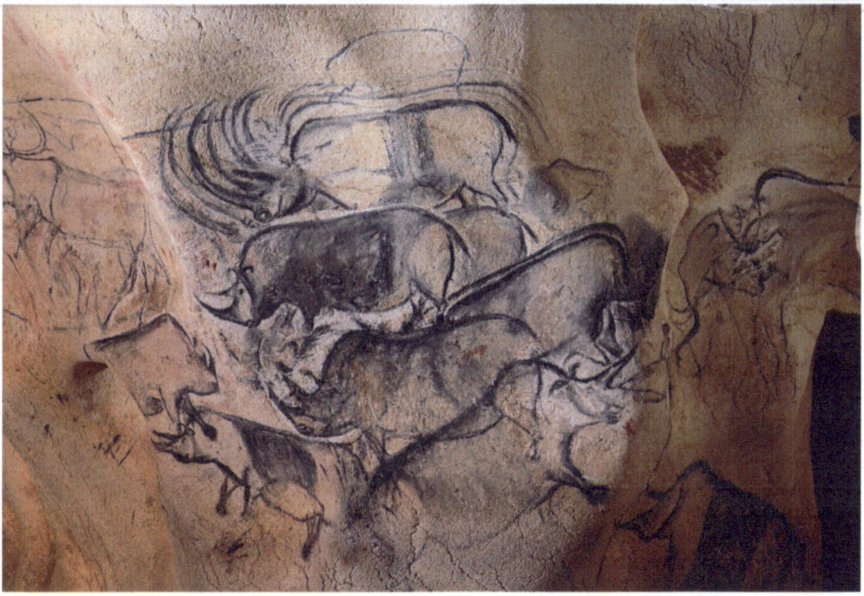

Fig. 6.9 Chauvet Cave. Rhinoceros panel in La Salle du Fond (Clottes & Le Guillou, 2001). (Photo and copyright Valérie Feruglio & Camille Bourdier/French Ministry of Culture, with permission)

cave lions and should be considered part of "pre-echoes"[2] of moving pictures (Azéma & Rivère, 2012). Different observers (Clottes & Azéma, 2005) converged on an interpretation of the depicted behavior of a pair of lions on the same wall (Fig. 4.4). From the nuances of the depiction, seers, including an animal ethologist with modern experience of lion behavior in Africa, could tell that the female lion was cozying up "to the male and probably soliciting mating" (Davidson, 2021, 25). Azéma's analysis of French art representations, from Chauvet Cave 35 thousand years ago and La Vache 15 thousand years ago, of the successive images of the head and neck of horses, the heads of cave lions, the images of lions in motion show model observation of these animals. In the case of the gait (Azéma & Rivère, 2012, Fig. 6), they could almost be illustrations from the photographs by Eadweard Muybridge (using a "zoopraxiscope") which opened the eyes of artists to the way in which animals (including humans) really moved (Shimamura, 2002). In the absence of a special camera, the observations by the cave artists were meticulous, even though the specific story of the scenes is lost. Like many other scenes in art we can tell from the interactions of the animals represented that there was a story that would give life to the meaning of the panel but we cannot now know any more than that.

We have the expectation of meaning, but the indeterminacy is only resolved because of the certainty of a story that accounts for the relationships. By the same

[2] I have no idea what, if anything, "pre-echoes" could possibly mean

token, the church (or priory), with a Saxon scene of fantastic animals and the geometric design (Fig. 6.10), is dedicated to Saints Mary and Hardulph, and who now can tell you who Saint Hardulph was and what s/he did (or even, for that matter, whether s/he was male or female)? Indeterminacy increases with distance from the original ritual, though the concept of sainthood has been retained (albeit with skepticism) in our culture.

One survey has gone about it differently: Culley (2021) has shown that painters generally did not represent scenes on the walls in caves in early European cave art, but when the art was produced on slabs of rock or on fragments of bone we can identify conventional scenes much more often. Valentín Villaverde (2021, Fig. 15.2b) illustrated this by using images from the small stone plaquettes from Parpalló (Fig. 6.11) which themselves seem to have played a role in the cognitive life of the artists (Davidson, 2012c).

Culley (2021) showed that different contexts of art production demanded different classes of art. This is, in a sense, already implicit in the art of Sulawesi where animal images, scenes and hand-stencils co-existed and may have had different contexts. European cave artists began the process of removing the indeterminacy of the meaning of images by representing images in different contexts (e.g., walls and

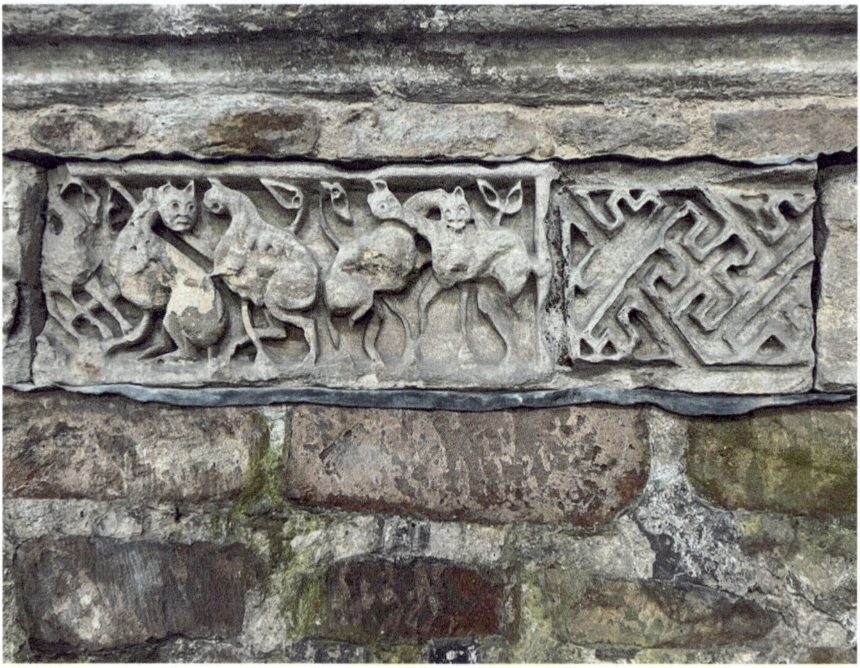

Fig. 6.10 Saxon carvings depicting animals in a frieze from the Church of Breedon on the Hill, Leicestershire. The seers can tell from the interaction of the animals on the panel that the producer represented a scene, but we cannot now recover the story of the meaning of the panel, any more than we know the code that unlocked the secrets of the geometric panel next to it. (Photo by Fiona Davidson and David Davidson with permission)

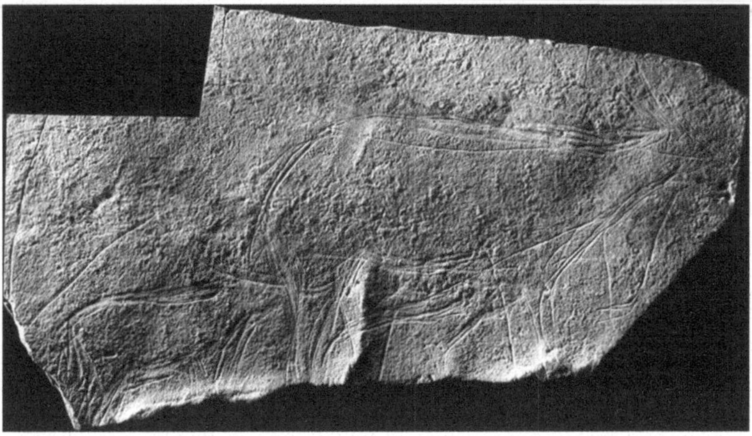

Fig. 6.11 Images of hind and young engraved on a stone plaquette from Parpalló (around 19–17 thousand years ago). You could argue that the animals appear to be standing on the edge of the plaquette which might be taken as the ground. (Villaverde Bonilla, 2021). (Photo by Valentín Villaverde)

stone plaquettes or fragments of bone) and in some of those (mostly the stone plaquettes or fragments of bone) by representing animals in scenes. Hand stencils can mean many things such as association of the people who made them and the stories they told, but seers can never tell which of those many things were involved. Indeed, it has been suggested that hand stencils might not be art (Forge, 1991), but I will counter that in Chapter Nine. Scenes reduced the indeterminacy.

Sometimes variation between different bodies of art in a particular region may be attributed to different chronologies. Villaverde demonstrated that in the sequence of painted cave art in Eastern Iberia, the scenes in the early art—particularly from Parpalló—involved animals interacting with each other but essentially in a rather static scene where they just were there with each other. In his words (Villaverde Bonilla, 2021, 229–30), "These are like snapshots of a well-known ethological reality, but they do not imply either past times or a relatively distant future…[They] refer to an 'eternal present'…" In the later Levantine Art tradition in the same region of eastern Iberia, Villaverde showed that some of the scenes had a serial character and often represented different phases of an action in a single narrative: this may include hunting scenes and scenes interpreted as warfare. Levantine Art, now generally accepted to be made by pastoralists and agriculturalists several thousand years later, often depicts hunting but was quite different in character from the Paleolithic art of the fisher-gatherer-hunter artists in the rest of Iberia. This difference is because of its frequent inclusion of narrative scenes and its depiction of people within them. Inés Domingo (2021, 254) argued that the new narrative style, in a region otherwise part of a world that, if anything, depicted relatively static scenes, could be the result of "signaling a new global change—to the human mind, in human visual communication patterns, and in the way humans conceptualize themselves in art."

When in later prehistory (and in art in the last few centuries, at least in Europe) images of people, animals and activities occurred in abundance, the issues for analysis become more complex. Craig Alexander et al. (2021) proposed a six-part classification of the combinations of images that might contribute to scenes in the northern Italian site of Valcamonica: solitary figures; figures in clusters; figures in arrays; figures in graphic designs; figures together making scenes; and narratives (Fig. 6.12). Christopher Chippindale (2004) had previously drawn attention to the different scales at which similar images at Monte Bego could be considered to work. Even when there is some overlap between categories and scales, for our purposes here, combinations of images in one form or another can contribute to the reduction of uncertainty of reading the meaning of the art.

Jerry Spangler surveyed the 1000-year-old rock art of Nine Mile Canyon systematically to show its spatial variation (Spangler & Davidson, 2021), demonstrating that scenes were often produced close to settlements, while both formal and informal isolated images were produced some distance above them. The two types of art might have had different functions. By now we are familiar with the notion that within a single tradition of art there is variation in the way art is used depending on the cultural context in which it was produced.

When we look at the way scenes were composed and used in other circumstances, the story is that archeological art is just as complex and rich as some of the art studied by art historians—we are just less familiar with the contexts of the painting. Carolyn Boyd's interpretation of 2000-year-old paintings from the Lower Pecos in Texas has carried the interpretation of rock art panels as scenes to an unprecedented level (Boyd, 2016). She achieved this by making the bold assumption that an analysis can be successful if it treats a whole rock art panel as a single composition (Fig. 6.13). Her vindication came when she invited a Huichol shaman, Matsihua, to

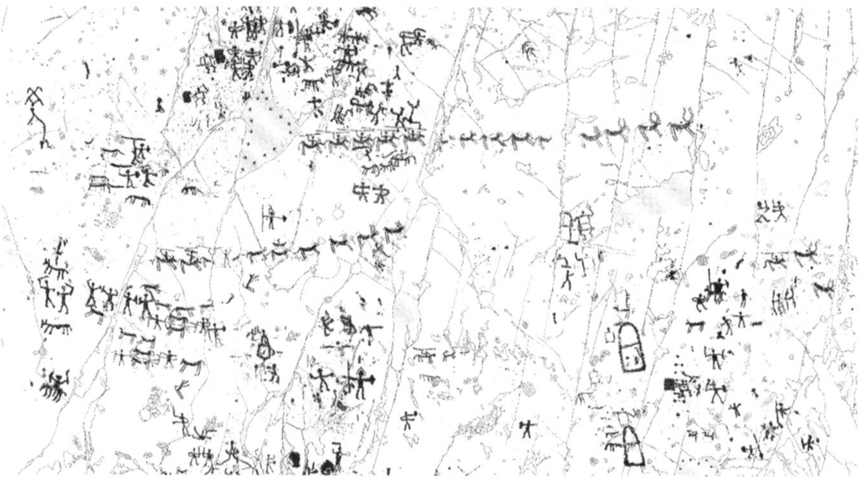

Fig. 6.12 Valcamonica. Figured surface from Seradina 12. (Copyright and tracing by Alberto Marretta, Parco Archeologico Comunale di Seradina-Bedolina (Marretta, 2018) (See also Alexander et al., 2021)

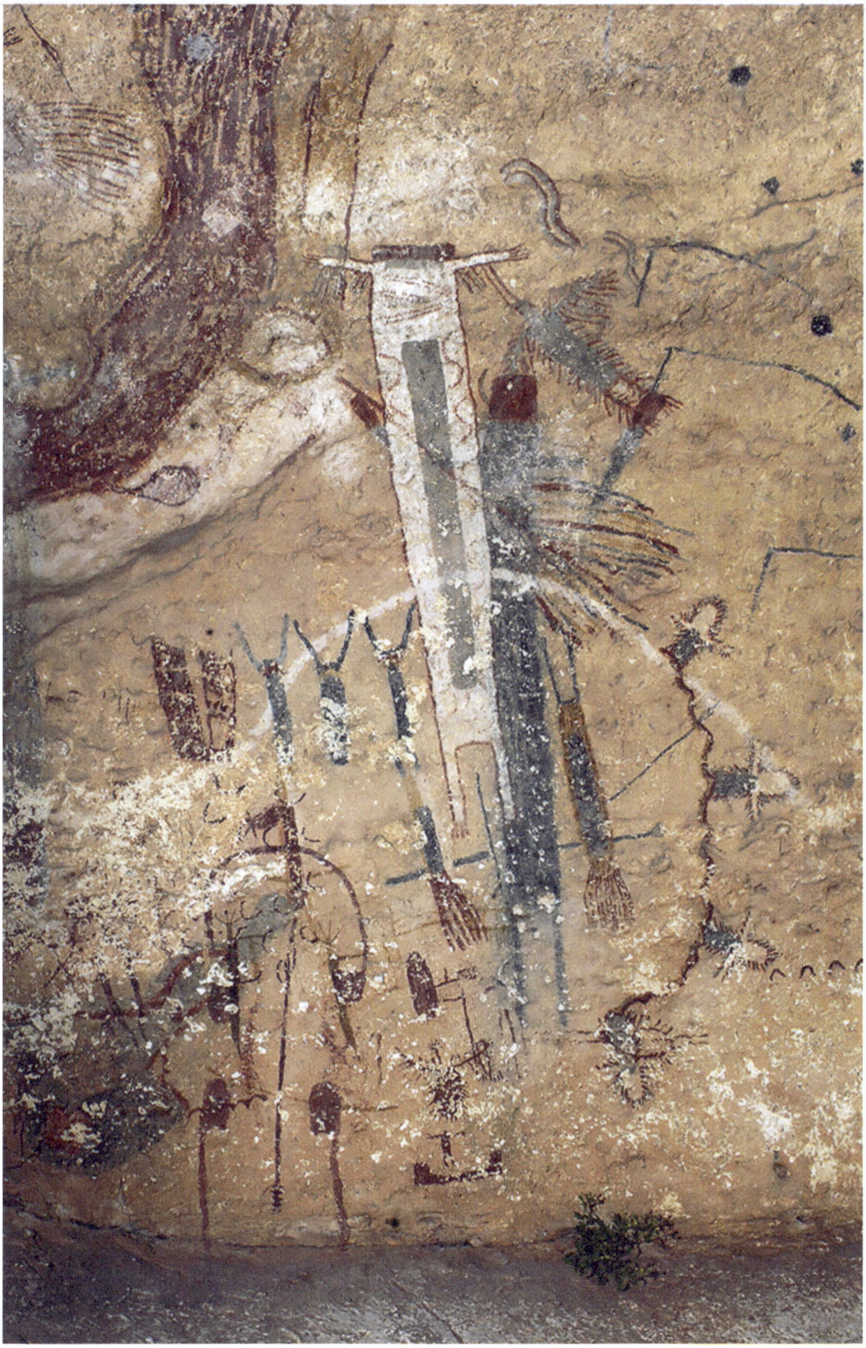

Fig. 6.13 Central portion of the White Shaman rock art panel. (Copyright Shumla Archaeological Research and Education Center and with permission from Carolyn Boyd)

travel 600 km to visit the shelter. On seeing and absorbing the pictures that he saw then for the first time, he proclaimed "These are my grandfathers, grandfathers, grandfathers. They are all here. All my grandfathers, all my ancestors, they are all here" (quoted in Boyd, 2021).

Boyd and Kim Cox analyzed the panel at White Shaman in meticulous detail and followed that by comparison to a variety of other sites in the region (Boyd, 2016). They showed that there are repeated patterns of images of anthropomorphs with attached antlers, often with peyote buttons on the tines—and peyote buttons are known hallucinogens (Boyd & Dering, 1996)—of deer impaled by feathered darts, of serpents which may form an archway through which a "shaman" must pass. Some of the so-called shaman figures in these panels may carry staff-like objects that seem to resemble the fruit of the powerful hallucinogen *Datura* (known as thornapple, or jimson weed). These themes occur in many of the painted rock shelters of the region, allowing the recognition that there is a consistent pattern of composition of the images at each site. Boyd then compared the recurring elements with ethnographic accounts of shamanic ideology and art still being practiced further south in Mexico (Boyd, 2021). This ideology relates animals, animal characteristics adopted by shamans, peyote and its association with deer, rainmaking, and *Datura*.

This case, taken together with the chain of inference required to work out the meaning of the imagery in and around Fig. 6.13, suggests that particular images in rock art can be interpreted in the light of a whole body of other images to allow inferences that would not be possible without the broader context. Scenes permit the understanding of individual examples of art, while understanding of them in the context of other regional sites allows the discussion of the way the storytelling about one site connected the people of whole regions to each other. In terms of the argument here, the recognition of a scene is part of the process of adding a determined meaning to the expectation of meaning. The analysis of the panel, together with understandings from comparison within a wide cultural context, allows some of the resolution of the indeterminacy of the images.

6.4 Essay 6 Pictures and Meaning (Mondrian Rubbish Bin from Rockport and Anonymous *Two Chairs*)

For much art, the significance derives not from the content of the image alone but from the statements of the artists about their work. So, the stories about the work become as important as the work. Two artists stand out for me as making marks that are incomprehensible, but command high prices: Piet Mondrian and Mark Rothko. In both cases, I think the works are pretty, but part of me wants to say that I cannot understand why they are considered art at all. Both seem to me to be on the verge of the design of great wrapping paper. Yet, I am reminded of the work in Australia about the meanings of a circle—a simple design with multiple meanings (Morphy, 1980). The point is that the significance, if any, derives in all four instances (Mondrian, Rothko, wrapping paper, and circles in art) not from the content of the image alone but from the statements of the artists about their work. (There is also the perennial statement "But my child could produce that", to which there is a simple response: "No they could not.")

Piet Mondrian's art style is one of the most recognizable in the world. It is well-exemplified by the image painted onto a rubbish bin in Rockport, Massachusetts, (Fig. 6.14 Left)—most distinctively white squares and rectangles defined by black lines with primary colors, red, blue and yellow filling in some of the shapes. Mondrian's web site[3] links him with Rembrandt (seventeenth century) and Van Gogh (nineteenth century) as great figures of Dutch art, but we should add Van Eyck (fifteenth century), Hieronymus Bosch (fifteenth to sixteenth century) and Vermeer (seventeenth century). As the web site tells us, Mondrian's original work was representational (Fig. 6.15), yet he showed the influence of Monet's Impressionism, Georges Seurat's Pointillism and Van Gogh's vibrant colors until in the twentieth century, influenced by the Cubism of Picasso and Braque, he simplified his forms

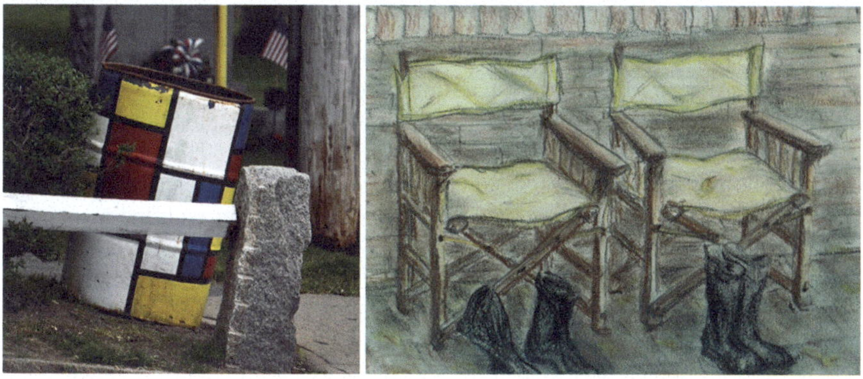

Fig. 6.14 Left: Mondrian rubbish bin, Rockport, Massachusetts 2009. (Photo Iain Davidson). Right: Two chairs, by unnamed artist. (Photo and permission Matt Cawood)

[3] https://www.piet-mondrian.org Accessed 25 November 2021.

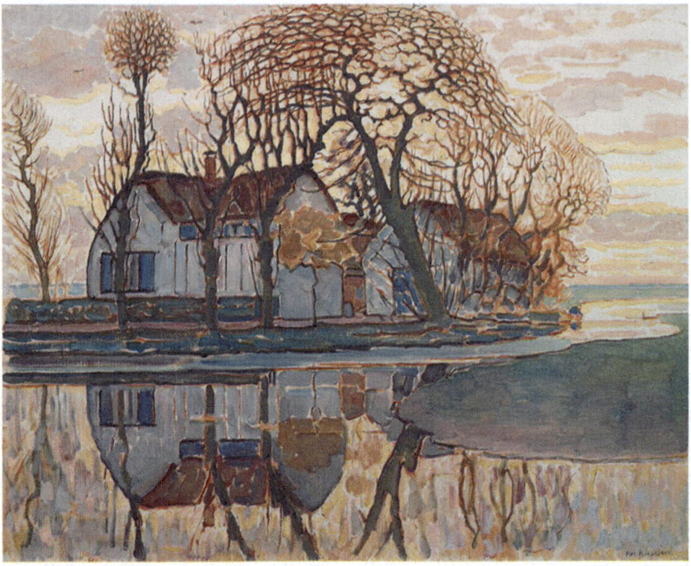

Fig. 6.15 Piet Mondrian (1872–1944), *Farm near Duivendrecht* (c 1916) (Wikimedia Commons PD Creative Commons CC0 1.0 Universal Public Domain Dedication)

first to show how light and shape interacted. He ended up with the strict geometry and bright but austere primary colors now associated with him.

To some the simple patterns and colors appear little different from the patterns on wrapping paper, an attitude exemplified when, in the mid-1960s, 20 years after the artist died, the Mondrian style inspired the designer, Yves St Laurent, for some of his Paris fashion garments. And from there the design was very rapidly taken over by cheap copies and the circular interaction between artist's and couturier's design and the economy ensured that a story could be told about St Laurent purchasing paintings by Mondrian and increasing their value as well as that of his dresses (Troy, 2018). Seeing original Mondrian works in the Dallas art museum (https://dma.org/art/collection/object/5073694) the paintings prove to be much more expressive when you see them than any prettily colored dress or a design on a rubbish bin.

What emerges from this is that art is neither mark-making nor using marks as representation by themselves, nor even doing that beautifully (by some criterion of aesthetics). We need to deal not only with "works of art" but with the associated narratives, and contexts, and the relationships between seers and the "work."

In essence, even in the modern world, the art is embedded within a culture that talks about, and tells stories about, the works. And the acceptance of *The Fountain* and Mondrian's or Rothko's paintings results from an expectation within our culture that the observer should be able to appreciate the work because it is presented as art, and it means more to the seers than what can be seen on the surface.

The pastel drawing in Fig. 6.14 Right is by an unnamed artist who has not attempted to sell any pictures, but it was produced in homage to a series of paintings by Van Gogh. One of that series, a painting of three pairs of boots is in the Art

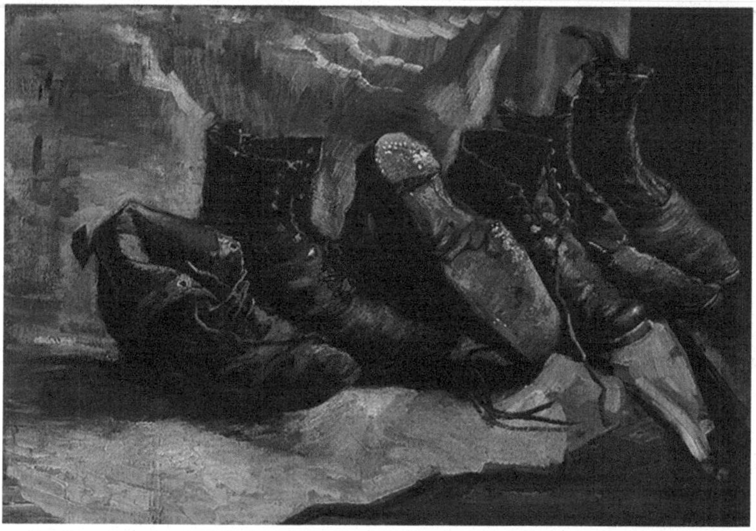

Fig. 6.16 Vincent van Gogh, Three Pairs of Shoes, Harvard Art Museums/Fogg Museum, Bequest from the Collection of Maurice Wertheim, Class of 1906, Photo President and Fellows of Harvard College, 1951.66

Gallery at Harvard University (https://harvardartmuseums.org/art/228744) (Fig. 6.16). An artist starts with marks, and hence art is always about marks, but some marks create memories through which we see other marks.

Whatever observers may say about the picture (Fig. 6.14 Right) as involving skill of execution, it involved a producer and a seer (anyone who sees it hung in its frame on the wall) and the producer has told the story about the significance of time at Harvard and appreciation of the boots picture in relation to Van Gogh. It also evokes a visit by artist and seer to the Hill-Stead Museum in Connecticut which now houses a collection of Impressionist paintings by Édouard Manet, Monet, Edgar Degas and Mary Cassatt purchased in Paris in the 1880s. The collectors at Hill-Stead, Alfred Pope and his daughter Theodate, purchased one of the Monet paintings from Van Gogh's brother Théo. The pastel meets all of the criteria for art, in my definition, and all it lacks is that it has never been sold or bought.

References Cited

Ahearn, L. M. (1999). Agency. *J Linguist Anthropol, 9*(1/2), 12–15.

Alexander, C., Marretta, A., Huet, T., & Chippindale, C. (2021). 17. Rules of ordering and grouping in the pitoti, the later prehistoric rock-engravings of Valcamonica (BS), Italy: From solitary figures through clusters, graphic groups, and scenes to narrative. In I. Davidson & A. Nowell (Eds.), *Making scenes: Global perspectives on scenes in rock art*. Berghahn Books.

Aubert, M., Brumm, A., Ramli, M., Sutikna, T., Saptomo, E. W., Hakim, B., et al. (2014). Pleistocene cave art from Sulawesi, Indonesia. *Nature, 514*(7521), 223–227. https://doi.org/10.1038/nature13422

Aubert, M., Lebe, R., Oktaviana, A. A., Tang, M., Burhan, B., Hamrullah, et al. (2019). Earliest hunting scene in prehistoric art. *Nature, 576*(7787), 442–445. https://doi.org/10.1038/s41586-019-1806-y

Austin-Broos, D. J. (2010). Translating Christianity: Some keywords, events and sites in Western Arrernte conversion. *The Australian Journal of Anthropology, 21*(1), 14–32.

Azéma, M. (2015). Animation and graphic narration in the Aurignacian. In R. White, R. Bourrillo, F. Bon (Eds.), *Aurignacian Genius: Art, Technology and Society of the First Modern Humans in Europe, Proceedings of the International Symposium* (pp. 256–279). New York University.

Azéma, M., & Rivère, F. (2012). Animation in Palaeolithic art: A pre-echo of cinema. *Antiquity, 86*(332), 316–324.

Baffier, D., & Feruglio, V. (2001). La galerie des Mégacéros. In J. Clottes (Ed.), *La Grotte Chauvet. L'art des origines* (pp. 118–127). Seuil.

Barron, A. B., Halina, M., & Klein, C. (2023). Transitions in cognitive evolution. *Proceedings of the Royal Society B: Biological Sciences, 290*(2002), 20230671. https://doi.org/10.1098/rspb.2023.0671

Bar-Yosef, O. (1997). Symbolic expression in later prehistory of the Levant: Why are there so few? In M. Conkey, O. Soffer, D. Stratmann, & N. Jablonski (Eds.), *Beyond art: Pleistocene image and symbol* (pp. 161–187). California Academy of the Sciences.

Bateson, G. (2000). *Steps to an ecology of mind: Collected essays in anthropology, psychiatry, evolution, and epistemology*. University of Chicago Press.

Botha, R. P. (2009). Theoretical underpinnings of inferences about language evolution: The syntax used at Blombos Cave. In R. Botha & C. Knight (Eds.), *The cradle of language* (pp. 93–111). Oxford University Press.

Boyd, C. E. (2016). *The White Shaman mural. An enduring creation narrative in the rock art of the Lower Pecos*. University of Texas Press.

Boyd, C. E. (2021). Images-in-the-making. Process and vivification in Pecos River-style rock art. In O. Moro Abadía & M. Porr (Eds.), *Ontologies of rock art: Images, relational approaches, and indigenous knowledges* (pp. 245–263). Routledge.

Boyd, C. E., & Dering, J. P. (1996). Medicinal and hallucinogenic plants identified in the sediments and pictographs of the Lower Pecos, Texas Archaic. *Antiquity, 70*(268), 256–275. https://doi.org/10.1017/S0003598X00083265

Breuil, L. A. H. (1930). 121. The Palaeolithic art of North-Eastern Spain, and the art of the bushmen: A comparison. *Man, 30*, 149–151.

Browning, R. (1845). *How they brought the good news from Ghent to Aix*. Retrieved from https://www.litcharts.com/poetry/robert-browning/how-they-brought-the-good-news-from-ghent-to-aix

Brumm, A., Oktaviana, A. A., Burhan, B., Hakim, B., Lebe, R., Zhao, J.-x., et al. (2021). Oldest cave art found in Sulawesi. *Science Advances, 7*(3), eabd4648. https://www.science.org/doi/10.1126/sciadv.abd4648

Carr, J. A. (2007). The Battle of New Orleans and the Treaty of Ghent. *Diplomatic History, 3*(3), 273–282. https://doi.org/10.1111/j.1467-7709.1979.tb00315.x

Cartailhac E, Breuil H (1906) La caverne d'Altamira à Santillana, près Santander (Espagne), Monaco

Chippindale, C. (2004). From millimetre up to kilometre: A framework of space and of scale for reporting and studying rock-art in its landscape. In C. Chippindale & G. Nash (Eds.), *The figured landscapes of rock-art. Looking at pictures in place* (pp. 102–117). Cambridge University Press.

Churcher, B. (2014). *Australian notebooks*. The Miegunyah Press.

Clottes, J. (Ed.). (2001). *La Grotte Chauvet. L'art des origines*. Seuil.

Clottes, J., & Azéma, M. (2005). Les images de félins de la grotte Chauvet. *Bulletin de la Société Préhistorique Française, 102*(1), 173–182. Retrieved from http://www.jstor.org/stable/27923872

Clottes, J., & Le Guillou, Y. (2001). La salle du Fond. In J. Clottes (Ed.), *La Grotte Chauvet. L'Art des origines* (pp. 128–148). Seuil.

Clottes, J., & Lewis-Williams, J. D. (1998). *The shamans of prehistory. Trance and magic in the painted caves*. Harry N. Abrams.

Conkey, M. W. (1987). New approaches in the search for meaning? A review of research in "Paleolithic Art". *Journal of Field Archaeology, 14*, 413–430.

Conkey, M. W. (1989). The structural analysis of Paleolithic art. In C. C. Lamberg-Karlovsky (Ed.), *Archaeological thought in America* (pp. 135–154). Cambridge University Press.

Cowling, E. (2002). *Interpreting Matisse Picasso*. Tate Publishing.

Culley, E. V. (2021). 12. A comparison of "scenes" in parietal and non-parietal Upper Paleolithic imagery: Formal differences and ontological implications. In I. Davidson & A. Nowell (Eds.), *Making scenes: Global perspectives on scenes in rock art*. Berghahn Books.

Da Vinci, L. (2012). *Leonardo on Art and the Artist*. Dover Publications, Mineola, NY.

David, B. (2017). *Cave art*. Thames & Hudson.

Davidson, I. (1997). The power of pictures. In M. Conkey, O. Soffer, D. Stratmann, & N. G. Jablonski (Eds.), *Beyond art: Pleistocene image and symbol* (Vol. 23, pp. 128–158). The California Academy of Sciences.

Davidson I (2012a) Symbolism and becoming a hunter-gatherer. L'art pléistocène dans le monde / Pleistocene art of the world / Arte pleistoceno en el mundo, Actes du Congrès IFRAO, Tarascon-sur-Ariège, septembre 2010, Symposium « Signes, symboles, mythes et idéologie… ». N° spécial de Préhistoire, Art et Sociétés, Bulletin de la Société Préhistorique Ariège-Pyrénées, LXV-LXVI, Book: pp 292–293, CD: pp 1689–1705

Davidson, I. (2012b). Variation in early paintings and engravings. In J. McDonald & P. Veth (Eds.), *A companion to rock art* (pp. 51–68). Wiley-Blackwell.

Davidson I (2012c) What a carry on? Portable art and changes of symbolic meaning. L'art pléistocène dans le monde / Pleistocene art of the world / Arte pleistoceno en el mundo, Actes du Congrès IFRAO, Tarascon-sur-Ariège, septembre 2010, N° spécial de Préhistoire, Art et Sociétés, Bulletin de la Société Préhistorique Ariège-Pyrénées, LXV-LXVI, Book pp 268–269, CD: pp 1559–1570

Davidson, I. (2013). Origins of pictures: An argument for transformation of signs. In K. Sachs-Hombach & J. R. J. Schirra (Eds.), *Origins of pictures. Anthropological discourses in image science* (pp. 15–45). Halem.

Davidson, I. (2016). Book review of "Religion at work in a Neolithic society: Vital matters". *Australian Archaeology, 82*(2), 192–195. https://doi.org/10.1080/03122417.2016.1182300

Davidson, I. (2017a). Images of animals in rock art: Not just "good to think". In B. David & I. J. McNiven (Eds.), *Oxford handbook of the archaeology and anthropology of rock art* (pp. 435–467). Oxford University Press.

Davidson, I. (2017b). Paleolithic art. In J. Jackson (Ed.), *Oxford bibliographies in anthropology*. Oxford University Press.

Davidson, I. (2020). Marks, pictures and art: Their contribution to revolutions in communication. *Journal of Archaeological Method and Theory, 27*(3), 745–770. https://doi.org/10.1007/s10816-020-09472-9

Davidson, I. (2021). 1. Scenes and non-scenes in rock art. In I. Davidson & A. Nowell (Eds.), *Making scenes: Global perspectives on scenes in rock art* (pp. 16–31). Berghahn Books.

Davidson, I. (2023). Humans making history through continuities and discontinuities in art. *Cambridge Archaeological Journal, 33*, 637–654. https://doi.org/10.1017/S0959774323000057

Davidson, I., & Noble, W. (1989). The archaeology of perception: Traces of depiction and language. *Current Anthropology, 30*(2), 125–155.

Davidson, I., & Nowell, A. (2021a). Introduction. Behind the scenes—did scenes in rock art create new ways of seeing the world? In I. Davidson & A. Nowell (Eds.), *Making scenes: Global perspectives on scenes in rock art* (pp. 1–15). Berghahn Books.

Davidson, I., & Nowell, A. (Eds.). (2021b). *Making scenes: Global perspectives on scenes in rock art*. Berghahn Books.

Davidson, I., Cook, N. D. J., Fischer, M., Ridges, M., Ross, J., & Sutton, S. A. (2005). Archaeology in another country: Exchange and symbols in North West Central Queensland. In I. Macfarlane, M.-J. Mountain, & R. Paton (Eds.), *Many exchanges: Archaeology, history, community and the work of Isabel McBryde* (pp. 101–128). Aboriginal History Inc.

Davis, W. (1986). The origins of image making. *Current Anthropology, 27*, 193–215.

Descola, P. (2006). La fabrique des images. *Anthropologie et Societes, 30*(2), 167–182.

Descola P (2013 [2005]) Beyond nature and culture (Par-delà nature et culture) (Trans. Jane Lloyd). The University Press of Chicago, Chicago

Dobrez, L. (2021). 2. The possible significance of depicted scenes for cognitive development. In I. Davidson & A. Nowell (Eds.), *Making scenes: Global perspectives on scenes in rock art* (pp. 32–50). Berghahn Books.

Domingo Sanz, I. (2021). 16. New insights into the analysis of Levantine Rock Art scenes informed by observations on Western Arnhem Land Rock Art. In I. Davidson & A. Nowell (Eds.), *Making scenes: Global perspectives on scenes in rock art* (pp. 240–258). Berghahn Books.

Duncan, C. (1995). The art museum as ritual. *The Art Bulletin, 77*(1), 10.

Forge, J. A. (1991). Handstencils: Rock art or not art. In A. Rosenfeld & P. G. Bahn (Eds.), *Rock art and prehistory* (pp. 39–44). Oxbow Press.

Fritz, C., & Tosello, G. (2007). The hidden meaning of forms: Methods of recording Paleolithic parietal art. *Journal of Archaeological Method and Theory, 14*, 48–80. https://doi.org/10.1007/s10816-007-9027-3

Fritz, C., Tosello, G., & Conkey, M. W. (2016). Reflections on the identities and roles of the artists in European Paleolithic societies. *Journal of Archaeological Method and Theory, 23*(4), 1307–1332.

Geertz, C. (1973). *The interpretation of cultures* (Vol. 5019). Basic Books, New York.

Gell, A. F. (1998). *Art and agency. An anthropological theory*. Clarendon Press.

Gosden, C. (2001). Making sense: Archaeology and aesthetics. *World Archaeology, 33*(2), 163–167.

Green, J. (2014). *Drawn from the ground. Sound, sign and inscription in Central Australian sand stories*. Cambridge University Press.

Guenther, M. G. (2002). The bushman trickster: Protagonist, divinity, and agent of creativity. *Marvels & Tales, 16*(1), 13–28.

Hahn, J. (1972). Aurignacian signs, pendants and art objects in Central and Eastern Europe. *World Archaeology, 3*(3), 252–266.

Henshilwood, C. S., d'Errico, F., Vanhaeren, M., van Niekerk, K., & Jacobs, Z. (2004). Middle Stone Age shell beads from South Africa. *Science, 304*(5669), 404.

Hiatt, L. R. (1996). *Arguments about aborigines: Australia and the evolution of social anthropology*. Cambridge University Press.

Hill, A. (1999). Phantom limb pain: A review of the literature on attributes and potential mechanisms. *Journal of Pain and Symptom Management, 17*(2), 125–142.

Ingold, T. (2018). One world anthropology. *HAU: Journal of Ethnographic Theory, 8*(1-2), 158–171.

Iriki, A., & Sakura, O. (2008). The neuroscience of primate intellectual evolution: Natural selection and passive and intentional niche construction. *Philosophical Transactions of the Royal Society B, 363*(1500), 2229–2241.

Kaur, A., & Guan, Y. (2018). Phantom limb pain: A literature review. *Chinese Journal of Traumatology, 21*(06), 366–368.

Kingma, E. (2019). Were you a part of your mother? *Mind, 128*(511), 609–646. https://doi.org/10.1093/mind/fzy087

Larkin, P. (2012). *The less deceived*. Faber & Faber, London.

Layton, R. H. (1991). Figure, motif and symbol in the hunter-gatherer rock art of Europe and Australia. In P. Bahn & A. Rosenfeld (Eds.), *Rock art and prehistory* (pp. 23–38). Oxbow Books.

Lenssen-Erz, T., Fäder, E., Vogels, O., & Mathiak, B. (2021). 6. Scenic narratives of humans and animals in Namibian rock art – a methodological restart with data mining. In I. Davidson & A. Nowell (Eds.), *Making scenes: Global perspectives on scenes in rock art* (pp. 90–106). Berghahn Books.

Levi-Strauss C (1964) Totemism (Trans. R. Needham). Merlin Press, London, UK
Lewis-Williams, J. D. (1981). *Believing and seeing: Symbolic meanings in southern San rock paintings*. Academic.
Lewis-Williams, J. D. (1987). A dream of eland: An unexplored component of San shamanism and rock art. *World Archaeology, 19*(2), 165–177.
Lewis-Williams, J. D., & Pearce, D. (2005). *Inside the Neolithic mind: Consciousness, cosmos and the realm of the gods*. Thames and Hudson.
Lui-Chivizhe, L. (2022). *Masked histories: Turtle shell masks and Torres Strait islander people*. Melbourne Univ. Publishing.
Malafouris, L. (2008). Beads for a plastic mind: The 'Blind Man's Stick' (BMS) hypothesis and the active nature of material culture. *Cambridge Archaeological Journal, 18*(03), 401–414. https://doi.org/10.1017/s0959774308000449
Malafouris, L. (2013). *How things shape the mind. A theory of material engagement*. MIT Press.
Marretta, A. (2018). *La Roccia 12 di Seradina I: documentazione, analisi e interpretazione di un capolavoro dell'arte rupestre alpina*. Edizioni del Parco di Seradina-Bedolina.
Meincke, A. S. (2022). One or two? A process view of pregnancy. *Philosophical Studies, 179*(5), 1495–1521.
Mitchell, W. J. T. (1996). What do pictures "really" want? *October, 77*, 71–82.
Morin, O., Kelly, P., & Winters, J. (2020). Writing, graphic codes, and asynchronous communication. *Topics in Cognitive Science, 12*(2), 727–743.
Moro Abadía, O. (2006). Arts, crafts and Paleolithic art. *Journal of Social Archaeology, 6*(1), 119–141.
Moro Abadía, O., & Gonzáles Morales, M. (2005). L'analogie et la représentation de l'art primitif à la fin du XIXe siècle. *L'Anthropologie, 109*, 703–721.
Moro Abadía, O., & Gonzáles Morales, M. (2004). 1864–1902: El reconocimiento del arte paleolítico. *Zephyrus, 57*, 119–123.
Moro Abadía, O., & González Morales, M. R. (2005). "El arte por el arte": revisión de una teoría historiográfica. *Munibe (Antropologia-Arkeologia), 57*, 5–15.
Morphy, H. (1980). What circles look like. *Canberra Anthropology, 3*(1), 17–36.
Morphy, H. (1991). *Ancestral connections: Art and an aboriginal system of knowledge*. University of Chicago Press, Chicago.
Morphy, H. (2009). Art as a mode of action: Some problems with Gell's Art and Agency. *Journal of Material Culture, 14*(1), 5–27.
Mulvaney, D. J. (2008). From Oxford to the bush: W.E. Roth, W.B. Spencer and Australian anthropology. In R. McDougall & I. Davidson (Eds.), *The Roth family, anthropology and colonial administration*. University College London Press.
Mulvaney, D. J., Morphy, H., & Petch, A. (Eds.). (1997). *"My dear Spencer": The letters of F.J. Gillen to Baldwin Spencer*. Hyland House.
Munn, N. (1973). *Walbiri iconography. Graphic representation and cultural symbolism in a Central Australian society*. Cornell University Press.
Munn, N. (1996). Excluded spaces: The figure in the Australian Aboriginal landscape. *Critical Inquiry, 22*(3), 446–465.
Nomade, S., Genty, D., Sasco, R., Scao, V., Féruglio, V., Baffier, D., et al. (2016). A 36,000-year-old volcanic eruption depicted in the Chauvet-Pont d'Arc cave (Ardèche, France)? *PLoS One, 11*(1), e0146621. https://doi.org/10.1371/journal.pone.0146621
O'Connor, S., & Fankhauser, B. (2001). Art at 40, 000 BP? One step closer: An ochre covered rock from Carpenter's Gap Shelter 1, Kimberley region, Western Australia. In A. Anderson, I. Lilley, & S. O'Connor (Eds.), *Histories of old ages. Essays in honour of Rhys Jones* (pp. 287–300). Pandanus Books, Australian National University.
Palacio-Pérez, E. (2010). Cave art and the theory of art: The origins of the religious interpretation of Palaeolithic graphic expression. *Oxford Journal of Archaeology, 29*(1), 1–14. https://doi.org/10.1111/j.1468-0092.2009.00337.x
Peters, J., & Schmidt, K. (2004). Animals in the symbolic world of Pre-Pottery Neolithic Göbekli Tepe, south-eastern Turkey: A preliminary assessment. *Anthropozoologica, 39*(1), 179–218.

Pigeaud, R. (2007). Determining style in Palaeolithic cave art: A new method from horse images. *Antiquity, 81*(312), 409–422.
Quine, W. V. (1970). On the reasons for indeterminacy of translation. *Journal of Philosophy, 67*(6), 178–183.
Robb, J. (2015). Prehistoric art in Europe: A deep-time social history. *American Antiquity, 80*(4), 635–654.
Rose, D. B. (2024). *Dreaming ecology. Nomadics and Indigenous ecological knowledge, Victoria River, Northern Australia*. Australian National University Press.
Ross, J. (2005). Rock art of the red centre. In M. A. Smith & P. Hesse (Eds.), *23°S. Archaeology and environmental history of the Southern Deserts* (pp. 217–230). National Museum of Australia Press.
Roth, W. E. (1897). In Facsimile (Ed.), *Ethnological studies among the North-West-Central Queensland Aborigines*. Government Printer.
Rubin, W. (1989). *Picasso and Braque. Pioneering cubism*. The Museum of Modern Art.
Sachs-Hombach, K., & Schirra, J. R. J. (Eds.). (2013). *Origins of pictures. Anthropological discourses in image science*. Halem.
Schmidt, K. (2010). Göbekli Tepe–the Stone Age Sanctuaries. New results of ongoing excavations with a special focus on sculptures and high reliefs. *Documenta Praehistorica, 37*, 239–256.
Shimamura, A. P. (2002). Muybridge in motion: Travels in art, psychology and neurology. *History of Photography, 26*(4), 341–350.
Solomon, A. (2013). The death of trance: Recent perspectives on San ethnographies and rock arts. *Antiquity, 87*(338), 1208–1213.
Solomon, A., & Bahn, P. (2023). Rites, wrongs and analogies: Religion and ritual as explanation of prehistoric rock art. In J. d'Huy, F. Duquesnoy, & P. Lajoye (Eds.), *Le gai sçavoir: Mélanges en hommage à Jean-Loïc Le Quellec* (pp. 274–288). Archaeopress.
Spangler, J. D., & Davidson, I. (2021). 18. Finding order out of chaos. A statistical analysis of Nine Mile Canyon rock art. In I. Davidson & A. Nowell (Eds.), *Making scenes: Global perspectives on scenes in rock art* (pp. 277–294). Berghahn Books.
Spencer, B., & Gillen, F. J. (1899). *Native tribes of Central Australia*. Macmillan.
Sutton, P. (2009). *The politics of suffering. Indigenous Australia and the end of the liberal consensus*. Melbourne University Press.
Taçon, P. S. C., Wilson, M., & Chippindale, C. (1996). Birth of the Rainbow Serpent in Arnhem Land rock art and oral history. *Archaeology in Oceania, 31*(3), 103–124.
Tattersall, I. (1998). *Becoming human. Evolution and human uniqueness*. Harcourt Brace & Company.
Troy, N. J. (2018). Mondrian's dress: Yves Saint Laurent , copying (and) the couture copy. *Bulletin of the German Historical Institute, 63*, 27–43.
Ucko, P. J., & Rosenfeld, A. (1967). *Palaeolithic cave art*. Weidenfeld and Nicolson.
Van Huyssteen, J. W. (2010). Coding the nonvisible: Epistemic limitations and understanding symbolic behavior at Çatalhöyük. In: I. Hodder (Ed.), *Religion in the emergence of civilization: Çatalhöyük as a case study* (pp. 99–121). Cambridge University Press, Cambridge.
Villaverde Bonilla, V. (2021). 15. Scenes in the Paleolithic and Levantine Art of Eastern Spain. In I. Davidson & A. Nowell (Eds.), *Making scenes: Global perspectives on scenes in rock art* (pp. 224–239). Berghahn Books.
Vinnicombe, P. (1976). *People of the Eland. Rock paintings of the Drakensberg Bushmen as a reflection of their life and thought*. University of Natal Press.
Viveiros de Castro, E. (2013). The relative native. *HAU: Journal of Ethnographic Theory, 3*(3), 473–502.
Warner WL (1964 [1937]) A black civilization. A social study of an Australian tribe. Harper Torchbooks, New York
Wickham, C. (1842). Notes on Depuch Island. *Journal of the Royal Geographical Society of London, 12*(ArticleType: research-article / Full publication date: 1842), 79–83. Retrieved from http://www.jstor.org/stable/1797981

Chapter 7
Painting by Numbers: Seven Types of Ambiguity in Art

Abstract By using an old phrase about ambiguity, I look at the different ways of looking at art without getting on board with the attitudes of others, particularly the privileged with their restricted cultural background. Leonardo said that art is a *cosa mentale* (rather as McLuhan did later), and as a result, art does not so much act on people as that people react to art in light of their experiences. Those experiences include various cognitive habits such as reactions to *trompe l'oeil*, pareidolia, and other gestalt processes. The reactions of people like us and of our various ancestors at different stages have been different, and the chapter discusses those characteristics at three different stages of the evolution of the production of art—it was from this that I originally identified the seven stages of ambiguity that form the core of the chapter.

I point out that some famous works of art involve breaking arbitrary rules, others pay homage to earlier works which they evoke—all belong to the last 300 years in the period that art historians recognize as one where people produced "Fine" Art which I call Art 1. In doing so, the artist appears to play with the concept of indeterminacy that underlies the concept of the expectation of meaning in the marks artists have always made. Various examples of rock art (which I call Art 2) show that reducing that indeterminacy was part of the way in which people determined a canonical message in the scenes they represented.

As we survey the world of art, it becomes clear how important the role of commerce has been in the last three or four hundred years, as artists have used new possibilities that resulted from social change (such as the damage to the unique power of the Catholic Church) and technological change (invention of paper and oil paint, use of canvas, as well as printing) and sought to make a living from their labors without being tied to patronage. At the same time, gate-keepers—particularly in museums, galleries, and new Academies of Art—determined cultural rules of the emerging market.

A survey of some of the early examples of art (which I have called and defined as Art 3) allows us to begin to understand some of the cognitive processes that were involved in our ancestors slowly beginning to produce meaningful marks and talk

I. Davidson, *Art or Scribbles? In the Eye of the Beholder: The Evolutionary Emergence of Visual Communication*, Interdisciplinary Evolution Research 11,
https://doi.org/10.1007/978-3-031-90458-5_7

about their meanings in the context of the cultural knowledge of people in the society. The chapter ends with an Essay about the Trinil Shell, an example of Art 3, and one of the anthropomorphs from Northwest Central Queensland, an example of Art 2, and a discussion of the social role of cultural context in interpreting them as art.

7.1 Why Ambiguity?

Nearly one hundred years ago, the critic William Empson (1947 [1930]) published a book about poetry called *Seven Types of Ambiguity* which referred to ways in which words were used to enhance meanings. I have shamelessly borrowed the apt phrase. I did not mean, as I do now, to refer to those situations in which the work of art of a producer brought to the mind of the seer something that may not have been exactly what the producer had in mind. That is, though, certainly one of the ways in which Empson made his readers think about poetry—with the poet as producer and the reader or hearer in the role of seer. Because my borrowing is informal, I do not want to go through the full set of Empson's types, nor his discussions of the precise meanings and intentions of ambiguity, but rather to use those that are appropriate to art as we have been discussing it here.

I chose the original title of this section in homage to Empson's original formulation without remembering much about it. On reflection, I discover that he was seeking to tackle variations in the way people looked at poetry in the face of the modernist changes that arose from T.S. Eliot's contribution to the cultural values of the early years of the last century (Eliot, 2010), and Sigmund Freud's imaginings about the mind (Empson, 1947 [1930], viii–ix). The comparison is appropriate: we are considering variations in ways of looking at art—including our cognitive attitudes, wherever we get them—without being beholden to the views of those who are interested in art. We are not overly concerned about the dominance of one way of looking at aspects of cultural behavior by one privileged group of people rather than another. I have previously extolled Berger's *Ways of Seeing* (Berger, 2008) as making massive changes to our understanding of art when originally broadcast in 1972. One commentator on the world into which it emerged stated "The art world was still dominated by cravat-wearing connoisseurs, aristocratic collectors and haughty curators" (Papastergiadis, 2024, 3).

One of Empson's leading examples of ambiguity is the use of metaphor. Once before my co-author and I argued in favor of the science of trying to understand what is behind the metaphor of mind and against the use of that metaphor as a means of seeming to understand the mind by avoiding the science (Davidson & Noble, 1997). I might be less strict about this now, given that the reflectivity which I have said is an important part of art (see also Barron et al., 2023) is actually a form of metaphor, but my task is a little different. Through the metaphor anything that looks like "understanding" derives from the cultural construction of such metaphors—we might almost call it their ambiguity. As Culley (2008, 69) says "metaphors are systematically structured by our physical experience" and always have been structured in that way by the people who use them. The understanding of metaphors will always depend on the particular cultural background of the individuals employing them. Moreover, "each language has its own set of metaphorical principles" (Morphy, 2023, 61) through which language and its conventions of meaning permeate all aspects of perception of the world.

I am trying to understand the whole topic of art, bringing you along with me, without engaging in much more than the seductive simplicities of the claim that art is a metaphor (or any other figure of speech) and hence is part of those perceptions. Likewise, I want to avoid straying into the topics of puns or irony which (for all their personal appeal to me) seem quite difficult to apply to visual art without first accepting an argument about how such devices depend on the conventions of particular cultural contexts. Arguably Duchamp's ready-mades (Fig. 6.3) such as *The Fountain* could be considered as examples of visual puns—the urinal that looks like a sculpture of a fountain when it is reversed. Such excursions would make my point that context is everything. You would not expect to see a urinal in the public spaces of an art gallery. Duchamp's pun may have been seen differently by men than by women. I could argue that Velázquez was using this sort of ambiguity in *Las Meninas* (Fig. 1.10 Left) precisely by leaving unstated—unrepresented—the painting on the other side of the canvas he depicted. Ten years later, in *The Art of Painting* Vermeer reversed Velázquez's point of view and showed the back of the artist, part of the picture on the canvas and the model who was the artist's subject (Fig. 7.1).[1] Robb (2017) gives the example of Warhol buying real Brillo soap pad boxes for an exhibition, because it was easier than making more of the exact replicas he had previously shown as art (for discussion and illustration of Warhol's Brillo boxes and Duchamp's The Fountain, see Davies, 2006, 42 and Figs. 2.1 and 2.2).

It remains true of paintings and engravings, though, that the mark is not all there is to art. The fact that you and I see the set of marks as a picture of something, is a demonstration of the metaphorical nature of art, and that sets up the expectation of meaning. Incomplete figures that are a famous component of Paleolithic cave art exemplify this perfectly (see, for example, the Rhinoceros in Figs. 6.7 and 7.2). All painted or sculpted figures are inevitably incomplete because the only complete version would be the thing itself, whereas the representation is often invoked by a small number of lines.

Any art is composed of a set of marks which we see as something more than just marks. All other animals make some sort of mark, but not all of them behave as if they can work out what the marks were. Humans can. Our ancestors made marks, worked out that they were signs of the activity that caused them, then realized that they could make marks for the sake of signaling something. Even if there were some changes in the human brain during evolution that made it possible to achieve this, there would still have been a process in which marks came to be seen as something

---

[1] We do not know for sure whether Vermeer ever saw the painting by Velázquez.

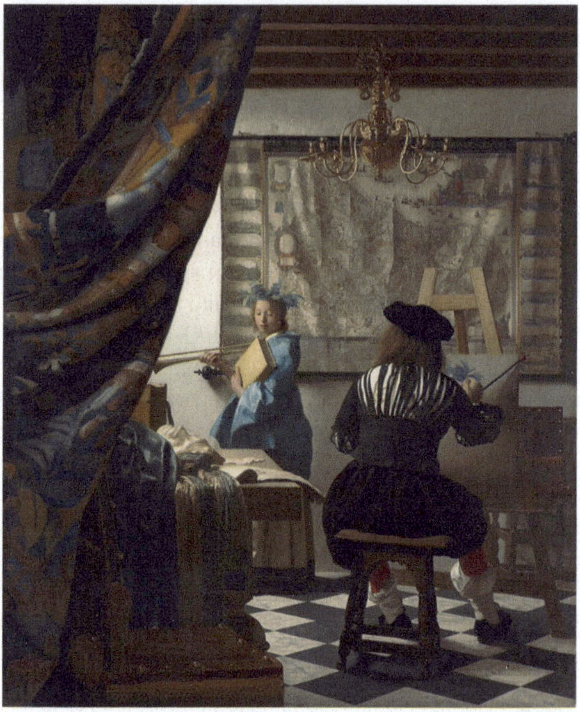

Fig. 7.1 Jan Vermeer 1666–68 *The Art of Painting*. (Wikimedia Commons PD)

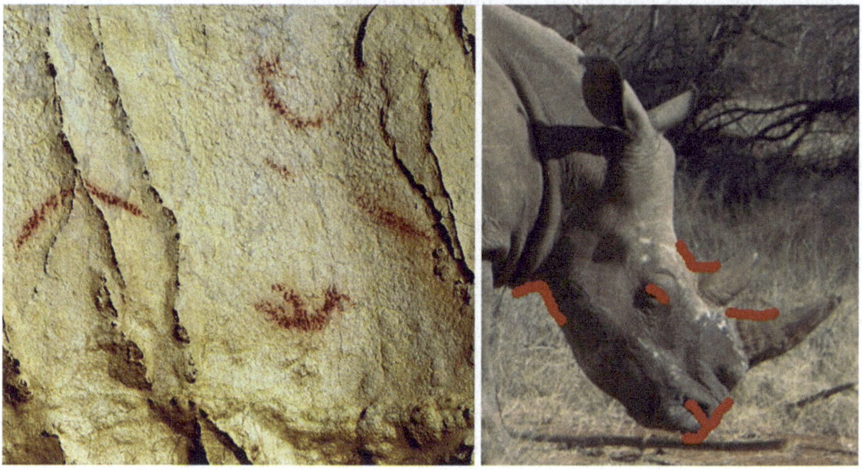

Fig. 7.2 Left: Marks on the wall of Chauvet Cave that can be formed up into the head of a Rhinoceros. (Photo Jean Clottes). Right: Head of an African Rhinoceros marked to show how the Chauvet image was composed. (Photo by Greg Laden with permission)

else. Some elements of how that transition might have been achieved can be seen in markings at Chauvet Cave.

I visited Chauvet Cave in 2002 as a member of the Chauvet International Advisory Scientific Committee invited to see that the research work was good—it was exemplary (Clottes & Geneste, 2012). In addition to the more celebrated paintings and petroglyphs at the site, there is a set of disconnected marks which by themselves do not form an image (Figs. 6.7 and 7.2) (Clottes, 2001, 74) but which can be connected in our minds to look like the head of a rhinoceros, similar to other images of the animal in the cave. This happens because of a mental process which constructs the head of a rhinoceros from those marks, rather like the Gestalt involved with the illusion of the Kanisza Triangle (Fig. 7.3) (Kanisza, 1976). The seer completes the picture by a sort of gestalt (for discussion of Gestalt principles see Guberman 2017). This sort of completion of the picture will be familiar to all of us through the drawn illusions, such as the Kanizsa Triangle, that are the stuff of psychological understandings of perception (e.g., Noble & Davidson, 1996, 120–126).

Arguably, now, anything goes in art—art is in the eye of the beholder. As Shiner (2001, 3) said "Today you can call virtually anything 'art' and get away with it." The whole history of art from the very beginning has been to set up expectations that observers will think about what was in the mind of the producer who offered a particular object (or, in effect nowadays, none) as a work of art. Leonardo said that painting is a *cosa mentale*—the historical and by now cognitive expectation that we, the seers, have been given something to think about. Robb, wrote about Gell's (1998) approach to agency by saying that "Art acts upon people." On the contrary, all agency in art is about one fact that is the reverse of that statement. To continue a theme that experts are often wrong, both Robb and Gell were wrong about this. People react to art. It is the people who are agents, not the art. It is not just that the marks on the wall are marks on the wall but that seers think either that they represent something they look like, or, as the seer reflects upon what they see, that they have a thought about those marks. We can turn that around to make "The marks" the subject of the sentence, but it is still the people doing the thinking. Because of that

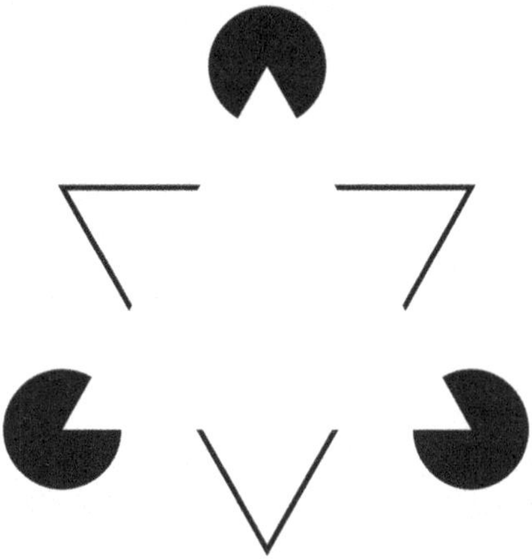

Fig. 7.3 Kanisza Triangle showing how, under certain circumstances, the eye completes the shapes that are not drawn. (Copied from Kanisza, 1976)

interaction between the seer and the marks, we tend to think that the marks have played a role but the only agency in the interaction is that of the seer. The nature of those reactions is determined by the cultural contexts people live in, and that includes their familiarity with art and its accompanying stories. It is an illusion to believe that art has agency, because that familiarity arises from the role of stories about art. Anything else is a trick of the writing.

As astute readers have probably picked up, it is worth noting that as I have been writing these remarks about agency I have repeatedly needed to recast a sentence because I originally wrote it with "art", or an equivalent, as the subject of the sentence. The idea that things have agency is primarily a product of the flexibility of language. People use language flexibly in a way that makes us think that things have agency. Express it how you wish. The idea of the agency of things results from the flexibility of language and a grammatical error.

There are various techniques in visual representation which entail ambiguity. Because art is an illusion in which individual marks create an image that is something previously unseen, there is a sense that all art is a metaphor at the simplest level, though it may have other levels of metaphorical meaning too (Stern, 1997). One example of this use of metaphor is the use of *trompe l'oeil* images, in which something is painted to create an illusion which tricks the eyes (the catalogue of illusions begins with *trompe l'oeil* in Seckel, 2004). The mural art from Punta Arenas, Chile, uses the features of the architecture (the vertical slab with the tv camera at the top) but adds to them by seeming to enhance them (Fig. 7.4). On the right of the vertical slab there seems to be a grand entrance to a house with one person looking out at the other young person sitting on the steps. But the entrance, window, steps and people do not exist.

Another example of such metaphors is pareidolia (Alais et al., 2021; Taubert et al., 2017), in which natural objects seem to represent faces (Fig. 7.5 and see Fig. 9.20). These are perfect examples of ambiguity. I previously suggested (Davidson, 2020) that the Makapansgat cobble[2] looks like a face to us because of the power of pareidolia (Fig. 7.5). The natural cobble (meticulously described by Bednarik, 1998) was found at the site of Makapansgat in South Africa in a breccia layer that is conventionally said to date to 2.5 million years. Yet, as it was not deliberately made but is an unmodified cobble, and we know nothing about the cognition of the pre-human ancestors, Australopithecines, who carried it 32 km to the site, we cannot say that it was anything more than a cobble, albeit a red one, to anyone before the modern human discoverers found it (Eitzman, 1958), and described it as a face (Dart, 1974).

The whole history and experience of art is about the expectation that when we see something deliberately made, we expect to have a reaction that the art is not just about the marks—it has, perhaps, some other "meaning". For deliberately made

[2] The technical boundary between pebbles and cobbles lies at 64 mm. This object, at 83 mm, is just larger than that boundary, so it is technically a cobble.

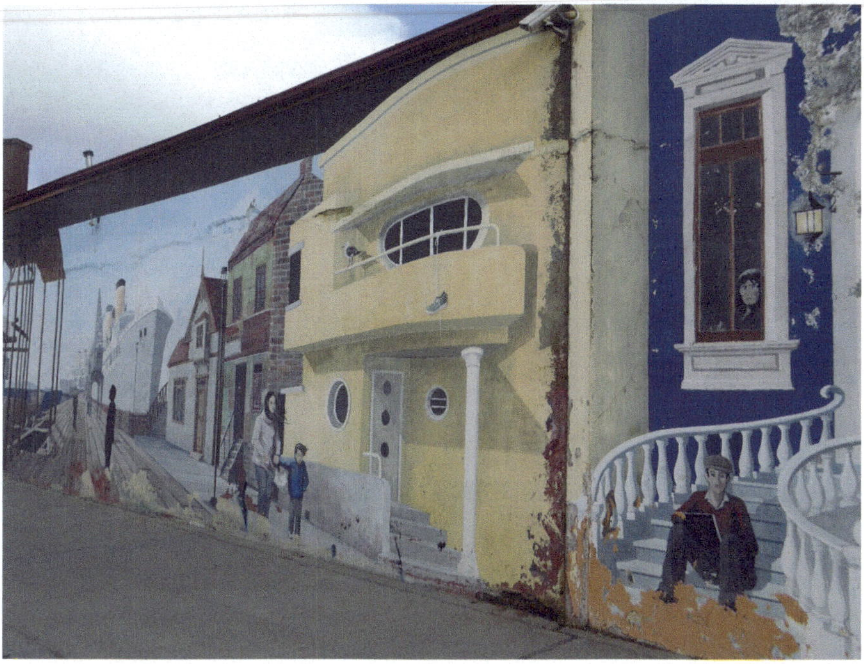

Fig. 7.4 Mural art in Punta Arenas, Chile, to illustrate *trompe l'oeil* painting. The part to the right of the vertical slab, clearly shows the use of painting to create the illusion of a woman looking out of the window beside the door at a person sitting on a staircase that sweeps up to it. (Artist unknown. Photo Iain Davidson)

objects and marks, it is what I call "the expectation of meaning." I would argue that is a more important principle than any discussion of beliefs about spirits, or power, or the presence of life or mind in an inanimate object. All beliefs like them depend upon the cultural networks that anthropologists can tease out of studies of the societies themselves. In the case of Makapansgat, we know it was not deliberately made, we know nothing (and have no evidence) about spirits or power, and we probably have some expectation that there were no networks among such Australopithecines that we might want to call cultural (Haidle & Conard, 2011). We humans, accustomed to the power of pareidolia, are the only ones who can, now, see this cobble as a representation of a face.

We could try to ask whether we should consider this cobble as art within the framework I have been setting up here. Let us consider (1) the producer, (2) their story about an object, which evokes something in the mind of (3) the seer about (4) a subject. If (1) the Australopithecine did not talk about anything (as I have argued, for example, in Davidson, 2003) then the chain of reasoning stops there. If (2) they talked but told no story—made no reference to anything that they had in their

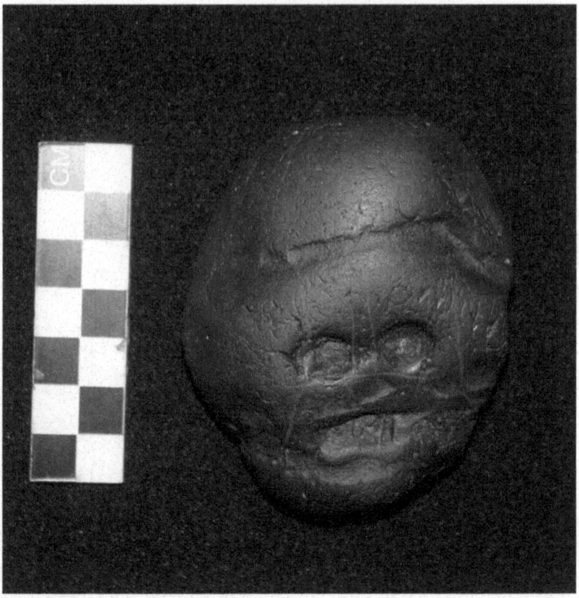

Fig. 7.5 Unmodified cobble, not deliberately made, from Makapansgat. (With kind permission of Bernhard Zipfel, curator of the cobble, and photographer)

minds—then the chain is cut.[3] If (3) the seer does not understand anything about how a producer can tell a story about art, then the chain stops. If (4) all known creatures that might look a little like the cobble were not human (as Australopithecines were not), then there is no case. Under all of these circumstances, all the typical circumstances of Australopithecines, there was no chance that the cobble could be spoken of or thought of as an image of a face.

On the other hand, if we (or Wilfred Eitzman, or Raymond Dart, or Bernhard Zipfel) see such a thing and say that it shows a face, then we (they) are entitled to say that it is—for us—art, but not that we know anything more about Australopithecines. The cobble tells us about us. Art is a mental phenomenon, a *cosa mentale.* We could try to reason in the opposite direction. If Australopithecines did not have minds like ours, it is difficult to pull off the trick of saying that this is evidence for their mindedness. If we argue that pareidolia would be a stimulus to the emergence of minds that could see a face in a cobble, then it might be part of an evolutionary process. For that to happen, there must be a chronological chain with appropriate links. There is none here. The natural cobble is an object unlike any other for 2 million more years. Sometimes in stating what we know about the past we have to be brutal about our own ignorance and wishful thinking.

Finally, in terms of the paradoxes and ambiguities, let us go back to my original citation of the concept of ambiguity by comparing the earliest art (Art 3) with modern art (Art 1) in the first and last columns of Table 7.1. We will discuss the

[3] There was once a series of discussions about whether Australopithecines could sing or perhaps sign (Falk, 1980).

Table 7.1 Seven types of ambiguity in Art (this started from the comparison between Art 1 and Art 3 in Table 1 of Davidson, 2020)

	Characteristics of Art 3	Characteristics of Art 2	Characteristics of Art 1
1	Pattern	Pattern with convention	Uniqueness
2	Repetition	Common style, often unique within it	Stands alone
3	Association of speech with marks	Marked objects are highly variable, some utilitarian, some associated with stories	Resemblance between object and subject
4	Producers engaged with seers	Producers engaged with seers for some but not all marked objects	Producers separated from seers
5	Persistence	Objects persist, but some were portable	Transferability—complex relations to money and markets
6	Social contexts of production	Social networks of interaction	Contexts open to all (who can pay)
7	Meanings given by relations between producer and seer	Meanings spread more openly within society	Meanings not necessarily limited by what can be seen in the work

comparisons between the different categories that get called art, in the next few sections, but we need to begin with the ambiguity between what we all recognize as Art 1 and the beginnings of art in Art 3. I introduced these columns in an earlier paper (Davidson, 2020, Table 1) to make the point that there are important differences between the early claims for art and art as it is recognized by people who are interested in art. But how different are they, really?

Let us look in a little more detail at some examples that show the ambiguity between Art 3 and Art 1.

7.1.1 *Pattern or Uniqueness*

When people make marks deliberately, the seers can work out that there was an intended meaning to the marks; the question is how anyone can identify that marks were made deliberately. In this, as an archeologist, I continue the argument that we recognize pattern through repetition of a particular set of marks. I think most non-archeologists use the same process of recognition. Sometimes the repetition seems to be no more than that repeated actions have been applied to an object. I am most familiar with it through the fact that archeologists recognize that fragments of stone were made deliberately because of the nature of the patterning, and that may not be so common. There is a problem, though, that repetitive behavior is also a feature of many other animals, and even natural processes. On one occasion a geologist claimed that I was trying to stop her mine because I identified flaked stones which she said were the natural products of "weathering" breaking up stones. The outcomes from the two processes are slightly similar but any trained archeologist can demonstrate that they are different. Natural processes of weathering of rock tend to produce fragments that reflect the operation of the processes of the weathering to produce some similar characteristics. They are almost all different from the patterning of the operation of human and hominin processes making flakes on rocks. We have all seen (hopefully less often than we used to) animals caged for display by themselves which engage in repetitive movements indicating their distress and we have seen the repetitive patterns animals engage in when courting—such as the pigeons on electric wires outside my study window. In terms of physical products, we also understand that the bears of Chauvet marked the walls by scratching the cave surface. Likewise, there are repeated marks elsewhere which we would not deem to be art (see the scratched abalone shells in Fig. 3.27 Left) and, it is possible, that some people regard marks that show repetition as art but incorrectly. The mine went ahead, but, at the insistence of First Nations people, the developers had to move their airstrip (because we could demonstrate that people had left cobble hammers at the stone flaking quarry after they had made stone tools with them) and the location of the worker accommodation (because we found piles of flaked stone marking campsites there).

What we do understand is that there is a reason why archeologists think the marked ochres from the South African cave site of Blombos (Henshilwood et al., 2002) (see Fig. 3.27 Right) and from some other South African sites, like Klein Kiphuis (Mackay & Welz, 2008), are significant (Fig. 7.6).

This is because there was a pattern to some of the marks which made parallel lines that crossed each other to produce a distinctive cross-hatching. At the later South African site of Diepkloof a different group of archeologists found fragments of ostrich eggshell marked with designs that include cross-hatching (Texier et al., 2010) (Fig. 3.28). We are much more inclined to recognize that these fragments of ostrich eggshell were deliberately marked because the cross-hatch pattern is often repeated, and there are other repeated patterns (e.g., delimited bands or zones of parallel lines).

On the other hand, for art since the eighteenth century, the prize has been uniqueness. Within a market for paintings (and other art works) the essence has always been that people will pay more money for works that are unlike any other—prints are routinely cheaper than paintings. They may share a stylistic similarity, and some of those who are interested in art have sought to define rules for what they would define as good art, such that style would involve sticking with those rules. There is a famous story that the eighteenth century English painter Joshua Reynolds pontificated (admittedly in a publication 8 years after the Gainsborough painting) (White, 2021), that blue was not a suitable color to dominate a painting (Gower, 1903, 78):

> It ought, in my opinion, to be indispensably observed, that the masses of light in a picture be always of a warm, mellow colour, yellow, red, or a yellowish white, and that the blue, the grey, or the green colours be kept almost entirely out of these masses, and be used only to support or set off these warm colours; and for this purpose, a small proportion of cold colour will be sufficient.

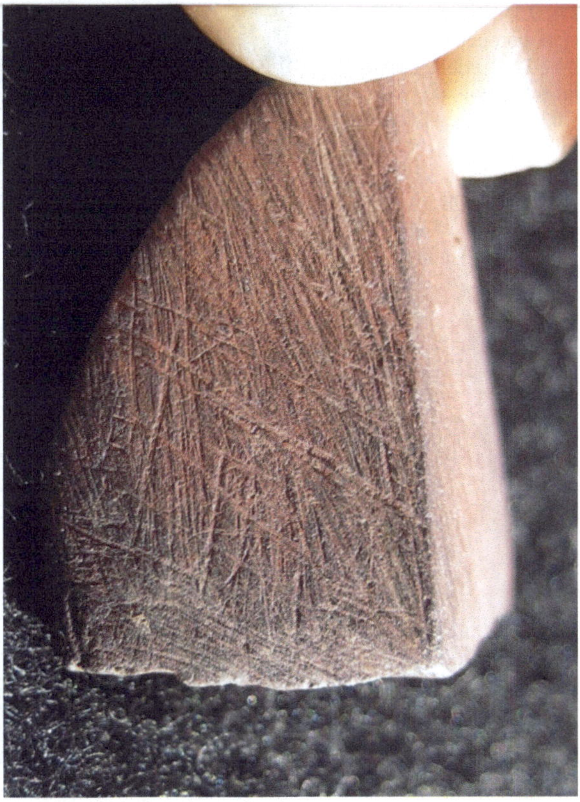

Fig. 7.6 Scored ochre from Klein Kiphuis, South Africa. This tiny fragment (it is being held upright by human fingers) shows that the ochre was scratched by grinding before being marked by the lines that are thought to be significant. The face to the right of the fragment is also ground. (Photo provided by Alexander Mackay)

In 1770, the equally English Thomas Gainsborough painted his *The Blue Boy*, to great acclaim (Fig. 7.7), and according to Ronald Gower it "in some respects is the finest portrait ever painted by an Englishman."

The Gainsborough portrait was much copied, according to Gower—even being used for the back of playing cards. This would not be surprising given the fame of the painting—but even in Gower's book, the reproduction is in monochrome. It is known that the copies were copies, not new originals.

The Reynolds edict is sometimes cited as a rule that Gainsborough was deliberately breaking, when instead it might have been a way of diminishing his painting (as would appear likely to judge by other comments on the rivalry in Hedquist, 2019, Chs 1 & 2). It may have been both, for Reynolds almost certainly stated his views earlier than their publication. Whichever it was, it demonstrates that art mavens may set rules but the combination of another producer and different seers can lead to different views. And, of course, such mavens also feel free to break their own rules (Fig. 7.8), as Reynolds himself seems to have done with a painting in even colder colors a mere 10 years after Gainsborough's success.

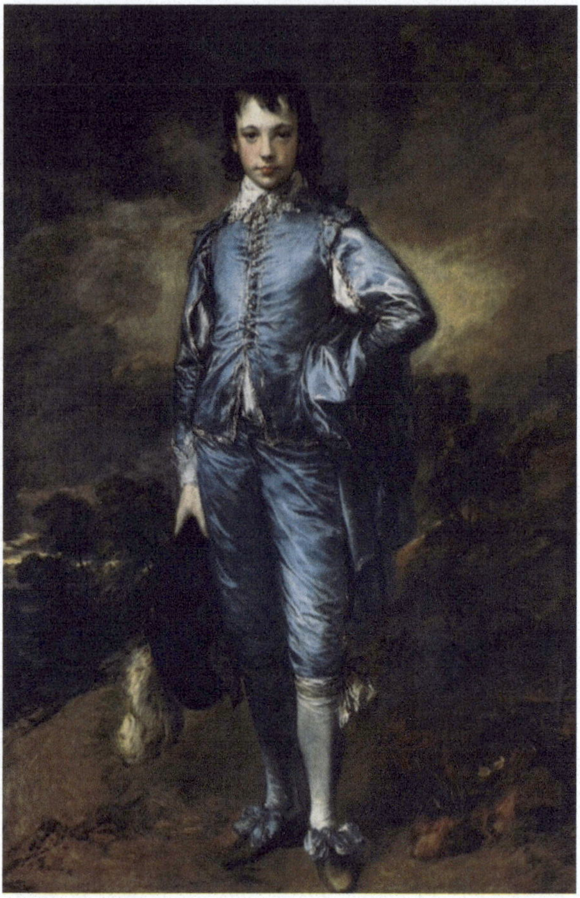

Fig. 7.7 Thomas Gainsborough 1770 *The Blue Boy*. (Public Domain. https://commons.wikimedia.org/wiki/File:Thomas_Gainsborough_-_The_Blue_Boy_(The_Huntington_Library,_San_Marino_L._A.).jpg)

7.1.2 Repetition or Stands Alone

The question of repetition in Art 3 is exemplified by the 350 thousand year-old marked bones from Bilzingsleben (Fig. 7.9) seen by some as representing very early attempts at art (Mania & Mania, 1988). Villaverde (2020, Fig. 10–13) assembled line drawings of examples from different sites in Europe and Africa from the Lower Paleolithic onwards to show that they were the basis of art and symbolism, which I discuss below in the section on Art 3 (Fig. 8.4). I have always argued that we tend to think that marks like this are significant because they remind us of old geometry lessons in school derived from Euclid, but that comparison might be irrelevant for those who were not brought up with the same values of geometry.

I realized the importance of this when engaged with the scratched bones from Bilzingsleben (Mania & Mania, 1988) (Fig. 7.9). I first saw a photo of these marks when I was asked to comment on a publication about them (Davidson, 1989). In this paper, the excavators Dietrich Mania and Ursula Mania had published magnificent

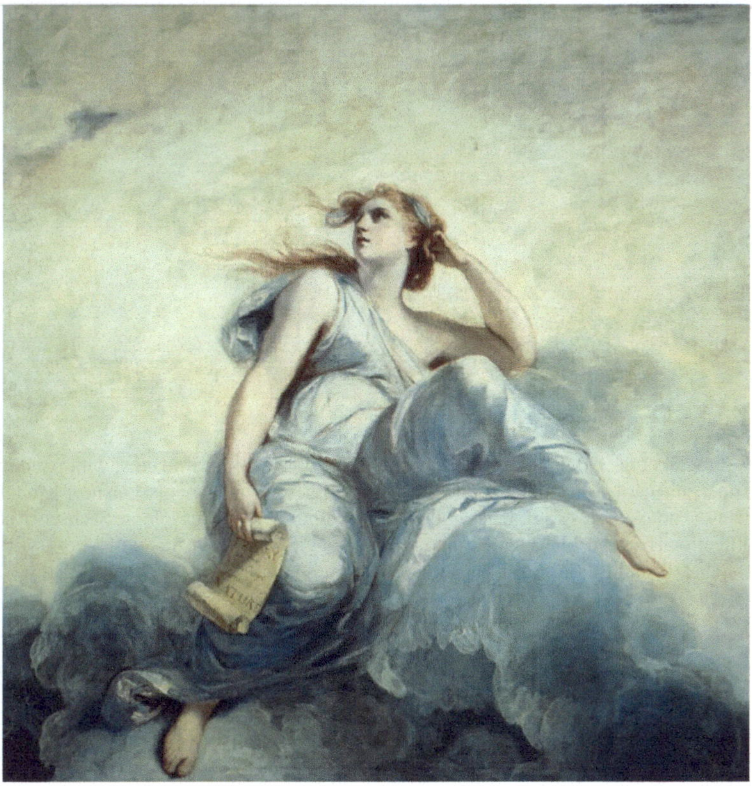

Fig. 7.8 Joshua Reynolds 1779–80 *Theory*. (CC-BY-NC-ND at the Royal Academy)

photographs and claimed that these select scratches were deliberate (Mania & Mania, 1988). At the time, my student Su Solomon was studying the way animals interact with bones and joined me in recognizing the marks of gnawing on the edge. She added that there were imprints of carnivore teeth coinciding with several of the lines. When she plotted them, they fell into the pattern of the teeth in a dog's jaw. I then took advantage of the opportunity to visit the site in East Germany just before the fall of the Berlin Wall in 1989. There the Manias showed me the many scratched bones from the site, as well as many of the bone artefacts. I also saw the small number of bones with scratch marks that the Manias thought were significant.

The bone in Fig. 7.9 is heavily eroded but it is marked both by very clear cuts—which appear parallel—and by the tooth marks and damage to the edges by gnawing. The cuts are single lines, not repeated scores—unlike the marks at Blombos or Diepkloof, and some recent finds (Płonka et al., 2024). I thought it impossible to know whether the marks were made by people using stone tools or, somehow, by the teeth, especially because the surface of the bone is clearly eroded. It is all too easy to ascribe significance to the marks while ignoring the eroded surface. In fact, it was impossible to tell whether the marks were deliberately made. Others have added

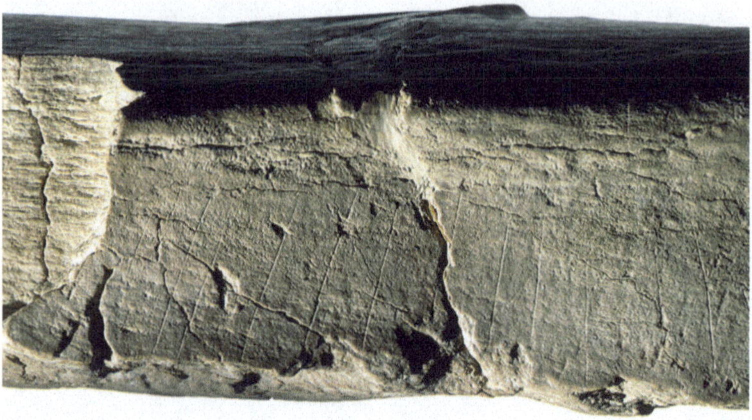

Fig. 7.9 Fragment of 350, 000 year old elephant tibia, gnawed by carnivores, with parallel lines marked on it, and carnivore tooth marks lying on those lines, from Bilzinglseben near Kindelbruck in Germany. (Photo courtesy of Landesamt für Denkmalpflege und Archäologie Sachsen-Anhalt—State Office for Heritage Management and Archaeology Saxony-Anhalt—with permission from Michael Schefzik and Bettina Stoll-Tucker at Landesmuseum für Vorgeschichte, Halle)

their expectation of significance by making their own drawings from the photos and leaving out the tooth marks and the signs of gnawing on the edges (one such drawing was by the person who edited the journal in which the original photos were published, Bednarik, 1995), but that does not solve the problem. I concluded that it was impossible to decide that the marks were significant, because there was, in addition, a large number of bones scratched with lines which did not have any geometric characteristics (Davidson, 1990). I am not convinced that there is anything more going on, cognitively for the producer, with these lines than the seers' recognition of patterns we have been taught to deem significant.

The question of whether Art 1 stands alone is also a little unclear. Most artists have sought to produce original works and most purchasers require it. I have already cited the example of Andy Warhol buying real Brillo boxes for an exhibition because it was easier than producing new examples of his art which reproduced Brillo boxes, but that might be a special case. One example concerns the same Joshua Reynolds who sparred with Gainsborough. It turns out that another artist Nathaniel Hone painted a picture called *The Pictorial Conjuror, Displaying the Whole Art of Optical Deception* (which had its own checkered history). For our purposes here, the story was that Hone mocked Reynolds for copying poses represented in other paintings (Newman, 1986; Stefanis, 2013) (Fig. 7.10).

It is a common feature of works of art that the producer pays homage to another work of art: Gainsborough is generally thought to have been paying homage to Antony Van Dyck paintings of a hundred years earlier when he painted *The Blue Boy*, and it is said that Beatrix Potter adopted aspects of *The Blue Boy* when naming and painting Peter Rabbit in his blue jacket (Paul, 2002). Tim Storrier is quite unequivocal about his Archibald Prize portrait, *The Histrionic Wayfarer (after*

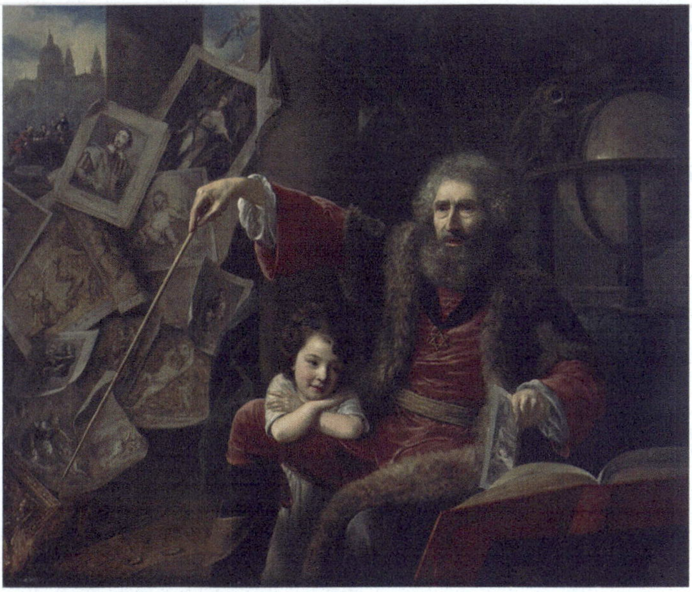

Fig. 7.10 Nathaniel Hone's *The Pictorial Conjuror, Displaying the Whole Art of Optical Deception 1775* (National Gallery of Ireland. Public domain, from Wikimedia Commons)

Bosch), being influenced by a Hieronymus Bosch painting from 500 years before, but inspection (not even close inspection) shows that the two paintings are very different, except in the angle of pose of the main figures (Fig. 7.11). Moreover, the Bosch painting can only be seen when the two side panels of a triptych painting, in the Museo del Prado in Madrid, are closed (and nowadays that is quite difficult). The full triptych, with the side panels open, shows *The Hay Wain* in which large numbers of people revel while the harvest is collected; the side panels show scenes from the early chapters of the Christian Bible on one side, and the descent into hell on the other.

The theme of reference to other works of art, continues with the Storrier piece because he also produced a sculpture, *Equine Impedimenta (The Artist's Mount)*, of a horse laden with artists' gear which he presented to the New England Regional Art Museum in 2023 (https://www.armidaleexpress.com.au/story/8191190/impressive-bronze-beast-with-connections-to-australian-racing-royalty/). NERAM runs a fund-raising art sale each year named after a Packsaddle that featured in the first such sale, and the horse, clearly, has such a pack saddle. (The print by Lloyd Rhys in Fig. 5.22 Left was purchased at that sale.) Some of the same gear carried by the figure in the painting is also visible on the sculpture (Fig. 7.12).

In addition, Storrier acknowledged: "Coincidentally after completing a group of works from this series, I was alerted to a small drawing by John Sell Cotman [Fig. 7.13], probably completed between 1800 and 1830, of an alert, slightly mythical pony loaded with the equipment used by an artist: palette, easel, canvases, brushes

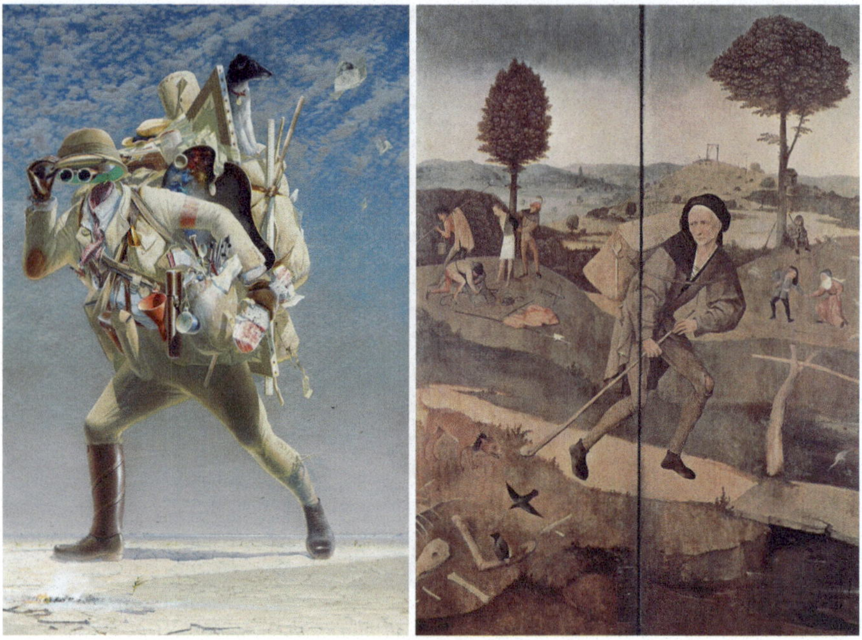

Fig. 7.11 Left, Tim Storrier 2012 *The Histrionic Wayfarer (after Bosch)*. (Original image from the artist reproduced with permission); Right, Hieronymous Bosch 1512–1515 *The Pedlar, closed state of The Hay Wain* (https://www.museodelprado.es/en/the-collection/art-work/the-haywain-triptych/7673843a-d2b6-497a-ac80-16242b36c3ce). (Public Domain from Wikimedia Commons)

and paper portfolio." Perhaps the point is that the indeterminacy of art allows the producer to be influenced by and evoke references to other works which may or may not be known by the seer. The producer weaves together through references known or unknown a whole canvas of meanings, precisely because works of art are not fixed in place but can move from one place to another through space and time, as well as being stored to be seen at other times by other artists and seers. Hone's mockery of Reynolds seems to be commentary on widespread practice in painting.

7.1.3 Marks and Stories or Resemblance

We cannot ever know about stories from a past before anything was written down. I am suggesting that whenever people make marks that persist, though, something important happens. It does not seem to apply to ephemeral traces of the presence of any animal, such as the scents they emit involuntarily. Whether we are archeologists or not, we have to produce arguments about the way stories might have been used to interpret persisting marks made in that remote time—constrained by our ability to use the evidence in front of us. My argument is that the deliberate making of marks produced an expectation of meaning—that the marks could be interpreted. Indeed,

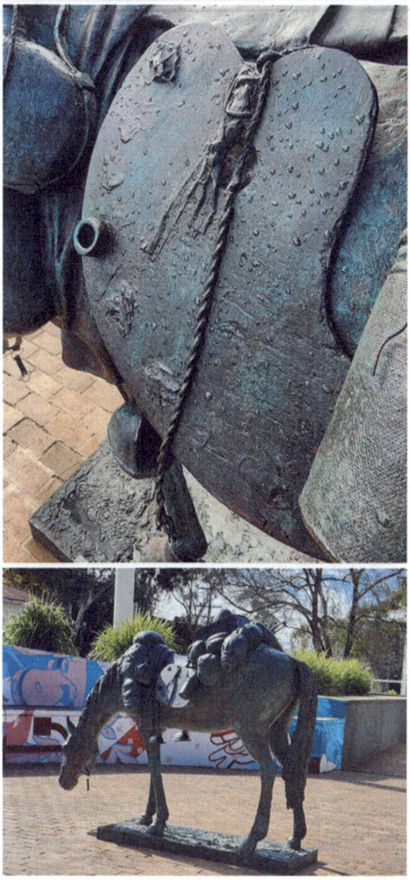

Fig. 7.12 Tim Storrier *Equine Impedimenta (The Artist's Mount)* at NERAM with detail showing a palette as also seen in Fig. 7.11 Left. (Photos Iain Davidson)

we can argue very straightforwardly that once the marks were made, they could only have led to an understanding of an original meaning if the mark-maker told a story or sang a song about them. Importantly, such meanings were inevitably indeterminate—we could only make an approach to the original reasons for producing the marks. Only stories about the marks can reduce the indeterminacy of meaning. The song or story had to be told either at the time, probably by the producers, or by some other means. Sometimes people needed a shared code about, say, geometric shape; sometimes through associations between original storytelling and particular marks.

In Fig. 3.27 Left I reproduced a picture of 100-thousand-year-old marks on the bottom of abalone shells from Blombos in South Africa which were made while stirring the contents of the shell (likely ocher-based paint) (Henshilwood et al., 2011). The marks were similar to each other in the two shells, yet no one is saying that the repetition means they have anything to do with the production of art. The pattern—the similarity between the marks in the two shells—derives from common actions, not from conventions. In discussing the characteristics of ritual, Rappaport

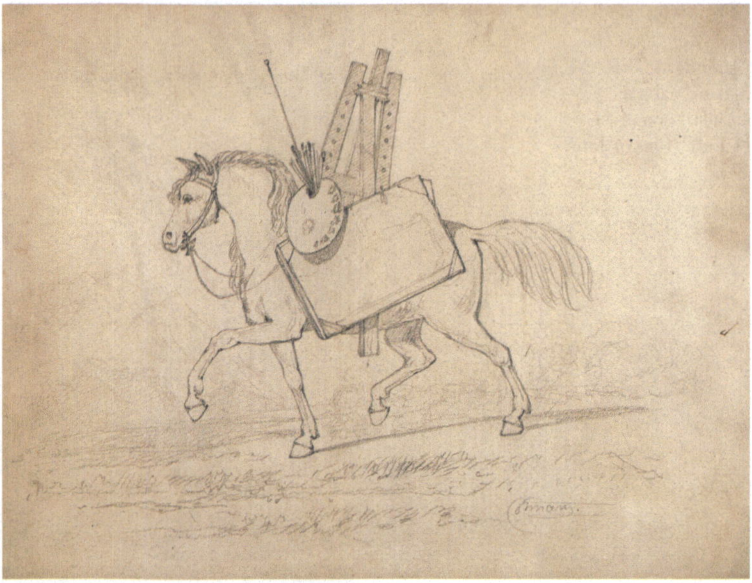

Fig. 7.13 John Sell Cotman, 1782–1842, British, *Horse Loaded with Artist's Paraphernalia*, undated, (Wikimedia Commons, Public Domain. Yale Center for British Art, Paul Mellon Collection, B2001.2.741)

(1999, 52–54) distinguished between two sorts of repeated actions; those that I watch each day, for example, of pigeons outside my window; and those from which I infer a deeper meaning. The first Rappaport calls "self-referential" and the actions reflect "the current physical, psychic or social states of individual participants, or of the body of participants as a whole, [or] are confined to the here and now." I think it is fair to say that the scratches from stirring paint in the abalone shells are of this type.

On the other hand, the examples that pervade the work of Munn (1973) (Fig. 6.5) and Green (2014), as well as the example of the songs by Mrs Hansen recounting her traditional stories to her daughter Isabel Tarragó (Fig. 3.24) demonstrated that, for certain categories of image production there are and were accompanying songs and stories. These are not an example from that distant time before written traditions, nor even from beyond the reach of oral histories, but they illustrate the alternative to the self-referential movements of pigeons or paint-makers. Within the Country of Mrs Hansen and her family, the Mulligan River People, there are known petroglyph sites (Fig. 7.14). Mrs Hansen showed me (and others) one sign of a circle with crossed lines inside it which was used by important people "when they wanted to travel unseen". That use was something we could not possibly have known without Mrs Hansen's story. The same goes for many of the marks in the sites of that region, and elsewhere. These were not scenes, but the art could not possibly have meant anything without the coding of meanings implied for geometric or non-representational signs.

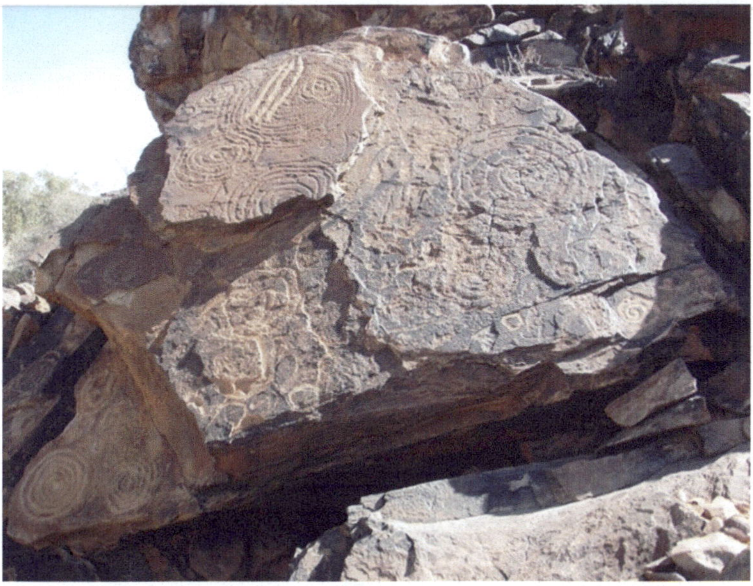

Fig. 7.14 One slab among many engraved at different times at Nganganterra, in the Country of Mulligan River People on the edge of the Simpson Desert, Australia. Signs are all geometric or not representational. The grazier Bill might call these marks "scribbles" but there were made at three different times (at least), and they are all clearly very structured and deliberate. (Photo Iain Davidson 2008) (Reproduced with permission of the Native Title holders of the Mulligan River People)

In the previous chapter, I referred to an analysis of scenes at the Iberian site of Parpalló that represent interaction between a small number of participants shown in the image (citing Fig. 6.11) in essentially static interactions (Villaverde Bonilla, 2021, Fig. 15.2b). Such patterns can be considered as self-referential in this way, though in their broader context, we will see that there is more to it than that. Villaverde went on to show that in much the same region of Iberia, Levantine Art that is separated from the plaquettes of Parpalló by many thousands of years and by dominant economy, there are paintings on the walls of the rock shelters which represent scenes that show much more activity than before. In one instance, at La Palla, "up to four stages are depicted: the sighting and firing; the following of the trail; the final shot to the injured prey animal; and its collapse, including the blood vomiting and agony." Another scene, at La Araña, shows people firing arrows from their bows at a herd of ibex, killing some of them (Fig. 7.15).

Villaverde (2021, 232) continued:

> In Levantine art, this continuous suggestion of past or future events, including both the background and the expected resolution of the actions, is repeated with sufficient consistency to make us think that this temporal narrative concept is part of a standard means of graphic expression on the part of their creators. The vividness of some of the scenes comes precisely from their representation of the passage of time: such scenes are not static representations of a moment, but actions that develop through time and contain within their structure the keys to reading them.

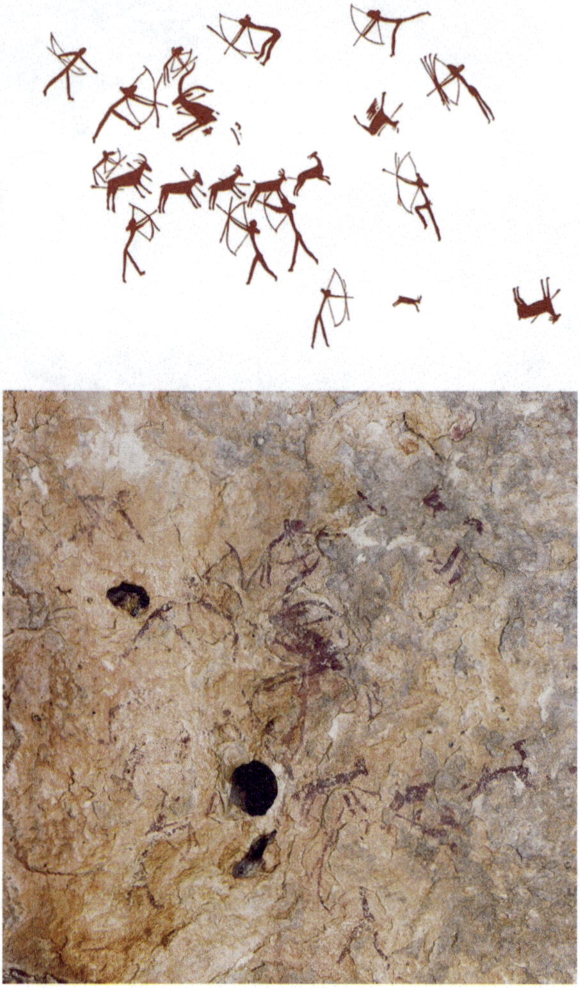

Fig. 7.15 Photo (lower) and tracing of a scene from the Levantine Art at La Araña, Bicorp, Valencia showing ibex being hunted by men with bows and arrows. (From Villaverde 2021 Fig. 15.4)

In Rappaport's (1999, 52–53) terms, this represents the second aspect of rituals that makes them distinctively human. The story of the scene is more than a juxtaposition of static elements but represents what he calls a canonical message about events linked by their depicted relationships to describe a past and (at least) a present. The message of the interactions shown in the painting is contained in the narrative story. The static painting has become a story.

In the previous chapter, I argued that the production of such scenes reduced the indeterminacy of marks, and most non-specialists in rock art are familiar with the idea that artists once depicted people hunting animals with bows and arrows, exactly as at La Araña. Such scenes are not as common as all that: indeed, Paul Bahn (2006, 575) stated, with the authority of someone who has seen much of European cave art,

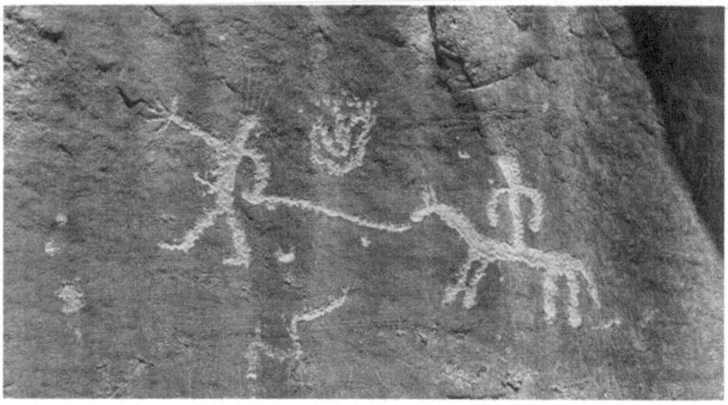

Fig. 7.16 Simplest scene at Nine Mile Canyon, Utah—one person (with headdress?) appears to be leading another on horseback. (Photograph Iain Davidson 2011)

"in the tens of thousands of images, there is not a single clear depiction of a hunt". The simplest version of a scene is when the producers represented an interaction, say between people and animals or between two animals, as in the relatively recent image (within the last 1000 years) from Nine Mile Canyon in Utah (Fig. 7.16) (as described in Spangler, 2013). At some level there is a representation of aspects of the things represented: the human figures may have arms and legs; the horse has legs and ears, and there is a line between the vertical human and the head of the horse such as you would expect if one was leading the other with a rope. As I have discussed before (Davidson, 2013), the resemblance to real aspects of the individual creatures represented is said to be an iconic resemblance in the semiotic terminology of Peirce (1986). In creating semiotics Peirce defined three classes of signs relevant to his study: Index, Icon and Symbol (and many things related to them) (although it is generally best to go back to the original, some derived works may be useful, Deacon, 1997; Kissel & Fuentes, 2017; Peirce, 1986). Peirce (1986) defined a sign as "something which stands to somebody for something in some respect or capacity," and taking a cue from Eco (1986), I suggest it is a thing you can see from which you can work out something about another thing you can't see.

Peirce (1992, 13) himself wrote: "An icon is a sign which stands for its object because as a thing perceived it excites an idea naturally allied to the idea that object would excite. Most icons, if not all, are *likenesses* of their objects" (italics in the original). Thus, Boas and Gombrich (in Chap. 3) both remarked that mutilation of images "excited" (though perhaps we would avoid that word here) ideas of harming real people. The individual agents in the rock art scene have some resemblance to what they represent—a walking person, a rope, a horse, and a person riding. But the resemblance can be quite limited—so this represents a person on a horse but neither The Lone Ranger nor Silver. It could be The Lone Ranger and Silver, but there is nothing about the marks which allows that identification rather than an identification as Roy Rogers and Trigger, and given the cultural specificity, a seer could tell a story about their choice, but it would not be constrained by the image itself. Iconicity of the named characters requires some further characteristics of the image to excite an allied idea, in the same way that the images of faces in "ikons" require something more to be seen as likenesses of the religious figures they purport to represent (Figs. 3.3 and 7.18).

My earlier point about art and communication—that there is convention in the relationships between people and the things that get to be called art—reflects the fact that people use conventional means of communication called "language" and that the sum total of those conventions makes up their culture (Davidson, 2013). Those conventions arise from the variety of ways in which people recognize and use signs. Peirce (1992, 228) wrote that a sign is "something which stands to somebody for something in some respect or capacity," thus emphasizing the centrality of "somebody" for the interpretation of all signs. For example, people recognize the marks made on the ground by different animals as they pass, often with great accuracy—naming species and age-class of animals, for example—and there is remarkably little evidence that any other animal, not even monkeys, do that. To be sure, there are reactions to scents left, for example, by canines marking fenceposts, but we cannot be sure that there is any more reflection than that the scent has been left. Those marks made by passing animals would be called indexical signs of the animals, as would the scratch marks made by bears on some of the painted panels of Chauvet Cave—they indicate that bears made marks on the walls that were subsequently painted by people. The scratch marks are both under and over paintings by people in the cave (Clottes, 2009). Such indexical signs are part of Peirce's (1992, 14) original definitions in semiotics: "an Index stands for its object by virtue of a real connection with it, or because it forces the mind to attend to that object." A sign left by a passing animal—such as a paw print, scratch mark, or a scat—would be an indexical sign because it would be quite specific to that animal, there would be a real connection. The apparent lack of other animals that interpret the presence or behavior of another animal from a sign probably shows that humans are needed to give meaning—to turn a sign into an indexical sign.

A sign that is both conventional and has a (somewhat) arbitrary relationship with the subject it refers to—like the marker on a toilet door or the crucifix worn by some believers—exemplifies the symbolic sign. Humans are needed to interpret the sign itself—that is an indication of a male human because it looks a bit, but not much, like a man—and most often a sign of that type is accompanied by a similar but different sign which can be taken as a sign of a female. They are also needed to understand the whole cultural context of the meanings the sign stands for—that symbol marks a room where men and not women go inside to do something. Importantly, we all have to learn the convention.

A symbolic sign is not the same as a religious sign—religions are special cases of the accumulation of conventions that enable symbolic signs to have meaning, but there are many other symbolic signs. Between these is the iconic sign,[4] which, like the human leading another on horse-back in Fig. 7.16, according to Peirce's definition "may represent its object mainly by its similarity"—the symbol on the toilet door resembles in some arbitrary way the outline of a human but conventions give it meaning about the nature of the room it is attached to. In the same way, the

[4] I prefer the phrase "iconic sign" because then it cannot be so easily confused with the images on a computer screen which have become known as "icons," nor with famous places or people who are called "icons" by the media. Those places or people are actually indexical of a particular place or reason for being famous. Nor can they be so easily confused with the religious images that get called "ikons" (Fig. 3.2).

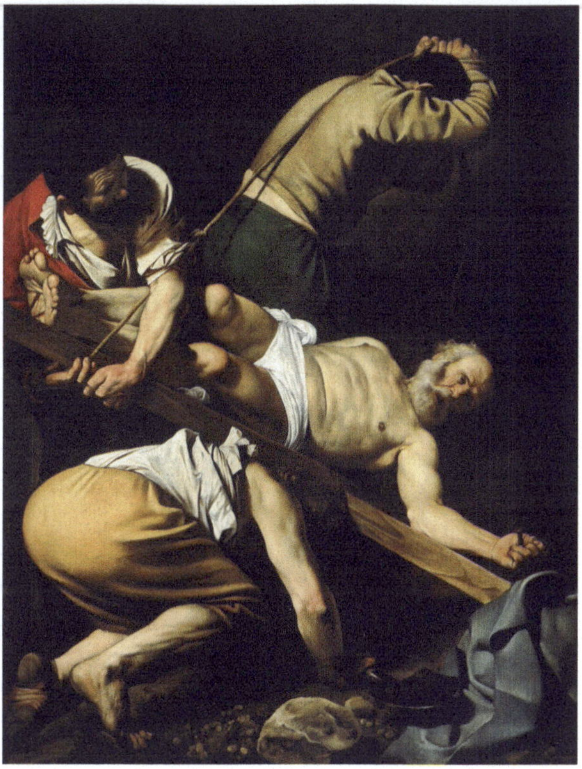

Fig. 7.17 Caravaggio *The Crucifixion of Saint Peter* (1601). Oil on canvas, 230 × 175 cm (90.5 × 68.8 in). Cerasi Chapel, Santa Maria del Popolo, Rome. (Wikimedia commons Public Domain)

crucifix worn by some adherents of a religion has a supposed (conventional) iconic resemblance to an original wooden cross used by Roman executioners in Palestine (as represented by Caravaggio in Fig. 7.17).

By convention a crucifix is an indexical sign of the person's adherence to one particular religion and symbolizes the probable beliefs of the wearer. In considering how mark-making turned into art, the objects did not belong only to one category of sign but necessarily moved between them as people saw them differently, or learned and internalized stories about them. Others have considered a more subtle version of semiosis by considering how objects might have functioned as signs, but signs cannot mean anything without people to provide either meaning or function (Kissel & Fuentes, 2017). People make signs mean. They do this sometimes directly by teaching others, learning from others, by scaffolding, or by experience (e.g., De Houwer et al., 2013; Scalise Sugiyama, 2011).

Importantly, all art can be considered as symbolic at some level. Thus, I would argue (against some opposition) that an iconic sign is really an example of a symbol. It certainly represents through its similarity, but there is an element of arbitrariness about choosing which characteristics of a subject represent it. In addition, for others to see any representation as an iconic sign requires that producer and seer share conventions about the selection. Religious icons—ikons—have some iconic

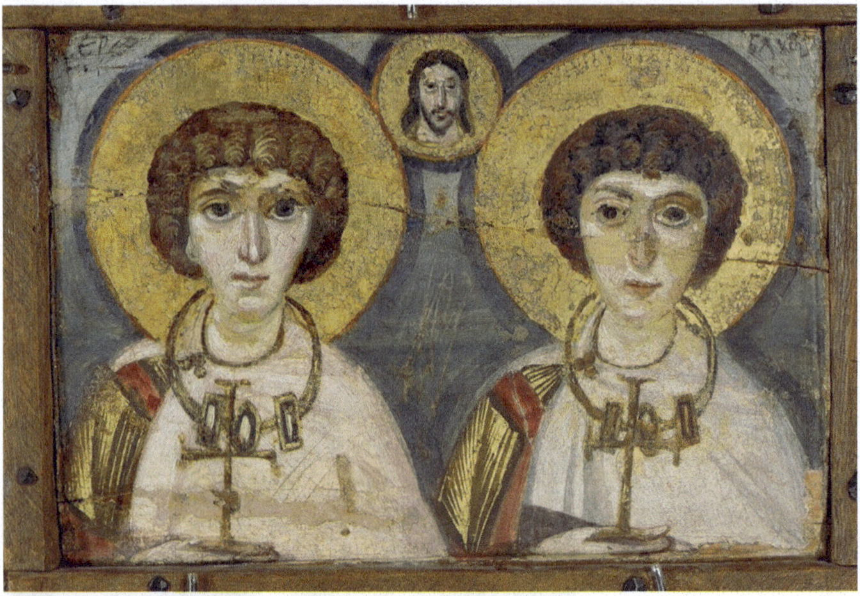

Fig. 7.18 Ikon painting of Saints Sergius and Bacchus from the Khanenko Museum, Kyiv. From seventh century Byzantium. Cold encaustic on pine, gold leaf. (https://web.archive.org/web/20191222175614/https:/khanenkomuseum.kiev.ua/en/collection/item/saints-sergius-and-bacchus). Public Domain

resemblance to a human, though the conventions about representation in that context mean that the resemblance is very stylized (Fig. 7.18). There is no specific information that enables a choice between The Lone Ranger and Roy Rogers, or in this case between Sergius and Bacchus.

Iconic signs therefore meet the criteria to be symbolic signs—an apparently arbitrary choice of which features to represent and convention about which choices to make. A representational image is a symbol and as the image becomes less similar to the thing it represents, so it must be more conventional to be understood as an iconic image. In the limiting case the image is just a mark with little iconic similarity to the subject. Any interpretation of that mark requires that seers be participants in the cultural conventions shared with other people. The rules of interpretation apply among the people of a culture but similar forms may vary across regions as people's culture varies. And rules can change through time, as different people learn to interpret signs differently and as signs change slowly and are opened to different interpretations.

As we have found with other types of ambiguity, there seems to be a case that there is actually an element of similarity between the earliest marks that could be called art, and other types of representation which most would accept are really art.

7.1.4 Producers Engaged with Seers or Separated

In this section, I want to begin at the other end of the story with an idea about, or caricature of, artists of the last couple of hundred years, that they suffered poverty for their art, and thus were set apart from the rest of society. The ambiguity would, then, be that these artists and their art—Art 1—were somehow separated from their seers, and in contrast Art 2 and Art 3 were produced in contexts that were more social and artists were fully integrated with their society. The truth is not necessarily easy to determine, though the image of the artist starving alone was undoubtedly romanticized in the nineteenth century as a version of bohemian life—as through the opera *La Bohème*. The classic stereotype is Van Gogh, whose paintings now command the highest prices, but who only sold one, *The Red Vineyards*, in his lifetime (Fig. 7.19), and there is a whole other history of how the painting was transferred from Anna Boch's hands to the Moscow Museum where it is now. In the absence of sales, Van Gogh depended on his brother Théo for money, which he mostly spent on paints and canvases (Naifeh, 2021, xxvii). Rather earlier "[d]espite the riches of the Dutch Republic, the wealth acquired through trade and speculation, many artists died in poverty … Rembrandt went bankrupt and spent the last decade of his life in reduced circumstances … Vermeer spent his last years in financial hardship…." (Stonard, 2021, 294). But these were not cases of starving artists in garrets but cases of misjudgments of economic conditions from positions of wealth, either because of family background or by earnings from the sales of their art.

On the other hand, it could be argued that successful artists have not been poor recluses but engaged in strong social contexts. Many of them seem to have been well-placed in society and the eighteenth century saw the foundation of Academies of Art, one of them founded by Joshua Reynolds (Shiner, 2001, 101) which provided a further social context for individuals to talk to others about their work. The Impressionists, of course, found the stranglehold of the members of the French

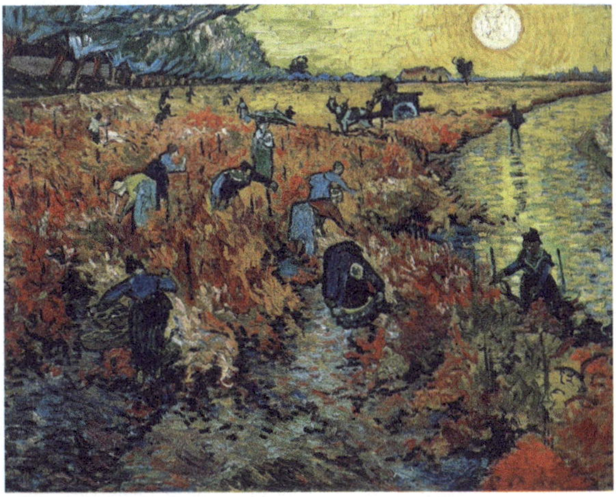

Fig. 7.19 Van Gogh *The Red Vineyards*. Originally sold to Anna Boch. Now in Pushkin State Museum of Fine Arts, Moscow. (From the Wikimedia Commons. Public Domain)

Academy so oppressive that they broke away. Even Van Gogh was not isolated, as he successfully persuaded Paul Gauguin to share his house in Arles. Picasso was from a relatively well-to-do background in Spain.

Determining the objective reality of the situation is not simple, given the abundant bias that can intervene between asking the question and finding evidence to provide an answer (Melchionne, 2007). One recent study, for example, looked for economic data in the US Census and concluded that the image of the starving artist did not correspond to reality (Filer, 1986). Of course, it could take no account of people who might have wanted to be artists but had given up on the idea rather than endure penury.

So, the ambiguity between the very beginnings of art—Art 3—and the art of the last 3–400 years—Art 1—is between marking objects, however we construe that, and a stereotype of the way recent art production is perceived by some. The extent to which this difference just derives from a caricature of Art 1 in some popular imagination depends on the important question of how marks became meaningful in any way in Art 3. At its simplest, people (seers) think that Art 1 has high value because of their own social conditioning—their culture means that they have absorbed the idea that certain sorts of production have artistic value where other sorts do not. Other marks are meaningless unless people learn that they are meaningful, and as a result stories are essential to making marks symbolic (see Nowell, 2023). If this were not the case, the bumps on the ceilings of European caves would also have meant something to the cave bears that saw them: artists at Altamira (Beltran & Saura Ramos, 1998) in northern Iberia made use of the bumps on the ceiling to give the appearance of 3-dimensional painting, but there were also clear marks made by now extinct cave bears (Breuil & Obermaier, 1935, Fig. 2) (Fig. 3.17); at Ekain (Barandiarán & Altuna, 1969), in the Basque country of the western end of the Pyrenees, complete skeletons showed that hibernating bears died in their beds not far from the paintings; people painted at Chauvet Cave (Philippe & Fosse, 2001), in eastern France, on walls that were already scratched by cave bears and the bears scratched the painted walls afterwards (Fig. 4.4). In Chauvet Cave, people also painted images of the cave bears (Fig. 7.20), and we are reasonably certain that

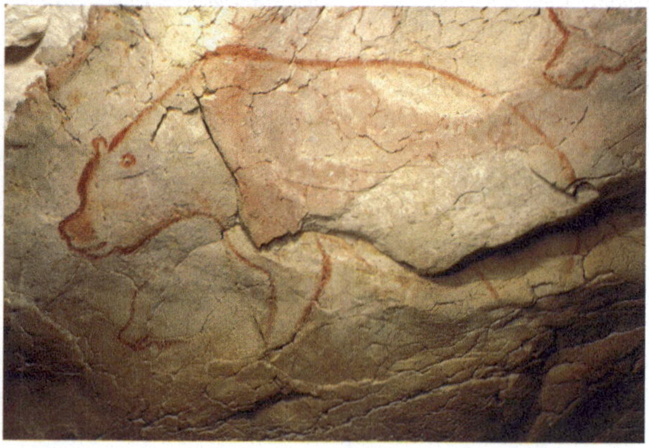

Fig. 7.20 Chauvet Cave. Painting of cave bear (though I am always tempted to say that this is a picture of the 3 Bears). (Baffier & Feruglio, 2001, Fig. 193). (Copyright granted by Valérie Feruglio/French Ministry of Culture)

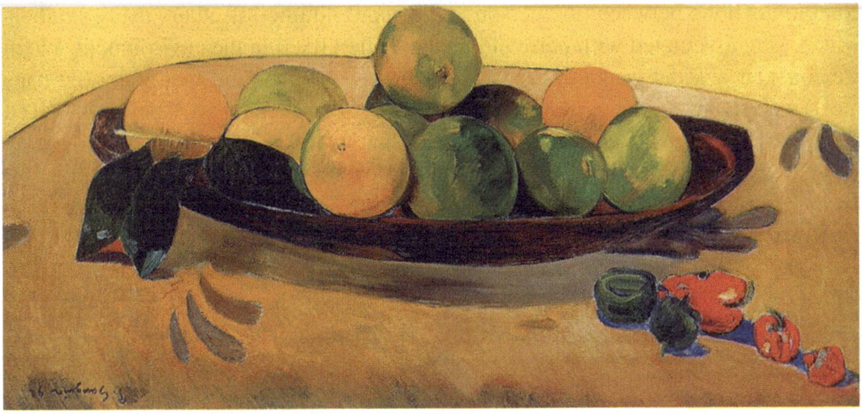

Fig. 7.21 Gauguin 1892 *Still Life with Oranges*. (Public Domain from WikiArt)

the bears did not paint the images. In reality, we can be more certain than that. Indeed, it is more likely that for a long time there was a practice of people butchering bears for their furs (Romandini et al., 2018).

As I have indicated in Chap. 4 (Table 4.1), the process through which people communicate by making marks entails mental processes both by the mark maker and by the seer of the marks. It is a *cosa mentale*. The semantic role of art is qualitatively different from that of stone tools, for example, because of that cognitive requirement—the stone tools leave a physical product which can be pondered (or used), but art leaves a material product whose nature is primarily mental. You can cut with the flakes left at a knapping floor, but you cannot catch an animal in the nets marked on the ochre of Klein Kliphuis (if you think that that is what the marks represent) (Fig. 7.6). But then you also cannot eat the fruit in a still life by Gauguin (Fig. 7.21). The art can be seen but the story that led to its creation cannot be heard.

In the end, the ambiguity may be less than the stereotypes lead us to expect. It remains the case that artist and representation must have been included within the daily lives of other people for all early art. It is certainly the case with Art 1. The difference, perhaps, is a result of the practice of setting places aside for seers to enjoy the works of the producers—museums and galleries to exhibit the works of artists and set them aside from society.

7.1.5 Persistence or Transferability

I have emphasized, particularly in Chap. 4, that Art 3 is often persistent in a particular place, and that that persistence has an effect on the interactions between people in a society. When we move to consider rock art, of course, we know about it precisely because it has persisted on the permanent surfaces onto which or into which it is marked. As I outlined before, early art, whether in Sulawesi or Swabia, involved communications between one person and a small number of others using symbols which were associated with particular places, often fixed in the environment. When we consider a wider range of archeological art, Art 2, we find that many (perhaps most) materials are either art objects themselves, or marked by art. They may indeed be portable, but they "belong" to the society that made them.

Something changed at the time generally known as the Renaissance (southern or northern). Earlier church art or monumental art was often persistent, though clearly some art was not. Interaction between that persistence and the other artistic things associated with it (such as the Gutenberg Bible illustrated in Fig. 5.15 or other illuminated works, or the Saxon sculpture in Fig. 6.10) depended in part on the other developments in communication that I have outlined in Chaps. 5 and 6.

One of the things that seems to have made a difference for Art 1, however, was the production of art in ways that allowed the art itself to be transferred from the producer to consumers and that was a product largely of changes in the technology of production of art. As I have noted in Chap. 3 these involved changes in the support, such as the invention of paper and the use of canvas for paintings; changes in the medium both through the invention of printing and the making of blocks to print pictures—with the associated expectations of multiple identical copies; through changes in paint technology, particularly through the use of oil paint; and through development of production techniques as a result of the scientific description of perspective and the addition of *chiaroscuro* which together enabled the trick of representing an appearance of three dimensions in two.

But these technological developments also permitted the production of art such that individuals were able to make their own art collections. This, in turn, required four things: (1) that art should be available for collectors to acquire; (2) that artists should be free to sell their work; (3) that there should be people in a position to purchase art; (4) that there should be places for collectors to keep their collections. For the first of these, the producer should have been free to make art which would then be available to purchasers. Leonardo's *Última cena*, was commissioned by Ludovico Sforza (an important person in Milan) and is not in private hands because it was painted on a persistent place, the wall of a church (Fig. 7.22). Copies of the painting were produced within 30 years of its completion, and these found their way into collections and were the source of images when it came to repainting the chapel. The history of commission, production, deterioration and repainting and appearance in recent works of popular culture is very sad (https://en.wikipedia.org/wiki/The_Last_Supper_(Leonardo)). By the time Vasari (1998, 289–91) wrote about it, in three short paragraphs, the original painting had already substantially deteriorated.

The second requirement was that the art was not already contracted by someone else, so that it was available for collectors to purchase it. The *Última Cena* painting was commissioned, so that even if it were not attached to the church wall it would not have been available for sale. The third condition depended on inequalities in wealth, which have been there for a very long time but in addition there was a requirement that people other than the titled could purchase the work rather than buying the labor of the producer through patronage, commissions or sponsorship.

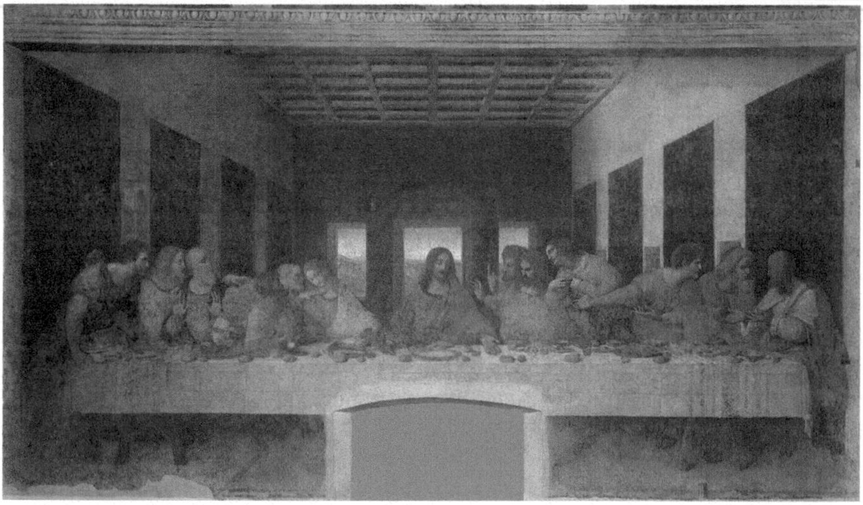

Fig. 7.22 Leonardo da Vinci 1495–98. *Última Cena* on wall of Chiesa di Santa Maria delle Grazie, Milan. (Public Domain from Wikimedia)

At first these inequalities were a product of the emergence of different classes, particularly through dealings in land and other property. That is one reason why so many works of Art 1 have a history in royal or religious collections, as in Italy, Spain, France or England, for example. Later inequalities involved the accumulation of wealth through the values that developed trade of other commodities, such as the spices that fueled early voyages of colonization. Works of art themselves also acquired value. In addition, once colonists travelled to distant places, the commodities included the bodies of other human beings who were traded to enormous profit into slavery. In some cases, those enriched by that trade in human misery used their wealth to invest in grand houses and art to put on the walls.

There are, of course, some exceptions (outstanding exceptions at that) about Art 1 being made on non-persistent surfaces—but in general that means that producers were not in a position to dispose of the work, and may never have intended to. In particular, one room in the Prado Museum in Madrid is devoted to the paintings by Francisco Goya of his so-called Black Paintings. He painted these towards the end of his life on the walls of his house outside Madrid and every account suggests he never intended for anyone else to see them. They remained on the walls for 50 years after Goya moved away from that house, at which stage, with the characteristic assessment of priorities by people who are interested in art, they were transferred to canvas and now are on the walls of the Prado (Figs. 3.25 and 9.21).

The third condition is that there be people who had acquired enough wealth (it is very unlikely that they had earned it) to purchase the works. This is really a matter for economic historians to comment about the class stratification that led to some having much wealth and much purchasing power, while others did not. In some sense, this was little different from the earlier situation in which people would commission art if they controlled the means to pay others to make it. The difference now was that the stratification of society according to the acquisition of wealth enabled those with money to purchase the works of artists. To this end, the eighteenth century saw the emergence both of sales of paintings and of academies which could pronounce on the worth of the works. In some respects, the Academies functioned as institutions separate both from the religious and from those who called themselves "aristocracy."

Once those conditions existed, the fourth requirement was that there should be museums or grand houses where Art 1 could be stored. Again, the economic historians could tell the story of the emergence (and toleration) of that sort of inequality. It is the case, though, that the British Museum was founded with works (Greek sculptures) accumulated by fair means or foul (Hamilakis, 1999) by Thomas Bruce and purchased for the British by Act of Parliament in 1816, as we will see in the next Chapter.

7.1.6 Context Within Society or Availability to Everyone (Who Can Pay)

It is easy to slip carelessly between the contexts for the production of the first things that might be art (Art 3) into a discussion of the first things that were self-consciously made with deliberate design, and which may vary from culture to culture (Art 2). "There are only artists," declared Gombrich, yet he omitted the fact that such artists live in communities and get the stories which give their work context from those around them—as integral parts of the culture of all of them. One of the points here is that the contexts pervade the society of the producers without seeming to do so. That is my understanding of what culture IS. Thus, for example, there is a discussion of the consequences of the widespread availability in 1870 of (slightly doctored) photographs of Michelangelo's *The Creation of Adam* (Steinberg, 1992) (see Fig. 7.23). Vasari (1998, 444–5), once apprenticed to Michelangelo, wrote an authoritative book about the lives of (all Italian) artists in the middle of the sixteenth century. He described *The Creation of Adam* in half a paragraph, with no illustration. We might ask what people who had not seen the ceiling actually knew of it before the publication of the photographs. Who knew of its colors until more recently than that? The publication of the color photographs was the first time in over 300 years that images had been available which allowed people to see the work at something like close quarters and inevitably discussion ensued about what the human figures might mean (especially since Adam, said to be the first human, had not quite been created). But there were no color photographs in 1870.

We need not detain ourselves with the details here (which are argued by Steinberg) except to note that the whole discussion was centered on a belief that the stories in the Bible (about God, Adam, Creation etc) were, somehow, true despite the contradictions, as well as other assumptions about the main players in the story (Eve,

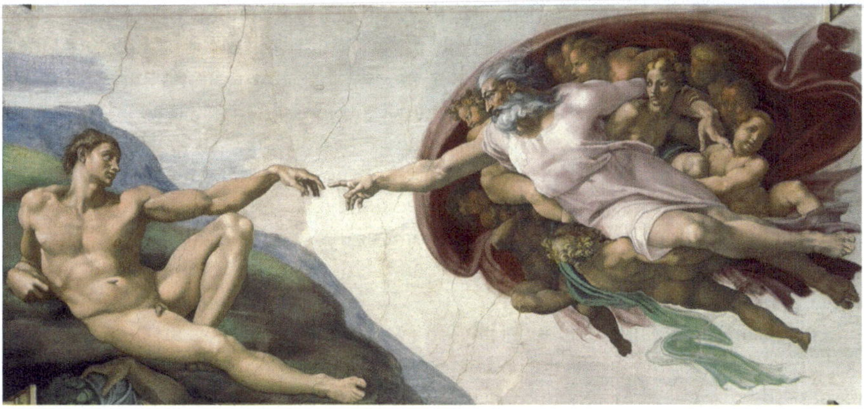

Fig. 7.23 Michelangelo 1508–1512 *The Creation of Adam* (cleaned version). (Wikimedia Commons Public Domain)

Lucifer, Michael, Jesus) and the way they were treated in art. These are what I referred to in Chap. 3 as unquestioned beliefs: they dominated the context of production of the paintings (Popes, Chapel, Michelangelo's personal cultural background). The whole discussion, for example by Steinberg, depends on assumptions about the biblical context that would have been well-known and accepted in Europe at Michelangelo's time, but may not be to the same extent nowadays. Funded by Japanese television, the paintings were cleaned and photographed in color more than quarter of a century ago (Fig. 7.23). The people who funded that act were from a cultural background—about biblical context, for example—that was completely different from Michelangelo's. The imagery became clearer so that people can now see either color photographs or the real thing and see all of the figures. Ironically, Japanese people and many others for whom the original cultural context is now unfamiliar can see images from a completely different cultural context.

Archeologists are very familiar with the importance of context in society when interpreting their finds, and that applies to art too, as we shall see when we concentrate on Art 2. But the important point here is about Art 3 before we are confident that the productions meet our expectations for things to be art. In Table 4.1 I pointed out that four of the eight case categories that define semantic roles for people engaged with art involve an entity that only exists in the minds of the people involved—it has no material substance. And the whole argument of this book is that mental aspects of semantic roles are determined in social interactions—we know the story of the whale because someone else told us about it, so that we shared knowledge about it. My thoughts about the whale and my memory of the place were determined by that social interaction. The issue here, of course, is that for Art 3 it does not seem to be the case that the sporadic occurrence of pieces that might be art (as at Blombos) led on to very much else. The story is very much one where archeologists fill in the gaps in a way that is not always the most revealing option (Davidson, 2023).

In both South Africa and in Europe, there are early examples of what has been called art, but they do not seem to provide any continuity either around the time they were produced or from that time to later times. Thus, for example, archeologists working in the cave site of Bruniquel in France observed and dated an arrangement of broken stalagmites to 176.5 thousand years ago (Jaubert et al., 2016) (Fig. 5.2). Analysis of the processes necessary to construct such an arrangement shows that at least a dozen operations were necessary to achieve it. Yet there is nothing even approaching this complexity in any contemporary or later material evidence of behavior for, perhaps, 125 thousand years. How do we explain such unique constructions?

The Bruniquel example is not straightforwardly considered as art by anyone, so perhaps a little justification is needed before we continue. Whatever else the arrangement is or was, it was an example of individuals making marks on the environment which could be seen, but only by others who entered the cave. In terms of the sorts of definition we have been using here, there were one or more producers, and an object, and presumably one or more seers—even if some of the seers were also producers. There does not seem to have been a subject, though, and there does not seem to be any pattern or repetition from which we can infer more than the existence of the thing. In Chap. 3, I suggested that marks acquire meaning "by the repeated associations between marks and subjects, sometimes including resemblance, and seers of the marks learned those meanings." There does not seem to have been any repeated association or resemblance in this case, though it is undoubtedly an example of one or several individuals making marks. In addition, in Chap. 3, I argued that "eventually those thoughts become part of the seer's ways of seeing—generally as unquestioned beliefs—as a result of conventions of mark-making." In the absence of the repetition, it is highly likely that whatever was involved mentally in the Bruniquel arrangements did not become part of the unquestioned cognition and beliefs of any others, because they lacked the social context for continuation of a convention.

There are several other examples in which archeologists have sought to claim early evidence for art as I have discussed previously (Culley & Davidson, 2021) (and will return to in the section on Art 3), most notably in Trinil (Figs. 3.7 and 7.28 Left), Berekhat Ram on the Golan Heights (d'Errico & Nowell, 2000) (Fig. 7.24—see discussion in Chap. 8), South African sites such as Blombos (Figs. 3.27 and 3.29) (Henshilwood et al., 2009) and Diepkloof (Fig. 3.28) (Texier et al., 2010), and among Neandertals in Iberia as at La Pasiega (Fig. 4.6) (Hoffmann et al., 2018) or the recently reported 57.3 thousand year old finger markings in La Roche Cotard in northern France which date to the time of Neandertals (Marquet et al., 2023). In all these cases and others, it is difficult to argue that such early examples led on to a widespread practice that started at those times and those places. Rather, these seem to be indications that between 176.5 thousand and 57.3 thousand years ago there were people (technically hominins, but we slip in to calling them humans-like-us) who were physically able to put together things which might, from later periods or contexts, be called "art", but who lacked something.

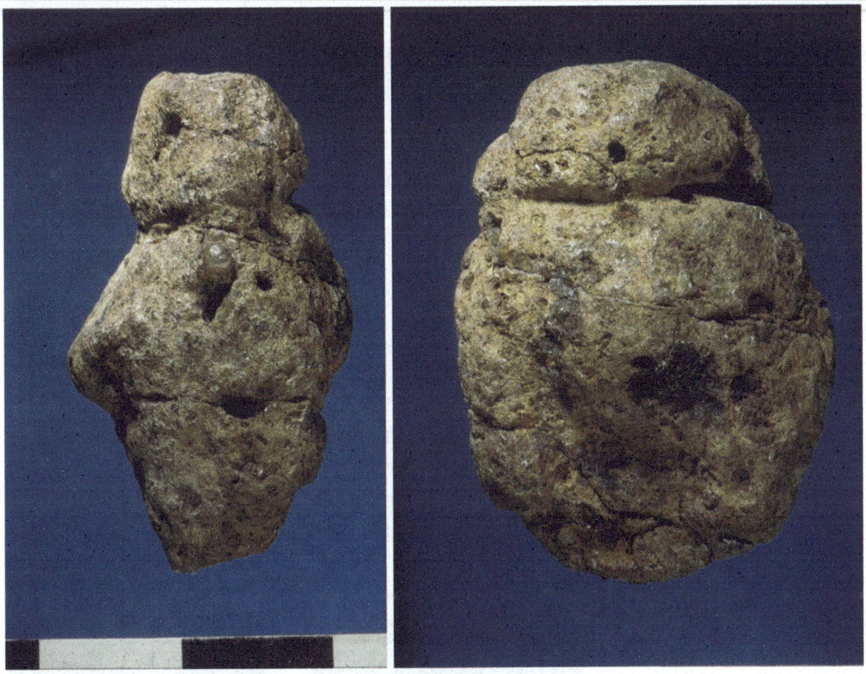

Fig. 7.24 Two views of a modified piece of 260-thousand-year-old volcanic rock, scoria, from Berekhat Ram oriented to look as if there is a head at the top. (Photo courtesy of Francesco d'Errico, Université de Bordeaux, and April Nowell, University of Victoria)

The key site here is Blombos, where the two well-known ochre slabs marked with cross-hatched lines can be dated by their discovery in dated layers at the site (Fig. 3.29). As I discussed in Chap. 3, before this there was a lot of scratching, for one reason or another, accidental or deliberate, but not much sense of pattern. These are probably what grazier Bill would have been justified in calling scribbles. But by the production of cross-hatched pattern, I would say that the people of Blombos had created the material through which it was possible to have an expectation of meaning.

A little later in time, at Diepkloof (Fig. 3.28) people marked ostrich eggshell (probably containers) with cross-hatching and with other patterned sets of scratches. Some other sites such as Klein Kliphuis (Fig. 7.6) had (small bits of) ochre with cross-hatched designs. I suggest that what is encapsulated in these widespread isolated single instances is the emergence of a mental capacity for an expectation of meaning-in-pattern as people in societies became aware of what could be achieved by individuals alone. That became a shared cognitive capacity once there was a social context in which people could share responses to external stimuli. I would say that it was NOT art, but has the appearance that it might become art. It was, I suggest, mark-making such as the world had never previously seen.

I would argue that the evolution of modern cognition probably happened in many places, and that there was the opportunity for it to emerge which was not always capitalized upon. The best-known example of this is the production among Neandertals of things that are now generally considered symbolic or called art, but actually have some important characteristics which suggest they could have become art in the course of evolution, but they did not. These things are the stalagmite arrangement in Bruniquel (Jaubert et al., 2016) (Fig. 5.2), paintings in caves (Hoffmann et al., 2018; Pike et al., 2012) (Fig. 4.6) and also engravings in one cave (Rodríguez-Vidal et al., 2014), use of feathers (Finlayson et al., 2012; Peresani et al., 2011), ochered marine shells (Peresani et al., 2013; Zilhão et al., 2010) and bird talons (Radovčić et al., 2015, 2020) (Fig. 7.25). Put together, they sound as if they might constitute a body of symbolic use of the environment of the sort represented by later art, but I suggest they are something more interesting. Leaving aside challenges to the dating of the paintings (and the credibility of the challenges) (Sauvet et al., 2017), one problem does arise from the dating which I have set out in Table 7.2. It looks as if many features which get called art, mostly Art 3, appear intermittently in the Middle Paleolithic—in Europe that means that they appear among Neandertals—but the dates suggest that there may not have been any

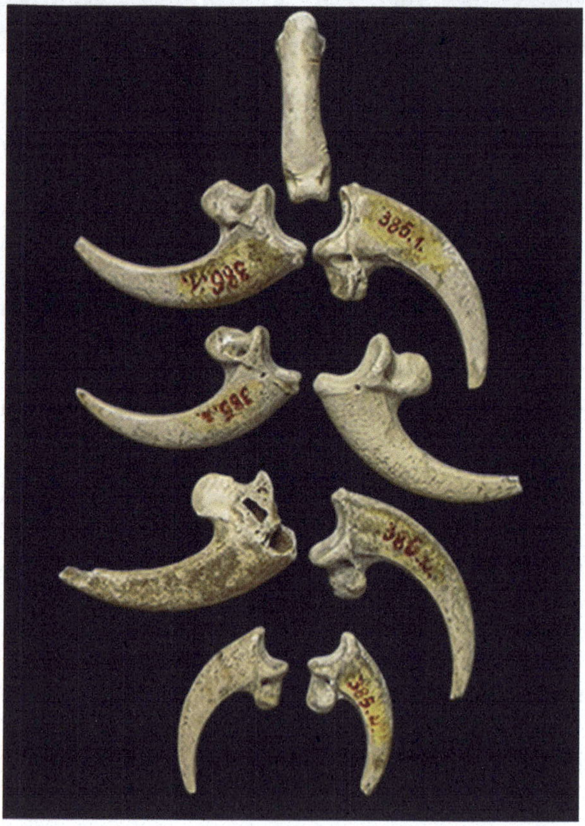

Fig. 7.25 Krapina, Croatia, presumed 130-thousand-year-old necklace consisting of the claws of white-tailed eagle carefully removed (Radovčić et al., 2015). (Permission given by photographer, Luka Mjeda 9 Oct 2023)

Table 7.2 Dates for Middle Paleolithic examples of Art 3 and earliest examples in Art 2

	Middle Paleolithic	Earliest Art 2 in SW Asia, Europe	Examples of Art 2 South Africa	Examples of Art 2 Wallacea and Australia
Stone/Stalagmite arrangements	Bruniquel 176.5			
Finger flutings and other marking	Roche Cotard 57.3	Castillo ~40; Chauvet 38–33		Koonalda 30
Sculpted stones or bones	Berekhat Ram 260	Hohlenstein Stadel 40; Hohlefels 36–35; Geissenklosterle 30–40		
Marked stones, shells, or bones	Trinil 500	Dziadowa Skala 130–115	Blombos 75; Diepklooof 60; Apollo XI 30	Carpenter's Gap Nose bone 45
Colored shells	Aviones 115; Fumane 47.6–45		Blombos 100	
Pierced bone or shell ornaments (Shuidonggou, China 31)	Bizmoune ~142; Krapina 130; Qafzeh 92	Sunghir 34	Blombos 72; Sibudu 72, 65, 47	Jerimalai 42–38; Mandu Mandu 34
Evidence of feather use	Qesem 420–200; Gorham's Cave	Fumane 44		
Wall paintings/ engravings	La Pasiega 64; Gorham's Cave 40	Fumane 40–37.5; Chauvet 38–33		Sulawesi 45.5

Dates in thousand years. The entries seem to show a lack of continuity between early sporadic examples (e.g., in the Middle Paleolithic of Europe or in South Africa) and later traditions

continuity with the varieties of art among the people of the Upper Paleolithic. Likewise, the South African evidence from Blombos and Diepkloof seems to appear early but then is not represented after about 60 thousand years ago, and more recent evidence of marked or modified objects does not re-appear until the Late Stone Age.

This is a slightly controversial view because there appears to be a model of cognition as something that is basically biological. On this model, human cognition suddenly appears—and there does not seem to be an evolutionary argument about how it appears gradually and is subject to natural selection—such that the people now suddenly all have modern human cognition. This view continues that with that cognitive ability they can "pop" out into a world producing art and other symbolic behaviors such as burial and other creative things. There is a problem that those who had not yet got that cognitive ability would not understand what had happened. If meaning can be indeterminate when everyone has the same cognitive ability, it would never be able to be determinate if no one else understands them because they have not got that "popping" cognition.

My model would have cognitive changes as a result of the storytelling that goes with the mark-making that we now see as art, not as a biological thing, at first, but as a learned thing. Thus, it would be possible to have either marking without storytelling or marking with storytelling. In the case of marking with storytelling, those who know that marks made by people often have "meanings" can share some of them with those who have not been through the whole process. The expectation of meaning can spread and with it the non-biological, "higher" cognition spreads. The first point is that deliberate marks are associated with stories. The storytelling acts as natural selection for the gradual biological changes that enabled the mental changes associated with modern human cognition (Nowell, 2023). The oral channel still has to do all the heavy lifting.

In my scenario, I think that the first Blombos pattern was little different from the scratches to get ochre powder off bigger slabs that have been found throughout the sequence (Fig. 3.29). Likewise, I'd say that the marks in Dinaledi in South Africa—if that is what they turn out to be—where several unusual skeletons were found deep in inaccessible caves, (Fuentes et al., 2023) are not necessarily "Abstract" marks. The use of the word abstract usually means that WE do not know their meaning because they do not look like anything else, but I think it is better to be explicit that such marks might acquire meaning from the telling of stories. In the absence of repetition, we can say that it is really unlikely that there was any storytelling. The same would be true of the paint marks made by Neandertals (Fig. 4.6) (Hoffmann et al., 2018) or the scratches on the Trinil shell (Fig. 3.7 or 7.28 Left) or the grinding of the Berekhat Ram scoria (Fig. 7.24).

We cannot understand such processes if we assume that art was suddenly created by someone when there was no one else who understood what art was, and we cannot assume that the necessary cognitive capacity emerged fully formed either. Both of these required gradual processes, and both involved the evolution of the combinations among people who were producers, storytellers and seers, rather than a sudden event which appeared in one place and then, as if by magic, among all humans.

The ambiguity in this particular type of comparison arises because unlike the very earliest things that are candidates to be art, the works of Art 1 are frequently said to be "Works of Art" ignoring the artists, their stories and the seers. Yet, as we have seen throughout this book, and especially in discussing *The Creation of Adam*, the work cannot actually be taken as an item by itself, but in every sense derives its importance from the social context in which it was created and in which it was and is presented to the seers. By the same token, those candidates for Art 3 acquire their possibility of meaning from the fact that the producers were members of a society. They were only art insofar as the producers could explain them to the seers in the context of the society of the time.

7.1.7 Meanings from Social Context or Meanings Unlimited

By now we have found a familiar pattern. Although I suggested in Table 7.1 that there seems to be a case for ambiguity between what defines Art 1 and the candidates for art in Art 3, the differences can be seen to be less than they seem. The last

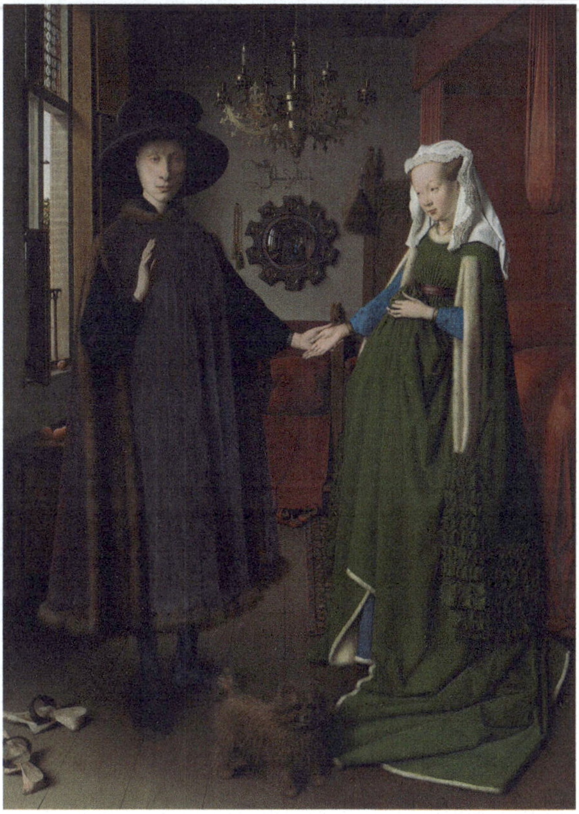

Fig. 7.26 Jan van Eyck, *The Arnolfini Portrait* (1434). Oil on oak panel, 82.2 × 60 cm (32.4 × 23.6 in). National Gallery, London. (Wikimedia Commons Public Domain)

type of ambiguity, then is one that has some similarity with the previous one. The items in Art 1 often acquire much greater significance than was originally intended. Thus, Van Eyck's 1434 *The Arnolfini Portrait* was one of the earliest paintings to use oil paint which transformed painting (Gombrich, 1995, 240–243) (Fig. 7.26). However much we admire the painting, the oil paint became the story, in addition, and arguably the mirror was referenced in Velázquez's *Las Meninas* (Fig. 1.10 Left).

Monet's *Boulevard des Capucines* (Fig. 7.27) was exhibited in 1874 at one of the very early shows by the group who became known as Impressionists. The show took place in rooms formerly used by a photographer, and the scene of this painting by Monet gives something of the impression of the first photograph that includes human figures taken by Daguerre in 1838 (Fig. 8.33). The exhibition broke away from some of the existing power structures that were dominated by well-known traditional painters such as Ingres and Delacroix (Stonard, 2021, 350). It is one of the foundation paintings of Impressionism—a movement that became as important as any individual painting.

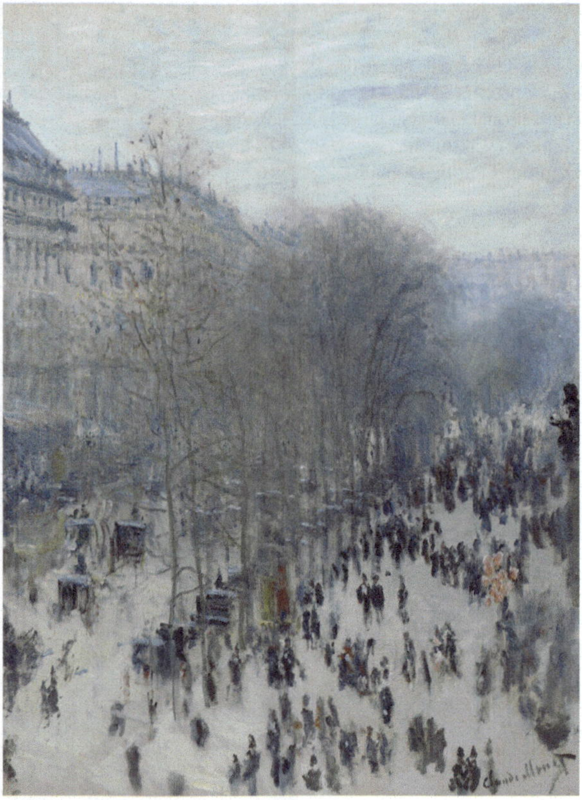

Fig. 7.27 Claude Monet, 1873–74, *Boulevard des Capucines*, oil on canvas, 80.3 × 60.3 cm, Nelson-Atkins Museum of Art, Kansas City. (Wikimedia Commons Public Domain)

Art 3, on the other hand, was coming from nowhere: it can only be understood now, if at all, by putting the materials into the context of the producers of the marks and their social relations with those who saw them. Archeologists and others tend to see the markings as indicative of art as they would see it today. That is why the objects are so often given artistic terms (such as "abstract") or names that evoke later examples of art (the Berekhat Ram "figurine"). The cave paintings of Sulawesi and other areas of South East Asia or the earliest cave paintings of Europe are later than the isolated or discontinuous evidence of Art 3. It is open to interpretation what status the examples of Art 3 represent among the catalogue of species names given to fossil hominins by people who study their bones (the options are outlined by Culley & Davidson, 2021). The function of the discussion in this book is not to make a stand about such issues of paleoanthropology, but to make an argument that early art is worthy of that name. It may be that such an argument would contribute to clarifying one aspect of the paleoanthropological debate, but that is not our main purpose here.

This discussion has been structured around the seven types of ambiguity between the earliest art and the art we are all familiar with. The ambiguity often involves repetition for archeologists to recognize objects that Art 3 is an intentional production determined, perhaps, by convention identifiable through the presence of patterns among the objects. In Art 1, the work is generally allowed to stand alone and is judged on those merits. For some (but not all), the skill in the production of Art 1 is shown through the resemblance between subject and object, while in Art 3 any meaning of the image was originally conveyed through speech acts by the producers. Pasztory (2005, 22) stated that: "With some exceptions, artists are notoriously nonverbal and react to art by making art and not words." My point throughout has been that speech acts by artists (and all of us) establish the frameworks for internalized cultural knowledge and values which may include the interpretation of particular sets of marks with the effect that they become art. Speech is not necessary at the time of making art but has been intrinsic to the cultures of the artists as they make it. The painter of the scene of Odysseus and the Sirens in Fig. 4.8 Right did not need to tell the story as he painted because it was a familiar part of Greek culture through the repetition of the myths of Homer's songs.

The question is how it came to be that people might have made marks that could be expected to have meaning, however indeterminate that meaning was at a time prior to the incorporation of marks or images into the stories that pervaded the culture of the time. An object in Art 1, by persisting as a material object, can stand alone so that seers can be separated from producers, even though at some stage seers from the same social context had the same cultural values as the producers.

Rock and cave art gain some of their force from being persistent in the environment, but in Art 1 much of the force derives from the object's transferability—it can be moved from one environment to another, and, most importantly it can be bought and sold—but generally within a social context where seers accept the cultural contexts of the subject matter and of the place where it can be seen. Art 3 required social interaction between producers and seers in order to generate an expectation that marks had meaning. Repetition became a context in which people could pass that expectation on to others and as such it can be, and usually was, an element of ritual. Art 1 is open to all (who can pay) often without any ritual related to the content of the image(s) but nevertheless involving the rituals to be observed in museums and galleries.

7.2 Essay 7 Types of Ambiguity (Trinil Shell and Northwest Central Queensland Anthropomorph)

In 2015 some large shells excavated in the 1890s provided evidence from half a million years ago in Java of patterned marks made by a non-human species which biological anthropologists call *Homo erectus*. Many of the shells have been punctured at a position which would have allowed the easy opening of the shell to extract the animal as food. In addition, one shell has scratch marks on it (Fig. 3.7), possibly made by a shark tooth, which include some lines that have no pattern and others that make an M-shaped zigzag (Fig. 7.28 Left). Making marks that were apparently meaningless, but nevertheless patterned, is something that looks as if it is on the way to symbolic use of marks (https://theconversation.com/marksonanancientshell-leadtoarethinkofhumanhistory34957.) We the observers from a long time afterwards are looking for pattern. As I have argued here, what is required for such a pattern to be given meaning—to become symbolic—is that it be repeated. I will elaborate about the semiotics of symbols in Chap. 9. At Blombos the patterned marks are repeated, but neither at Trinil nor at Berekhat Ram was there any repetition.

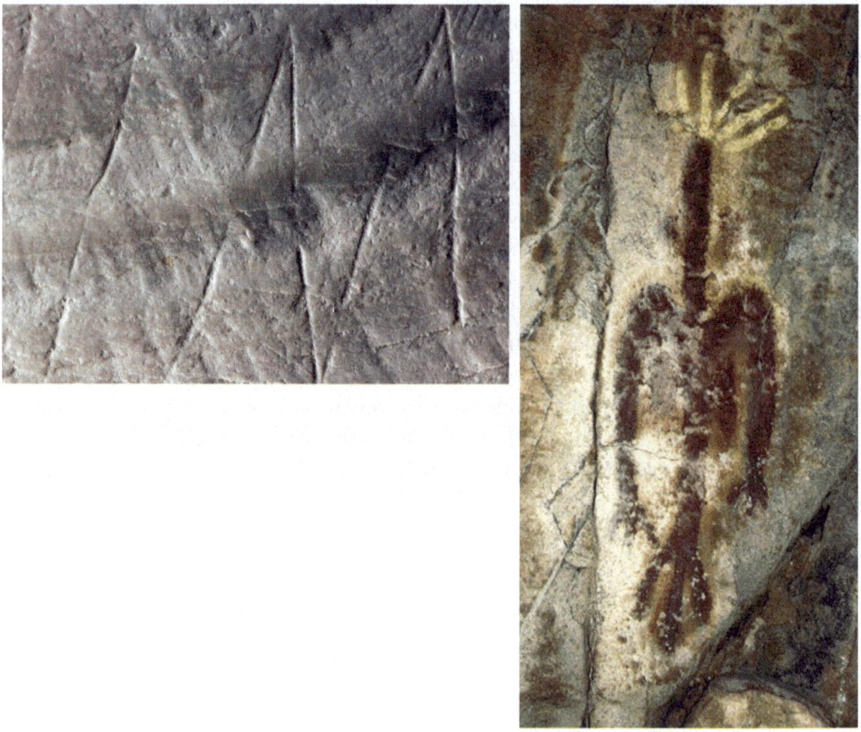

Fig. 7.28 Left: Engraved mussel shell from Trinil, Indonesia, about 0.5 M years old. (Courtesy of Wim Lustenhouwer, VU University Amsterdam, and Naturalis Biodiversity Center where the shell is curated). Right: "Anthropomorphic image" NW Queensland. (Photo Iain Davidson)

No one expected to find such a marking on such an artefact from such an early date, and it was only found because the materials from the excavations of the 1890s were conserved and looked at again 120 years later. The insidious point is that there may be other finds lying in drawers from old excavations in other places in the world, equally unexpected because they did not fit the version of the past that was then current.

A work of art typical of an art museum is unique and can be appreciated on its own. Before the twentieth century, a painting generally resembled the subject (unless, of course it was imaginary, as, say a mythological scene—but even then, the characters in the picture were very familiar). Producers often worked alone—the "starving artist in a garret". The work has sometimes been available to be transferred from one institution to another, and now is subject to copyright and ownership but other than that there are no barriers to who can view it (except cost restrictions). The meaning of the work is not confined to the object as seen in the work but can be interpreted by the seers. I call this Art 1, and it is exemplified by the artists I mentioned in Chap. 3. The linkage between these examples, I suggest, is the transferability of ownership, usually involving money.

At the beginning of art, the very recognition of things as probably Art at all depended on the producer making patterns in which a convention can be observed, usually from the repetition of the marks. I call this Art 3. The necessary existence of the convention means that we can infer that speech was associated with the marks (creating what Peirce called the "factitious habits of their interpreters") and that the producer probably interacted with the seers—by telling the stories—while engaged in the marking. In some way, the marks must have persisted (at least they must survive for us to make the argument) and it is likely that ownership of the marks could not be transferred—at least not for an intermediary material token. They depended on understandings within a society of the producer and the seers, such that any meaning, at the time, and for some time afterwards, depended on that social relationship.

A feature of this contrast is that the characteristics of Art 3 are needed for archeologists to be able to identify materials as art in the first place. Yet the characteristics also contrast in all respects with the characteristics of Art 1. At the same time, I argue, the characteristics that archeologists need to identify early art objects as Art 3, are the same characteristics that turned those ambiguous objects into regular art—particularly repetition and persistence. But in archeological art, Art 2, the imagery was or seems likely to have been necessarily accompanied by a narrative known in the society. In turn, Art 1, as identified, for example in art museums and art markets, was a product of the changed social circumstances of art sponsors and markets. It could have been produced without the seers knowing an accompanying story. These contrasts define the ambiguity in the identification of art.

Figure 7.28 Right shows another anthropomorphic figure from northern Queensland akin to that in Fig. 2 Left. The rocks are less than 7 km apart. The headdress is radial rather than branched. The hands, three fingers, and legs, if I have named them correctly, are all anomalous. Images that have a short projection between the legs are often interpreted as anthropomorphic and male; those where the projection is longer than the legs, are interpreted as having a tail, such as a lizard. In symbolic images, elements do not necessarily derive from iconic conventions of representation. This image has some similarities with a bird: the hand becomes a claw, the headdress becomes the crest, the rounded, elevated "shoulders" become the outline of wings and the three lower appendages become tail feathers. In such a symbolic tradition, it is relatively unproblematic that the claw should be a continuation of the wings, because it is not straightforward iconic representation. Modern oral testimony of the people of the region in the tradition of the painters, lacks detail because of vicious disruption of life in north-west central Queensland, but informants talk about a hawk as fundamental to their spiritual beliefs. There is, therefore, a tantalizing possibility that the painted imagery illustrates a symbolic world that related humans and animals.

References

Alais, D., Xu, Y., Wardle, S. G., & Taubert, J. (2021). A shared mechanism for facial expression in human faces and face pareidolia. *Proceedings of the Royal Society B: Biological Sciences, 288*(1954), 20210966. https://doi.org/10.1098/rspb.2021.0966

Baffier, D., & Feruglio, V. (2001). La salle Brunel. In J. Clottes (Ed.), *La Grotte Chauvet. L'art des origines* (pp. 64–73). Seuil.

Bahn, P. G. (2006). Sex and violence in rock art. *Nature, 441*(7093), 575–576. https://doi.org/10.1038/441575a

Barandiarán, J. M. d., & Altuna, J. (1969). *La cueva de Ekain y sus figuras rupestres*. Sociedad de ciencias naturales Aranzadi.

Barron, A. B., Halina, M., & Klein, C. (2023). Transitions in cognitive evolution. *Proceedings of the Royal Society B: Biological Sciences, 290*(2002), 20230671. https://doi.org/10.1098/rspb.2023.0671

Bednarik, R. G. (1995). Concept-mediated marking in the Lower Palaeolithic. *Current Anthropology, 36*(4), 605–634. http://www.jstor.org/stable/2744248. Accessed 20 May 2025.

Bednarik, R. G. (1998). The 'Australopithecine' cobble from Makapansgat, South Africa. *South African Archaeological Bulletin, 53*, 4–8.

Beltran, A., & Saura Ramos, P. A. (Eds.). (1998). *The cave of Altamira*. Harry N. Abrams.

Berger, J. (2008). *Ways of seeing*. Penguin UK.

Breuil, H., & Obermaier, H. (1935). *The cave of Altamira at Santillana del Mar, Spain*. Junta de las Cuevas de Altamira.

Clottes, J. (Ed.). (2001). *La Grotte Chauvet. L'art des origines*. Seuil.

Clottes, J. (2009). Sticking bones into cracks in the Upper Palaeolithic. In C. Renfrew & I. Morley (Eds.), *Becoming human: Innovation in prehistoric material and spiritual culture* (pp. 195–211). Cambridge University Press.

Clottes, J., & Geneste, J. M. (2012). Twelve years of research in Chauvet Cave: Methodology and main results. In J. McDonald & P. Veth (Eds.), *A companion to rock art* (pp. 583–604). Blackwell.

Culley, E. V. (2008). Supernatural metaphors and belief in the past: Defining an archaeology of religion. In K. Hays-Gilpin & D. S. Whitley (Eds.), *Belief in the past: Theoretical approaches to the archaeology of religion* (pp. 67–83). Left Coast Press.

Culley, E. V., & Davidson, I. (2021). Art, sign, and representation. In N. Gontier, A. Lock, & C. Sinha (Eds.), *The Oxford handbook of human symbolic evolution* (Second ed., pp. 331–368). Oxford University Press.

d'Errico, F., & Nowell, A. (2000). A new look at the Berekhat Ram figurine: Implications for the origins of symbolism. *Cambridge Archaeological Journal, 10*(1), 123–167.

Dart, R. A. (1974). The waterworn Australopithecine Pebble of many faces from Makapansgat. *South African Journal of Science, 70*(6), 167. https://doi.org/10.10520/AJA00382353_4093

Davidson, I. (1989). Comment on "Deliberate engravings on bone artefacts of *Homo erectus*". *Rock Art Research, 5*, 100–101.

Davidson, I. (1990). Bilzingsleben and early marking. *Rock Art Research, 7*, 52–56.

Davidson, I. (2003). The archaeological evidence of language origins: States of art. In M. H. Christiansen & S. Kirby (Eds.), *Language evolution* (pp. 140–157). Oxford University Press.

Davidson, I. (2013). Origins of pictures: An argument for transformation of signs. In K. Sachs-Hombach & J. R. J. Schirra (Eds.), *Origins of pictures. Anthropological discourses in image science* (pp. 15–45). Halem.

Davidson, I. (2020). Marks, pictures and art: Their contribution to revolutions in communication. *Journal of Archaeological Method and Theory, 27*(3), 745–770. https://doi.org/10.1007/s10816-020-09472-9

Davidson, I. (2023). Humans making history through continuities and discontinuities in art. *Cambridge Archaeological Journal, 33*, 637–654. https://doi.org/10.1017/S0959774323000057

Davidson, I., & Noble, W. (1997). Making up the mind. *Cambridge Archaeological Journal, 7*(2), 275–277.

Davies, S. (2006). *The philosophy of art*. Wiley Blackwell.

De Houwer, J., Barnes-Holmes, D., & Moors, A. (2013). What is learning? On the nature and merits of a functional definition of learning. *Psychonomic Bulletin & Review, 20*, 631–642.

Deacon, T. W. (1997). *The symbolic species*. W.W. Norton.

Eco, U. (1986). *Semiotics and the philosophy of language*. Indiana University Press.

Eitzman, W. I. (1958). Reminiscences of Makapansgat limeworks and its bone-breccial layers. *South African Journal of Science, 54*(7), 177–182. https://doi.org/10.10520/AJA00382353_542

Eliot, T. S. (2010). *Notes towards the Definition of Culture*. Faber & Faber.

Empson, W. (1947 [1930]). *Seven types of ambiguity* (2nd ed.). Chatto and Windus.

Falk, D. (1980). Language, handedness, and primate brains: Did the Australopithecines sign? *American Anthropologist, 82*, 72–78.

Filer, R. K. (1986). The "Starving artist"—Myth or reality? Earnings of artists in the United States. *Journal of Political Economy, 94*(1), 56–75.

Finlayson, C., Brown, K., Blasco, R., Rosell, J., Negro, J. J., Bortolotti, G. R., et al. (2012). Birds of a feather: Neanderthal exploitation of raptors and corvids. *PLoS One, 7*(9), e45927. https://doi.org/10.1371/journal.pone.0045927

Fuentes, A., Kissel, M., Spikins, P., Molopyane, K., Hawks, J., & Berger, L. R. (2023). Burials and engravings in a small-brained hominin, Homo naledi, from the late Pleistocene: contexts and evolutionary implications. *bioRxiv*, 2023.2006.2001.543135. https://doi.org/10.1101/2023.06.01.543135.

Gell, A. F. (1998). *Art and agency. An anthropological theory*. Clarendon Press.

Gombrich, E. H. (1995). *The story of art* (16th ed.). Phaidon.

Gower, R. S. (1903). *Thomas Gainsborough*. G. Bell and Sons.

Green, J. (2014). *Drawn from the ground. Sound, sign and inscription in Central Australian sand stories*. Cambridge University Press.

Haidle, M. N., & Conard, N. J. (2011). The nature of culture. Synthesis of an interdisciplinary symposium held in Tübingen, Germany, 15–18 June 2011. *Mitteilungen der Gesellschaft für Urgeschichte, 20*, 65–78.

Hamilakis, Y. (1999). Stories from exile: Fragments from the cultural biography of the Parthenon (or 'Elgin') marbles. *World Archaeology, 31*(2), 303–320.

Hedquist, V. (2019). *Class, gender, and sexuality in Thomas Gainsborough's Blue Boy*. Routledge.

Henshilwood, C. S., d'Errico, F., Yates, R., Jacobs, Z., Tribolo, C., Duller, G. A. T., et al. (2002). Emergence of modern human behavior: Middle Stone Age engravings from South Africa. *Science, 295*(5558), 1278–1280. https://doi.org/10.1126/science.1067575

Henshilwood, C. S., d'Errico, F., & Watts, I. (2009). Engraved ochres from the Middle Stone Age levels at Blombos Cave, South Africa. *Journal of Human Evolution, 57*(1), 27–47.

Henshilwood, C. S., d'Errico, F., Niekerk, K. L. v., Coquinot, Y., Jacobs, Z., Lauritzen, S.-E., et al. (2011). A 100,000-Year-Old ochre-processing workshop at Blombos Cave, South Africa. *Science, 334*, 219–222.

Hoffmann, D. L., Standish, C. D., García-Diez, M., Pettitt, P. B., Milton, J. A., Zilhão, J., et al. (2018). U-Th dating of carbonate crusts reveals Neandertal origin of Iberian cave art. *Science, 359*(6378), 912–915. https://doi.org/10.1126/science.aap7778

Jaubert, J., Verheyden, S., Genty, D., Soulier, M., Cheng, H., Blamart, D., et al. (2016). Early Neanderthal constructions deep in Bruniquel Cave in southwestern France. *Nature, 534*(7605), 111–114. https://doi.org/10.1038/nature18291

Kanisza, G. (1976). Subjective contours. *Scientific American, 234*(1), 48–53.

Kissel, M., & Fuentes, A. (2017). Semiosis in the Pleistocene. *Cambridge Archaeological Journal, 27*(3), 397–412.

Mackay, A., & Welz, A. (2008). Engraved ochre from a Middle Stone Age context at Klein Kliphuis in the Western Cape of South Africa. *Journal of Archaeological Science, 35*, 1521–1532.

Mania, D., & Mania, U. (1988). Deliberate engravings on bone artefacts of Homo erectus. *Rock Art Research, 5*, 91–109.

Marquet, J.-C., Freiesleben, T. H., Thomsen, K. J., Murray, A. S., Calligaro, M., Macaire, J.-J., et al. (2023). The earliest unambiguous Neanderthal engravings on cave walls: La Roche-Cotard, Loire Valley, France. *PLoS One, 18*(6), e0286568. https://doi.org/10.1371/journal.pone.0286568

Melchionne, K. (2007). Why artists starve. *Philosophy and Literature, 31*(1), 142–148.

Morphy, F. (2023). *When two worlds meet: "Translation" and its dilemmas*. Paper presented at the Madayin: Eight Decades of Australian Aboriginal Bark Paintings from Yirrkala, University of Virginia in partnership with the Buku-Larrŋgay Mulka Centre, February 4–May 14, 2023, American University Museum, Katzen Arts Center, Washington, DC (February 3–July 21, 2024 on display at The Fralin Museum of Art in Charlottesville, VA).

Munn, N. (1973). *Walbiri iconography. Graphic representation and cultural symbolism in a Central Australian society*. Cornell University Press.

Naifeh, S. (2021). *Van Gogh and the artists he loved*. Random House.

Newman, J. (1986). Reynolds and Hone: 'The Conjuror' unmasked. *Reynolds*, 344–354.

Noble, W., & Davidson, I. (1996). *Human evolution, language and mind*. Cambridge University Press.

Nowell, A. (2023). Oral storytelling and knowledge transmission in upper paleolithic children and adolescents. *Journal of Archaeological Method and Theory, 30*(1), 9–31.

Papastergiadis, N. (2024). *John Berger and me*. Giramondo Publishing.

Pasztory, E. (2005). *Thinking with things: Toward a new vision of art*. University of Texas Press.

Paul, L. (2002). Beatrix Potter and John Everett Millais: Reproductive technologies and coolhunting. In M. Mackey (Ed.), *Beatrix Potter's Peter rabbit: A children's classic at 100* (Vol. 1, pp. 53–75). The Scarecrow Press.

Peirce, C. S. (1986). Logic as semiotic: The theory of signs. In R. E. Innis (Ed.), *Semiotics. An introductory anthology* (pp. 1–23). Hutchinson.

Peirce, C. S. (1992). *The essential Peirce, volume 2: Selected philosophical writings (1893–1913)* (Vol. 2). Indiana University Press.

Peresani, M., Fiore, I., Gala, M., Romandini, M., & Tagliacozzo, A. (2011). Late Neandertals and the intentional removal of feathers as evidenced from bird bone taphonomy at Fumane Cave 44 ky B.P., Italy. *Proceedings of the National Academy of Sciences, 108*(10), 3888–3893. https://doi.org/10.1073/pnas.1016212108

Peresani, M., Vanhaeren, M., Quaggiotto, E., Queffelec, A., & d'Errico, F. (2013). An Ochered fossil marine shell from the Mousterian of Fumane Cave, Italy. *PLoS One, 8*(7), e68572. https://doi.org/10.1371/journal.pone.0068572

Philippe, M., & Fosse, P. (2001). Les ossements des animaux sur le sol de la grotte. In J. Clottes (Ed.), *La Grotte Chuavet. L'art des origines* (pp. 51–57). Seuil.

Pike, A. W. G., Hoffmann, D. L., García-Diez, M., Pettitt, P. B., Alcolea, J., De Balbín, R., et al. (2012). U-series dating of Paleolithic art in 11 caves in Spain. *Science, 336*(6087), 1409–1413. https://doi.org/10.1126/science.1219957

Płonka, T., Wiśniewski, A., Marciszak, A., Ziółkowski, G., Lipecki, G., Diakowski, M., & Serwatka, K. (2024). A Middle Palaeolithic incised bear bone from the Dziadowa Skała Cave, Poland: The oldest marked object north of the Carpathian Mountains. *Journal of Archaeological Science, 166*, 105971. https://doi.org/10.1016/j.jas.2024.105971

Radovčić, D., Sršen, A. O., Radovčić, J., & Frayer, D. W. (2015). Evidence for neandertal jewelry: Modified white-tailed eagle claws at Krapina. *PLoS One, 10*(3), e0119802. https://doi.org/10.1371/journal.pone.0119802

Radovčić, D., Birarda, G., Sršen, A. O., Vaccari, L., Radovčić, J., & Frayer, D. W. (2020). Surface analysis of an eagle talon from Krapina. *Scientific Reports, 10*(1), 6329. https://doi.org/10.1038/s41598-020-62938-4

Rappaport, R. A. (1999). *Ritual and religion in the making of humanity*. Cambridge University Press.

Rodríguez-Vidal, J., d'Errico, F., Pacheco, F. G., Blasco, R., Rosell, J., Jennings, R. P., et al. (2014). A rock engraving made by neanderthals in Gibraltar. *Proceedings of the National Academy of Sciences, 111*(37), 13301–13306. https://doi.org/10.1073/pnas.1411529111

Romandini, M., Terlato, G., Nannini, N., Tagliacozzo, A., Benazzi, S., & Peresani, M. (2018). Bears and humans, a Neanderthal tale. Reconstructing uncommon behaviors from zooarchaeological evidence in southern Europe. *Journal of Archaeological Science, 90*, 71–91.

Sauvet, G., Bourrillon, R., Conkey, M., Fritz, C., Gárate-Maidagan, D., Rivero Vilá, O., et al. (2017). Uranium–thorium dating method and Palaeolithic rock art. *Quaternary International, 432*(Part B), 86–92. https://doi.org/10.1016/j.quaint.2015.03.053

Scalise Sugiyama, M. (2011). The forager oral tradition and the evolution of prolonged juvenility. *Frontiers in Psychology, 2*. https://doi.org/10.3389/fpsyg.2011.00133

Seckel, A. (2004). *Masters of deception: Escher, Dalí & the artists of optical illusion*. Sterling Publishing Company, Inc.

Shiner, L. (2001). *The invention of art: A cultural history*. The University of Chicago Press.

Spangler, J. D. (2013). *Nine mile canyon. The archaeological history of an American treasure*. The University of Utah Press.

Stefanis, K. (2013). Nathaniel Hone's 1775 exhibition: The first single-artist retrospective. *Visual Culture in Britain, 14*(2), 131–153.

Steinberg, L. (1992). Who's who in Michelangelo's creation of Adam: A chronology of the picture's reluctant self-revelation. *The Art Bulletin, 74*(4), 552–566.

Stern, J. (1997). Metaphors in pictures. *Philosophical Topics, 25*(1), 255–293.

Stonard, J.-P. (2021). *Creation: A fully illustrated, panoramic world history of art from ancient civilisation to the present day*. Bloomsbury Publishing.

Taubert, J., Wardle, S. G., Flessert, M., Leopold, D. A., & Ungerleider, L. G. (2017). Face pareidolia in the rhesus monkey. *Current Biology, 27*(16), 2505–2509.e2502. https://doi.org/10.1016/j.cub.2017.06.075

Texier, P.-J., Porraz, G., Parkington, J., Rigaud, J.-P., Poggenpoel, C., Miller, C., et al. (2010). A Howiesons Poort tradition of engraving ostrich eggshell containers dated to 60,000 years ago at Diepkloof Rock Shelter, South Africa. *Proceedings of the National Academy of Sciences, 107*(14), 6180–6185.

Vasari, G. (1998). *The lives of the artists*. Oxford University Press.

Villaverde, V. (2020). *La mirada neandertal. Origenes del arte visual*. Universitat de Valencia.

Villaverde Bonilla, V. (2021). 15. Scenes in the Paleolithic and Levantine Art of Eastern Spain. In I. Davidson & A. Nowell (Eds.), *Making scenes: Global perspectives on scenes in rock art* (pp. 224–239). Berghahn Books.

White, K. (2021). *Thomas Gainsborough's "Blue Boy" was once the world's most famous painting--here are 3 surprising facts abut it*. Retrieved from https://news.artnet.com/art-world/three-things-thomas-gainsborough-blue-boy-2048661

Zilhão, J., Angelucci, D. E., Badal-García, E., d'Errico, F., Daniel, F., Dayet, L., et al. (2010). Symbolic use of marine shells and mineral pigments by Iberian Neandertals. *Proceedings of the National Academy of Sciences, 107*(3), 1023–1028. https://doi.org/10.1073/pnas.0914088107

Chapter 8
But Is It Art: Let Me Count the Ways

Abstract The beginnings of art involved several different processes: the emergence of, recognition of, and marking of individuality; mark-making such that humans at the time expected there to be meaning even though it was indeterminate; and the first faltering steps in the removal of that indeterminacy of meaning.

I have shown that we can tease out several crucial features that characterize art ever since. Art is closely involved in overcoming three tyrannies: of distance, of place and of time. Art is used to define individuality; its engagement in ritual and the storytelling that allows teaching; the involvement of people and groups to locate power to themselves; and the first steps in the use of scenes to reduce the indeterminacy of the marks—steps that would, ultimately, make available to seers the mysteries of art which individuals otherwise used as resources of power. Art was involved, however slightly, in the process of overcoming the powers of institutions, but once those were overcome, professionals in the field set about creating new institutions of power especially for art, and the more recent history of art has been about overcoming the new power constraints.

8.1 Beyond Ambiguity—The 7 States of Archeology and Art

The art of the modern world to be found in museums and galleries I call Art 1. It is also found in the dwellings of the rich, and I might also use the label for that on the walls of motels. Some is difficult to tell apart from wrapping paper—grazier Bill would probably approve—and some wrapping paper, and garments (see Fig. 3.8), reproduce the art that is found on the walls of others. Sometimes people express their view about such judgments because they have attitudes derived from their membership of particular social groupings.

I give the label Art 2 to the abundance of different art forms discovered by archeology. This labelling began with Robb using different fonts for different types of art (Robb, 2017), but then I realized it was difficult to speak about the two sorts of art in different fonts and switched to numbers. I added a third form for earlier items that

I. Davidson, *Art or Scribbles? In the Eye of the Beholder: The Evolutionary Emergence of Visual Communication*, Interdisciplinary Evolution Research 11,
https://doi.org/10.1007/978-3-031-90458-5_8

do not necessarily reach the status of Art 2 and called it Art 3. That form of mark-making has special characteristics that are important in understanding the evolution of art and of cognition. Using photographs, Culley and Davidson (2021) discussed many of the earliest candidates in Art 3, which might or might not be art and put them in the context of a brief summary of Art 2. Some people would have no doubt that Art 3 represents the earliest examples of art. The real question is how we can establish that the archeological context allows us to infer conditions which do meet my criteria for art.

I have found it useful, in thinking about all the different things that can be given the name "art", to consider how archeologists identify it using the characteristics of the earliest deliberate mark-making (Table 7.1). In the previous chapter, I considered the seven apparent ambiguities between Art 3 and Art 1 and found that although there are nuanced differences, particularly in the popular imagination, there is much in common. It all hinges on the ability of the original mark-makers in Art 3 to be able to persuade those who saw them that the marks were meaningful. I have suggested that, at very least, the marks in a site such as Blombos set up what I have called the expectation of meaning. That expectation is one of the fundamental features of human cognition, and it is a major reason why some people consider Art 3 to be art. The crucial next development was the recognition of indeterminacy of those meanings, together with attempts to reduce that indeterminacy, some of which reduced as a result of the values and meanings of marks infiltrating society so that they became intrinsic to its culture. This infiltration came partly through ritual, partly through being representative of the real world, and partly through the incorporation of elements of the real world into stories and myths told at the same time as, or after, the performance and production of the art (see, for example, Nowell, 2023a). These two concepts—the expectation of meaning and the indeterminacy of meaning—are, together, the reason for the power of art in the human imagination.

What were the changes that took place in moving from the earliest signs to archeological art (Art 2), as well as in the transition from Art 2 to Art 1? Shiner (2001, 5–10) argued that Art 1 includes a segment that is known to specialists as "Fine"Art. Following Kristeller (1951), he said that Fine Art was invented in the eighteenth century. Those specialists subsequently dropped the important qualifier "fine" and pretend that only "their" stuff is "art"—what, as I indicated in Chap. 3, Collingwood would have called "proper art". Labels are often about the culturally specific contexts of things, such as academic disciplines, and the defense of disciplinary boundaries of all sorts. That leads to questions like "what is art?" So, what happens if we adopt the same principle about (fine) art?

I welcome the fact that "those who are interested in art"—to use the phrase I adopted in Chap. 2—drop "fine" as an adjective, not for the self-serving reason they have, but because they should be (and actually are) interested in much more than the art of the last 250 years (or so) which they call "fine". My argument is that such Fine Art is continuous with earlier works and calling it "fine" is not because of any characteristic of the art, so much as the culturally specific contexts of the interests of recent artists and art historians in those works. The timing also goes together with the emergence of a widespread commerce in art—art which (sometimes) gets the label "fine" is easier to sell. People who use the label think that, as a consequence of liking a certain sort of art, they too are "fine". Without saying anything about the majority of Art 2, who could not accept an argument that there was much art that was "fine" from before the eighteenth century (The Parthenon, Pompeii, Iron Age gold torcs, the Mezquita de Cordoba, Sutton Hoo, Leonardo, Dürer, Caravaggio, Artemisia Gentileschi, Rembrandt, Velázquez), and following the same logic also at the same times beyond Europe? I left out the pinnacles of Art 2 to be found in the Egyptian pharaonic period such as Nefertiti, the plaque of Narmer or the mask of Tutankhamun but all of them need to be included in Fine Art (Fig. 8.1). The study of the archeology of art is of relevance to all who study the human sciences.

I summarized some evidence in Chaps. 5 and 6 to the effect that art has been an expression which began to overcome the tyrannies of time and place. I sought to show how some of the consequences of changes in technology of communication affected the dual expression of graphic communication. These in turn, first led to, then benefitted from, the expectation of meaning and the successive efforts within cultures to reduce the indeterminacy of meaning. The same sorts of consequences are true of other aspects of the long history of art. I, therefore, introduce an unconventional 7-State classification of the past within this framework, recognizing the fact that each part might be subdivided according to individual specialties and interests. I used the word "Stage" initially because there was a chronological element in some places. I changed to "State," on advice, because the idea of stages implies some sort of progression, which was not my intention:

State A, Creating the expectation of meaning—Before modern human behavior appeared more than 100 thousand years ago—some marking creating the expectation of meaning. This state was experienced all over the world;

State B, Emergence of one species—Around the time of earliest modern human behavior between 100 thousand and 12 thousand years ago. There was only one species in the world at 12 thousand years ago, before that it is less certain. In this State, people used ritual to establish the importance of place;

State C, Settling down—12 thousand years ago to the appearance of writing. In some parts of the world, writing did not "appear." It was introduced. In others it appears to have been invented in towns and villages, first appearing about 5 thousand years ago—the appearance of formal teaching and representation of formal narratives;

State D, Writing—From the appearance of writing about 5000 years ago to the appearance of printing with moveable type 500 years ago—fixing the story. The timing and conditions of this State are based on events in parts of Europe, though writing was much more widespread than that;

State E, Printing—From printing 500 years ago to the use of semaphore to communicate at a distance about 250 years ago—releasing the story from the institutions of control. Again, the chronology is principally based on events in parts of Europe. Although Islam introduced paper and paper-making to Iberia from the East, Islam did not adopt printing until long after the rest of Europe;

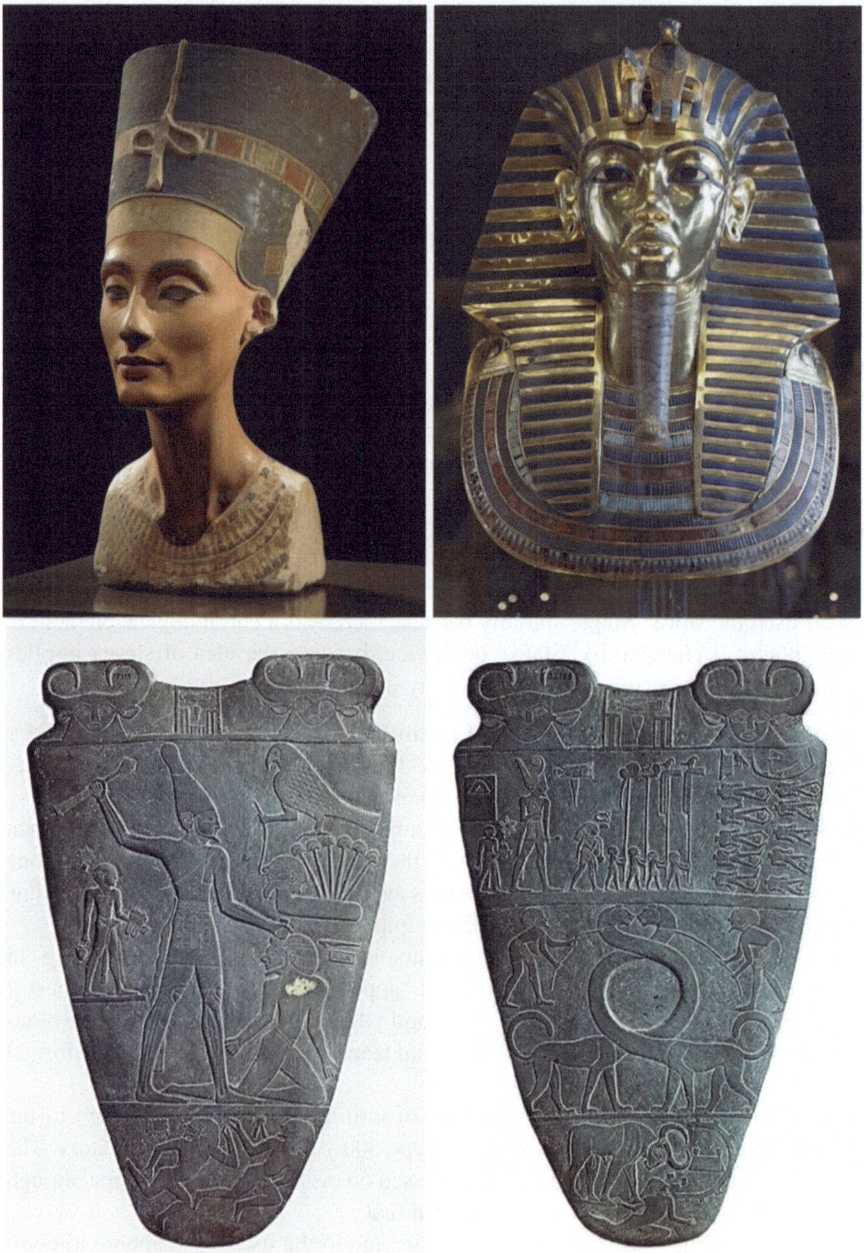

Fig. 8.1 Examples of art from Egypt that should be called Fine: Top Left, Bust of Nefertiti by sculptor Thutmose, about 3340 years ago, currently held in Neues Museum, Berlin (Philip Pikart, CC BY-SA 3.0, via Wikimedia Commons); Top Right The funerary mask of Tutankhamun, about 3300 years ago (Roland Unger, CC BY-SA 3.0, via Wikimedia Commons); Bottom, both sides of the Narmer Palette, about 5000 years ago (Wikipedia Commons Public Domain)

State F, "Fine" Art—From semaphore to the widespread use of photography around 1900—spreading the story rapidly to overcome the tyranny of time. The label "Fine Art" seems to have been principally a European invention, related to European commerce;

State G, Art is not a photograph—Let me tell you what you see.

I suggest this 7-State classification is a suitable alternative to the description of several different Ages defined for the first time more than 150 years ago (Lubbock, 1865) which offers little about how people used the resources that defined those ages to exert power over each other. I have used this classification to order the sorts of works of art I have already illustrated here and the ways in which they compare to what was known in periods with different characteristics (Table 8.1). Art 2, has three States, and Art 1 three more, but the time periods are not commensurate. Each of the Sates ends with a significant development in the evolution of communications, in the manner I discussed in Chap. 5.

I will address each of these States in turn. One applies to Art 3, and three each to Art 2 and Art 1.

8.2 Art 3: The Earliest Art of State A

How does Art 3 involve the four elements of the art-system, and what is it about them that makes it art? All art, from the very beginning, consists of marks made by a person, made by adding pigment (as in painting—hence the importance of La Pasiega, dated to State B, in Fig. 4.6) or removing part of the surface of the material (as in sculpting—hence the importance of Berekhat Ram in Fig. 7.24). In that process, it can become art if there is storytelling either in its creation or surrounding it. There is nothing about those individual marks alone that generates the meanings or understandings that people bring to art objects—people are the agents of that generating. Art did not suddenly appear fully formed from the minds of all humans. So how, in the course of the evolution of human behavior, did marks get seen as something more than marks—more, indeed, than scribbles? Understanding that will increase our understanding of what constitutes art.

The sequence of events in human evolution, which is the context in which people did or did not produce art, has been described in many different ways. Some concentrate on changes in the skeleton and the names that paleoanthropologists make up (observing some rules written long ago) according to the patterns of features they discern among the bones (e.g., Wood & Boyle, 2016); others concentrate on the impressions of brains on the inside of the skull known as endocasts (Davidson, 2007; Falk, 1992; Ponce de León et al., 2021); still others relate bodies and brains to environmental variables (Will et al., 2021). I will avoid the business—the academic industry—of classification of the fossils of our ancestors by any means for fear that the taxa and relationships preempt arguments about what the species were capable of (Botha, 2024). These classifications and interpretations may, or may not,

Table 8.1 Art at 7 different States of its emergence

State	When (years ago)	**1. Markers of personal identity.** e.g., beads	**2. Persistent visible marks.** e.g., rock shelter art	**3. Persistent marks needing education or initiation.** e.g., Cave art	**4. Marks on useful portable objects.** e.g., Pottery	**5. Non-utilitarian marked objects.** e.g., Sculpture and Art 1	**6. Scenes.** On any medium
Creating the expectation of meaning—Art 3							
A—Art 3	**>100,000**	*.Some beads*		*.Bruniquel stalagmite circle*	**Creating the expectation of meaning** *.Ostrich Eggshell*	**Creating the indeterminacy of meaning**	
Emergence of one species—ritual and the importance of distance							
B—Art 2	**100,000-12,000**	**Defining the self** *.Sunghir* **Overcoming the tyranny of distance** *.Willendorf*	*.La Pasiega* *.Gibraltar*	**Ritual and teaching** *.Altamira* *.Les Trois Frères*	**Locating Power** *.Marks on bones* *.Plaquettes*	*.Swabia*	**Reducing the indeterminacy** *.Leang Bulu' Sipong 4* *.Lascaux*
Settling down—overcoming the tyranny of place							
C—Art 2	**12,000–5000**		**Overcoming the tyranny of place**	**Formal teaching** *.Babylonia*	*.Ceramics*	**Sacred places** *.Göbekli Tepe* *.Çatalhöyük* *.'Ain Ghazal*	**Sequential narrative** *.Levantine Art*
Writing—fixing the story							
D—Art 2	**5000–500**		**Fixing the story** *.Valcamonica* *.Duccio* *.Giotto*	**Written texts in public places** *.Pyramids* *.Bisitun* *.Mezquita*	**Fixing the story** *.Library of Nineveh* *.Odysseus* *.Moche*	**Religious art** *.Angkor Wat* **Monumental story display** *.Teotihuacan*	**Narratives and stories** *.The Book of the Dead* *.Gilgamesh* *.Genesis* *.The Iliad* *.Mahabarata* *.Confucius*

Printing—freeing the story from institutions of control							
E— Art 1	**500–250**				**Freeing the story from institutions of control** *.Mainz* *.Wittenberg*		
"Fine" Art—overcoming the tyranny of time							
F— Art 1	**250–120**				**Overcoming the tyranny of time** *.Semaphore* *.Telegraph*	*.Wedgewood*	
Art is not a photograph—let me tell you what you see							
G— Art 1	**<120**					**Let me tell you what you see** *.Impressionism* *.Cubism* *.Abstract Expressionism*	

have anything to do with the first appearance of art and its subsequent incorporation into the behavioral repertoire of humans. We need to be clear that at some stage our ancestors, however we group their bones together and whatever the name we choose for those groupings and whatever the name they went by, had several ranges of behaviors. Those ranges both required and provided the selective context for cognitive abilities that evolved towards ours. Art is related to cognitive ability, but cognitive ability is not simply related to brain size or skull shape, and particular abilities did not suddenly appear without any selective context. I have laid out previously a scheme for this, which culminated with the required cognitive ability to engage in one task in order to achieve another (e.g., Davidson, 2019, 91–2). We should also always remember that the basic building block of evolution is variation inherited by one generation from another and that "we" too have variable behavior.

8.3 State A Creating the Expectation of Meaning

State A is either characterized by the production of no art at all or by Art 3. I suggest that it is not possible, at the moment, to resolve the question of whether the individual marked objects of this state were part of anything that was given any meaning by any of the makers or to any of the observers. That is to say that in almost all cases, the marked objects are unlike any other marked objects contemporary with them however similar they may be to objects from other times and places. I analyzed some of the key individual objects in the previous chapter.

In State A, I have sought not to get sidetracked by discussion of the emergence of language which must have taken place in at least one of the *Homo* species that lived during this state (Davidson, 2020), and there may have been four or more species according to the names given by paleoanthropologists (Smith & Wood, 2017). Suffice to say that not many players are suggesting that any people communicated using language early in State A (for an argument about how something like language emerged early, see Dediu & Levinson, 2013), but by the end of the State, language was certainly present among *Homo sapiens*, whatever your definition of language (Botha, 2020). A review of recent work on Neandertal brains claims that they had less cognitive flexibility than modern humans, while modern humans have greater capacity for "self-awareness, creativity, prosocial behavior, and healthy longevity" (Nowell, 2023b, 155)—though these characteristics cannot be discerned from aspects of the brain so much as from the evidence of their behavior.

The question of the emergence of an awareness of self has been well argued by Iriki and Sakura (2008) on the basis of their experimental teaching of tool use to monkeys. Their work shows that learning tool-use results in changes to the brain and that one of outcomes has been that tool-users come to recognize a separation between their bodies and the material that makes up the tool. It led to what Iriki and Sakura called the capacity to "objectify the self." As I indicated, the importance of "self" was not just a brain thing, nor a material thing, but a combination of these with the behavioral things that people engaged in. I suggest that there was a transitional state where individuals had an awareness of the objective self, without the recognition of the social consequences of that "self." It could be argued that prior to the emergence of the species we recognize as *Homo sapiens*, the markers for self-awareness did not get actualized to provide an awareness of self in the agency of those people.

This variation between the two species is something that natural selection could have worked on, bearing in mind that even within one species there is variation in those capacities: there are people with more or less self-awareness (so far as I know); some present evidence that they are more creative than others (I imagine); we are all aware that some people are more prosocial than others (from charity workers even so far as to some business people); and young people die as well as old and all ages in between.

My take is that, as with the early appearance of marked objects in South Africa, the evidence from Neandertals in Europe looks like the beginning of marking, but that there is not yet (perhaps) enough evidence for us to suggest that cognitively the people of Blombos, Neandertals, and modern humans were all on a par with each other. Nowell (2023b, 156) suggests that natural selection acted on the sorts of differences between Neandertals and modern humans. I would go further and point to the role of things we call art in the process by which selection operated on associated behavior and led to cognition and brain functions differentiating. This was the emergence of an awareness of self, which, became marked by the use of personal ornaments and marking. We will see some more indications when discussing State B.

I previously offered the simple notion: "Stories make marks art." By that definition, language must precede art, though in reality there is widespread understanding that they are closely related to each other (Fig. 8.2). But there is more to it than that. If a person can tell a story about "any old mark"—such as some of the scratches on the Trinil shell (Fig. 3.7 and 7.29 Left)—then it is the seers' memory of the context of the story of the marks that allows them to recognize or repeat the association between the mark and the story, together with the mark-maker's own credibility as a storyteller. If the mark is repeated and patterned, though, any seer can ask the question about why the mark-maker made the mark in a pattern and told the story. Without repetition and patterning, the mark is almost incidental, and merges in with all the other marks that people leave in their ordinary passage on the surfaces of the world. It is through the telling of stories, their repetition, and their association with the marks, that the marks become art as part of the culture of the producers and the seers. Along the way, storytelling became part of the way in which mark-makers acquired credibility. Kelly (2015, 7) quotes Ross and Davidson (2006, 308) as claiming that there is no ethnographic information of people making rock art while engaged in ritual—but that does not preclude past cases, and most importantly, does not preclude ritual behavior being enacted at places where art had been previously made.

In an attempt to solve this archeological problem of identifying the context for meaning-making, I enunciated a Convention Criterion which must be met for

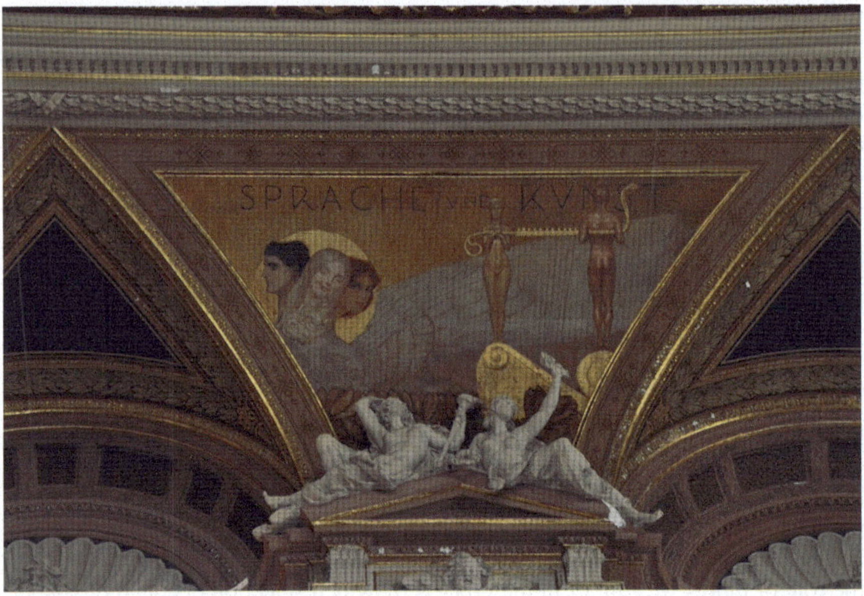

Fig. 8.2 Spandrel celebrating Language and Art, in the Main Ceremonial Hall of the University of Vienna, probably painted by Gustav Klimt, though most comment about his paintings there, is about those which were deemed too pornographic to be shown and were destroyed in 1945. (Photo by Iain Davidson)

archeologists to recognize that marks were being made to carry meaning (Davidson, 1992, 2003, 153):

> We can infer a convention or code through which meaning might be recognized if we find depictive or non-depictive marks on objects in repeated patterns, restricted in time and distribution.

With the benefit of the archeologists' hindsight, some objects seem to be conventionalized yet they do not meet the Convention Criterion because they were not restricted in time and distribution. I argue, therefore, that restricted repetition is essential for there to be a pattern, and for there to be a convention. Hidden away from plain sight is the fact that the cultural context that is the fabric of people's lives (that gives rise to that variation) is created by those very repetitions. For all of us, the culture we live in is unique to us, but insofar as there is commonality from one person to another, it is made up of those shared behaviors, beliefs, storytelling and other products of our practices that reflect the shared daily routines of our lives (Davidson, 2016). Archeologists can and do interpret the material products of behaviors, beliefs, and practices through the understanding of the repetitive, shared products that are left in the archeological record.

If we are to say that a type of mark had meaning, then that meaning would initially be restricted in time or place. These characteristics of Art 3 are also necessary for archeologists to be able to identify, long after the event, that marks were deliberately made. Without that, they were not art at all. It would be impossible to recognize them as art unless the marks were repeated in some way, but even more impossible if they were not repeated so as to make a recognizable pattern "restricted in time and distribution." This is one of the reasons why the Makapansgat cobble (Fig. 7.5) does not make it on to my list of very early art objects. We cannot infer anything about the behavior of the non-human Australopithecines that brought it to Makapansgat, because we have no evidence that any other such creature did anything similar at similar times or places. A similar restriction would apply to the claim for a much later block from La Roche Cotard in France which seems to have had some alterations, but which is (at the moment) similarly completely unique (Pettitt, 2003). It actually makes no difference that this is the same site where finger markings have recently been reported after a cave entrance was revealed by removing layers of sediment dated back more than 57 thousand years (Marquet et al., 2023). All people who offer interpretations that claim otherwise are actually making statements about their own beliefs about art objects, about their own role in society, and about their own beliefs about the neurology of the brains of other creatures (e.g., Watson, 2011, 90–91), but they are not saying anything about the past, or about how the events of the past made us human.

In a book with Bill Noble (1996, 75–77), I previously applied the same logic to early claims for a modified piece of volcanic rock, scoria, from Berekhat Ram (Fig. 7.24) (for a detailed discussion see Culley & Davidson, 2021) in the Golan Heights (about 100 km SW of Damascus in Syria) dated about 260 thousand years ago (Goren-Inbar, 1986). In this case, it was shown subsequent to our previous skepticism that the cobble was modified in specific ways (d'Errico & Nowell, 2000; Marshack, 1997). On the assumption (and it is quite a big one) that the lump of rock could be seen to have something like the head and neck of a human figurine, those modifications appear to isolate the head and neck. So far so good—the modification might mean that there was an attempt to get a seer to identify a head. But the modification leaves untouched the projection said to be the chest of the figurine, which remains unseparated like the appearance of a well-clothed woman of the nineteenth century (Fig. 8.3). Although the examination has shown that the modifications were undoubtedly deliberate there is still considerable doubt about the identification of the anatomical parts. Its status as a figurine is still implausible after the identification (d'Errico & Nowell, 2000; Marshack, 1997). It is also the case that there are no other modified rocks like this one: it remains unique. The expression I used before, which applies under the Convention Criterion is that "We cannot be seduced by the promise of iconicity into accepting the plausibility of the original identification. Even an accumulation of implausible claims does not strengthen the claim for any one of them" (Davidson, 2003, 153).

What about non-figurative marks—what grazier Bill would have called scribbles? Two candidates are the marked mussel shell from Indonesia that has been dated to about half a million years ago (Fig. 3.7 and 7.28 Left), and the 350,000-year-old scratched bones from Bilzingsleben in Germany (Fig. 7.9). The first consists of marks on a mussel shell from Trinil in Indonesia (Fig. 7.28 Left), the site where the fossilized bones of the early human ancestor, *Homo erectus*, were first found over

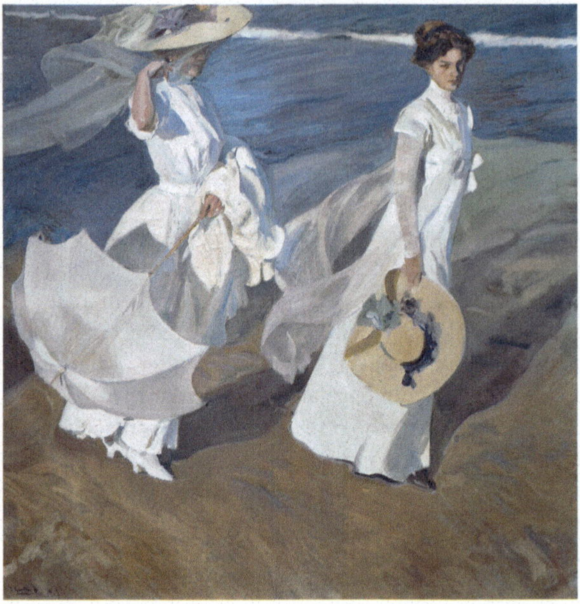

Fig. 8.3 Joaquín Sorolla Strolling along the sea-shore 1909. Sorolla Museum, Madrid. From Wikimedia Commons Public Domain

120 years ago (Joordens et al., 2015). At the time, a collection of large mussel shells was removed from the layers at the archeological site, and these were kept but not carefully examined. When the people who recently rediscovered them in a museum undertook a detailed examination, they found that several of the shells had puncture marks which implied patterned skilled behavior to open the bivalve. In addition, on one shell in the group they found scratch marks, some of which seemed patterned (numbered in the original drawing) though some marks did not. For some, the lines seem to show patterned behavior by the human ancestor who made them, but there is no sign that such lines were ever made anywhere else for many tens (or hundreds) of thousands of years. There was no repetition. I discussed this in detail in Chap. 7.

Villaverde (2020, 171–174) sought to show the earliest marking around the world that constitutes the beginning of art (of which I have selected one of his sets of assembled examples in Fig. 8.4).

Despite the fact that these specimens were selected to make a point about similarity, it is not clear to me that it is a very strong one. Even if you judge that the similarities were strong, there are clearly two groups, as Villaverde recognized—those from Africa and those from Europe. Yet those from southern Africa (as far away from Europe, do not forget, as you can get in Africa) are from dates that do not overlap each other—clearly within State A—and from a different set of times from the specimens from Europe—which are from State B. This does not sound like a convention to me, but we archeologists, as well as you readers, have the advantage of being able to make comparisons across time and space so that things look like part of a common tradition when in reality, perhaps, we should think of them as isolated scribbles.

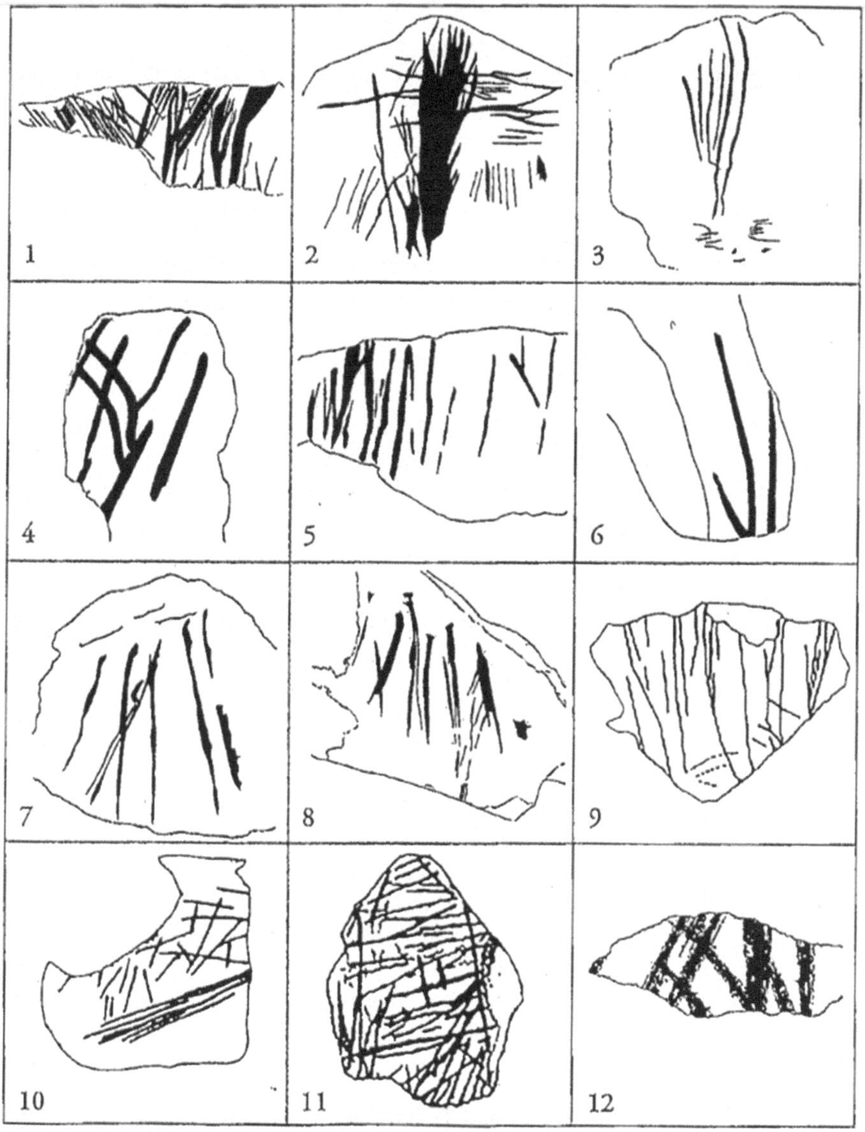

Fig. 8.4 Villaverde's comparison of engraved motifs from the European Lower and Middle Paleolithic (ELMP) and the Southern African Middle Stone Age (SAMSA). The specimens are SAMSA: *1–4, 12* Blombos (100–75 thousand years ago); *5* Klasies River (100–185 thousand years ago); *6* Sibudu (77–29 thousand years ago); *9* Wonderwerk (it turns out it is Holocene, and should not be with this group). ELMP: *7* Pesturina (95–64 thousand years ago); *8*. Kiik-Koba (about 32 thousand years ago); *10* Fumane (MIS 5–3, i.e. between 130–30 thousand years ago); *11* Champlost (MIS3 i.e. 57–30 thousand years ago)

I have, however, argued that, in State B, Blombos does give us some clue to a convention, because of the repetition of the cross-hatched pattern in two archeological layers quite close to each other in time (Fig. 3.29) around 72.7 thousand years ago. The pattern of cross-hatching appears to be the same as that from Gorham's Cave, Gibraltar (from my State B) (Rodríguez-Vidal et al., 2014) from just before the appearance of modern humans there and dated perhaps to near 39 thousand years ago—perhaps 33 thousand years and a continent away from similar things from Blombos. Again, they cannot be a part of a single cultural convention. From about the same time as the image in Gorham's Cave, in the southwest region of Germany called Swabia, but associated with modern humans, several bone or ivory figurines had cross-hatching (Fig. 8.5). This appears as two sets of parallel lines crossing each other where the parallel lines have similar spacing to each other as in other figurines from sites in the region (Figs. 2.14 or 2.16). I do not doubt that the Swabian figurines are part of State B and Art 2, and it is partly by association with these marks that the Gorham's Cave markings are said to be art.

There are several possible reasons for the appearance of this simple pattern in early art from three separate regions, in which the two European regions (almost 2000 km apart) are separated from Blombos by the Straits of Gibraltar and the whole of Africa (at least 10,000 km). Most obviously, it is a simple pattern with minimum requirements for recognition, or it is a pattern that is somehow generated within the visual system of people (and? not other animals?). The third option is the one I was taught as a student more than 50 years ago; such a pattern may represent aspects of the coat of animals (though that would not account for Blombos or Gorham's Cave except by abstracting the convention from the realism of the sculpture—which would itself imply cognitive abilities unknown from other evidence of the time). My preference is for the first of these options with the added interpretation that the appearance that we would call cross-hatching is part of the production of what I have called "the expectation of meaning". At very least the producer, or someone, needed to tell a story about what they were doing, and it is this necessary inference that there was a story here which makes this into art. It was the reason for the meaning, invisible though it may now be.

The expectation of meaning among humanly made marks derives from the recognition that other humans made marks in a pattern. On that line of reasoning, it

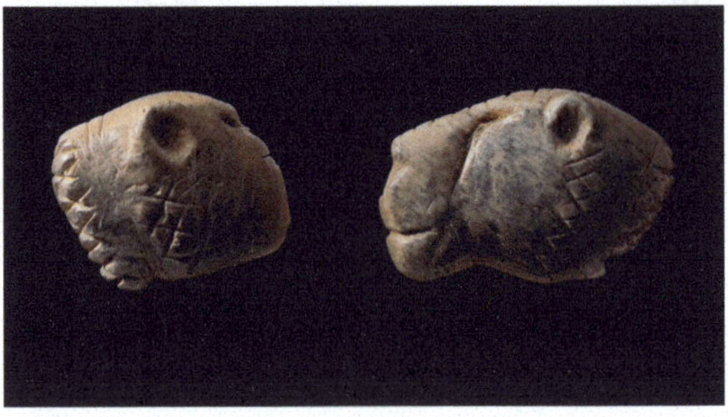

Fig. 8.5 Two views of the mammoth ivory lion's head from Vogelherd with cross-hatching and rounded ears like the almost contemporary paintings from Chauvet Cave. Between 30–40 thousand years old. (Conard & Floss, 2013)

does not matter why the pattern occurs, simply that people have the capacity to recognize it as a pattern. I would argue that the earliest examples in Art 3 demonstrate nothing more than the capacity of the producers to arrange marks in a pattern, which led others (including us) to recognize the arrangement and develop an expectation. For later marks, once we have established that people make patterns to which they ascribe meaning by telling stories, the meaning derives from the stories and their place in the culture and not from the marks alone.

I think it is quite close to a confirmation of an old argument by Wobst (1977, 326) who argued that:

> …given a category of material culture, stylistic messaging is either absent altogether or it is all-pervasive. … it argues for the sudden appearance of stylistic form in material culture, instead of the gradual incremental evolution often anticipated: a state of no-stylistic-messaging should suddenly be replaced by a state in which stylistic form has pervaded at least one (or more) categories of material culture.

I tend to summarize this colloquially as "When one thing has meaning, then all things have meaning," but on that reading, the critical question is "What is a 'thing'?" Wobst was arguing about the appearance among people with modern cognition of behavior which would have "stylistic form". I am suggesting that what came first was the expectation that patterns carry meaning, and then the evolution of modern cognition, through which there are several avenues for reducing the indeterminacy of the meaning. Art 3 is the earliest art which may have been the beginnings of art (Fig. 3.27 Right), but it did not appear all at once. In South Africa at least, it appeared over a documentable sequence of 30 thousand years (Fig. 3.29), and the two pieces of ocher with cross-hatched lines at Blombos occurred right at the end of that time. Remember that the length of time between now and the earliest tradition of cave art in Sulawesi and in Europe is about 40 thousand years. It would be quite appropriate to argue that the ambiguities allow those who are interested in art to dismiss the claims of those looking for the beginning of art but, as we will see, their claim may not be appropriate for Art 2.

What was the process of getting from Art 3 to Art 1? I think that the answer is contained within Table 4.1 which includes: a column for Semantic Roles among humans and chimpanzees—the animals were (surprisingly) similar in the range of roles (Byrne et al., 2004); the column of the Semantic Roles involving stone tools from Davidson and McGrew (2005) entails that there was a physical product which differentiated humans from apes, and which, by persisting, might be thought about; and the column of Semantic Roles for art production shows that for some of the semantic cases, which we know to have existed because they were present both in apes and in early stone toolmakers, the object (which was an idea) must be in the mind of the artist when it was produced (Davidson, 2013a)—*cosa mentale*. There is a literature about the cognitive ability of apes (Tomasello & Herrmann, 2010), and from that the question has been why apes do not produce art? The reasons begin with the middle column of the table: there needs to be a physical and persistent product of the actions. Thinking about those products, then, produces a) permanence and b) the need for individuals to account for—tell stories about—what they have done. That second aspect inevitably involves and encourages the collaboration that Michael Tomasello (and others) suggest is one of the key features of human, but not ape, cognition.

Combine that with the rest of the table and you have the question of the emergence of pattern—and the mental ability that that entails. I need to acknowledge that the stone tools known as handaxes have a role in that evolutionary process (though I have been arguing against the extreme common case of it in Davidson, 2002) (Fig. 8.6). The really important point is that handaxes are an identifiable form, can emerge in trials of making stone tools when choices of striking platforms were nothing other than random, as Mark Moore and Yinika Perston (2016) have shown, and that appears an identifiable number of times. That is to say that out of the routine of producing stone flakes with sharp edges by breaking stones for cutting in one form or another some random processes could become habits of action to produce recognizable patterns (Davidson, 2019). What was needed then was for such knappers to recognize that such a form was a desired outcome—that the pattern has a function. They would do this if they found that such forms were good sources for further knapping to produce flakes (or even that they could be used) and then that the form could be flaked. Either the form was coarse, in which case the tool maybe could not be used, or it was pretty in which case it could. Once the knappers realized that it

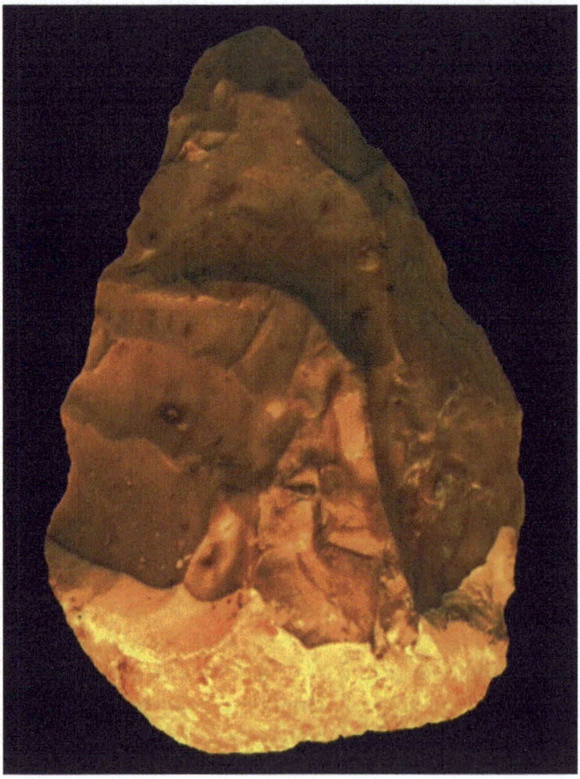

Fig. 8.6 Handaxe from Jelinek's excavations at Tabun. A core from the same site has flakes removed which are almost identical to those that have been removed from this rock (Davidson, 2002), Fig. 9.4). (Photo Iain Davidson)

could be pretty as well as functional, then the way was open for something like art (Wynn & Berlant, 2019). The point is that the prettiness is a construct in the mind of the knapper (as well as the archeologist).

Art 3 is the first stage in the appearance around the world of marking that we might now recognize as art. I suggest that neither the "pretty" form of handaxes, nor the production of patterns of marks on objects, which we might want to interpret as art nowadays, should really be considered as art. Instead, I suggest that what our ancestors did through leaving marks of their accidental or deliberate actions was to set up the expectation of meaning, a meaning that was essentially indeterminate.

And yet… I do not want to get bogged down with arguments about precursors to modern humans and their ability to make art, but just as there are materials from Blombos early in State B which have a somewhat uncertain status as art, so there are increasingly large numbers of things found at the time of the Neandertals of Europe which might lead on to art, yet do not appear to. Philip Barnard (2019) would account for such convergence and intermittent appearance of "things which might lead to art" in terms of cognitive evolution. In addition to things we have discussed such as the Trinil and Berekhat Ram pieces and the arrangement of stalagmites from Bruniquel, these include the talons of predatory birds from sites like Krapina in Croatia, which must be construed as being personal ornaments from as long ago as 130 thousand years (Radovčić et al., 2015) (Fig. 7.26). It is a perfect illustration of the discontinuities that other examples of likely suspended talons are from sites much later in date and from far away (see Table 7.2). So other examples are from France and 50 thousand years later, or elsewhere in France but 30 thousand years more recent still. One group of ochered marine shells are from Southern Iberia more than 110 thousand years ago, another from alpine Italy more than 60 thousand years later. Several other sites now have similar types of indications of personal ornament, including others with pierced shells, dating back to about 142 thousand years ago in Morocco (Sehasseh et al., 2021). It is our propensity to consider the overview of a hundred thousand years which makes them seem like a tradition.

As I have argued before (Davidson, 2023), archeologists make up stories "as if there was originally some sort of continuity between the different elements of archaeological evidence." This is convenient for one sort of storytelling, which relies on a continuity with (usually modern ethnographic) evidence that sounds familiar. So, the zigzag on the Trinil shell calls to mind other examples of the use of the motif in art—as, for example, in South American cave paintings like those from Cueva de las Manos (Aschero, 2021) (Fig. 8.7). Yet these are separated by half a million years and half the world.

In fact, the continuity may not mean that we should think of a long unbroken set of practices between the almost familiar and the familiar. Rather, we should take the opportunity to question what the underlying reasons for superficial similarities might have been as in Barnard's (2019) mental modelling. I suggest that we are dealing with three things in State A: the emergence of recognizing and marking individuality; mark-making such that the contemporary humans expected there to be meaning that was indeterminate; and the first faltering steps in the removal of that indeterminacy of meaning. But there was no tradition.

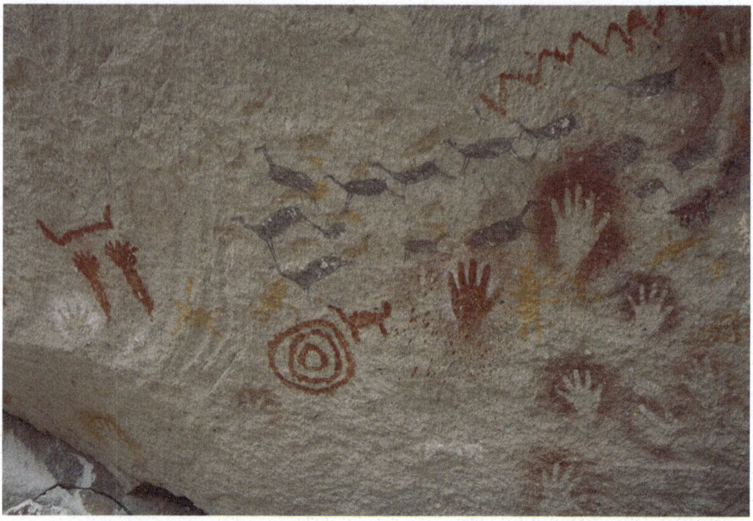

Fig. 8.7 Painting from Cueva de Las Manos, Argentina with zigzag (Photo and permission from Carlos Aschero)

8.4 Art 2: Archeological Art

Art 2 seems to be the product of the daily lives of the people who saw it and appreciated it, such that the cultures of the users were fully engaged with the producers and especially their stories and materials that went with them. Art 1, on the other hand, is different because of the importance of the market in Art which created an independence from producers, stories and the day-to-day culture of users. By the end of State B, people all over the world had begun to live in settlements they built from stone or clay, and when the walls were smooth or plastered, they painted on them, as at the 6 thousand year old site of Teleilat Ghassul in Jordan (e.g., Drabsch & Bourke, 2014) (Fig. 8.8), and in some instances included sculptures within them.

Robb has worked hard at expanding our understanding of the relationship between archeological art (Art 2) and society. In particular, Elizabeth DeMarrais and Robb (2013), stated that:

> Arguing that art is participatory as well as visually affecting, we first suggest that art creates sites of activity for shared interaction. Second, we discuss the varied ways that people use art to create and assert representational models for social relations. Third, we consider the varied roles of art as cultural capital, marking out members of society through shared forms of knowledge or access to art. Finally, we document the ways that art serves as a medium of exclusion and as a means for resisting authority or challenging power relations. We highlight the layered meanings inherent in many artworks.

This list includes many of the characteristics that I am saying define other works of art, including persistence, social relations between producers and seers, embedding of stories within society and incorporation of storytellers and learners jointly into the same set of cultural values, the use of art to maintain boundaries of entitlement

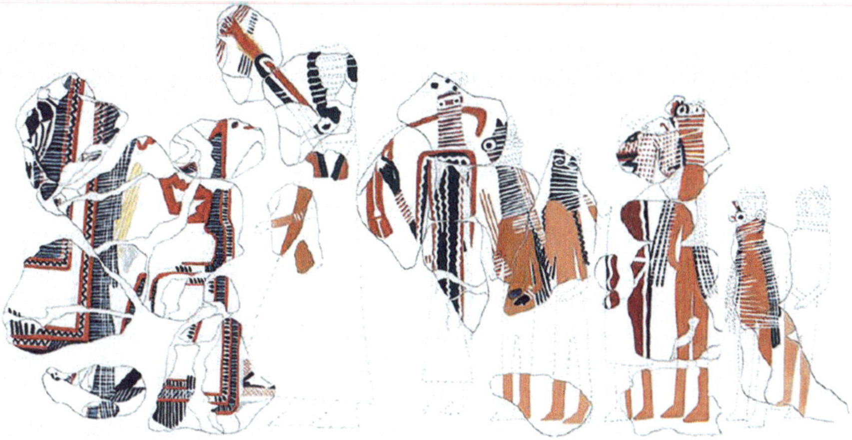

Fig. 8.8 Reconstructed wall painting of procession scene from Teleilhat Ghassul, Jordan (6700–5700 years ago) (Drabsch & Bourke, 2014) (Permission from Dr. Bernadette Drabsch)

and snobbery. Note, however, that I introduced some of my themes while discussing Art 1, and DeMarais and Robb are talking about these themes in Art 2. There seems to be no reason to produce an artificial boundary in this regard between Art 1 and Art 2, except in terms of those same criteria of entitlement and snobbery. Indeed, we go back to grazier Bill saying "Not that I would call them art, so much as scribbles" or to the cliché "I know what I like." People use different criteria: that is why I have been trying to broaden our thinking to connect the familiar inventory of Art 1 (which art historians might prefer) with the earliest art, Art 3, and its evolution to become archeological, Art 2. Otherwise, art, and the label "art," is no more than a thing people use to reinforce their own prejudices about class, race or gender. Any judgment about whether something is art or scribbles tells us more about the judge than about the thing.

Robb (2015) sought to break the mold of the story of Art 2, archeological art, of Europe by assembling a database of evidence from 211 art traditions across the continent from the earliest sites then known before about 2000 years ago, with the aim of an analysis which moved from "'what does art mean?' to 'what does art do?'" (p. 636). Art, Robb said, "encompasses figurines, cave paintings, rock carvings, statuary, megalithic tomb decorations, metalwork, modeling in metal and clay, and various figured objects." One of the objectives of my discussion has been to provide an understanding derived from the evolutionary emergence of art in the archeological past, which would encompass Robb's work.

Nearly 30 years ago, I attempted to understand the earliest art of the Upper Paleolithic, between 40 thousand years and 10 thousand years ago (Davidson, 1997). I concluded (p. 153) that "it signifies the way humans first worked out their relations with land and with each other once they had discovered language. It means the opposition between fisher-gatherer-hunters and agriculturalists and thus it means the largest themes of human history: the conflict between different ways of life and dispossession through colonization."

In a later paper, Robb (2017) mentioned as archeological art (Art 2) the Lascaux painted horses (probably painted about 21 thousand years ago) (Fig. 2.11), Neolithic figurines (Fig. 1.3), Celtic designs, Pueblo pottery, Iroquoian pipes, and Andean geometrical designs. He went on to attempt a sequential account of European art (Robb, 2020) which he characterized as moving from the second half of my State C "Neolithic iconic art which uses heavily encoded imagery, often schematic geometric motifs to a Bronze/Iron Age narrative art [my State D], which increasingly involves imagery of identifiable people, animals and objects." When he analyzed those data, he showed that "[art] encompasses many things: more or less formalized ritual paraphernalia, body ornaments, decorations that make a place appropriate for a given purpose, objects signifying the material embodiment of supernatural beings, and traces of action (for instance, a rough sketch made as a gesture emphasizing a narrative)."

In some sense, this whole book is about understanding how all of these things can be different but at the same time are, nevertheless, art. I will introduce in each State, though, some archeological art equivalent to those pieces mentioned by Robb.

I have tried to summarize what appears in this book according to a set of subdivisions which suits the argument. In particular, Art 2 was an element of the move from ritual to formal teaching, from indeterminacy of meaning to the invention of agents to give meaning to things by the use of narrative, from art defined by a time and place to art that transcends both of them, and from an art that was controlled by the institutions of power obtained by other means to an art that could be independent of such social structures.

8.5 State B The Emergence of One Species—Ritual and the Importance of Distance

By the end of State B, our species, *Homo sapiens*, was the only one surviving on the planet. I have sought to avoid the knotty problems of the art status of any of the other ancestors in this state by lumping them all together, with the caveat that some "art behavior" may have appeared without the cognitive implications it had by the end of State B. It would be controversial to argue that art had a role in the dominance our species has acquired over others, and I want to avoid that controversy here. If it is true, then I would argue that it was the interaction between art and human cognitive ability which had that role, and not any art object itself. It could be argued that the expansion of those the roles of individuals to include ritual, personal ornaments, decorations, representations of imaginary beings was a product of the capacity to produce persisting and repeated material culture as hooks for the creation and diffusion of narratives. But that would require another book.

Langley (2013, 624) suggested that "The act of landscape socialisation may have remained solely a Modern Human characteristic with Neanderthal populations solving problems of navigation, access to resources and interaction between groups in a manner which did not require the extensive development of stories (and consequent group identities) tied to specific landscapes." This is not the place to question whether this summary of a major difference is a result of or was caused by what Langley calls "The lower population densities during the Middle Palaeolithic" which "resulted in a lesser need for portable symbolic material culture, and a 'story-less' landscape" which "would account from [sic] the lack of landscape marking and infrequent long-distance transport of distinctive raw materials." By the middle of State B, however, the transition had taken place and Neandertals were extinct, landscapes were full of stories, and an important element of this was the increased transport of raw materials over larger distances than before (Davidson, 2016, Fig. 10.2). The story-tellers had won.

Early in State B, people were using paint, perhaps on their bodies (d'Errico et al., 2023; Henshilwood et al., 2011), as well as personal ornaments which initially consisted of pierced shells (Henshilwood et al., 2004) around Africa (Sehasseh et al., 2021) and later were finely made of ostrich eggshell using which they may have signified social networks (Miller & Wang, 2022) as part of the process of indicating their "symbolically mediated self and personal identity" (Nowell & Cooke, 2021). In some instances, it appears, shells with natural perforations were strung and sometimes stained with ocher (as at Qafzeh, not far from the East Mediterranean coast, dated to about 92 thousand years ago) (Bar-Yosef Mayer et al., 2009).

The most spectacular example of the use of beads for personal ornament was found at the Russian 34-thousand-year-old archeological site of Sunghir near Moscow (Trinkaus et al., 2014) (Fig. 8.9). From their disposition around the bodies these beads seem to have been sewn on to fitted clothing.

This process of identifying individuality was an important means of defining membership of one group rather than another (Davidson & Noble, 1992, 139)—something I have argued in this book is an important aspect of art behavior. It turns out that defining the self to others—giving the individual's personal perception an external marker—is one of the constants from the very beginning of art, and the reason is very simple. Within the four-part definition of art as an art-system—the producer, their subject, the object they produce and the seer of the object—if neither of the two people identifies their distinctiveness and cannot be seen to have it, then it would be difficult to produce art that depends on producer and seer as distinct roles. Stiner (2014) has argued that the use of beads had two important functions: as a way of communicating personal identity or membership of a particular group (cf Davidson & Noble, 1992); and, as an indicator of the spatial dispersal of the social network of the individual. This use of ornaments, therefore, likely contributed to overcoming not the tyranny of distance, so much as the perception that individuals could overcome it in a social network.

The most remarkable other example of this use of body ornaments also introduces one of the most famous examples of sculpted figures in the world—I once referred to it as "the most famous woman in the world"—the figurine from Willendorf (Fig. 8.10). This 30-thousand-year-old sculpture is distinctive for a number of reasons, including her lack of facial features, which are covered by what seems to be elaborately arranged hair (or a head-dress), and lack of feet. She shares the lack of facial features and of feet with many contemporary figurines which also

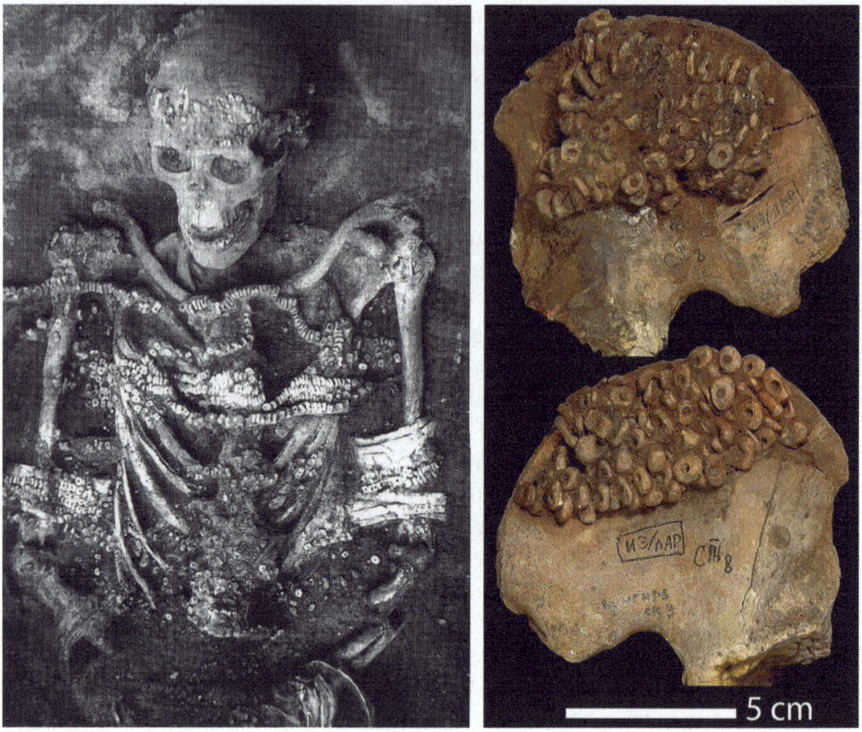

Fig. 8.9 Left: Superior view of the Sunghir 1 skull, axial remains, and arms in situ (from O.N. Bader 1998: Plate 2) (Original not seen). Right: External view of the Sunghir 3 ilia with the mammoth ivory beads still attached to it and the ocher staining of the left ilium and beads (above) (Trinkaus et al., 2014). (Both photos and permission courtesy of Erik Trinkaus)

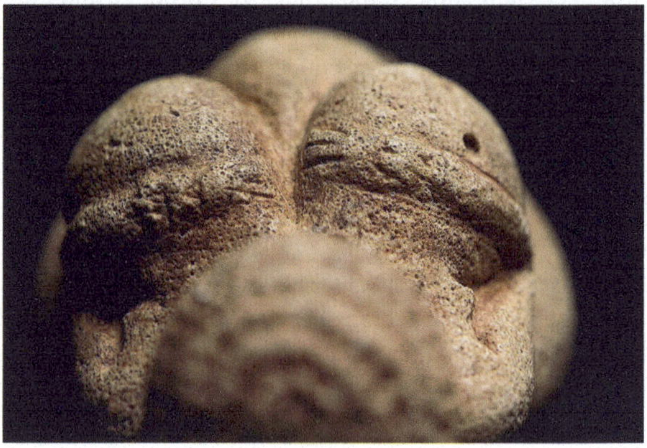

Fig. 8.10 Venus of Willendorf showing detail of the ornaments on her wrists from above, with out-of-focus view of the elaborately arranged hair. (Courtesy Caroline Posch and Lammerhuber, Natural History Museum Vienna)

mostly model women if they have any distinctive sexual features (Bowdler & Balme, 2010; Conroy, 1993). She also shows a zig-zag ornament on her right wrist. Soffer et al. (2000) have argued that several of these figurines seem to mark aspects of their, probably organic, body adornment, clothing even. The figurine is carved in limestone which probably came from 730 km away on the other side of the Alps (Weber et al., 2022). It seems plausible that the distance between the source of the stone and the findspot of the figurine gives some idea of the extent of the spatial dispersal of the social network of those who were connected with the art. They had overcome the tyranny of distance.

Some art sites were fixed points in the world of mobile people, and these need to be considered in the context of networks of information exchange (Whallon, 2006). Ironically such information exchange could best be analyzed by studying art that was not fixed. Conkey pointed out long ago that it was possible to analyze the marked bones in cave sites such as Altamira and its surroundings and show that there was a pattern to the distribution of stylized marking on the bones. Many bones had different marks on them which could be found in a variety of different sites but that all of the marks could be found on marked bones at Altamira (Conkey, 1980). On that basis, she argued that different parts of the community dispersed at different times to some of the other sites, but that the whole community came together at Altamira. What has emerged as archeologists have developed ways of identifying the movement of raw materials from sources to find-spots, as in the case of the Willendorf figurine (Weber et al., 2022), is that some raw materials were transferred over very long distances late in State B, but not before (Davidson, 2016, Fig. 10.2; Whallon, 2006). In doing so, people appear to have overcome the tyranny of distance through their willingness to obtain materials, either directly or by exchange, from far away.

People also marked caves sporadically with fingermarks (Nowell & Van Gelder, 2020), engraving (Norbert Aujoulat et al., 2002) or painting (Brumm et al., 2021; Pitarch Martí et al., 2021). In Chap. 5, I discussed the importance of ritual and of the word of mouth that is associated with it. I would go further and suggest that this marking almost always required repeated storytelling to make it possible for seers to understand and internalize the marks. Repetition is an important component of rituals which drives the process of narrative and its embedding in the context of a society of collaborating individuals. I would argue that ritual and teaching go and went hand-in-hand, and on the same basis, that many of the places where isolated exceptional instances of marking have been reported were probably not embedded into the mental understanding through the use of ritual (as appears to be the case, for example, among the Neandertal paint marks studied by Hoffmann et al., 2018). It may be that we should not consider them as art at all.

To leap ahead to the present, Morphy (1991) described how the initiated taught uninitiated individuals to understand the rules of society by making marks and paintings. The point is that producers tell stories (or are capable of telling them) about the relations between the object they produce and the subject of it. In the case of the sand drawings I witnessed, the images illustrated the story (Fig. 3.24), and while there was no formal teaching of the sand drawings, they were part of

Mrs Hansen's teaching of the story to her daughter, Isabel. In the case of European cave art, the contexts are such, some of them extreme cases, that the argument is incontrovertible that some teaching went on (Mithen, 1988)—and it is most likely that it involved stories (Davidson, 2013a; Nowell, 2021, Ch. 5; 2023a). Ritual enhanced the stories, generating what Rappaport called canonical messages as I described in the previous chapter. In the fullness of time, and by such devices, the stories and their accompanying pictures became part of the culture of the whole society.

But this teaching cannot be considered "formal teaching" in the sense that I will discuss it in the next section, but it is clear that there must have been some social learning both for language and for ritual (Davidson, 2016, 101–103). Such learning might be considered observational learning acquired during joint attention to a common task and not requiring formal teaching (leaving aside the arguments among pedagogues about the best methods of teaching). As I pointed out in the 2016 paper, all aspects of culture involve learning, but the values associated with individual societies and particular sections within society involve different aspects of learning. At some stage, teaching enhances the acquisition of skills that lead to such differentiation. What I call "formal teaching" involves the specialized social system which is designed to ensure that all members of particular sections of society have the opportunity to acquire those skills from more experienced elders.

The extent to which knowledgeable people had responsibility for sharing the stories, told during ritual, implies a social inequality that probably underlay differentials of power. I argued that the extraordinary concentration at Parpalló of plaquettes with paintings and engravings implied that there was a differential about the sharing of information which concentrated power at Parpalló (Davidson, 1989a, 2012b). Different individuals doubtless were able to exploit different resources of power by revealing the secrets of the variation in the art in the caves and their control of access to information about how the variation was represented conventionally.

Even close to the Western European region with cave art, there was such variation: in Swabia, there were mostly animal figurines of bone or ivory like those from Hohlenstein-Stadel (Fig. 2.14), Geissenklosterle (Fig. 2.16) and Vogelherd (Fig. 8.5); and human figurines (including one from Swabia at Hohlefels shown in Fig. 8.11) (Conard, 2009). The variation was also represented by the figurines of which the one from Willendorf (which is stone) was an example (Fig. 8.10) which were distributed over large regions of northern and eastern Europe (Bowdler & Balme, 2010; Conroy, 1993) but to all intents and purposes not to Spain, Italy and Greece. At very least, the lack of faces and feet on these figurines was a convention about how to represent people.

Linda Conroy (1993) pointed out that the figurines could be divided into two groups, one with representations of females, and another with figurines which were not female, but were generally not male. This opposition suggested that this was the earliest example of gender that was socially defined rather than mapping on to biological sex (which would have given us female and male figurines) but representing female and non-female. Sandra Bowdler and Jane Balme (2010) later suggested that the figurines could represent women who were past child-bearing age.

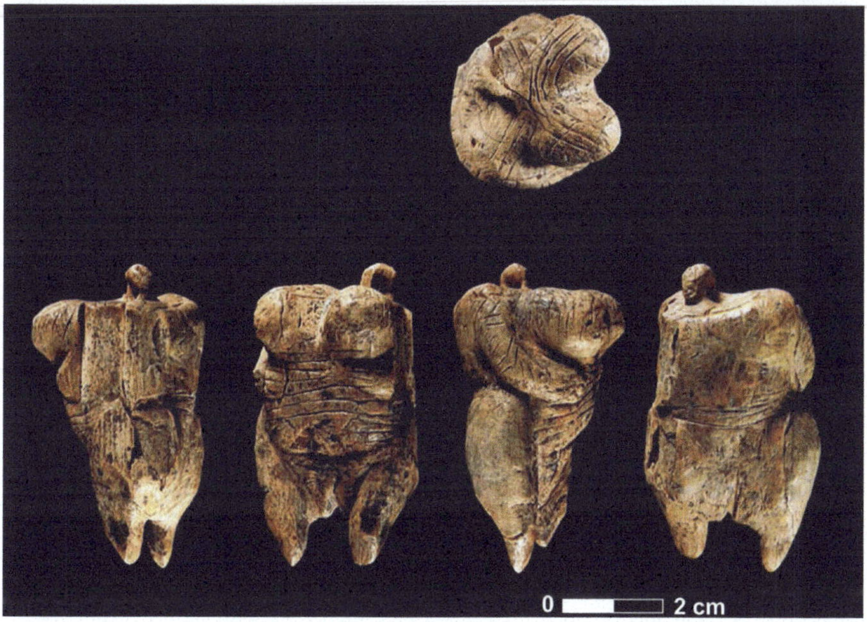

Fig. 8.11 Four views of the ivory figurine from Hohlefels dated older than 35 thousand years ago (Conard, 2009). (Permission from University of Tübingen, Photo: M. Malina)

The idea of a social convention is supported by the gender interpretation and the idea that the figurines represented a common socially important state. They were probably not sexual objects and little to do with conventional beliefs about fertility (Nowell & Chang, 2014). The essence was that the people who made the figurines worked within a convention and told stories about what was represented to seers who themselves had faces and feet, probably included more than one gender, as well as people who were not beyond reproductive age. The figurines themselves and the limits to their distribution imply variation in the art that was expressed in non-utilitarian marked objects around 30 thousand years ago.

Sometime during State B, people used watercraft to move from Borneo into Sulawesi, and painted rock walls have been found on both sides of the crossing, in Borneo (which was part of Sunda, the continent now split into island Southeast Asia), and Sulawesi (which is one of the islands in Wallacea between Sunda and Sahul) (Aubert et al., 2018; Brumm et al., 2021). From there after several sea-crossings between islands they reached Sahul, the continent now split into Australia and New Guinea by the sea rising 14 thousand years ago (Shipton et al., 2021). They certainly made landfall in Australia by 65 thousand years ago (Clarkson et al., 2017) and the evidence from Madjedbebe shows they were already grinding ocher and using grindstones for plant food and stone tools (Hayes et al., 2022). In all probability, body painting (for there is no evidence that ground ocher was used to paint anything else) became part of the process of identification of individuals (perhaps including the groups they belonged to) and that permitted engagement in ritual. It seems likely that a similar interpretation must be applied to the discovery of two abalone shells from 100 thousand years ago used as paint pots in Blombos cave (Fig. 3.27 Left) at the start of State B. A painted rock fragment which had fallen from the ceiling was discovered at the cave of Nawarla Gabarnmang in northern Australia dated to 28 thousand years ago (David et al., 2013), but, although it is currently the oldest evidence in Sahul that rocks were painted, it cannot be construed as an image of anything.

At a later time towards the end of State B, people moved into the Americas, and, although it was initially thought they walked across a dry Bering Strait, the evidence for the use of watercraft makes it more uncertain when and how they made their move (Braje et al., 2020; Davidson, 2013b). The evidence of early art in the Americas is very slight, but there is at least one quite good image of a Mammoth engraved on bone,[1] though how it survived is much more open to question. A battery of scientific tests established that the engraving is old (Purdy et al., 2011), and we know that people left their spear points in Mammoth skeletons (Grayson & Meltzer, 2003) before the species became extinct so that the image must have been made before that date before 13 thousand years ago (Buck & Bard, 2007).

One of the features of Art 2 from State B, more than 12 thousand years ago, in rock-shelters and caves, is that in the west of the Mediterranean region, there was art produced deep inside caves in France and Spain, as well as art produced on bones and stone plaques excavated from archeological sites often just outside those caves. The art inside the caves was different from the portable stuff—there were more straightforward scenes in the portable art than on the walls of caves (Culley, 2021). At the same time in the countries in the Eastern Mediterranean region, there was almost no art either in caves or in archeological sediments (Davidson, 2012a).

There is an aspect of the claim about variation that is often overlooked. We have known for a long time that the cave art (Art 2) of Western Europe (from State B) was not widespread outside France and Spain but that in those two places it was quite abundant. It could be, of course, that it was widespread elsewhere, but not in the same ways (on bone or wood that has perished, or on stone that has not been found). At the same time, there was very little art of any sort visible in the abundant sites of the east Mediterranean until right at the end of the period (Bar-Yosef, 1997; Davidson, 1997). Erella Hovers and Anna Belfer-Cohen (2022), though, argue it may simply be that symbolism was present but manifested by narratives which left different material traces. One figurine, the 11 thousand year old 'Ain Sakhri figurine from a Palestinian cave near Bethlehem, does appear to be representational—a sexual act (Fig. 8.12) (some caution about the dating may be needed according to Boyd & Cook, 1993).

The important point I take from this is that in archeological art during State B there is variation both within regions and between regions: "The differences between the uses of painting and engraving in the east and west Mediterranean regions

[1] My negotiations to obtain an image to reproduce provided the information that the piece had been sold into private hands and could not be displayed—a perfect example of the selfish impact of the commercialization of art.

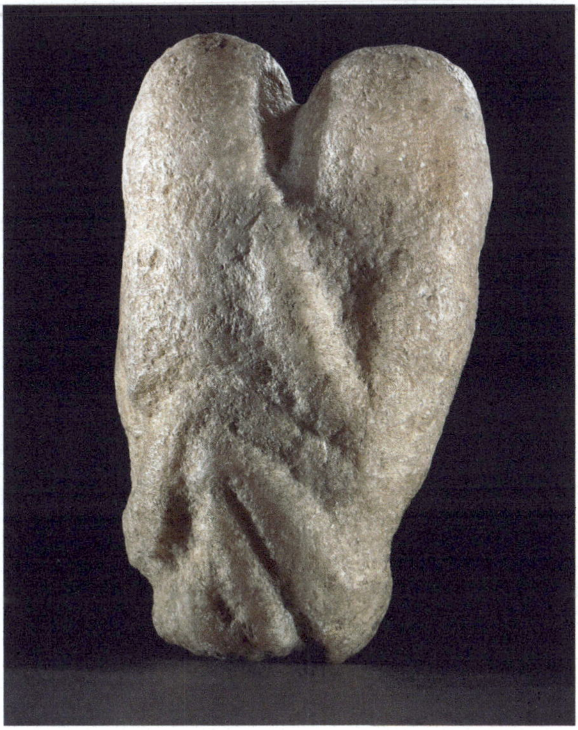

Fig. 8.12 Limestone figurine of two people apparently in a sexual embrace, probably 11 thousand years old from 'Ain Sakhri near Bethlehem (Boyd & Cook, 1993). (Photo obtained from British Museum Image ID: 00045676001)

probably reflect ontological, ritual and religious differences between the societies in those regions" (Davidson, 2012a). There were different outcomes in State A of the early experiments with making meaning through marks and they grew and changed in unique ways so that people in different regions produced art with different patterns and conventions of variation from the time of State B.

Much art in State B probably involved ritual. June Ross defined how to identify ritual using criteria developed by Roy Rappaport (1999; Ross & Davidson, 2006) and I used those criteria for the plaquette art of eastern Iberia (Davidson, 2012b). Art is an important characteristic of the lives of the people who made it and of the people in their societies who are consumers of art. And we can even identify that it has the three characteristics that Warner derived from his study of the Murngin—ritual, human relationships with (parts of) nature, and human relationships with other people.

Finally, I should note the pattern of use of scenes in State B. So far as we know, many of the scenes in cave art in Europe involved rather static representations of individual relationships, say between a hind and a fawn (Villaverde Bonilla, 2021), though there were undoubtedly scenes which imply a more dynamic relationship, such as that in the Shaft at Lascaux (Fig. 8.13) (Davenport & Jochim, 1988).

The date for Lascaux is commonly given as 21.5 thousand years ago (Ducasse & Langlais, 2019). The hunting scene from Leang Bulu' Sipong 4 in southern Sulawesi

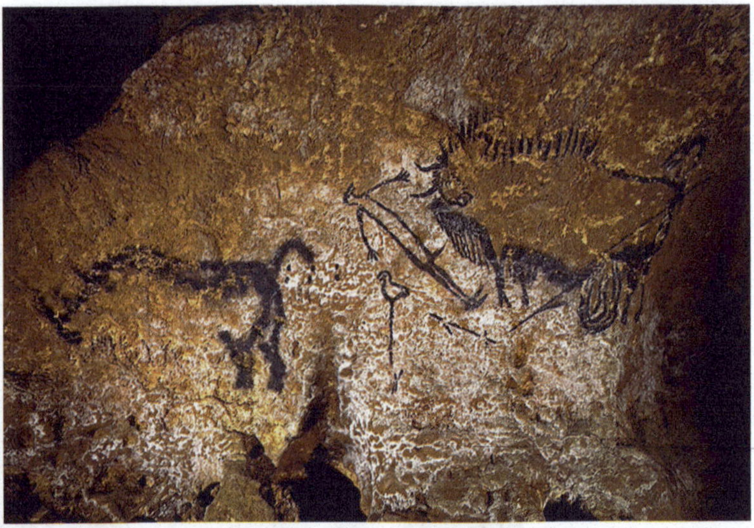

Fig. 8.13 The shaft scene at Lascaux (Aujoulat, 2004, 160–61). (Photo and permission supplied by Sophie Goedert, Ministère de la culture, Direction générale des patrimoines et de l'architecture, Sous-direction de l'archéologie, France)

shows humans apparently dressed in masks hunting wild pigs and dwarf bovids more than 43.9 thousand years ago (Aubert et al., 2019). So, the art of Sulawesi is twice as old as the most famous French painted cave and showed the social convention and storytelling that must accompany the use of masks.

The great French archeologist Gabrielle de Mortillet commented in 1883 on the engraved bones excavated from European sites that were then the first art known to show images of animals extinct in that part of the world (see Fig. 2.17): "They are the childhood of art. But they are not the work of children." The saying was taught to me when I was an undergraduate in the 1960s complete with misquotation, as it was to others (e.g., Pettitt, 2003), and until writing this I did not seek a source. I am very grateful to Valérie Feruglio for pointing me to de Mortillet (1883, 416) whose words were:

> Ce qui caractérise surtout les oeuvres de cette époque, c'est une extrême naïveté. Nous sommes là en présence de l'enfance de l'art mais d'un art très vrai, très réel. Cette enfance de l'art est loin d'être de l'art d'enfant. Il y a loin, bien loin des oeuvres de cette époque aux ébauches informes qui parfois couvrent les murs autour de nos écoles.

Feruglio offered me the following gloss on this: "'Enfance de l'art' in French is used to say that something is simple, is basic, something easy to learn and to apply. So I guess that de Mortillet used it as a play on words. He considered this art as naive but with a certain skill far from the shapeless scribbles done by children on the walls of their school…" (in Email 11 Sept 2022).

The remarks probably referred to apparently simple images such as that in Fig. 2.17. The piece was described in some detail including the corroboration of the

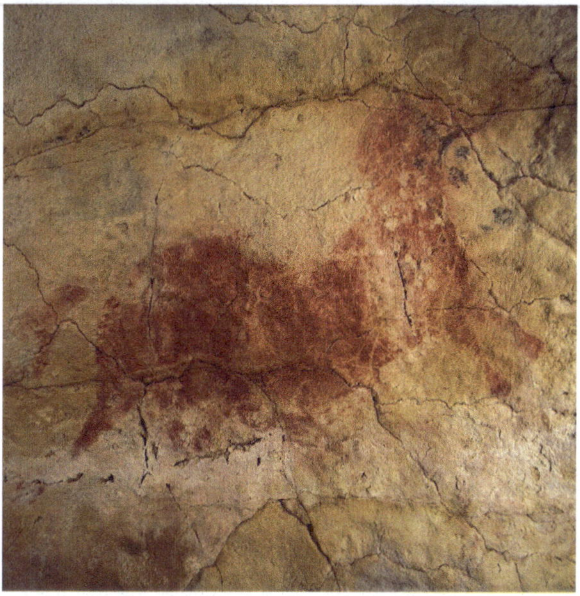

Fig. 8.14 Painting of horse on the ceiling of Altamira, first reported in 1880 after supposedly being discovered by the daughter of the excavator Marcelino Sanz de Sautuola with animals painted on bosses (Fig. 3.17) which created a 3-dimensional effect. (From Beltran & Saura Ramos, 1998, 63) (Photo by P. A. Saura Ramos, permission from Adriana Chauvin Grandela, Museo de Prehistoria y Arqueología de Cantabria, Santander)

identification of species by half a dozen paleontologists. This was deemed necessary to confirm "the coexistence of Man [*sic*] with the fossil Elephant" (Lartet & Christy, 1875). The original cited in this work was published in 1865, only 6 years after the first demonstration of that coexistence by the discovery of a flaked stone "handaxe" in the gravels of Amiens with such bones[2] (Gamble & Kruszynski, 2009).

According to Moro Abadía (2006), the cave art of Altamira in northern Spain was first reported by Quiroga and Torres Campos (1880) (Fig. 2.10) so de Mortillet did not live to see the acceptance of rock and cave art as being of great antiquity and could not have been commenting on the most famous examples of cave art (Fig. 8.14) from Altamira in northern Spain (Beltran & Saura Ramos, 1998).

The acceptance of the Spanish claim awaited French finds and French scholars and the publication by Cartailhac (1902) of his famous "Mea culpa d'un sceptique." According to Moro the other reason for initially not accepting that the paintings at Altamira were genuine was the general disbelief that hunter-gatherers could make art. Engraved bones had been known for at least 45 years before Altamira was first described (see Fig. 2.1) (Bahn, 1998, 57–60), but they were deemed to be "craft" in the rhetoric of nineteenth century assessment of the products of different groups in society (Moro Abadía, 2006). Abolish the elitism of groups and maybe "craft" would disappear.

In addition, the delightful image of the mammoth in Fig. 2.17 (Lartet & Christy, 1875) required "scientific reportage" in order to establish that the animal

[2] Mark Moore (personal communication 14 June 2024) pointed out that de Perthes identified sculptures of animal shapes among the handaxes at Amiens. Such things are not now given much credence.

represented was likely to have been drawn in antiquity and represented an animal type that existed long ago; not an imaginary creation of a falsifier—a person who had embellished their knowledge of an existing type. If that was true, then it really established that proving humans and extinct animals had coexisted, was a job for an archeologist excavating the site. "People who are interested in art" or historians can know that the picture of a mammoth incised into bone was left at a particular place in France, and that might be enough of a narrative for some. A far richer narrative can follow an archeological analysis: when the particular picture was made; that people were contemporary with the animals; those people were also drawing what they saw (unlike Dürer); and for many "people who are interested in art," what they drew was also aesthetically pleasing—it was art. Without this analysis (or its reverse as shown by the North American mammoth engraving), the narrative is almost meaningless—it may or may not be true—and the critics might make it important as art, but the art is much more important than that. It is an element of a story about people inhabiting their world, how they lived, and how they depicted a world that was full of creatures that were alive around them and filled them with wonder, not imaginary fantastic creatures.

State B, therefore, was marked by the characteristics that give art its special nature. We have known this throughout the last 140 years. But in this analysis, I have shown that we can tease out several crucial features that characterize art ever since. These are its use to define individuality; its engagement in ritual and the storytelling that facilitates teaching; the involvement of people and groups to locate power to themselves; and, paradoxically, the first steps in the use of scenes to reduce the indeterminacy of the marks—steps that would, ultimately, make available to seers the mysteries of art which individuals otherwise used as resources of power.

8.6 State C Overcoming the Tyranny of Place—Settling Down

There are two features of human behavior during State C that pick-up aspects of the story of art as I have been unfolding it. The first is what, in Table 8.1, I call "Overcoming the tyranny of place"—the creation of places, some of them living spaces, that were identifiable with their producers. This involves what I call, following Gary Feinman and Jill Neitzel (2023), "settling down"—the establishment of permanent structures where art could be produced, and teaching could be undertaken—and then the isolation of particular structures within the settlement. The second is the institution of formal teaching. This is a complex topic which has been addressed in terms of the limited teaching abilities of non-human animals (Boesch, 1991; Caro & Hauser, 1992), in relation to the production of stone tools (Gärdenfors & Högberg, 2017) and of more general aspects (d'Errico & Banks, 2015). The emergence of writing systems at the end of the state entailed the emergence of formal teaching systems which had probably not been seen earlier.

At the boundary between States B and C, the conventional story is that agriculture (including pastoralism) emerged in the "Fertile Crescent", of Western Asia, based on sheep and goats (Daly et al., 2021; Stiner et al., 2022), wheat and barley (Fuller et al., 2023), but doubtless including cattle, pigs and horses (Cucchi & Arbuckle, 2021), and several other plant species (Kabukcu et al., 2021) as well as an earlier domestication of the dog (Perri et al., 2021; Range & Marshall-Pescini, 2022; Shipman, 2021). But agriculture emerged independently elsewhere too, both in East Asia (Fornasiero et al., 2022), New Guinea (Shaw et al., 2020) and in the Americas (Milner & Boldsen, 2023). In all likelihood, none of these involved teaching or direct involvement with art, but it seems likely that art was involved with the cultural values that allowed the particular attitudes to nature that were less directly predatory and involved protection and appropriation instead.

Hunter- gatherer lifestyles initially dominated the behavior of people in all parts of the world (e.g., Kuhn, 2020), though there was probably enormous variation from one region to another, with different contributions from hunting (of a variety of species of different sizes), from gathering (of different parts of plants growing above or below ground) and fishing (which has been seriously underestimated by referring only to hunter-gatherers) (Binford, 2001). As Milner and Boldsen (2023): "The early development of agricultural economies was variable within and between regions, and it did not follow an unwavering unilinear path toward ever-greater reliance on domesticated plants and animals. For much of this time, people cannot be neatly separated into those who relied on foraging or farming, because food-acquisition strategies featured elements of both." Recent literature in Australia has emphasized that many economies might be compared to agriculture (Pascoe, 2014), but this seems to me to involve a caricature of the simplicity of the process in other parts of the world and of the characterization of groups of people as either one or the other. It underestimates the nuances about Australian Aboriginal management of resource acquisition (Davidson, 1989b; White, 1971), using fire (Hallam, 1975; Jones, 1969), exploitation of grasses (Allen, 1974), cycads (Beck, 1992), and other plants (Gott, 1982, 1983), eel management (Lourandos, 1997, 66, 219–20), and the social control of resource management (Lourandos, 1983). These economic subtleties ultimately involved the regionalization of rock art styles (David & Lourandos, 1998).

In the Eastern Mediterranean, one of the supposedly earliest centers, the general narrative is that peoples continued hunter-gatherer lifestyles until farming was introduced (Rowley-Conwy & Layton, 2011) and they began to live in permanent settlements. We should not think, however, that there was a sharp break between mobile hunter-gatherers and settled farmers. In reality there was much overlap. There are two important aspects of the transition, however: first, in ecological terms "Agriculture is not just a further intensification of the hunter–gatherer niche but is a new and transformed niche of its own. … major long-term exploitation of wild cereals and animals does not necessarily lead to agriculture; and … was based on species that played relatively minor roles in the Epipalaeolithic economy" (Rowley-Conwy & Layton, 2011). Second, in social terms, as David Graeber and David Wengrow (2021) pointed out as a central feature of their analysis of their "new history of humanity," social anthropologists Marcel Mauss, among the Inuit of the Arctic, and Claude Levi-Strauss, among the Nambikwara of Brazil, both, in the first half of the twentieth century, described modern societies which were sedentary for part of the year and dispersed for other parts (Graeber & Wengrow, 2021, 99–100, 107). Among the Kwakwaka'wakw (called Kwakiutl by Boas) of the NW Coast of Canada there was similar seasonal complementarity but, in that instance, after a warmer season engaged in fishing, dispersed groups came together in settlements of massive wooden structures built for the winter and endowed with carved images such as those characterized in Boas's (1955) book *Primitive Art* (Fig. 8.15).

One of the key features of life before people settled down was that they led mobile lives which they managed according to the availability of food, water and other people in their environments (e.g., Binford, 2001; Grove, 2009; R. L. Kelly, 1983; Whallon, 2006) but had few long-lasting settlements. Very little has been said about the role of art in such mobility, and yet throughout Western Asia, the sites around the time of the earliest appearance of agriculture and pastoralism were characterized by art or monumental buildings, as at Jericho (plastered skulls) (Fig. 8.16), Çatalhöyük (wall-paintings and mounted cattle skulls as horn core displays), Beidha (sanctuary) (Fig. 8.19), 'Ain Ghazal (large plaster sculptures) (Fig. 1.3), Göbekli Tepe (carved standing stones) (Fig. 6.2) and many others (e.g., Kornienko, 2009).

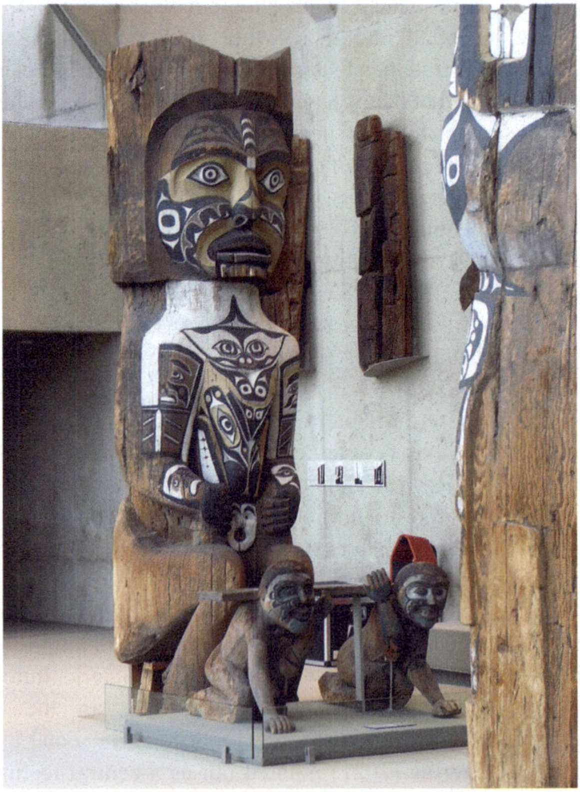

Fig. 8.15 Large NWC First Nations House Post from University of British Columbia Museum, (Object ID A50009 d-f), made by George Nelson about 1906. (Photo by Iain Davidson, permission given from Museum of Anthropology at UBC, Vancouver, Canada). http://collection-online.moa.ubc.ca/search/item?keywords=post&category%5Bculture%5D%5B0%5D=1596399&yearstart=&yearend=&row=14

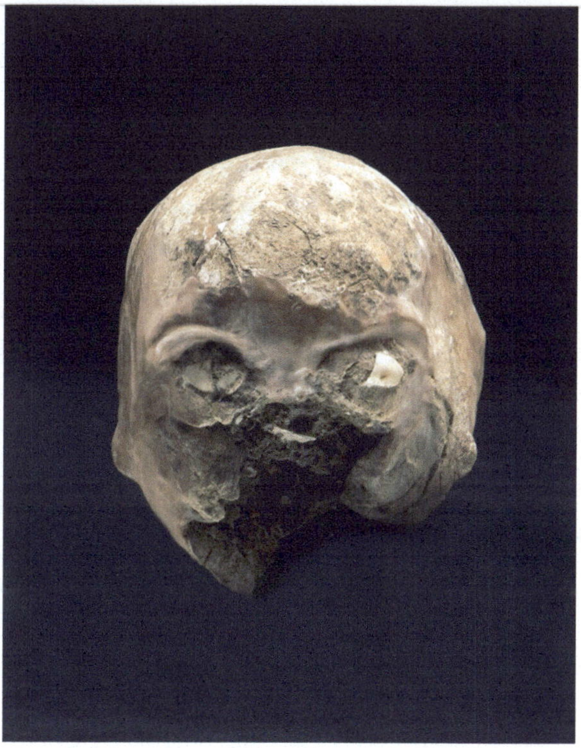

Fig. 8.16 Plastered skull of a woman from the Pre-Pottery Neolithic B of the Palestinian site of Jericho. (Cambridge Museum of Archaeology and Anthropology Accession #1957.159. Creative Commons license (CC BY-NC-ND 4.0)

In Levantine Iberia, the art of early agricultural societies was very different from the art of the caves (e.g., Domingo Sanz, 2021), even though, just as in many parts of the world, the rock art associated with the first pastoralists and agriculturalists was predominantly of hunting animals (Davidson & Nowell, 2021). As we saw in the previous chapter, unlike cave art which had almost no representations of people, and no scenes, Levantine art had both (Fig. 7.15).

It would appear that the art was not just holding up a mirror to the world around the producers and the seers but involved stories about their world. In other words, we must recognize that within any body of behavior that associated people with art there was variation from the beginning, what we might call a *cosa mentale*, at least as early as the time there was a tradition. Moreover, there was never any unity of art style around the world. From the moment art was produced, or, at least since art was expressed by more than just beads and personal ornaments (Miller & Wang, 2022), the fact that it was not mirroring what was simply visible in the world meant there were differences between the cultures and styles in different regions.

At the end of State B and during State C, many people came to live in settlements with dwellings constructed of stone or mud brick and to rely on cultivated plants and animals (Fig. 8.17), whatever ecological conditions preceded or followed the interactions (Davidson, 1980). Although such settlement was not universally true, it

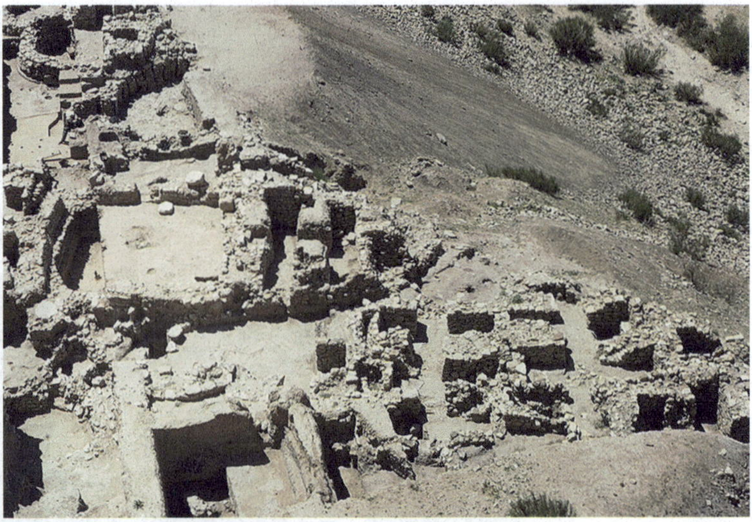

Fig. 8.17 The western side of the remains of the settlement of Beidha, Jordan, from above. The remains date from about 8.5 thousand years ago and feature in some discussions of the appearance of an agricultural or pastoral way of life (Kirkbride, 1968). (Photo Iain Davidson 1967)

did make a difference to the ways in which people used art and interacted with each other.

There has always been an issue, often ignored, of the mobility of the people who lived in such settlements and recent analysis of what is involved in "settling down" has emphasized the social networks involved in mobility, settlement and the interactions between them (Feinman & Neitzel, 2023). The issue here is not about the emergence of settlement (as in the discussion of Beidha, for example, by Byrd, 1994) but, in the terms I am using, about the way in which people with some degree of mobility also recognized the importance of particular places—such as those with important buildings. I suggest that this was achieved by concentrating on ritual and other importance in the mind, and hence the artistic questions and issues of communication. People used artificial, long-lasting shelter of stone or (mud-)brick as a new medium on which they could represent actions in stories using pictures and constructed special spaces within which they could conduct rituals that were accompanied by works of art (Verhoeven, 2002). At Beidha, the ritual area was not, however, accompanied by anything—the whole complex was empty of any other materials, rather as that arrangement of stalagmites made by Neandertals at Bruniquel about 170 thousand years before had been. The difference was that this ritual space was there for all to see without entering a cave. At the ritual area of Beidha, huge grinding dishes re-used in the construction (Fig. 8.18), suggested a previously unfamiliar use of ground food.

On the other hand, Ian Hodder and his teams have constructed an elegant argument that people at the early agricultural site of Çatalhöyük in Turkey built their houses as "history houses" by placing them successively one on top of another with

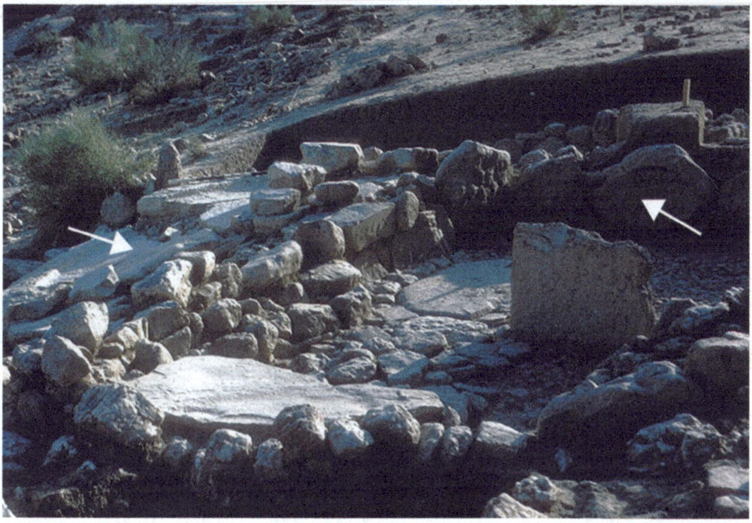

Fig. 8.18 Sanctuary or ritual area at Beidha. One grinding dish forms part of the "wall" behind the standing stone and in front of the archeological post, and another one is broken at mid left of the photograph, just below the standing column of sediment with a wooden peg on top. (Photo Iain Davidson May 1967)

a variety of repeated—probably ritual—elements in each successive build (Lercari, 2017). Such agency may well have been part of the "trick" of the Çatalhöyük "history houses" where repeated activities in the same location saturated the cultural lives of the inhabitants (Hodder & Pels, 2010) including horn core displays and burials of human bodies, sometimes subsequently disinterred and reburied, with or without their heads (Hodder, 2014, 16–17). I would argue that such processes contributed to the reduction of the indeterminacy associated with marking by diverting attention from the immediate uncertainties and reifying the set of cultural values through the invention of spirits or the supernatural. The agency is a product of the way language is used, rather than the reality of the supernatural. It is ultimately indeterminate even if it sounds as if it is not. Around this time, there were many sites in what is now southern and eastern Türkiye which included massive public displays of art (especially at Göbekli Tepe shown in Fig. 6.2), ritual, feasting and symbolic treatment of dead people (Atakuman, 2014). A similar argument could be run about megalithic structures in many places (e.g., Bradley, 2009) (Fig. 8.19).

One of the fundamental things to emerge in State C was the creation of portable objects which could be decorated with things we might want to call art, especially pottery. As State C was also a period in which people adopted agriculture and pastoralism very widely in both the Old and New Worlds, there was also an explosion of pottery within which such foods could be stored, prepared and cooked. With the appearance of pottery, there was also another ubiquitous medium on which people could express their visual marking. Some of the most spectacular were from northern Iraq and the Halafian culture (Fig. 8.20).

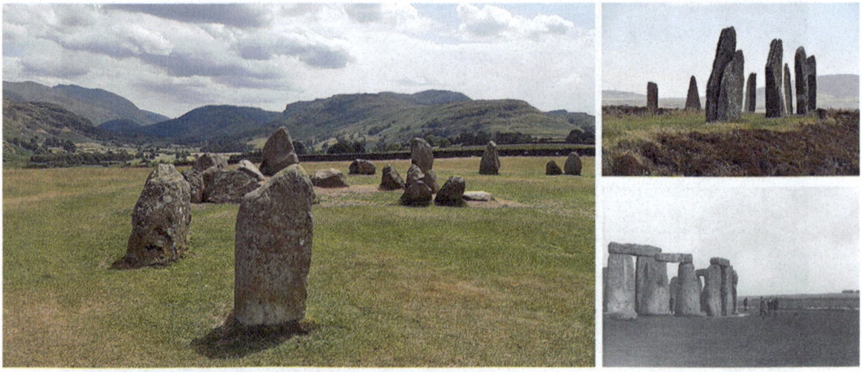

Fig. 8.19 Parts of stone circles in UK (Clockwise from left): Castlerigg (Lake District); Ring of Brodgar (Orkney); Stonehenge, during a field trip from a University where most of us had a sense of privilege and entitlement. (Photos Iain Davidson)

8.7 State D Fixing the Story with Writing—"Painting the Word and Speaking to the Eyes"

The most important element of State D was the fixing of the story of the societies relevant to any particular location. This happened in a variety of ways: in Valcamonica, in northern Italy, between 3000 and 10,000 years ago, vast panels of rock art were made in the open-air which at their most complex involved narrative scenes (Alexander et al., 2021, 272–4) (Fig. 6.12). Even in the many cases where something could be classified in a way other than as a narrative, the artist needed to tell a story about the art, or the composition was part of already-known stories. By the end of the State, art was being produced in the fixed locations of cathedrals and elsewhere (Fig. 8.31) or the massive sculptural sites such as Angkor Wat (Fig. 1.4) and the Cathedral of Ghent (Fig. 8.21).

Apart from such obvious examples of religious monumentalism, there were many others all over the world from the earliest parts of the State. Most famously, some of the Pyramids of Egypt date from the very beginning of the State. They were often tombs in which the bodies were not only embalmed but buried with astonishing funerary ceremony (as in the Tomb of Tutankhamun in Fig. 8.1, Top Right). The Pyramids of Teotihuacan in Mexico date from 1700 years ago and do not seem to have been for the burial of important people (Fig. 8.22), though the archeology of that is changing.

Many of the surrounding buildings in Teotihuacan included wall paintings and some had elaborate sculptures. At the opposite end of the Avenue of the Dead from the Pyramid of the Moon was the Temple of the Plumed Serpent where the sculptures were elaborate (Fig. 8.23). Art seems to have accompanied the construction of important and less-important places.

In State D, the invention of writing was marked by several significant events. Producers of Art sometimes included the text of the story that was represented in

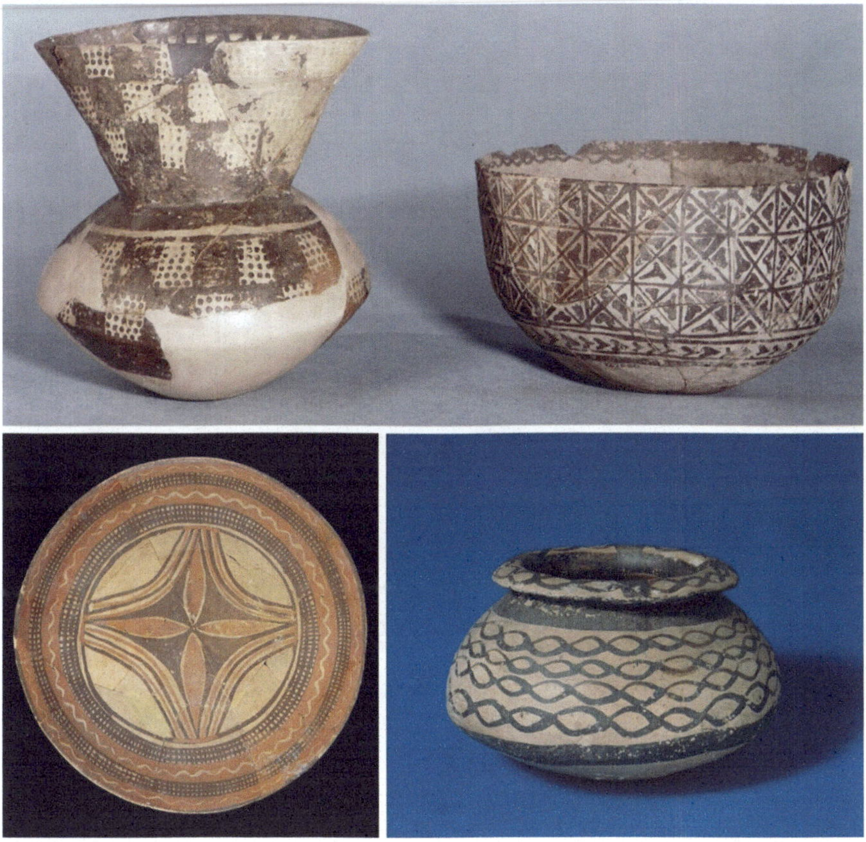

Fig. 8.20 Four Halafian ceramics. British Museum catalogue numbers clockwise from top: 247739001; 581,794,001; 144,655,001. Date given as 8000–7000 years ago. The ceramics were excavated by Max Mallowan and obtained by the Museum through the division of finds which was a condition of funding the research (Mallowan & Rose, 1935). There is a discussion of more recent excavations and dating at the site (Hijara, 1980). (Trustees of the British Museum: Creative Commons Attribution-NonCommercial-ShareAlike 4.0 International (CC BY-NC-SA 4.0) license)

pictures: the works were part of a rich culture that included multicolor paintings and generally included writing (Fig. 8.24).

The combination of pictures and text is reminiscent of the Duccio work where it was necessary to make observers aware of the relevant text. The reasons may have been different but the intent was the same—either to tell the story that goes with the pictures, or to make the story more vivid by illustrating it. Moreover, the effect was, like the use of religious imagery in religious buildings everywhere, to fix the story within the visible mythology of the society. The written texts reinforced that there was more than just the picture and the possibly variable interpretations of it. The tradition has continued with the stylized Kufic text around the Mihrab, and many other places, mostly from the Qur'an (for translations see Khoury, 1996), in the

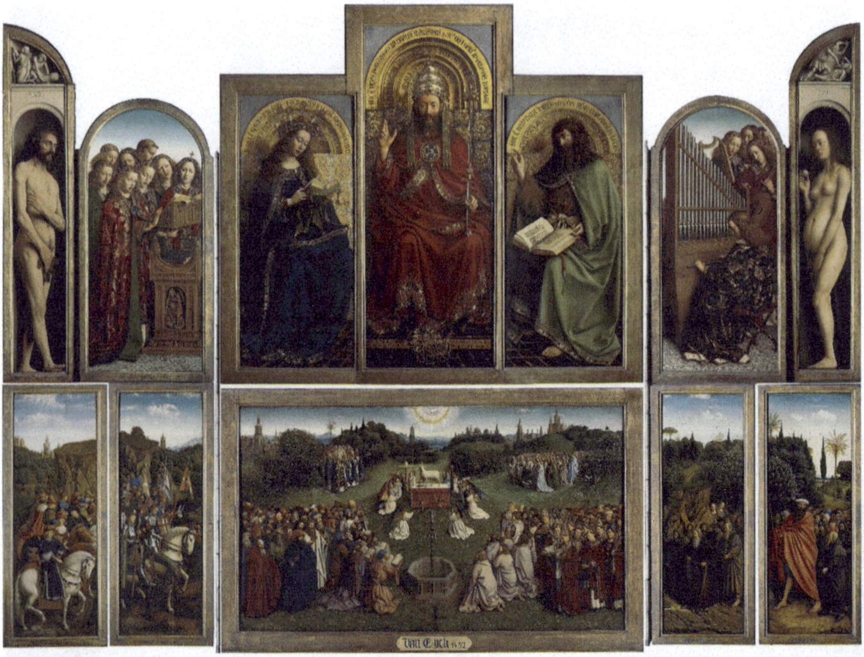

Fig. 8.21 Ghent Altarpiece by Jan Van Eyck and his brother Hubert, 1432. (Wikimedia Commons, Public Domain)

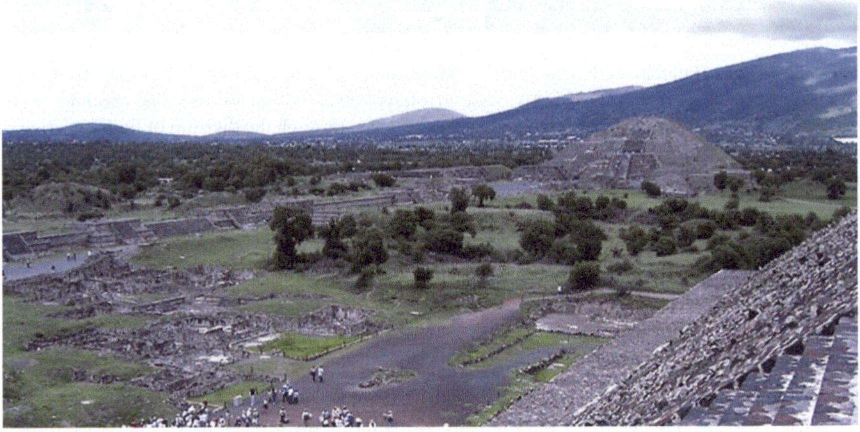

Fig. 8.22 The Pyramid of the Moon taken from halfway up the Pyramid of the Sun showing the array of structures along the Avenue of the Dead in Teotihuacan, Mexico. The human figures on both monuments indicate the scales. (Photo Iain Davidson)

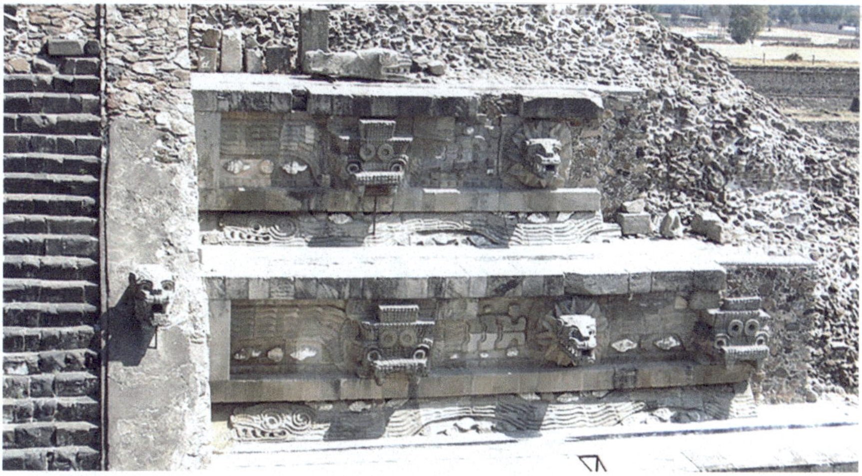

Fig. 8.23 Monumental sculptures in the Temple of the Plumed Serpent, Teotihuacan. (Photo Iain Davidson)

Fig. 8.24 Limestone plaque of Akhenaton and Nefertiti seated, holding 3 of their daughters, under the rays of the sun god Aton giving Ankh-symbols to them. Berlin, Neues Museum ÄM 14145. (Hans Ollermann, Flickr. Permission given 14 Nov 2023)

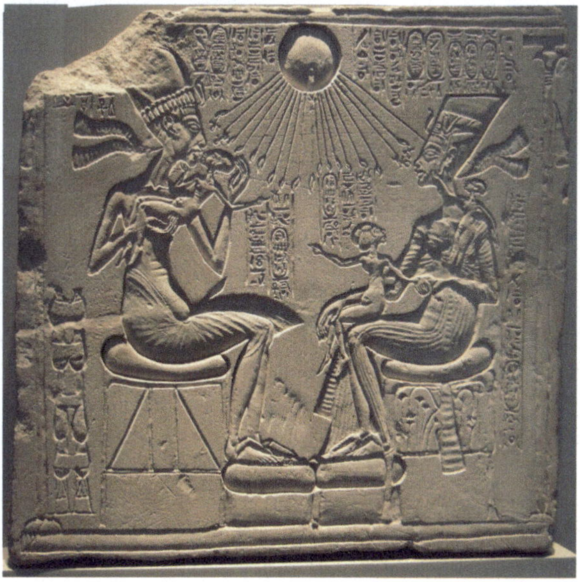

Mezquita de Cordoba (Fig. 5.9) where the text itself is almost irrelevant but has become an object of reverence.

People could write things down such that (almost) everyone in the society knew about them—that was one point of the Code of Hammurabi (Fig. 5.7) and the 2500-year-old writing on the wall on Behistun (Fig. 8.25) which was one of the trilingual texts that permitted archeologists (many years later) to decipher cuneiform script.

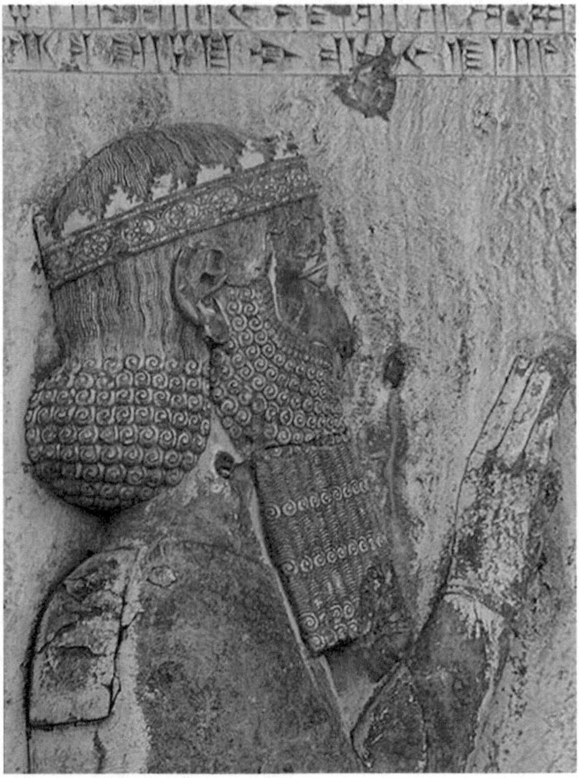

Fig. 8.25 Detail of sculpture of Darius the Great from the Behistun carved rock wall showing a tiny fragment of the original cuneiform inscription. (Wikimedia Commons. Creative Commons CC0 1.0 Universal Public Domain Dedication)

In addition, there was a need for people to teach and learn to use the new medium and they eventually took to it with the usual enthusiasm associated with the classroom, as shown from the analysis of teaching tablets from Nippur, Iraq (Paulus, 2023). From these it is possible to get "an emphatic insight into long-ago childhood … There is something particularly moving about holding a piece of clay, shaped by tiny fingers millennia ago…" (Robson, 2023). The example from 3700 years ago was a mature expression of formal teaching (and frustrated learners). We can surmise that formal teaching itself was older than this.

Writing also allowed people, or successions of people, to preserve great oral works by writing them down—the *Epic of Gilgamesh* (George, 1999) and the written Torah (Schiffman, 2012), the Egyptian *Book of the Dead* (Berisha & Pervorfi, 2022), the *Mahabarata* (for an account of the use of the Mahabarata in modern literature leading to its wider dissemination see Setyami & Faruk, 2023), the poems of Homer (the *Iliad* and the *Odyssey*) (for an account of the uncertainty about the authorship, showing that the poems penetrated the whole of Greek culture, see West, 1999), the *Analects* of Confucius (Chin, 2014), the New Testament, and the Qur'an. Such formal texts became embedded in the culture, so that, as I indicated in the Preface, the work of Homer continued in use for teaching Greek for at least 1000 years after it was written down (Horsley, 2000, 62). This list includes some of the foundational works of several world religions that all seem to have been established at about the same time and generated many works of art to illustrate the texts.

Some people came to call the few hundred years in which these texts from all over the world were written down "the Axial Age" (Boy & Torpey, 2013). I do not suggest we link these texts together in the way Karl Jaspers originally intended—in terms of the emergence of ongoing religion: "Religion was rendered ethical, and the majesty of the deity was thereby increased. The myth, on the other hand ... was turned into parable" (Jaspers, 1953, 3 as quoted in Mullins et al., 2018). Such attitudes seem to me to misunderstand the nature of anthropological descriptions of religion which suggest that all (modern) societies have religious beliefs, but not all of them wrote them down (Durkheim, 1972, Ch. 11). Moreover, Ingold (2000, 316) cites Edmund Leach as suggesting that "Ritual behaviour...is essentially communicative, and serves to convey information, in a symbolic code, about group membership or social identity." The existence of written texts enabled the words of the document—literally, the scripture—to be memorized and quoted in a way that created a symbolic code that all could refer to. Under these circumstances, the emergence of significant texts must have followed the emergence of writing in any society, in ritual context. The public display of texts such as the Code of Hammurabi from at least 3750 years ago was long before that axial age (Fig. 5.7) (Slanski, 2012). The penetration of the stories from oral histories into the artistic domain indicates that those texts which get lumped together as "axial" need not be related to each other except through the pervasiveness of oral traditions, the rituals associated with them and the presence of writing. "These 'breakthroughs' occurred in different civilizations at different times and to different degrees under different conditions and with different consequences" (Assmann, 2012, 398). In all of them we may imagine that what was involved was "painting the word and speaking to the eyes" as de Joucourt (1755) put it, making the oral visual, as I described in Chap. 5.

I should include one other work with this list of great texts. That is Plato's (Rowe, 2005) work *Phaedrus* in which, following the pattern of his own writings, he transcribed comments by Socrates about writing, in particular, the well-known line that: "You have invented an elixir not of memory, but of reminding; and you offer your pupils the appearance of wisdom,[3] not true wisdom, for they will read many things without instruction and will therefore seem to know many things, when they are for the most part ignorant." As we all rely on Internet search engines to augment our memories, it is very easy to think this was just a simple statement of truth. People who rely on oral traditions and histories have astonishing memories (L. Kelly, 2015, 2017)—how could anyone commit to memory the whole of the *Iliad* or the *Epic of Gilgamesh*? What Socrates missed in defending the old ways, as old people always do, was that writing removed the story from its context, as Kelly implied: it fixed the story but overcame the tyranny of place. Writing was doing something new so that there could be a new process by which the producer of art was required to tell stories to get seers to understand the subject of the images. The stories could be

[3] One of my colleagues told me that this reminded him of the then-fashionable Artificial Intelligence.

well-known through the written text and their illustration could give the art the appearance of something independent of storytelling.

Two things followed from this liberation of art from the immediacy of storytelling. First the art was only as informative as the duration of the story, perhaps as part of a narrative, like the story of Odysseus (from State D) referred to in Chap. 4 (Fig. 4.8 Right). Hence everyone would know the story entailed in the "vase" painted with a picture of Odysseus listening to the Sirens while his crew rowed on without hearing. Importantly, for the argument here, the mythology was probably well-known 2500 years ago either through the oral singing or through the written preservation of the story.

I can illustrate this with an elongated bowl from Peru, of a type used in blood sacrifices, decorated with bas-relief images of human-like creatures and a reed boat with feline heads (Fig. 8.26) which would date from State D. Aside from saying that it probably relates to Ai Apaec, the heroic ancestor of the Moche people, the specifics of the story in this image seem not to be known to anyone, though boats figure largely in Moche narratives (Quilter, 1997).

The point is that the narrative of the bird-headed being and the reed boat has not been preserved in writing, so the story of the art is lost. We can only put it together through the comparative work that archeologists engage in.

Likewise, the identification of the people who figured in the Duccio picture (see Chap. 3) has to be augmented by texts (some of which may also no longer be known), just as, without the Internet, we now have no idea who St Hardulph might have been (Fig. 6.10). Second, the stories achieved an independence from the storytelling, such that it was possible to imagine that there was an agent for storytelling who was not actually identifiable—the storyteller became a supernatural but absent agent.

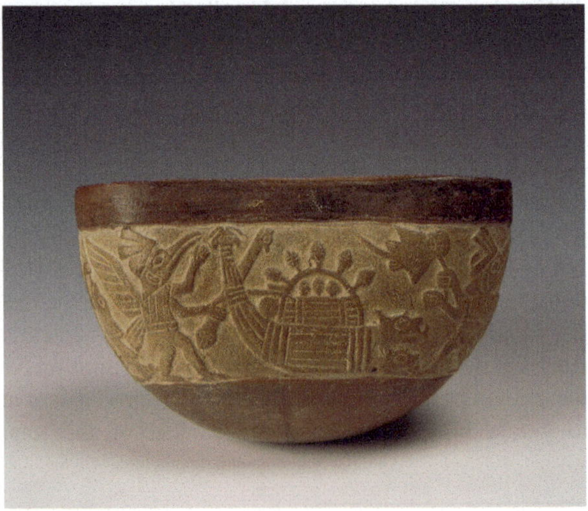

Fig. 8.26 Bowl from Museo Larco, Lima, Peru about 2000 years old (ML 002908). Showing a supernatural being, a reed boat and sacred animals. (Permission from Museo Larco, Lima, Peru 29/09/2023)

8.8 Art 1: The Art of Museums, Galleries (And Motels)—States E, F and G

In this book I have illustrated many works of Art 1 that seem to be in extreme contrast to the stone tools with which I ended a previous section. These works include those by Picasso, Matisse, Anish Kapoor, Lionel Lindsay, Olley, Streeton, Adnate, Goya, Caneletto, Turner, Lloyd Rhys, Duchamp, Braque, Mondrian, Van Gogh, Gainsborough, Reynolds, Horne, Storrier, Cottman, Gauguin, Sorolla, and Monet. These are all Art according to the criteria of the art historians as they were produced after the middle of the eighteenth century. There are also images by Rembrandt, Velázquez, Frans Hals, Botticelli, Dürer, Vermeer, Caravaggio, Leonardo da Vinci, Michelangelo, Giotto, van Eyck and an unnamed Ikon painter. The art in this second group was produced earlier than the eighteenth century and according to the view I have been criticizing would not be called Art 1. It is not clear to me that the works of the people in the first list are necessarily different in any important respect from those by people in the second list. What keeps them apart are the boundaries defended by those who say that art was invented in the eighteenth century (Kristeller, 1951; Shiner, 2001, 5), and that could explain one reason why a particular sort of Englishman might write a book about *Civilisation* and leave out all Spanish art (Clark, 1969). We tend to use the word "race" to define the way in which such elitists react to people from other countries, but the word and the concept have no support from biological characteristics, they are really about socially determined attitudes of one group towards another—in that case the attitudes of an English, man, of a certain class and education, with particular opinions derived from that education about the achievements in a place about which he was not so well educated. Art and social context are bound up together.

The importance of what is in the minds of the people who make decisions about how to define art or artefact types (and their own position in relation to those definitions) can be seen in the history of Art 1, for example, as exposed by Procter (2020) in writing about the relationship between colonialism and museums. The most famous example is the notional purchase of marble sculptures which, for the previous 2300 years, had been an intrinsic part of the temple of the goddess Athena in Athens, the Parthenon (see Fig. 8.31). The sculptures—works of art as well as part of the religious dressing of the temple—were removed from their position in the temple by the British Thomas Bruce, better known by the arbitrary title the elitist British gave him. He subsequently sold them to the British Museum, with instructions, in an act of arrogance, to use the name of his title, as examples of sculpture made in the Athens of Pericles. As is to be expected the history of the different contexts in which people were able to interact with the marble sculptures is convoluted and involves the history of Greece and the people who occupied it, their ideologies, religious and supernatural, which included attitudes to the art of the classical past by a plague of European tourists and travelers, particularly of wealthy classes (Hamilakis, 1999).

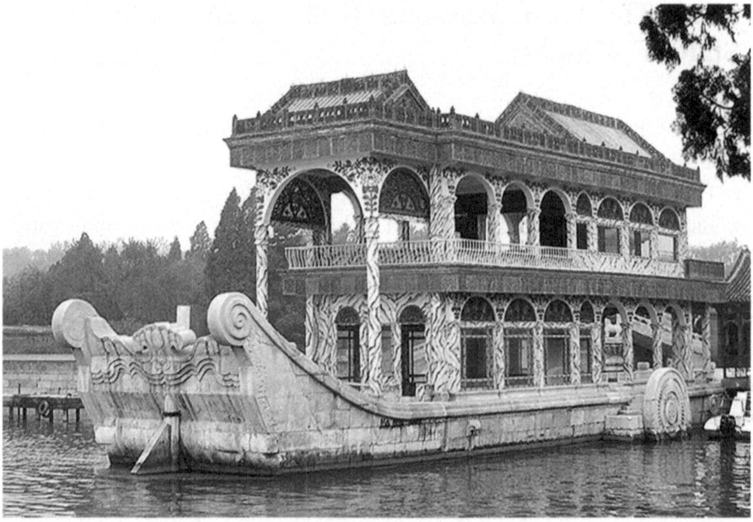

Fig. 8.27 Marble Boat in the Summer Palace near Beijing (in characteristic haze). (Photo Iain Davidson)

The Greek marble sculptures are not the only example (Davidson, 2015), nor even the first, even if they are one of the most famous. Bruce's title indicated his privilege and sense of entitlement in British society, characteristics which were passed on to his son, James, in the way such things happen. The son was instigator (with some French people) of the destruction of the Old Summer Palace in Beijing and the looting of its portable treasures—but not the marble. Some treasures remained which could neither be removed nor completely destroyed, such as the Marble Boat (Fig. 8.27), although the boat's superstructure had to be reconstructed after Bruce burned the original.

Victor Hugo (1985 [1861]), asked to comment on the sacking, burning and looting, one year later, wrote a furious and eloquent letter:

> One of the two victors [France and England] filled his pockets: when the other saw he filled his coffers. And back they came to Europe, arm in arm, laughing away. Such is the story of the two bandits.
>
> We Europeans are the civilized ones, and for us the Chinese are the barbarians. This is what civilization has done to barbarism.

The younger Bruce and his fellow vandals were responsible for the distribution of objects from their Chinese origins into more than 40 art collections (according to UNESCO) before they were gradually returned to China. The horse's head in Fig. 8.28 was part of a clock comprising heads of zodiac animals which each spouted water for 2 h. The irony is that the sculpture was designed by an artist, Giuseppe Castiglione, who was painter for the Qing Dynasty court in the eighteenth century (https://www.chinahighlights.com/beijing/attraction/old-summer-palace.htm) and himself an Italian. Who, then, were the barbarians? Such ironic attitudes to art are

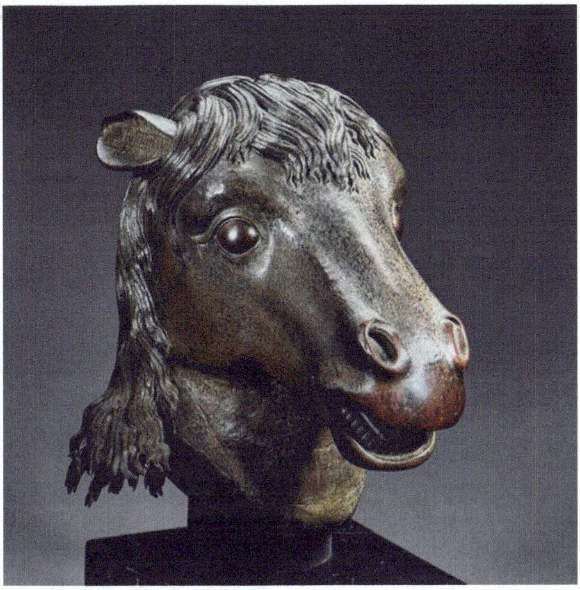

Fig. 8.28 Sculpture by Giuseppe Castiglione of horse's head in bronze that once was part of an installation of sculptures that formed a clock. It was looted from the Summer Palace in Beijing by the son of the man who looted the sculptures from the Parthenon in Athens. (http://images.china.cn/site1007/2019-11/14/422925d2-a943-408a-a447-fc6a9333b103.jpeg downloaded 13 Sept 2022)

quite general when they are held by one group of people with their own sense of their importance relative to people from other societies.

The Benin bronzes were looted from Nigeria in a British marauding expedition in 1897. They are generally called bronzes, but many are actually brass, using metal from brass rings used as currency during the slave trade (Skowronek et al., 2023). Felix Roth took part in the military raid and his brother Henry Roth recovered enough from his notes (McDougall, 2008) to write a book about Benin (Roth, 1968 [1903]). These Roths were both brothers of Walter E. Roth who wrote about the North West Central Aborigines of Queensland, who will feature in the next chapter.

Recently Dan Hicks (2020) writing about the Benin Bronzes used a succinct expression: "The theft of an object by a European museum is a negative act." Of course, the British troops regarded the Benin bronzes as the spoils of warfare, as if that gave legitimacy to the theft of art by the imperialists. You can still hear people seeking to justify the theft on the grounds that the art objects would have been destroyed anyway by subsequent civil wars, although no one has sought to evaluate the extant of destruction in different places. For what it is worth, in the very region where the Benin bronzes were stolen, art continued despite the influences introduced by colonialism (Nicklin, 1987; Sieber & Hecht, 2002). Hicks was mounting a sustained argument for the return of the bronzes: it may be a positive act for the thief but will always be negative for the victim. As always, the nature of the outcome

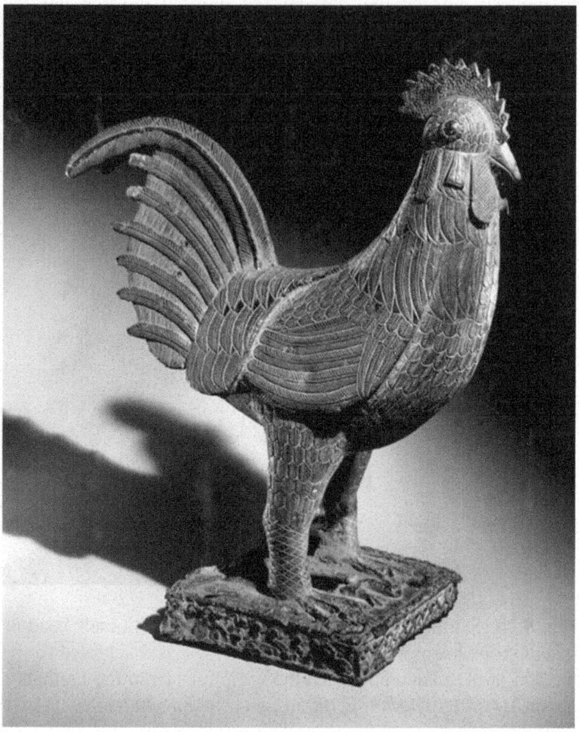

Fig. 8.29 Brass rooster from Benin City, Nigeria currently located in the Field Museum, Chicago. From the Fuller Collection. (Photograph The Field Museum, Image No, A99490, Cat. No. 210278 with permission)

depends on the context of the observer. Just as importantly, the thieves chose to loot objects which people used in ritual and they ended up in the art museums of Europe where the rituals were quite different. What they took depended on their own class-based ideas about what was worthy of exhibiting: it depended, in fact, on what they thought was art. The *Brass Rooster* (Fig. 8.29) is one such example, a work of art so fine that an almost identical piece was once on display in Cambridge University, England, which was presided over by one of the leaders in the fight to stop looting of archeological materials. The rooster can be seen in the movie *Page Eight*, in a scene shot in the Hall of Jesus College, Cambridge (https://en.wikipedia.org/wiki/Page_Eight).

As a result, colonizers chose to steal the very art objects that proved them wrong. They might have paused to reflect that there really was art in Africa. McDougall (2008, 87) suggested that Henry Roth certainly thought so, when he concluded about *Great Benin* written from his brother's notes: "only one conclusion" … was possible: the Benin bronzes were "a form of real native art" (Roth, 1968 (1903), 234). Military might and money do not entitle even those with recognition by their own governments to use acquired possession of art as a weapon for any reason (nor even to wilfully destroy it). Such attitudes of entitlement and snobbery run through the history of Art 1. In the absence of trade involving money by which the art could be seen in Academies, museums or the homes of the privileged, looted art became one of the valued spoils of war.

As I have predicted, the judgments about the relative worth of peoples were based on the sorts of prejudices that we often associate with class difference, and often call racial difference, and the position of the individual judge in such classificatory schemes. I have previously implied that the people who enjoy the Grand Ole Opry in Nashville, Tennessee and those who enjoy the Sydney Opera House might have contrasting cohorts despite both having a devotion to Opry/Opera music. And in the examples of the Greek sculptures, the art from the Chinese Palace, and the Benin bronzes, the story is of thefts by interlopers who thought themselves entitled to respect in a way that entitled them to not respect the Indigenous people of those places.

It so happens, that in all of these cases, the thieves violated art that had been or was part of the source of power for the elites whose treasures were being robbed. This sort of thing had happened before in a way that seems less violent and offensive to us when Gutenberg invented the printing press and could print Bibles which could be distributed beyond the church (Fig. 5.15) and enabled Luther (and others) to overthrow the stranglehold that the Church held over its adherents (Figs. 4.1 and 5.13) by undermining the source of their finances.

After making a case that the eighteenth century saw the emergence of a new "Fine" Art which has influenced the appreciation and location of art works ever since, Shiner (2001, 7) commented:

> …a central belief of the modern system of art has always been that money and class are irrelevant to the creation and appreciation of art. But to elevate some genres to the status of mere utility and their producers to fabricators is more than a conceptual transformation. And when genres and activities chosen for elevation and those chosen for demotion reinforce race, class, and gender lines, what once looked like a purely conceptual change begins to look like an underwriting of power relations as well.

It is difficult to disagree that what was realized at that time and became dominant thereafter has involved the use of works of art in commerce and in power (through the say-so of the Academies of art and the Museums where it was displayed) in the interests of those of the race, class and gender that Collingwood would have regarded as "proper" (as I argued in Chap. 3). The point of the last two chapters has been that these power struggles through art have not only been confined to Western Europe and its hangers on. As Morphy (1991) demonstrated (and we will see in the next chapter), Aboriginal art in eastern Arnhem Land has long been about the power of the painters, the stories they tell, the pictures they make and the initiation of seers into their society through their knowledge of story, picture and power maintained within the culture and the social structure (see the discussion of this system in Davidson, 1995). Myers (2002) has described in detail how Pintubi paintings from Central Australia became part of "high" art: one particular painting "moved from the bush to the Aboriginal Arts Board to world recognition in overseas exhibition." It seems that this system is part and parcel of what I described in Chap. 3 as "unquestioned attitudes." It is not just that the conceptual transformation reinforces statuses of race, class and gender and the associated unquestioned attitudes, but that through

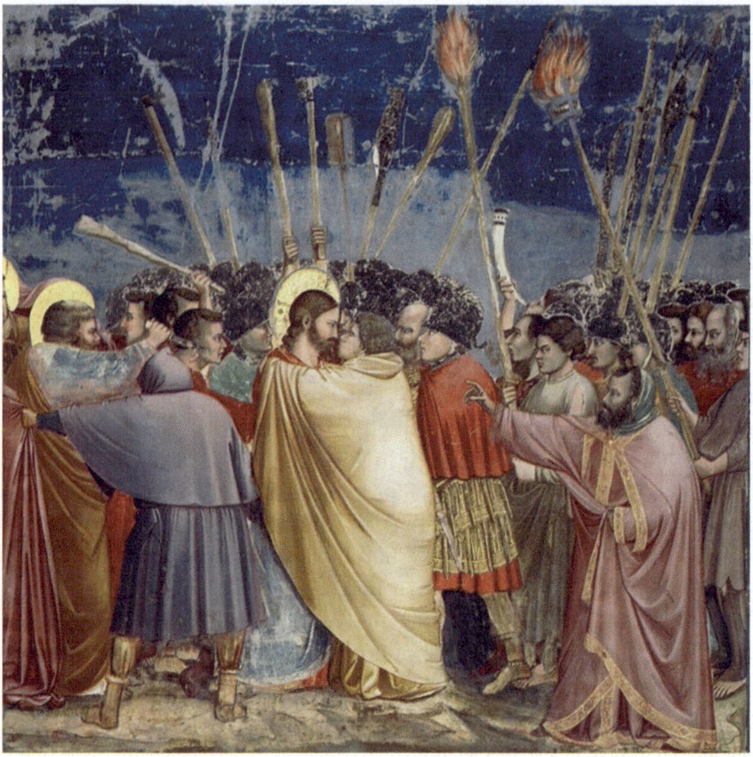

Fig. 8.30 Giotto Kiss of Judas (1304–1306, nearly 900 years ago), fresco in Scrovegni Chapel, Padua (https://upload.wikimedia.org/wikipedia/commons/e/ef/Giotto_-_Scrovegni_-_-31-_-_Kiss_of_Judas.jpg Public Domain)

the whole structure of the system of elevated arts, these attitudes have been created or reinforced. The status of individuals at different parts of the system also changed.

In the case of the 900-year-old Giotto fresco in the Scrovegni Chapel in Padua, persisting in the place where it was painted, (Fig. 8.30) the good people can be identified by their haloes which identify that they belong to a certain exclusive group and thus give a clue to the story associated with the scene. Much early art, therefore, has the quality of being part of the place where it was seen—but even then was emphasizing social divisions. Such persisting works are often defined by the beliefs associated with the place where they are still found. Is it Art? Of course it is.

Art 1 is a label attached to the art of museums and galleries. For most people there is no problem identifying what it is, and I included some examples in Table 3.2. Much other material culture with artistic qualities was more ephemeral and not fixed to a particular place, however. Some of that had religious connotations, such as illuminated volumes of bibles, prayer books, or books of hours (Binski & Zutshi, 2011), some of which were illuminated in gold and color. In the catalogue of early manuscripts in the Cambridge University Library, Paul Binski and Patrick

Zutshi say that these "can fairly be described as 'high art'" though most of them pre-date the eighteenth century.

It may well, therefore, be worth assessing the extent to which the eighteenth century revolution in art was really about a cultural capacity to produce works which were not of a religious nature and were not fixed to religious or other places. I have previously referred to the changes in technology of paint and support, and changes resulting from printing and its impact on the spread of knowledge, including, ultimately the development of the capacity to spread that knowledge widely and rapidly, and, in the end, with pictures. All of these things allowed individuals and institutions to make art the subject of an expanding commerce, divorcing it, to some extent, from the people who produced it, the story they told, and the general impact of such stories on the culture of the people who saw it.

John Berger famously suggested that one of the key events in ways of seeing was the invention of (or in his view the public access to) the camera (Berger, 2008, 18). As he put it: "what you saw depended on where you were when," and this changed the way painters represented the world. Note how we tend to see the importance in that statement of "where" and "when". For Impressionists "the visible became fugitive" and their role was to fix the visible to give some idea of the flux of that view; for Cubists "the visible was … the totality of possible views taken from points all around the object (or person) being depicted." We need to look again at Berger's statement and point out that "what you saw" depended on who "you" were. People from one culture will see things differently from people in another culture, because that culture is defined by the totality of prior knowledge due to education and experience, and they, in turn depend on the observer's class, race and gender. The experience is often determined by the visual images that surround us.

8.9 The Transition from Archeological Art (Art 2 and Art 3) to Art 1

What is the relationship between the earliest art and the much later art that hangs on the walls of galleries, museums, the dwellings of the wealthy, the doctor's office (Robb, 2015) and motels? There is a strong whiff of class about what is considered art and what is not. If class could account for enough, then when considering the issue of the evolution of the production and the recognition of art we would need also to consider the origins and evolution of class. What we can be sure of, from archeological evidence, is that the earliest things that get called "art" appeared long before there was any means by which class could be produced or identified, so class will not be enough. As Culley and Davidson (2021) suggested, "Prior to and during the nineteenth century, artistic works were not typically created for esthetic enjoyment, but commissioned by wealthy patrons and institutions to document, convey, and glorify cultural and religious ideals, historic events, and the wealth and power of individuals, groups, and states."

I now augment Table 7.1 to include a column for Art 2 the remainder of archeological art. That, of course, means that we are no longer talking about a strict ambiguity, because there are three things in the contrasts, but there are still seven characteristics. In Chap. 3, I indicated that part of the aim of this book is to show how the earliest stuff that gets called art, Art 3, became the stuff that people deal with when they are interested in art, Art 1. Now it is time to return to that theme, through a discussion of the three different art types: Art 1, Art 2 and Art 3, and the characteristics that make it possible to recognize them. Some of those characteristics have already emerged and I have indicated those characteristics in Table 7.1.

Meaning is not necessarily limited by what can be seen in an artwork, but, because the art and the artist in Art 1 (the art of museums, galleries, and motels) become separated as the art persists. The seer may bring to the work any interpretation they wish, though their wishes are constrained by the cultural knowledge they already have and bring to their viewing. At the time of production of Art 3 (the earliest objects claimed to be art), any meaning was given by the producer in their social interaction with the observer. Only under special circumstances did the persistence of the object in the shared culture of producer and seer, together with the nature of the marks lead to the possibility of the interpretation of meaning by the observer.

Consider how art has been used at different times and places. We are all comfortable with the idea that the ancient Greeks produced art because we know about the sculptures removed 220 years ago from the Parthenon (Fig. 8.31). In Athens, the sculptures formed an integral part of the, mostly ritual, buildings that still survive in

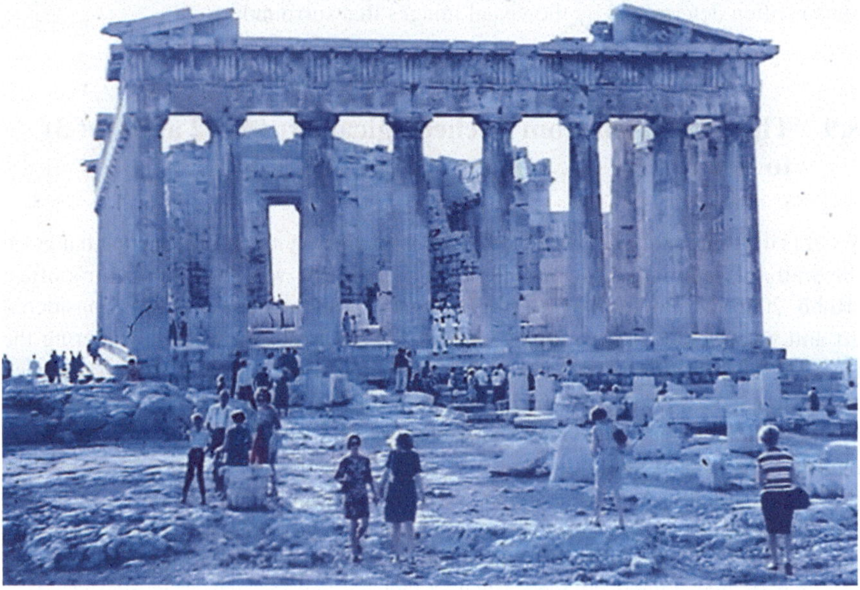

Fig. 8.31 The Parthenon in Athens, 1962, from which Thomas Bruce removed the most important sculptures 220 years ago. He took them back to Britain and sold them to the British Museum in London. (Photo Iain Davidson 1962)

some form in the center of Athens. They were not portable objects and had a proper place in the context of those buildings. Likewise there were paintings on the walls of some of the buildings. They were fixed there and but for greedy foreign intervention should still be there.

Much early art, then, was attached to the building in which it was seen, as with the frescoes from Pompeii (Fig. 1.8), or any of the religious art of the early contributors to Southern European Mediaeval Art, such as Giotto (Fig. 8.30) or Duccio. It was only the production of transportable art that led to it acquiring a monetary value, such that it could be sold to a museum or a wealthy individual or corporation.

The history of art, then, appears to have been a complex interplay between a number of competing factors. At the beginning, our ancestors had to create the expectation that marks were meaningful, but that the meaning was indeterminate. There was then the interplay of individual interests, ritual and storytelling, acquisition of power to individuals or institutions, all of which were involved in reducing or controlling the indeterminacy of meaning. Then after establishing that distance was no problem for the acquisition of raw materials, the next state was to establish that particular places had special importance. It was such places that could become sacred, partly through the use of particular art images, some of which showed the narratives that could be told around art.

Writing produced several fundamental changes. First, it enabled the storytellers—who we have defined as the initiators of acts of art—to fix their story, often including the teaching of the story in quite formal arrangements. That process came to a head with the explosion of written texts of oral traditions in, at least, the Middle East, Greece, India and China. These written texts became central to particular religions—indeed, they came to be revered as if they were the unchangeable words of religions—such that the pictures that went with the storytelling could often refer to such fixed texts and appear to be given by some absolute authority. All of these things allowed powerful people to control the dissemination of the text through institutions of control, and all of that was reinforced by the connection between the art and the stories in the texts.

Printing, however, whether by accident or design, began an erosion of the institutions of control since more and more people had access to the fixed texts and could see their way through them, and the press could be used in an uncomplicated way to produce stories that contradicted each other, thus reducing the power of the institutions which had had control. In effect the process liberated the stories from the institutions of control, and artists were much less beholden to stories that were already known. The impact was felt beyond the story allowing people to create art that was not subject to the control of the origin-set of powerful institutions. The reaction was almost immediate as art objects entered into a commerce which put value on the work rather than the sponsor or the patron. The response seems to have been a regulation of the market through the establishment of Academies of Art which sought to control attitudes to the productions of artists (Shiner, 2001, Ch. 6).

The Academies, then, and the Salon in Paris, had their favorite artists and exerted their power through the taste of the audience. Some, such as Gainsborough, exerted their independence by going against the apparent edicts of the head of the British Academy, Joshua Reynolds (Fig. 7.7). As a result of the power games, the young generation who came to be known as Impressionists exhibited independently as the Batignolles group. They included Monet's *Boulevard des Capucines* (Fig. 7.27) in the first exhibition, with predictable reactions (Stonard, 2021, 350), which proved that the people who were interested in art were actually interested in the power structures not the paintings.

In light of the importance of photography to later conceptions of the role of painting, I have wondered whether there might be some connection between Monet's picture and two coincidental facts. The exhibition was held in a place formerly used by a photographer, and Monet's picture had a very slight similarity with features of the first photograph to include people, taken 36 years earlier (Fig. 8.32). More likely we are dealing with coincidences due to the fact that both were street scenes from high windows in Paris.

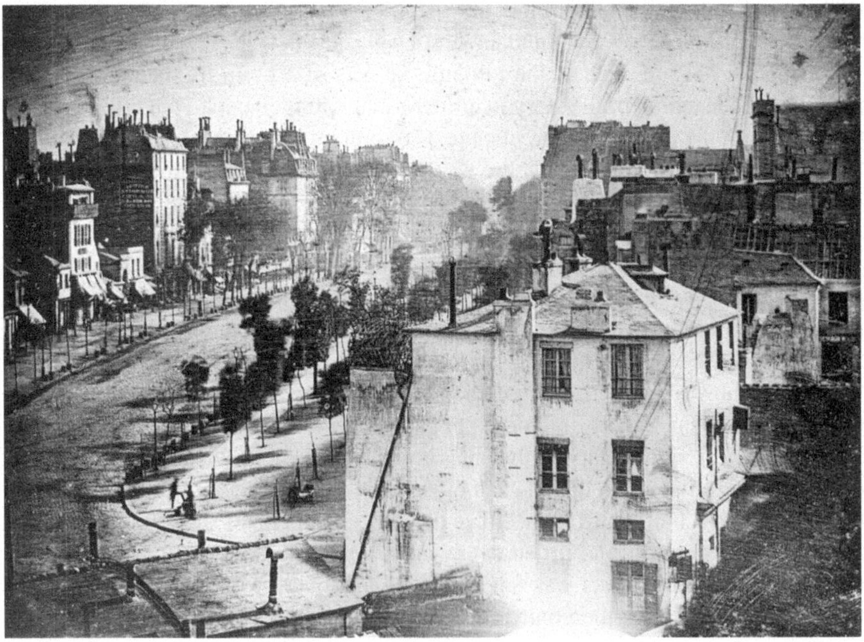

Fig. 8.32 Boulevard du Temple, Paris, third arrondissement, Daguerreotype. Made in 1838 by inventor Louis Daguerre, this is believed to be the earliest photograph showing living people. It is a view of a busy street, but because the exposure lasted for 4–5 min (see shutter speed Daguerre photo explained) the moving traffic left no trace. Only the two men near the bottom left corner, one apparently having his boots polished by the other, stayed in one place long enough to be visible. (Wikimedia Commons. Public Domain)

That said, both Daguerre and Monet emphasized the importance of the light in structuring the image and Monet went on to play with that concept, almost to the exclusion of the story of the subject he was painting. Photography, it would seem, opened up the world of art to free it somewhat from the storytelling I have been emphasizing hitherto. The Impressionists opened the world for other views of the subject. And that was the liberation that Picasso and the Cubists took up with enthusiasm, opening the way for the Abstract Expressionists and other artists of the last 120 years. Ironically, where art began with the storytelling by the artist about the subject it turned full circle at this stage so that there was not so much of a story—it is not so much what it means as my friends in MOMA would have said—as an account of how we see the world. The issue became "Let me tell you how you see" rather than just what you see.

Shiner (2001) quite properly began his book "Today you can call virtually anything 'art' and get away with it." What I have been arguing here is that the attitude towards the outputs of the art-system is correct. It only seems to be challenging if we adopt elitist stances to the things we look at. There is no need to invent a new concept such as "aesthetics" so that we may follow Kant and recognize that something is good. Instead we are free to say that we like something or do not like it (Dobrez, 2022).

8.10 Essay 8 Scenes (White Shaman, Texas)

Scenes involve sets of images in spatial proximity to each other from which, without any knowledge other than the images themselves, an observer can infer actions taking place among the agents represented in the images. This is the definition that we used for our book on *Making Scenes* (Davidson, 2021). It has always been said that there are few scenes in the Paleolithic cave art, but that unsubtle view is being broken down by new discoveries in Sulawesi (Brumm et al., 2021), more detailed reading of some cave images (Fritz et al., 2016), and by a shift in emphasis, in Western Europe, to the engraved and painted bone and stone portable pieces rather than just the art concealed within caves (Culley, 2021). Even so, Villaverde Bonilla (2021) has pointed out that there is a substantive difference between scenes that describe a moment, as at Parpalló, and those that can be attributed to an ongoing event, as in Levantine Art. And some careful readings have shown the possibility of representation of scenes inside a small number of caves.

I have previously offered a narrative (Davidson, 2013a) that traces the steps from hominins leaving signs of their presence, to making marks, to using marks to convey meaning through pattern and repetition, to the use of repetition in ritual. My sequence continued by moving from making repeated marks to creating the circumstances for recognising iconicity in what were otherwise primarily indexical marks. The two final stages to get from iconicity to Art 1 (the modern sense) were the production of iconic marks in a way that showed the agency of the creatures iconically represented in scenes and then the move for observers to identify the agency of the artist as well. By representing scenes, the producer could free the observer from the need to be told what individual images meant. This untethering of the image from the social context of its production left the possibility that the scene could stand alone and be interpreted in various ways. That is what post Renaissance art developed—the delicious uncertainty of interpretation, which I have referred to as the expectation of meaning combined with the indeterminacy of that meaning.

When we look at the way that scenes were composed and used, the story is that Art 2 is just as complex and rich as Art 1. Carolyn Boyd's interpretation of 2000-year-old paintings from the Lower Pecos in Texas has carried the interpretation of rock art panels as scenes to an unprecedented level (Boyd, 2016). Boyd achieved this by making the bold assumption that an analysis can be successful if it treats a whole rock art panel as a single composition. Her detailed analysis of the panel at White Shaman, near Langtry in Texas, followed by comparison to a variety of other sites in the region, show that there are repeated patterns of images of anthropomorphs with attached antlers, often with peyote buttons on the tines, of deer impaled by feathered darts (Fig. 8.33), of serpents which may form an archway through which a 'shaman' must pass. Some of the so-called shaman figures in these panels may carry staff-like objects that seem to resemble the fruit of the powerful hallucinogen

Datura (Boyd & Dering, 1996). These themes occur in many of the painted rock shelters of the region, allowing the recognition that there is a consistent pattern of composition of the images at each site. Boyd then compared the recurring elements with the ethnographic accounts of shamanic ideology and art still being practiced further south in Mexico as described from a visit by a Huichol shaman, Matsihua, from 600 km away (Boyd, 2016). This ideology relates animals, animal characteristics adopted by shamans, peyote and its association with deer, rainmaking, and *Datura*.

This case, taken together with the chain of inference exemplified by the image in Fig. 8.33, suggests that particular images in Art 2 can be interpreted in the light of a whole body of images to allow inferences which would not be possible without the broad context. Scenes permit the understanding of individual examples of art, while understanding of them in the context of other regional sites allows the discussion of the way the storytelling about one site connected the people of whole regions together. Boyd (2016, 162) ended her book about the White Shaman saying: "In the Lower Pecos, these remains engendered images of simple foragers engaging simple tools in an often harsh and unforgiving landscape. Without the art, they are yet another little understood and little regarded Archaic population eking out a meager existence as best they could. But with the art, worlds change and wonder begins."

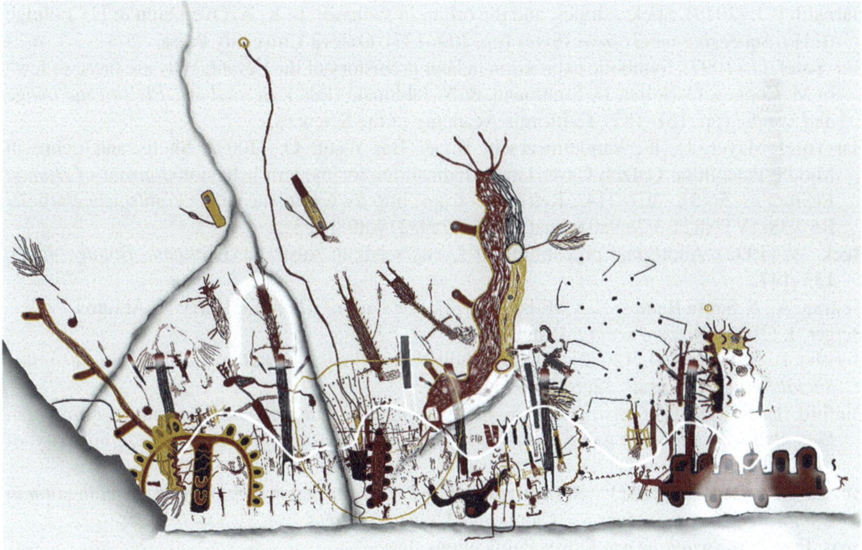

Fig. 8.33 Rendering of the White Shaman panel, near Comstock, Texas, USA. Created by Carolyn Boyd and supplied with permission. (Boyd, 2016)

Bibliography

Alexander, C., Marretta, A., Huet, T., & Chippindale, C. (2021). 17. Rules of ordering and grouping in the pitoti, the later prehistoric rock-engravings of Valcamonica (BS), Italy: From solitary figures through clusters, graphic groups, and scenes to narrative. In I. Davidson & A. Nowell (Eds.), *Making scenes: Global perspectives on scenes in rock art*. Berghahn Books.

Allen, H. (1974). The Bagundji of the Darling Basin: Cereal gatherers in an uncertain environment. *World Archaeology, 5*(3), 309–322. https://doi.org/10.1080/00438243.1974.9979576

Aschero, C. A. (2021). *Cueva de las Manos, Patagonia, Argentina. Patrimonio de la humanidad/ World Heritage Site*. AsociaiciónAmigos del Museo Nacional de Bellas Artes.

Assmann, J. (2012). 15 cultural memory and the Myth of the Axial Age. In *The axial age and its consequences* (pp. 366–408). Harvard University Press.

Atakuman, Ç. (2014). Architectural discourse and social transformation during the early neolithic of Southeast Anatolia. *Journal of World Prehistory, 27*(1), 1–42. https://doi.org/10.1007/s10963-014-9070-4

Aubert, M., Setiawan, P., Oktaviana, A. A., Brumm, A., Sulistyarto, P. H., Saptomo, E. W., et al. (2018). Palaeolithic cave art in Borneo. *Nature, 564*(7735), 254–257. https://doi.org/10.1038/s41586-018-0679-9

Aubert, M., Lebe, R., Oktaviana, A. A., Tang, M., Burhan, B., Hamrullah, et al. (2019). Earliest hunting scene in prehistoric art. *Nature, 576*(7787), 442–445. https://doi.org/10.1038/s41586-019-1806-y

Aujoulat, N. (2004). *Lascaux. Le geste, l'espace et le temps*. Seuil.

Aujoulat, N., Geneste, J.-M., Arhambeau, C., Delluc, M., Duday, H., & Henry-Gambier, D. (2002). La grotte ornée de Cussac-Le Buisson-de-Cadouin (Dordogne): Premières observations. *Bulletin de la Société Préhistorique Française, 99*(1), 129–137.

Bahn, P. G. (1998). *The Cambridge illustrated history of prehistoric art*. Cambridge University Press.

Barnard, P. J. (2019). Sticks, stones, and the origin of sapience. In K. A. Overmann & F. Coolidge (Eds.), *Squeezing minds from stones* (pp. 102–127). Oxford University Press.

Bar-Yosef, O. (1997). Symbolic expression in later prehistory of the Levant: Why are there so few? In M. Conkey, O. Soffer, D. Stratmann, & N. Jablonski (Eds.), *Beyond art: Pleistocene image and symbol* (pp. 161–187). California Academy of the Sciences.

Bar-Yosef Mayer, D. E., Vandermeersch, B., & Bar-Yosef, O. (2009). Shells and ochre in Middle Paleolithic Qafzeh Cave, Israel: Indications for modern behavior. *Journal of Human Evolution,* 56(3), 307–314. Retrieved from http://www.sciencedirect.com/science/article/B6WJS-4VTN82T-3/2/c9a01eb6af357a6d97a9a2140f98b9c3

Beck, W. (1992). Aboriginal preparation of *Cycas* seeds in Australia. *Economic Botany, 46*(2), 133–147.

Beltran, A., & Saura Ramos, P. A. (Eds.). (1998). *The cave of Altamira*. Harry N. Abrams.

Berger, J. (2008). *Ways of seeing*. Penguin.

Berisha, L., & Pervorfi, L. (2022). Literary tradition and the Egyptian book of the dead. *Technium Social Sciences Journal, 32*, 691–697.

Binford, L. R. (2001). *Constructing frames of reference. An analytical method for archaeological theory building using hunter-gatherer and environmental data sets*. University of California Press.

Binski, P., & Zutshi, P. (2011). *Western illuminated manuscripts: A catalogue of the collection in Cambridge University Library*. Cambridge University Press.

Boas, F. (1955). *Primitive art*. Dover Publications, Inc.

Boesch, C. (1991). Teaching among wild chimpanzees. *Animal Behaviour, 41*, 530–532.

Botha, R. (2020). *Neanderthal language: Demystifying the linguistic powers of our extinct cousins*. Cambridge University Press.

Botha, R. (2024). Did homo erectus have language? The seafaring inference. *Cambridge Archaeological Journal, 1–17*, 21–37. https://doi.org/10.1017/S0959774324000118

Bowdler, S., & Balme, J. (2010). Gatherers and grannies: Further thoughts on the origins of gender. *Australian Feminist Studies, 25*(66), 391–405.

Boy, J. D., & Torpey, J. (2013). Inventing the axial age: The origins and uses of a historical concept. *Theory and Society, 42*, 241–259.

Boyd, C. E. (2016). *The White Shaman mural. An enduring creation narrative in the rock art of the Lower Pecos*. University of Texas Press.

Boyd, B., & Cook, J. (1993). *A reconsideration of the 'Ain Sakhri' figurine*. Paper presented at the Proceedings of the Prehistoric Society (Vol. 59, pp. 399–405).

Boyd, C. E., & Dering, J. P. (1996). Medicinal and hallucinogenic plants identified in the sediments and pictographs of the Lower Pecos, Texas Archaic. *Antiquity, 70*(268), 256–275. https://doi.org/10.1017/S0003598X00083265

Bradley, R. (2009). *Image and audience. Rethinking prehistoric art*. Oxford University Press.

Braje, T. J., Erlandson, J. M., Rick, T. C., Davis, L., Dillehay, T., Fedje, D. W., et al. (2020). Fladmark+ 40: What have we learned about a potential Pacific Coast peopling of the Americas? *American Antiquity, 85*(1), 1–21.

Brumm, A., Oktaviana, A. A., Burhan, B., Hakim, B., Lebe, R., Zhao, J.-X., et al. (2021). Oldest cave art found in Sulawesi. *Science Advances, 7*(3), eabd4648. https://www.science.org/doi/10.1126/sciadv.abd4648

Buck, C. E., & Bard, E. (2007). A calendar chronology for Pleistocene mammoth and horse extinction in North America based on Bayesian radiocarbon calibration. *Quaternary Science Reviews, 26*(17–18), 2031–2035.

Byrd, B. F. (1994). Public and private, domestic and corporate: The emergence of the southwest Asian village. *American Antiquity, 59*(4), 639–666.

Byrne, R. W., Barnard, P. J., Davidson, I., Janik, V. M., McGrew, W. C., Miklósi, Á., & Wiessner, P. (2004). Understanding culture across species. *Trends in Cognitive Sciences, 8*(8), 341–346.

Caro, T. M., & Hauser, M. D. (1992). Is there teaching in nonhuman animals? *The Quarterly Review of Biology, 67*(2), 151–174.

Cartailhac, E. (1902). Les cavernes ornées de dessins. La grotte d'Altamira, Espagne Mea culpa d'un sceptique. *L'Anthropologie, 13*, 348–354.

Chin, A. (2014). *The analects*. Penguin Publishing Group.

Clark, K. (1969). *Civilisation*. John Murray.

Clarkson, C., Jacobs, Z., Marwick, B., Fullagar, R., Wallis, L., Smith, M. A., et al. (2017). Human occupation of northern Australia by 65,000 years ago. *Nature, 547*(7663), 306–310. https://doi.org/10.1038/nature22968

Conard, N. J. (2009). A female figurine from the basal Aurignacian of Hohle Fels Cave in southwestern Germany. *Nature, 459*(7244), 248–252. Retrieved from. https://doi.org/10.1038/nature07995

Conard, N. J., & Floss, H. (2013). Early figurative art and musical instruments from the Swabian Jura of Southwestern Germany and their implications for human evolution. In K. Sachs-Hombach & J. R. J. Schirra (Eds.), *Origins of pictures. Anthropological discourses in image science* (pp. 172–200). Halem.

Conkey, M. W. (1980). The identification of prehistoric hunter-gatherer aggregation sites: The case of Altamira. *Current Anthropology, 21*(5), 609–630.

Conroy, L. P. (1993). Female figurines of the Upper Palaeolithic and the emergence of gender. In H. du Cros & L. Smith (Eds.), *Women in archaeology. A feminist critique* (pp. 153–160). Australian National University.

Cucchi, T., & Arbuckle, B. (2021). Introduction. Animal domestication: From distant past to current development and issues. *Animal Frontiers, 11*(3), 6–9. https://doi.org/10.1093/af/vfab013

Culley, E. V. (2021). 12. A comparison of "scenes" in parietal and non-parietal Upper Paleolithic imagery: Formal differences and ontological implications. In I. Davidson & A. Nowell (Eds.), *Making scenes: Global perspectives on scenes in Rock Art*. Berghahn Books.

Culley, E. V., & Davidson, I. (2021). Art, sign, and representation. In N. Gontier, A. Lock, & C. Sinha (Eds.), *The Oxford handbook of human symbolic evolution* (2nd ed., pp. 331–368). Oxford University Press.

Daly, K. G., Mattiangeli, V., Hare, A. J., Davoudi, H., Fathi, H., Doost, S. B., et al. (2021). Herded and hunted goat genomes from the dawn of domestication in the Zagros Mountains. *Proceedings of the National Academy of Sciences, 118*(25), e2100901118. https://doi.org/10.1073/pnas.2100901118

Davenport, D., & Jochim, M. A. (1988). The scene in the Shaft at Lascaux. *Antiquity, 62*(236), 558–562. https://doi.org/10.1017/S0003598X00074718

David, B., & Lourandos, H. (1998). Rock art and socio-demography in northeastern Australian prehistory. *World Archaeology, 30*(2), 193–219.

David, B., Barker, B., Petchey, F., Delannoy, J.-J., Geneste, J.-M., Rowe, C., et al. (2013). A 28, 000 year old excavated painted rock from Nawarla Gabarnmang, northern Australia. *Journal of Archaeological Science, 40*(5), 2493–2501. https://doi.org/10.1016/j.jas.2012.08.015

Davidson, I. (1980). Transhumance, Spain and ethnoarchaeology. *Antiquity, 54*, 144–147.

Davidson, I. (1989a). Freedom of information: Aspects of art and society in western Europe during the last ice age. In H. Morphy (Ed.), *Animals into art* (pp. 440–456). Unwin Hyman.

Davidson, I. (1989b). Is intensification a condition of the fisher-hunter-gatherer way of life? *Archaeology in Oceania, 24*, 75–78.

Davidson, I. (1992). There's no art -- To find the mind's construction -- In offence. *Cambridge Archaeological Journal, 2*, 52–57.

Davidson, I. (1995). Paintings, power, and the past: Can there ever be an ethnoarchaeology of art? (Review of *Ancestral Connections: Art and an Aboriginal system of knowledge*. [H. Morphy]). *Current Anthropology, 36*(5), 889–892.

Davidson, I. (1997). The power of pictures. In M. Conkey, O. Soffer, D. Stratmann, & N. G. Jablonski (Eds.), *Beyond art: Pleistocene image and symbol* (Vol. 23, pp. 128–158). The California Academy of Sciences.

Davidson, I. (2002). The "Finished Artefact Fallacy": Acheulean handaxes and language origins. In A. Wray (Ed.), *Transitions to language* (pp. 180–203). Oxford University Press.

Davidson, I. (2003). The archaeological evidence of language origins: States of art. In M. H. Christiansen & S. Kirby (Eds.), *Language evolution* (pp. 140–157). Oxford University Press.

Davidson, I. (2007). As large as you need and as small as you can'--Implications of the brain size of *Homo floresiensis*. In A. Schalley & D. Khlentzos (Eds.), *Mental states: Evolution, function, nature* (pp. 35–42). John Benjamins.

Davidson, I. (2012a). Symbolism and becoming a hunter-gatherer. *L'art pléistocène dans le monde /Pleistocene art of the world / Arte pleistoceno en el mundo, Actes du Congrès IFRAO, Tarascon-sur-Ariège, septembre 2010, Symposium « Signes, symboles, mythes et idéologie… ». N° spécial de Préhistoire, Art et Sociétés, Bulletin de la Société Préhistorique Ariège-Pyrénées, LXV–LXVI*, Book: p. 292–293, CD: p. 1689–1705

Davidson, I. (2012b). What a carry on? Portable art and changes of symbolic meaning. *L'art pléistocène dans le monde/Pleistocene art of the world/Arte pleistoceno en el mundo, Actes du Congrès IFRAO, Tarascon-sur-Ariège, septembre* 2010, *N° spécial de Préhistoire, Art et Sociétés, Bulletin de la Société Préhistorique Ariège-Pyrénées, LXV-LXVI*, Book p. 268–269, CD: p. 1559–1570

Davidson, I. (2013a). Origins of pictures: An argument for transformation of signs. In K. Sachs-Hombach & J. R. J. Schirra (Eds.), *Origins of pictures. Anthropological discourses in image science* (pp. 15–45). Halem.

Davidson, I. (2013b). Peopling the last new worlds: The first colonisation of Sahul and the Americas. *Quaternary International, 285*(0), 1–29. https://doi.org/10.1016/j.quaint.2012.09.023

Davidson, I. (2015, October 14). Heritage, the ship of theseus and the song of homer. *EnviroSociety*. www.envirosociety.org/2015/10/heritage-the-ship-of-theseus-and-the-song-of-homer

Davidson, I. (2016). Stone tools: Evidence of something in between culture and cumulative culture? In M. N. Haidle, N. J. Conard, & M. Bolus (Eds.), *The nature of culture* (pp. 99–120). Springer.

Davidson, I. (2019). Evolution of cognitive archaeology through evolving cognitive systems. In K. A. Overmann & F. Coolidge (Eds.), *Squeezing minds from stones* (pp. 79–101). Oxford University Press.

Davidson, I. (2020). Chapter 2. IMHO: Inventing modern human origins. In M. Porr & J. Matthews (Eds.), *Interrogating human origins. Decolonisation and the deep past* (pp. 35–55). Routledge.
Davidson, I. (2021). 1. Scenes and non-scenes in rock art. In I. Davidson & A. Nowell (Eds.), *Making scenes: Global perspectives on scenes in rock art* (pp. 16–31). Berghahn Books.
Davidson, I. (2023). Humans making history through continuities and discontinuities in art. *Cambridge Archaeological Journal, 33*, 637–654. https://doi.org/10.1017/S0959774323000057
Davidson, I., & McGrew, W. C. (2005). Stone tools and the uniqueness of human culture. *The Journal of the Royal Anthropological Institute, 11*(4), 793–817.
Davidson, I., & Noble, W. (1992). Why the first colonisation of the Australian region is the earliest evidence of modern human behaviour. *Archaeology in Oceania, 27*, 135–142.
Davidson, I., & Nowell, A. (2021). Introduction. Behind the scenes—did scenes in rock art create new ways of seeing the world? In I. Davidson & A. Nowell (Eds.), *Making scenes: Global perspectives on scenes in rock art* (pp. 1–15). Berghahn Books.
Dediu, D., & Levinson, S. C. (2013). On the antiquity of language: The reinterpretation of Neandertal linguistic capacities and its consequences. *Frontiers in Psychology, 4*. https://doi.org/10.3389/fpsyg.2013.00397
DeMarrais, E., & Robb, J. (2013). Art makes society: An introductory visual essay. *World Art, 3*(1), 3–22. https://doi.org/10.1080/21500894.2013.782334
d'Errico, F., & Banks, W. E. (2015). The archaeology of teaching: A conceptual framework. *Cambridge Archaeological Journal, 25*(4), 859–866.
d'Errico, F., & Nowell, A. (2000). A new look at the Berekhat Ram figurine: Implications for the origins of symbolism. *Cambridge Archaeological Journal, 10*(1), 123–167.
d'Errico, F., van Niekerk, K. L., Geis, L., & Henshilwood, C. S. (2023). New Blombos Cave evidence supports a multistep evolutionary scenario for the culturalization of the human body. *Journal of Human Evolution, 184*, 103438. https://doi.org/10.1016/j.jhevol.2023.103438
Dobrez, L. (2022). Aesthetic response, kant, and rock art. *American Indian Rock Art, 48*, 155–166.
Domingo Sanz, I. (2021). 16. New insights into the analysis of Levantine rock art scenes informed by observations on Western Arnhem Land Rock Art. In I. Davidson & A. Nowell (Eds.), *Making scenes: Global perspectives on scenes in rock art* (pp. 240–258). Berghahn Books.
Drabsch, B., & Bourke, S. (2014). Ritual, art and society in the Levantine Chalcolithic: The 'Processional' wall painting from Teleilat Ghassul. *Antiquity, 88*(342), 1081–1098.
Ducasse, S., & Langlais, M. (2019). Twenty years on, a new date with Lascaux. Reassessing the chronology of the cave's Paleolithic occupations through new 14C AMS dating. *PALEO Revue d'archéologie préhistorique, 30-1*, 130–147.
Durkheim, E. (1972). *Emile Durkheim: Selected writings*. Cambridge University Press.
Falk, D. (1992). *Braindance*. Henry Holt.
Feinman, G. M., & Neitzel, J. E. (2023). The social dynamics of settling down. *Journal of Anthropological Archaeology, 69*, 101468.
Fornasiero, A., Wing, R. A., & Ronald, P. (2022). Rice domestication. *Current Biology, 32*(1), R20–R24. https://doi.org/10.1016/j.cub.2021.11.025
Fritz, C., Tosello, G., & Conkey, M. W. (2016). Reflections on the identities and roles of the artists in European Paleolithic societies. *Journal of Archaeological Method and Theory, 23*(4), 1307–1332.
Fuller, D. Q., Denham, T., & Allaby, R. (2023). Plant domestication and agricultural ecologies. *Current Biology, 33*(11), R636–R649. https://doi.org/10.1016/j.cub.2023.04.038
Gamble, C., & Kruszynski, R. (2009). John Evans, Joseph Prestwich and the stone that shattered the time barrier. *Antiquity, 83*(320), 461–475.
Gärdenfors, P., & Högberg, A. (2017). The archaeology of teaching and the evolution of Homo docens. *Current Anthropology, 58*(2), 188–208.
George, A. (1999). *The Epic of Gilgamesh. A new translation*. Allen Lane the Penguin Press.
Goren-Inbar, N. (1986). A figurine from the Acheulian site of Berekhat Ram. *Mitukefat Haeven, 19*, 7–12.
Gott, B. (1982). Ecology of root use by Aborigines of Southern Australia. *Archaeology in Oceania, 17*(1), 59–67.

Gott, B. (1983). Murnong-Microseris scapigera: A study of a staple food of Victorian Aborigines. *Australian Aboriginal Studies, 2*, 2.

Graeber, D., & Wengrow, D. (2021). *The dawn of everything: A new history of humanity*. Allen Lane.

Grayson, D. K., & Meltzer, D. J. (2003). A requiem for North American overkill. *Journal of Archaeological Science, 30*, 585–593.

Grove, M. (2009). Hunter–gatherer movement patterns: Causes and constraints. *Journal of Anthropological Archaeology, 28*(2), 222–233.

Hallam, S. J. (1975). *Fire and hearth: A study of Aboriginal usage and European usurpation in south-western Australia*. Australian Institute of Aboriginal Studies.

Hamilakis, Y. (1999). Stories from exile: Fragments from the cultural biography of the Parthenon (or 'Elgin') marbles. *World Archaeology, 31*(2), 303–320.

Hayes, E. H., Fullagar, R., Field, J. H., Coster, A. C. F., Matheson, C., Nango, M., et al. (2022). 65,000-years of continuous grinding stone use at Madjedbebe, Northern Australia. *Scientific Reports, 12*(1), 11747. https://doi.org/10.1038/s41598-022-15174-x

Henshilwood, C. S., d'Errico, F., Vanhaeren, M., van Niekerk, K., & Jacobs, Z. (2004). Middle Stone Age shell beads from South Africa. *Science, 304*(5669), 404.

Henshilwood, C. S., d'Errico, F., Niekerk, K. L. V., Coquinot, Y., Jacobs, Z., Lauritzen, S.-E., et al. (2011). A 100,000-Year-old ochre-processing workshop at Blombos Cave, South Africa. *Science, 334*, 219–222.

Hicks, D. (2020). *The Brutish Museums: The Benin Bronzes, colonial violence and cultural restitution*. Pluto Press.

Hijara, I. (1980). Arpachiyah 1976. *Iraq, 42*(2), 131–154.

Hodder, I. (2014). The vitalities of Çatalhöyük. In I. Hodder (Ed.), *Religion at work in a Neolithic Society: Vital matters* (pp. 1–32). Cambridge University Press.

Hodder, I., & Pels, P. (2010). History houses: A new interpretation of architectural elaboration at Çatalhöyük. In I. Hodder (Ed.), *Religion in the emergence of civilization: Çatalhöyük as a case study* (pp. 163–186). Cambridge, UK.

Hoffmann, D. L., Standish, C. D., García-Diez, M., Pettitt, P. B., Milton, J. A., Zilhão, J., et al. (2018). U-Th dating of carbonate crusts reveals Neandertal origin of Iberian cave art. *Science, 359*(6378), 912–915. https://doi.org/10.1126/science.aap7778

Horsley, G. (2000). Homer in Pisidia: Aspects of the history of Greek education in a remote Roman province. *Antichthon, 34*, 46–81.

Hovers, E., & Belfer-Cohen, A. (2022). A pleistocene record of making symbols. In: T. Wynn, K. Overmann, F. Coolidge Eds.), *The Oxford Handbook of Cognitive Archaeology*. Oxford University Press, Oxford, UK.

Hugo, V. (1985 [1861]). The sack of the summer palace *The UNESCO Courier,* 38(11), 14–16

Ingold, T. (2000). Chapter 16. Society, nature and the concept of technology. In T. Ingold (Ed.), *The perception of environment. Essays in livelihood, dwelling and skill* (pp. 312–322). Routledge.

Iriki, A., & Sakura, O. (2008). The neuroscience of primate intellectual evolution: Natural selection and passive and intentional niche construction. *Philosophical Transactions of the Royal Society B: Biological Sciences, 363*(1500), 2229–2241.

Jones, R. (1969). Firestick farming. *Australian Natural History, 16*, 224–228.

Joordens, J. C. A., d'Errico, F., Wesselingh, F. P., Munro, S., de Vos, J., Wallinga, J., et al. (2015). Homo erectus at Trinil on Java used shells for tool production and engraving. *Nature, 518*(7538), 228–231.

Kabukcu, C., Asouti, E., Pöllath, N., Peters, J., & Karul, N. (2021). Pathways to plant domestication in Southeast Anatolia based on new data from aceramic Neolithic Gusir Höyük. *Scientific Reports, 11*(1), 2112. https://doi.org/10.1038/s41598-021-81757-9

Kelly, R. L. (1983). Hunter-gatherer mobility strategies. *Journal of Anthropological Research, 39*(3), 277–306.

Kelly, L. (2015). *Knowledge and power in prehistoric societies: Orality, memory, and the transmission of culture*. Cambridge University Press.

Kelly, L. (2017). *The memory code: Unlocking the secrets of the lives of the ancients and the power of the human mind*. Atlantic Books.

Khoury, N. N. N. (1996). The meaning of the Great Mosque of Cordoba in the tenth century. *Muqarnas, 13*, 80–98.

Kirkbride, D. (1968). Beidha 1967: An interim report. *Palestine Exploration Quarterly, 100*, 83–89.

Kornienko, T. V. (2009). Notes on the cult buildings of northern Mesopotamia in the Aceramic Neolithic period. *Journal of Near Eastern Studies, 68*(2), 81–102.

Kristeller, P. O. (1951). The modern system of the arts: A study in the history of aesthetics part I. *Journal of the History of Ideas, 12*, 496–527.

Kuhn, S. L. (2020). Moving on from here: Suggestions for the future of "Mobility Thinking" in studies of Paleolithic Technologies. *Journal of Paleolithic Archaeology, 3*(4), 664–681. https://doi.org/10.1007/s41982-020-00060-7

Langley, M. C. (2013). Storied landscapes makes us (Modern) human: Landscape socialisation in the Palaeolithic and consequences for the archaeological record. *Journal of Anthropological Archaeology, 32*(4), 614–629.

Lartet, E., & Christy, H. (1875). *Reliquiae Aquitanicae: Being contributions to the archaeology and palaeontology of Périgord and the adjoining provinces of southern France*. Online at https://www.forgottenbooks.com/en/books/Reliquiae_Aquitanicae_1000698892. : Williams and Norgate

Lercari, N. (2017). 3D visualization and reflexive archaeology: A virtual reconstruction of Çatalhöyük history houses. *Digital Applications in Archaeology and Cultural Heritage, 6*, 10–17. https://doi.org/10.1016/j.daach.2017.03.001

Lourandos, H. (1983). Intensification: A late Pleistocene-Holocene archaeological sequence from southwestern Victoria. *Archaeology in Oceania, 18*, 81–94.

Lourandos, H. (1997). *Continent of hunter-gatherers: New perspectives in Australian prehistory*. Cambridge University Press.

Lubbock, J. (1865). *Pre-historic times* (1st ed.). Williams & Norgate.

Mallowan, M. E. L., & Rose, J. C. (1935). Excavations at Tall Arpachiyah, 1933. *Iraq, 2*, 1–178.

Marquet, J.-C., Freiesleben, T. H., Thomsen, K. J., Murray, A. S., Calligaro, M., Macaire, J.-J., et al. (2023). The earliest unambiguous Neanderthal engravings on cave walls: La Roche-Cotard, Loire Valley, France. *PLoS One, 18*(6), e0286568. https://doi.org/10.1371/journal.pone.0286568

Marshack, A. (1997). The Berekhat Ram figurine--a late Acheulian carving from the Middle East. *Antiquity, 71*(272), 327–337.

McDougall, R. (2008). Chapter 4. The making of *Great Benin*: Felix Norman Roth and Henry Ling Roth. In R. McDougall & I. Davidson (Eds.), *The Roth family, anthropology and colonial administration* (pp. 73–92). Left Coast Press.

Miller, J. M., & Wang, Y. V. (2022). Ostrich eggshell beads reveal 50,000-year-old social network in Africa. *Nature, 601*(7892), 234–239. https://doi.org/10.1038/s41586-021-04227-2

Milner, G. R., & Boldsen, J. L. (2023). Population trends and the transition to agriculture: Global processes as seen from North America. *Proceedings of the National Academy of Sciences, 120*(4), e2209478119. https://doi.org/10.1073/pnas.2209478119

Mithen, S. J. (1988). Looking and learning: Upper Palaeolithic art and information gathering. *World Archaeology, 19*(3), 297–327.

Moore, M. W., & Perston, Y. (2016). Experimental insights into the cognitive significance of early stone tools. *PLoS One, 11*(7), e0158803. https://doi.org/10.1371/journal.pone.0158803

Moro Abadía, O. (2006). Arts, crafts and Paleolithic art. *Journal of Social Archaeology, 6*(1), 119–141.

Morphy, H. (1991). *Ancestral connections: Art and an Aboriginal system of knowledge*. University of Chicago Press.

de Mortillet, G. (1883). *Le Préhistorique, Antiquité de l'Homme*. C. Reinwald.

Mullins, D. A., Hoyer, D., Collins, C., Currie, T., Feeney, K., François, P., et al. (2018). A systematic assessment of "axial age" proposals using global comparative historical evidence. *American Sociological Review, 83*(3), 596–626.

Myers, F. (2002). *Painting culture: The making of an aboriginal high art.* Duke University Press.

Nicklin, K. (1987). GI Jones; *The art of Eastern Nigeria.* x, 230 pp. Cambridge, etc.: Cambridge University Press, 1984.£ 30. *Bulletin of the School of Oriental and African Studies, 50*(1), 202–204.

Noble, W., & Davidson, I. (1996). *Human evolution, language and mind.* Cambridge University Press.

Nowell, A. (2021). *Growing up in the ice age: Fossil and archaeological evidence of the lived lives of plio-pleistocene children.* Oxbow Books.

Nowell, A. (2023a). Oral storytelling and knowledge transmission in upper paleolithic children and adolescents. *Journal of Archaeological Method and Theory, 30*(1), 9–31.

Nowell, A. (2023b). Rethinking Neandertals. *Annual Review of Anthropology, 52*(1), 170. https://doi.org/10.1146/annurev-anthro-052621-024752

Nowell, A., & Chang, M. L. (2014). Science, the media, and interpretations of Upper Paleolithic figurines. *American Anthropologist, 116*(3), 562–577.

Nowell, A., & Cooke, A. (2021). Culturing the Paleolithic body: Archaeological signatures of adornment and body modification. In N. Gontier, A. Lock, & C. Sinha (Eds.), *The Oxford handbook of human symbolic evolution* (2nd ed.). Oxford University Press.

Nowell, A., & Van Gelder, L. (2020). Entanglements: The role of finger flutings in the study of the lived lives of Upper Paleolithic peoples. *Journal of Archaeological Method and Theory, 27*, 585–606. https://doi.org/10.1007/s10816-020-09468-5

Pascoe, B. (2014). *Dark emu. Black seeds: Agriculture or accident?* Magabala Books.

Paulus, S. (Ed.). (2023). *Back to school in Babylonia.* Institute for the studiy of ancient cultures of the University of Chicago.

Perri, A. R., Feuerborn, T. R., Frantz, L. A. F., Larson, G., Malhi, R. S., Meltzer, D. J., & Witt, K. E. (2021). Dog domestication and the dual dispersal of people and dogs into the Americas. *Proceedings of the National Academy of Sciences, 118*(6), e2010083118. https://doi.org/10.1073/pnas.2010083118

Pettitt, P. (2003). Is this the infancy of art? Or the art of an infant? A possible Neanderthal face from La Roche-Cotard, France. *Before Farming, 2003*(4), 1–3.

Pitarch Martí, A., Zilhão, J., d'Errico, F., Cantalejo-Duarte, P., Domínguez-Bella, S., Fullola, J. M., et al. (2021). The symbolic role of the underground world among Middle Paleolithic Neanderthals. *Proceedings of the National Academy of Sciences, 118*(33), e2021495118. https://doi.org/10.1073/pnas.2021495118

Ponce de León, M. S. M. S., Bienvenu, T., Marom, A., Engel, S., Tafforeau, P., Warren, J. L. A., et al. (2021). The primitive brain of early *Homo. Science, 372*(6538), 165–171. https://doi.org/10.1126/science.aaz0032

Procter, A. (2020). *The whole picture: The colonial story of the art in our museums & why we need to talk about it.* Cassell.

Purdy, B. A., Jones, K. S., Mecholsky, J. J., Bourne, G., Hulbert, R. C., Jr., MacFadden, B. J., et al. (2011). Earliest art in the Americas: Incised image of a proboscidean on a mineralized extinct animal bone from Vero Beach, Florida. *Journal of Archaeological Science, 38*(11), 2908–2913. https://doi.org/10.1016/j.jas.2011.05.022

Quilter, J. (1997). The narrative approach to Moche iconography. *Latin American Antiquity, 8*(2), 113–133.

Quiroga, F., & Torres Campos, R. (1880). La cueva de Altamira. *Boletín de la Institución Libre de Enseñanza, 90*(161163), 258–268.

Radovčić, D., Sršen, A. O., Radovčić, J., & Frayer, D. W. (2015). Evidence for Neandertal Jewelry: Modified white-tailed eagle claws at Krapina. *PLoS One, 10*(3), e0119802. https://doi.org/10.1371/journal.pone.0119802

Range, F., & Marshall-Pescini, S. (2022). Comparing wolves and dogs: Current status and implications for human 'self-domestication'. *Trends in Cognitive Sciences, 26*(4), 337–349. https://doi.org/10.1016/j.tics.2022.01.003

Rappaport, R. A. (1999). *Ritual and religion in the making of humanity.* Cambridge University Press.

Robb, J. (2015). Prehistoric art in Europe: A deep-time social history. *American Antiquity, 80*(4), 635–654.
Robb, J. (2017). 'Art' in archaeology and anthropology: An overview of the concept. *Cambridge Archaeological Journal, 27*(4), 587–597.
Robb, J. (2020). Art (pre) history: Ritual, narrative and visual culture in Neolithic and Bronze Age Europe. *Journal of Archaeological Method and Theory, 27*(3), 454–480.
Robson, E. (2023). Foreword. In S. Paulus (Ed.), *Back to school in Babylonia* (pp. ix–x). Institute for the studiy of ancient cultures of the University of Chicago.
Rodríguez-Vidal, J., d'Errico, F., Pacheco, F. G., Blasco, R., Rosell, J., Jennings, R. P., et al. (2014). A rock engraving made by Neanderthals in Gibraltar. *Proceedings of the National Academy of Sciences, 111*(37), 13301–13306. https://doi.org/10.1073/pnas.1411529111
Ross, J., & Davidson, I. (2006). Rock art and ritual: An archaeological analysis of rock art in arid Central Australia. *Journal of Archaeological Method and Theory, 13*(4), 305–341. https://doi.org/10.1007/s10816-006-9021-1
Roth, H. L. (1968 [1903]). *Great Benin: Its customs, art and horrors*. Barnes and Noble
Rowe, C. J. (2005). *Phaedrus*. Penguin.
Rowley-Conwy, P., & Layton, R. (2011). Foraging and farming as niche construction: Stable and unstable adaptations. *Philosophical Transactions of the Royal Society B: Biological Sciences, 366*(1566), 849–862.
Schiffman, L. H. (2012). The Term and Concept of Torah. In: K. Finsterbusch, A Lange (Eds.), *What is Bible*. Peeters, Leuven, pp 173–191.
Sehasseh, E. M., Fernandez, P., Kuhn, S., Stiner, M. C., Mentzer, S., Colarossi, D., et al. (2021). Early Middle Stone Age personal ornaments from Bizmoune Cave, Essaouira, Morocco. *Science Advances, 7*(39), eabi8620. https://doi.org/10.1126/sciadv.abi8620
Setyami, I., & Faruk, F. (2023). *Epic of Mahabharata in Kuntowijoyo's Novels: Aesthetic response analysis*. Paper presented at the international joint conference on arts and humanities 2022 (IJCAH 2022)
Shaw, B., Field, J. H., Summerhayes, G. R., Coxe, S., Coster, A. C. F., Ford, A., et al. (2020). Emergence of a Neolithic in highland New Guinea by 5000 to 4000 years ago. *Science Advances, 6*(13), eaay4573. https://doi.org/10.1126/sciadv.aay4573
Shiner, L. (2001). *The invention of art: A cultural history*. The University of Chicago Press.
Shipman, P. (2021). *Our oldest companions: The story of the first dogs*. Harvard University Press.
Shipton, C., O'Connor, S., & Kealy, S. (2021). The biogeographic threshold of Wallacea in human evolution. *Quaternary International, 574*, 1–12. https://doi.org/10.1016/j.quaint.2020.07.028
Sieber, R., & Hecht, B. (2002). Eastern Nigerian art from the Toby and Barry Hecht collection. *African Arts, 35*(1), 56–79.
Skowronek, T. B., DeCorse, C. R., Denk, R., Birr, S. D., Kingsley, S., Cook, G. D., et al. (2023). German brass for Benin Bronzes: Geochemical analysis insights into the early Atlantic trade. *PLoS One, 18*(4), e0283415. https://doi.org/10.1371/journal.pone.0283415
Slanski, K. E. (2012). The law of Hammurabi and its audience. *Yale Journal of Law and the Humanities, 24*(1), 97–110.
Smith, R. J., & Wood, B. (2017). The principles and practice of human evolution research: Are we asking questions that can be answered? *Comptes Rendus Palevol, 16*(5–6), 670–679. https://doi.org/10.1016/j.crpv.2016.11.005
Soffer, O., Adovasio, J. M., & Hyland, D. C. (2000). The "Venus" figurines: Textiles, basketry, gender, and status in the Upper Paleolithic. *Current Anthropology, 41*(4), 511–537.
Stiner, M. C. (2014). Finding a common bandwidth: Causes of convergence and diversity in Paleolithic beads. *Biological Theory, 9*(1), 51–64. https://doi.org/10.1007/s13752-013-0157-4
Stiner, M. C., Munro, N. D., Buitenhuis, H., Duru, G., & Özbaşaran, M. (2022). An endemic pathway to sheep and goat domestication at Aşıklı Höyük (Central Anatolia, Turkey). *Proceedings of the National Academy of Sciences, 119*(4), e2110930119. https://doi.org/10.1073/pnas.2110930119
Stonard, J.-P. (2021). *Creation: A fully illustrated, panoramic world history of art from ancient civilisation to the present day*. Bloomsbury Publishing.

Tomasello, M., & Herrmann, E. (2010). Ape and human cognition: What's the difference? *Current Directions in Psychological Science, 19*(1), 3–8.

Trinkaus, E., Buzhilova, A. P., Mednikova, M., & Dobrovolskaya, M. V. (2014). *The people of Sunghir: Burials, bodies, and behavior in the earlier Upper.* Oxford University Press.

Verhoeven, M. (2002). Ritual and ideology in the pre-pottery Neolithic B of the Levant and south-east Anatolia. *Cambridge Archaeological Journal, 12*(2), 233–258.

Villaverde, V. (2020). *La mirada neandertal. Origenes del arte visual.* Universitat de Valencia.

Villaverde Bonilla, V. (2021). 15. Scenes in the Paleolithic and Levantine Art of Eastern Spain. In I. Davidson & A. Nowell (Eds.), *Making scenes: Global perspectives on scenes in rock art* (pp. 224–239). Berghahn Books.

Watson, B. (2011). The eyes have it: Human perception and anthropomorphic faces in world rock art. *Antiquity, 85*(327), 87–98.

Weber, G. W., Lukeneder, A., Harzhauser, M., Mitteroecker, P., Wurm, L., Hollaus, L.-M., et al. (2022). The microstructure and the origin of the Venus from Willendorf. *Scientific Reports, 12*(1), 1–10.

West, M. L. (1999). The invention of Homer. *The Classical Quarterly, 49*(2), 364–382.

Whallon, R. (2006). Social networks and information: Non-"utilitarian" mobility among hunter-gatherers. *Journal of Anthropological Archaeology, 25*(2), 259–270.

White, J. P. (1971). New Guinea and Australian prehistory: The "Neolithic problem". In D. J. Mulvaney & J. Golson (Eds.), *Aboriginal man and environment in Australia* (pp. 182–195). Australian National University Press.

Will, M., Krapp, M., Stock, J. T., & Manica, A. (2021). Different environmental variables predict body and brain size evolution in Homo. *Nature Communications, 12*(1), 4116. https://doi.org/10.1038/s41467-021-24290-7

Wobst, H. M. (1977). Stylistic behavior and information exchange. In C. E. Cleland (Ed.), *For the director* (pp. 317–342). Museum of Anthropology, University of Michigan.

Wood, B., & Boyle, E. K. (2016). Hominin taxic diversity: Fact or fantasy? *American Journal of Physical Anthropology, 159*(S61), S37–S78. https://doi.org/10.1002/ajpa.22902

Wynn, T., & Berlant, T. (2019). The handaxe aesthetic. In K. Overmann & F. Coolidge (Eds.), *Squeezing minds from stones: Cognitive archaeology and the evolution of the human mind* (pp. 278–303). Oxford University Press.

Chapter 9
Art and Archeohistory of Sahul (the Continent of Greater Australia and New Guinea)

Abstract In this chapter I seek to apply to the art of Sahul the principles I outlined for art in the previous chapter, beginning with some essential features of the continent of Sahul. I then look at the conditions applicable to the principles of art that I explored for the rest of the world: the expectation of meaning and the indeterminacy of that meaning; how individuals can define themselves through personal ornaments; the evidence of ritual and teaching; ways of reducing the indeterminacy of marking; overcoming the tyranny of distance and place and the importance of the art persisting in places.

These principles are derived from the history of communications around the world and end with three principles which I derived from behavior that did not exist in Aboriginal Australia or Sahul prior to colonization by Europeans: fixing the story through writing; freeing the story from institutions of control as a result of the invention of printing; and overcoming the tyranny of time as expressed by the introduction of the telegraph. I show that features of Australian art can be compared to what happened elsewhere despite the lack of writing, printing, and the telegraph.

9.1 The Art and the First People of Sahul

In 2010, I wrote (Davidson, 2010a):

> The relationships between art and the beliefs are complex, and cannot be understood simply from the images (or the sites) themselves, even when talking to people embedded in the belief system (Merlan, 1989; Morphy, 1991). A consequence of this is that, even in a situation where people can talk about the various meanings of non-iconic signs (Munn, 1973), the system can adapt to new meanings as appropriate, making it impossible to infer meanings or associated beliefs from images that were made long before the memory embedded in oral histories.
>
> In many respects, this is the key to continuity and change among Australian hunter gatherers—the system is structured to appear unchanging so that changes can be easily incorporated. The archaeological record documents change more faithfully, but, as with the

I. Davidson, *Art or Scribbles? In the Eye of the Beholder: The Evolutionary Emergence of Visual Communication*, Interdisciplinary Evolution Research 11,
https://doi.org/10.1007/978-3-031-90458-5_9

Dreaming, it is not a constant. It can only acquire relevance to Aboriginal people whose past it represents through successive revelation.

The grazier Bill told us, right at the beginning of this book, that he did not consider the rock art in the shelters of the Selwyn Ranges in Northwest Queensland as "art, so much as scribbles." I have established in the previous chapter some of the criteria by which we can assess the claims for art and its achievements in the earliest stages of human archeohistory. Now is the time to consider whether grazier Bill's contention is reasonable for the rock art he was describing. Is Australian rock and other art merely an example of scribbles, or is it art by the criteria we have established for other art?

A conventional approach to a version of art history that includes archeological evidence—the archeohistory of art of Australia and New Guinea—might have looked at the sequences of art of different dates in different regions to offer a more-or-less familiar culture history of one set of cultural values following on to another (it seems invidious to cite anyone here, but this is one guide: Flood, 1997). That would be instructive, if a little stereotyped, and open to the criticism that this is just another old white man's[1] approach to First Nations history, rooted in the approach to cultural history that has dominated the ideas of histories written by those of us who are not First Nations people. I want to adopt a slightly different approach deriving from the chapters that have gone before. I can never change the fact that I am white, a man, of a certain class and education, but this is not a conventional approach to the archeohistory of Australian art.

The important point for this book is that the first people who came to Sahul were capable of making and using art. After landfall somewhere in the north, they colonized a continent that was empty of people from the Bird's Head in New Guinea to the modern location of the d'Entrecasteaux Channel in Tasmania, from Barrow Island off Western Australia (which is normally part of the mainland) to what are now the d'Entrecasteaux Islands east of New Guinea (which were also generally part of the mainland). As people colonized they made their mark on the land by using art. The continent of Sahul is now composed of three large islands—Tasmania, New Guinea (from the Bird's Head to the d'Entrecasteaux Islands) and the large, unnamed, island that is most of Australia—but for 85% of the time, Sahul (made up of these three islands and the sea between them) was a single landmass. The landmass changed shape as the sea-level fluctuated cyclically (Norman et al., 2024) but for only about 15% of the time during the climatic changes of the last half million years or so,[2] higher sea-levels isolated the three principal islands, as well as isolating Barrow Island and the d'Entrecasteaux Islands, and inundating the d'Entrecasteaux Channel. Sea-levels are rising beyond the previous cyclical changes during and after my lifetime as a result of the climatic catastrophe people have produced, yet we are already living at a time of sea-level that is high within those

[1] I am aware the Josephine Flood is female!

[2] As shown for the last six cycles which included sea-level as high as it is now, for example, in (Mimura, 2013, Fig. 3).

cycles. The climate change we have produced will not restore the dry land of Sahul but continue to reduce the amount of it. These considerations, of course, are a product of non-Aboriginal definitions of land and sea and the boundary between them, and do not take account of different conceptions of this relationship by the Aboriginal populations most concerned (Morphy & Morphy, 2009).

Sahul was first colonized, by the ancestors of the people of its First Nations about 65 thousand years ago (Clarkson et al., 2017), although there are strong arguments that the real date may be more recent than that (O'Connell et al., 2018). This was about 20 thousand years before the earliest painted caves in Sulawesi (Brumm et al., 2021) and began the process of creating what I have defined as Art 2 using the ocher found in those earliest sites, including marking the walls of caves, carving bone and involving art in other ways in the lives of their communities. It was 25 thousand years earlier than modern people moved into Europe with their more famous cave art.

How did the First Sahulians use the symbolism of art to define their relationships to each other and to the (changing) land? Art of many different types is abundant throughout Sahul, well-known both for its rock art (Fig. 9.1) (and see, for example Morwood, 2002) and the recent explosion of paintings on bark, canvas and board (Fig. 9.2) (Morphy, 1998; Mulvaney, 2003; Myers, 2002).

In addition, there is abundant more recent art made by Aboriginal artists which is not closely connected to either rock art or the more famous acrylic paintings derived from sand drawings (Green, 2014), such as the print by Badger Bates made using techniques which would not have been available until recently (Fig. 9.3).

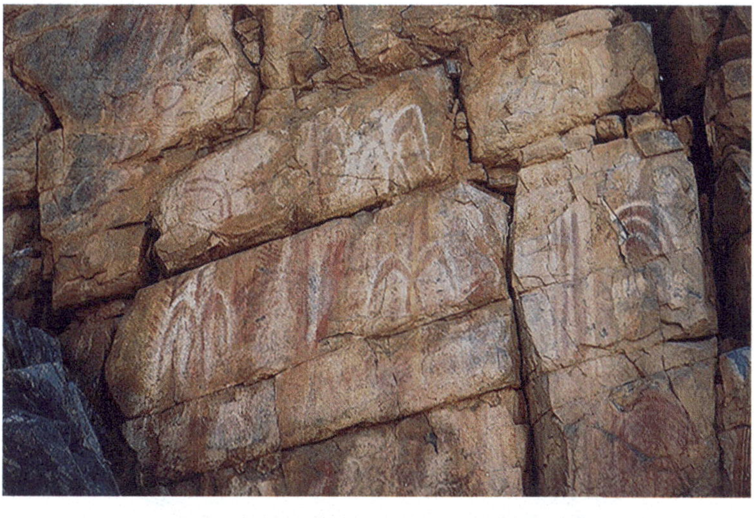

Fig. 9.1 A rock art image of a Northwest Central Queensland Anthropomorphs in Mitakoodi Country south of Cloncurry. I think it is easy to see how some images might fall into grazier Bill's "scribbles" category and be compared with Fig. 2 Right, while others are more difficult to describe in that way and have more in common with Figs. 2 Left and 7.28 Right. (Photograph Iain Davidson, permission from Pearl Connelly and Fred Turner of the Mitakoodi People)

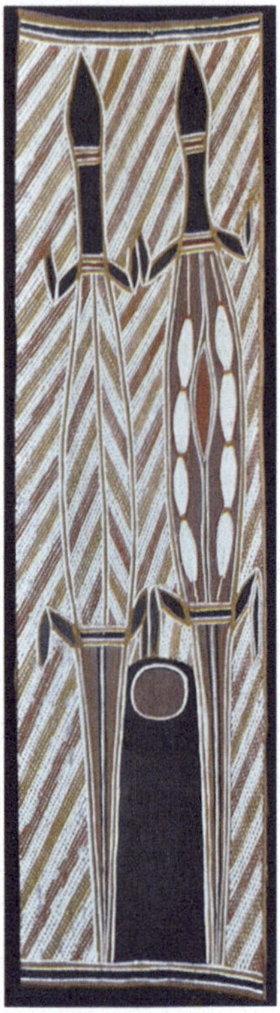

Fig. 9.2 Bark painting from Ramanginning by Fred Nanganarralil. (Photo by Matt Cawood. Permission from Milingimbi Art and Culture Aboriginal Corporation)

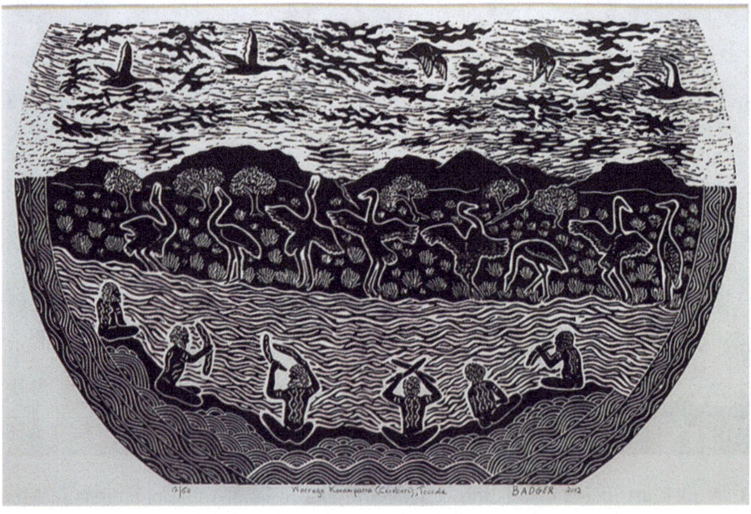

Fig. 9.3 Warrego Karamparra (Coroberi), Toorale by Barkantji elder Badger Bates. (Photo Matt Cawood. Permission granted by Badger Bates)

The islands that are the constituents of Sahul are also well-known for art of many types in perishable media, such as body ornamentation including body art, beads and pendants (Langley et al., 2018). The art of Papua New Guinea is very diverse and complex and is generally not discussed in the context of art in the rest of Sahul. Unsurprisingly, given that Rappaport's (1999) study of ritual derived from his years of experience in the Highlands of New Guinea, social interactions in Papua New Guinea involve much ceremonial activity. Individuals adorn their bodies (Sillitoe, 1988) with paint, feathers and items of material culture. Woodcarving is abundant and remarkable. In Chap. 3, I mentioned Ingold's argument about the way people learned to make the distinctive bags called *bilums* in the context of richly decorated house boards; the discussions in Chap. 6 about agency in art include the prows of the Trobriand Islands to the north east of the mainland; and, also in Chap. 6, I mentioned that people transform their social relationships in Torres Strait through their use of masks. In the Auwim region of northern New Guinea people made carved wood and stone artefacts, and some structures were ornamented with paintings in distinctive designs on sago bark, which encoded mythology of the society (Gabriel & Gorecki, 2014, 31–42). There is also rock art, some of which still has ethnographic oral traditions and links mainland Papua New Guinea to the islands of the Torres Strait (Tsang et al., 2022b), and elsewhere much of the rock art involves stencils of hands, material culture, shell ornaments, and cassowary body parts (Gabriel & Gorecki, 2014). One recent study concluded that rock art was "embedded in a network of relationships that involve oral traditions, place-making strategies and emotional responses such as overcoming fear" (Tsang et al., 2021). In the islands northeast of New Guinea in New Britain, large numbers of boulders were marked with mostly geometric signs which seem to have long histories going back thousands of years (Specht et al., 2021) (Fig. 9.4). Art, therefore, has a long tradition of importance and is still an important part of human life in Papua New Guinea, which we might call Northern Sahul. Incorporating these materials into the same framework as that I am applying to mainland Australia would probably require another book, and this one is already too long, so I do not detail anything here.

Paul Taçon (2000) has written that in Australia "Rock paintings and engravings can be found almost everywhere suitable surfaces occur. At least thirty key regions are identifiable on the basis of style, geography, and cultural history". This is complementary to the view that often the only thing that examples of rock art have in common is that they were painted on rocks (and that the boundaries of the stylistic regions are made up of regions with hardly any rock). Similar things could be said, though, of other categories of art (church art, mural art, museum art etc). Church art, for example, can be found almost everywhere there are suitable churches. For the art of modern Aboriginal communities, I have heard it said that almost every community has produced at least one artistic genius. Art has pervaded the society of the original inhabitants of Sahul since people first arrived here.

This chapter explores the principles that apply to art in a manner similar to those I outlined in the previous chapter: Expectation and indeterminacy of meaning; Defining the self; Ritual and teaching; Reducing the indeterminacy; Overcoming the tyranny of distance and place—the importance of persistence; Fixing the story;

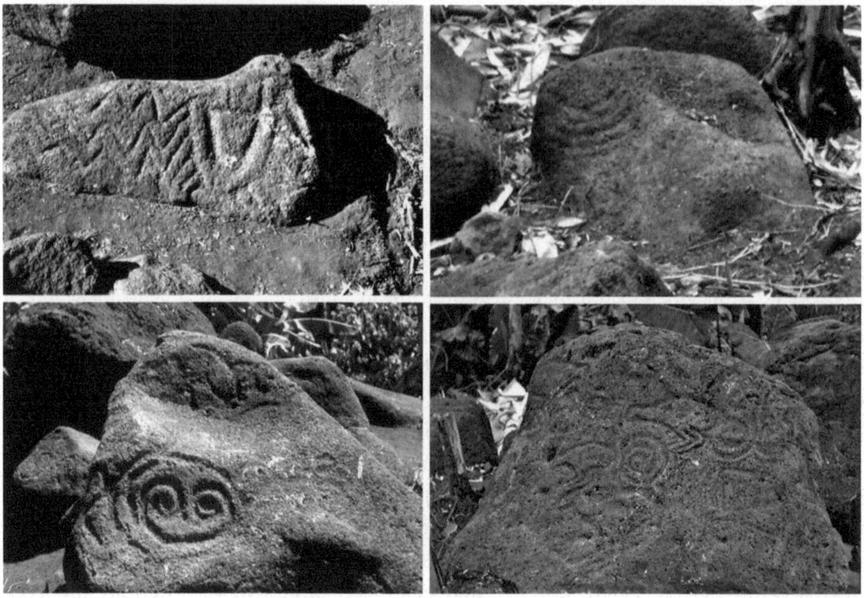

Fig. 9.4 Collage of images from New Britain in Papua New Guinea. (Specht et al., 2021. Permission from Jim Specht)

Freeing the stories from institutions of control; overcoming the tyranny of time. I am not saying that these are somehow built-in features of human behavior but that they are behaviors that were made possible by moving the symbolic capacity outside our minds (a concept associated with Donald, 1991) (in this case by making marks on the environment) to the products of the art-system.

9.2 Background

The first evidence for people here (and it does not matter for this argument whether it was about 65 thousand years ago or closer to 50 thousand), from Madjedbebe in Arnhem Land, includes evidence of the use of grinding stones to process ocher (Hayes et al., 2022). To get here, people had to cross from another continent, Sunda, which, like Sahul, is split up into islands on the brief occasions when sea-level is high. In conditions familiar to us, the continental region of Southeast Asia we call Sunda is divided for those same short periods into islands, including Sumatra, Java and Borneo. Sulawesi was never connected to the continent of Sunda, but always separated by the Makassar Strait, yet geographers name it as one of the Greater Sunda Islands.

After leaving Sunda, people had to make several more sea-crossings between the islands that lie between Sunda and Sahul in the region known as Wallacea. It was always necessary to make several sea-crossings (at a minimum, about ten) between the Wallacean islands to get from Sunda to Sahul. There was never a land bridge. The crossings were achieved successfully by people who had personal ornaments and shell fish-hooks, but not, initially, any sign of transfer of materials (such as obsidian) between islands (Norman et al., 2021; O'Connor et al., 2023). Noble and I previously argued that the first crossing to Sahul, with its several episodes of construction and use of watercraft, was the first unequivocal evidence for the mental capacity we called "uniquely human" and a product of the reflectivity consequent upon the emergence of language (Davidson & Noble, 1992). For those who choose the Working Memory argument about the emergence of modern humans, it also involved engagement in one activity while thinking of another (Davidson, 2010b).

Bradshaw and colleagues have used simulation modelling to show what might have been required for a successful first colonization of Sahul (Bradshaw et al., 2019):

> Using the best data available (while acknowledging that having more is desirable), our models estimate that as few as ~650 females (representing ~1,300 individuals in total) arriving in small groups (~130 each) over as many as 700–900 years would be sufficient to avoid a high probability of extinction during the probable environmental conditions that dominated northern Sahul between 65 and 50 ka.
>
> These numbers of people—whether ~130 arriving at semiregular intervals (Malaspinas et al., 2016) over a long period or 1,300 at one time—are substantively larger than ethnographically observed Aboriginal hunter-gatherers, except during larger ceremonial gatherings (Gould, 1969).

Some of the numbers that result from this simulation are not only implausible, but also remarkably similar to the numbers in another simulation elsewhere (Bradshaw et al., 2024). I think there are some unrealistic assumptions about the numbers of people who could travel in each watercraft (I illustrated several examples all with small numbers of occupants in Davidson, 2013b, Fig. 1) and the numbers of boats at each crossing (to say nothing about what averages out as one woman a year coming to Sahul).

We might begin by wondering about the small groups and the boats. Let us suppose that the boats could carry 13 people as indicated in some rock art (though none of those that represent boats seem to be ancient: O'Connor & Arrow, 2008; Oktaviana, 2018). On the figures we have been given, ten such watercraft would be needed to get 130 people across each of the ten sea-crossings—100 watercraft-voyages. To get 1300 people here that must have been repeated either (at least) 10 times, that is to say once every 70–90 years (say, 3 generations), meaning 1000 individual watercraft-voyages. The models represent one set of behaviors that are an alternative to sheer blind luck as an explanation for the successful presence of people in Sunda—but for what it is worth, I suspect a lot of luck was needed anyway, even if archeologists no longer consider accident to be sufficient (Allen & O'Connell, 2020; Bird et al., 2019). It was an astonishing achievement. As a comparison, the first European invasion of Australia 240 years ago began when 11 sailing ships in the First Fleet landed about the same number of people on the shores of what is now Sydney after sailing from England (Atkinson, 1997).

Support for large numbers of initial colonists can be found by examining the evidence from genomics which shows at least three Mitochondrial DNA haplotypes (which are transmitted only from mother to daughter) in Australia and four or five in Sahul (Allen & O'Connell, 2020, 6). This argument carries with it assumptions that the initial population must have been quite large. Nevertheless, however implausible the figures in the model calculations, the numbers of people who must first have entered northern Sahul across Wallacea were quite small and they certainly spoke few languages (there is a review of the arguments in Davidson, 2014a, 374–376). According to European records, the most recent survey lists more than 30 language families and 490 spoken languages recorded for mainland Australia alone (more than 500 when sign languages and creoles are included) (Bowern, 2023; McConvell & Bowern, 2011). No one is proposing that initial colonist numbers allowed for such large numbers of languages at the very beginning. We can say with certainty that many of them emerged uniquely within Sahul by processes that involved social differentiation which was probably fostered by the use of art.

From what is known and what has been reconstructed, one language family, Pama-Nyungan, covered 90% of the Australian mainland—almost the whole island except for the Northwest. It is said to have first appeared after 5 thousand years ago (Evans & McConvell, 1998) and to contain hardly any trace of earlier languages from that huge region. Such a history would make it really unlikely that the First Nations people of Tasmania spoke a Pama-Nyungan language, since sea-level separated the island from mainland Australia about 14 thousand years ago (Adeleye et al., 2024; Lambeck & Chappell, 2001) almost 10 thousand years before the spread of Pama-Nyungan languages. Unfortunately (for many reasons) far too little has been recorded of Tasmanian languages (Bowern, 2012).

The languages of New Guinea are more numerous in terms of language families and of languages. One language family alone has more than 400 languages. As a result, during those 65 thousand years, the population increased to support over 1000 languages (some estimate there were 1500) across Sahul, far more than the number of people or languages that entered the continent. At least we can point to one group of more than 400 Trans New Guinea (TNG) languages, with probable origins back more than 10,000 years to the Pleistocene (Pawley, 2005). Austronesian languages are more recent in New Guinea and arrived with people from the north bringing with them pottery, pigs, chickens, and possibly dogs, as well as other cultural features visible and invisible archeologically (Spriggs, 2007). Part of the process of first colonization, then, was the differentiation of the increasing population into groups that could be defined symbolically. My view of symbols is that art of some sort played a role both in that differentiation and its ongoing definition.

Early rock art has been found in Borneo dating to at least 40 thousand years ago (Aubert et al., 2018b) and a large body of art, some of it older, has been dated on the Wallacean island of Sulawesi (Aubert et al., 2014, 2019; Brumm et al., 2021; Langley et al., 2020; van den Bergh et al., 2016). For the moment, the oldest art in Sulawesi is 51.2 thousand years old (Oktaviana et al., 2024).

The early dates in Sunda point to the possibility that the rock art of Sahul was not independently invented (Langley et al., 2019)—there may be yet earlier art to be dated on Sunda and in Wallacea—otherwise the fact that the earliest evidence for people in Sahul/Australia is currently earlier (Clarkson et al., 2017) could mean that art was independently invented here. Indeed, it is conceivable that art was independently invented in all three regions, but no one seems to be contemplating the possibility that art was invented by the first people who came to Sahul and that they then took it back to Sulawesi and Sunda.

The first people on Sahul arrived technically able to construct watercraft (many times) (Balme, 2013; Balme et al., 2009; Veth et al., 2011) and I would argue that that fact means they were capable of making art as well. The first inhabitants in Sahul were cognitively equipped (Davidson, 2010b) both to expect that marks would have meanings and to construct and navigate in those watercraft. Both art and technology imply that. Further, the meanings of marks could be more easily understood, would be less indeterminate, as a result of the stories the makers (the artists) or other people told about them.

Many stories in Australia relate people to the landscape and to the other occupants of it. Such stories, which would be included in "The Dreaming" cosmology of Aboriginal people (see an account in Davidson et al., 2021) must have been developed within Sahul as people came to know the landscape and locations of other people; they are still told in such a way that they can be identified on the landscape. In a previous publication (Davidson, 2023, 639), I pointed out that:

> the memory of places (Gibson, 2013; Hercus et al., 2009; Russell, 2012) is absolutely central to the way people constructed their version of what is important in the landscape for the people who inhabit it (e.g., Hayward et al., 2018). Mythical accounts of the landscape begin with passing on oral accounts of something observed or imagined in the recent past.

The non-Aboriginal ecologist, Alan Newsome, for example, identified the relationships between red kangaroo mythology and important places in Central Australia as a result of working on the ecology of red kangaroos (Newsome, 1980). As Deborah Bird Rose (1996, 51) wrote: "One of the best documented examples of [Dreaming sites as] sanctuaries is along the Dreaming track of the red kangaroo in Central Australia. This track traverses some of the toughest desert country in the world, and sites coincide with the most favoured areas for kangaroos. As the sites are protected, so too are the kangaroos in these sites." The important point here is that histories and memories were needed before there could be any knowledge of landscape, facts and knowledge of locations of other people, memories of places and experiences of things to be observed or imagined in such places. We can add to that the experience of short term and longer term climatic variations and their impacts on the species of the environment.

Just over a hundred years ago, at the time Europeans started writing about the stories, the many groups of First Nations people were connected along routes of common ritual storytelling now called "Songlines" (for an introduction see Neale & Kelly, 2020), what Rose (and others) called "Dreaming tracks" (Rose, 1996, 51). For all the antiquity of these Songlines, linking places and people across the country, sometimes including their art, the lines and the belief system must have emerged in Sahul and be unique to this Country, but they did change. Bruno David has argued that the set of beliefs that would be recognizable as the Dreaming might have emerged relatively recently within the last 3.5 thousand to 1.5 thousand years ago (David, 2002). They would not have emerged fully formed but were worked out over many generations.

The earliest recording of rock art from the continent was in 1803 from Chasm Island in the Northern Territory (Bahn, 1998, 31–32) in a region described by Morphy (2007, 29–31). In 1842, after sailing on *The Beagle* six years after Darwin's voyage in the same ship, Wickham and Stokes reported rock art at Depuch Island off Western Australia (Wickham, 1842). Stokes (1969, 171) wrote (Fig. 9.5):

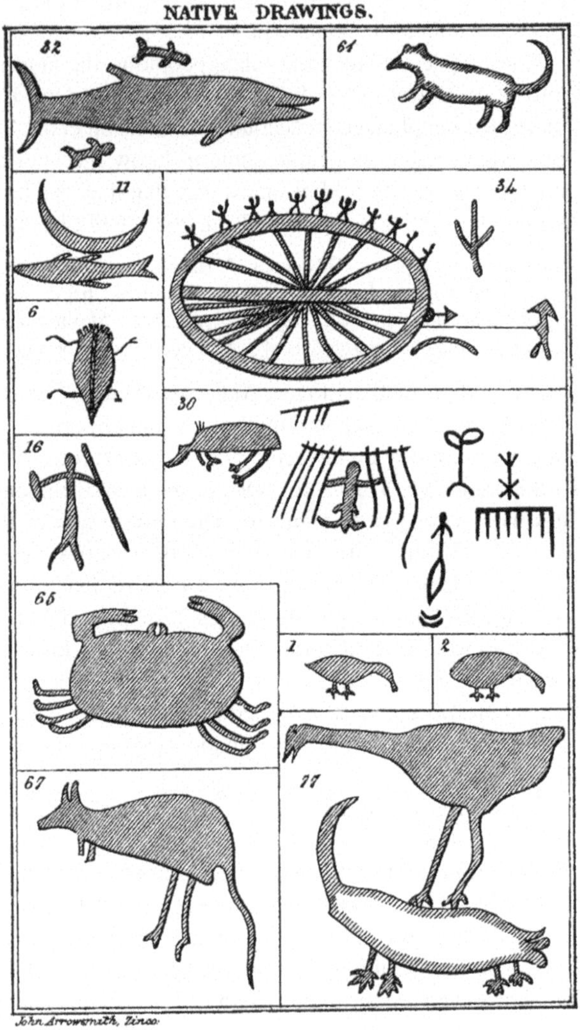

Fig. 9.5 "Impression of the copies made by Captain Wickham of the native drawings on Depuch Island" from Stokes (1969, 170). Some of these images appeared in the 1906 volume about Altamira (Fig. 6.6)

> No doubt they expended on their works of art as much patience and labour and enthusiasm as ever was exhibited by a Raphael or a Michael Angelo [sic] in adorning the walls of St. Peter or the Vatican; and perhaps the admiration and applause of their fellow-countrymen imparted as much pleasure to their minds as the patronage of popes and princes, and the laudation of the civilized world, to the great masters of Italy.

That was 3 years before the French acknowledged that an engraved bone from Le Chaffaud was made by fisher-gatherer-hunters during the Paleolithic (in my State B) (Fig. 2.1).

Some of the same images from Depuch Island were reprinted by Cartailhac and Breuil (1906, Fig. 151) (Fig. 6.6) as part of their argument that hunter-gatherers around the world could make art. Given that was true, then there was no argument against the art at Altamira (Figs. 2.10, 3.17, and 8.14) being painted by the hunter-gatherers of Europe. It might even have been made by people from a time sufficiently long ago to depict animals now extinct, such as bison or mammoth. Even before the official acceptance of Paleolithic hunter-gatherers making art in caves, Mark Twain (1989 [1897]) wrote: "…an Australian aboriginal's [sic] pictures of animals are nicely accurate in form, attitude, carriage; and he put spirit into them, and expression…As an untaught wielder of the pencil it is not likely that he has had his equal among savage people." Thus Twain's appreciation of their skill as artists echoed the judgment of Stokes (they were good artists), yet in referring to First Nations people as savages, he also showed the attitudes of his time towards people of other cultures, and showed attitudes to the possibility of women as artists. A few years later, Spencer compared the line work of Aboriginal painters near Oenpelli/Gunbalanya "with much the same freedom and precision as a Japanese or Chinese artist doing his more beautiful wash-work with his brush. They are so realistic, always expressing the characteristic features of the animal drawn…" (Spencer, 1928, 792–3). While Stokes and Twain claimed the artistry of Aboriginal rock art from seeing only the product, Spencer described it by detailing aspects of the production of the images as he saw them. Those barks contributed significantly to the perception of Aboriginal paintings as "art" in a series of studies and exhibitions from 1929 onwards (Goldhahn et al., 2021, 48–49).

Australia now has a vibrant Aboriginal "high" or "fine" art industry, which has created value from works that were originally produced for other purposes (Morphy, 2007, xi). By starting from his observations in 1973, Myers (2002) has copiously documented and argued the process by which the acrylic paintings of Central Australia emerged from less-commercial beginnings. For some it has been tempting to seek to draw conclusions about art processes elsewhere in the world on the basis of knowledge of the art of modern people in Australia (as, for example, Ingold, 2000), but I want to avoid that approach here.

The archeohistory of art in Sahul should tell us whether there is any merit to considering art in the steps I have isolated in the earlier chapters—trying to avoid inferences only from modern art—when we think about the ways in which people of the past used art in the process of discovering the landscapes and environments of Sahul. Can we tell anything about art from the way in which the First Peoples of Sahul used it? (Such an approach has been adopted, for example, for a comparison of social process affecting rock art in Levantine Spain and in Arnhem Land, by Domingo Sanz, 2021). I want you to join me in an attempt to think about the ways in which the First Peoples in Sahul arrived at the patterns of diversity recorded in recent times across the several parts of the continent (see, for example, Ross, 2013).

By the time Europeans arrived, maybe 65 thousand years after first landfall, and described those they had not killed, the First Nations inhabitants of Sahul had great diversity of material culture and social organization. According to Donald Davidson (1937)—no relation—they made visual art on rocks, wooden objects, netting and basketry, spears, watercraft, boomerangs, and personal decoration (summarized in Davidson, 2014b, 251–252; Satterthwaite & Arthur, 2005). Ross (2013) isolated some of the symbolic diversity between groups in different parts of Australia by looking at the evidence of the rock art in each region. She showed that the archeohistory of the continent is about the emergence of this symbolic diversity and its use to distinguish one group of people from another. That diversity depended, I argue, on the fact that art is a visible sign of the symbolic values of the society that makes it, and the symbolic values depend on conventions that are shared among members of the society through the stories that were fixed, at least in part, by making them part of the art that was left behind. The gradual changes in the conventions and the accumulation of differences led to the emergence of different societies, and with them the diversity of languages and material culture seen at contact.

9.3 Expectation of Meaning and the Indeterminacy of Meaning

Some people say the exhibition, *Dreamings*, brought the world's attention to the rich field of (modern) Aboriginal art (Morphy, 1998, 376–8; Sutton, 1989). Many of the paintings in the exhibition were acrylic paintings of dots, circles and tracks created by the Pintubi people of the Central Desert northwest of Alice Springs, encouraged by a non-Aboriginal schoolteacher (Bardon, 1991). Myers (2002), who worked with those people from 1973, was of the opinion that "Aboriginal art emphatically acquired the status of Fine Art with a much-publicized show in 1988." In the book that arose from the exhibition, Peter Sutton (1989, 13) recounted his experience with Aboriginal people on the Kirke River in Cape York Peninsula who were interacting with dingoes and totems, naming and land: "we would say that everything has meaning. In a traditional Aboriginal sense, the world is made up of signs. One may not know more than a fraction of the meanings, and not all of the meanings are of equal significance…" There was no "art" involved in this experience, no work that could be collected by a museum, framed, or hung on a wall. Yet, in telling this story of the experience in the bush and recording the accounts of Aboriginal men making incidental marks on the ground—semiotics would call them "indexical signs"—of the telling of stories and their appreciation by an audience, Sutton captured the essence of the ideas of the expectation of the meaning of signs. Signs were all around people in that apparently natural landscape—they were marks of human presence. Sometimes the marks persisted at particular places; at other times the marks were ephemeral. Sutton pointed out that that expectation went together with a certainty that any meaning of those signs was indeterminate—the signs alone did not speak. As he recounted the occasion, he had to ask for the meanings from the Aboriginal men with whom he was talking. The Aboriginal people made the meanings. Elsewhere, the Yolngu man, Narritjin, speaking to Morphy complained about one area of a bark painting with a pattern of red and black dots (Morphy, 1991, 143): "[I]t means blood and maggots, sand and worms, itchy red spots and rotten flesh ... One small dot, too many meanings." The multiplicity of meanings ensures that it is not the dots themselves that produce the meaning, but the people making them.

The most obvious example of rock art marks with meanings are those that involve human hands. We will look at them in detail when we consider what reduces indeterminacy. For the moment, all we need to remember is that the marks create a certain expectation of meaning, and most outsiders looking at hand stencils speculate about them as marks of a particular person (hence both determinate and a means of defining the self). In reality they are very much less determinate than that (Fernández Navarro et al., 2024) and do little more than define that a place was marked in particular ways.

Hand stencils have been found and dated in Europe (commonly from 37.3 thousand years ago in Spain, Pike et al., 2012). In addition, at the site of Maltravieso, in southern Spain, stencils have been dated as older than 66.7 thousand years ago so that, if that is its true date, they were probably created by Neandertals (Hoffmann et al., 2018a; Standish et al., 2025). Nothing about the early Maltravieso stencils, except the date, enables us to identify that it was made by a different species. That is the problem with hand-stencils—their associated meanings are indeterminate.

9.4 Defining the Self

In the previous chapter, I argued that, once individuals had established the autonomy of their sense of self through the use of material culture (Iriki & Sakura, 2008; Malafouris, 2013, 164–6), a fundamental feature of State B was the social definition of self through the use of individual markings or ornaments. This probably began with body painting at 100 thousand years ago in southern Africa (Henshilwood et al., 2011), but possibly with the use of pierced shells earlier in State A in Morocco in North Africa (Sehasseh et al., 2021). Balme and Morse (2006, 801) suggested that "Beads are a form of body ornament that, like other body ornaments including body modification, are a form of non-verbal communication indicating social value or difference." We should expect, therefore, that individual ornamentation and beads would have appeared fairly early in the first colonization of Australia, and we know that they were present almost everywhere when the material culture of Sahul was first described (McAdam & Davidson, 2018). In addition to arrays of more local distribution, there were also wide distributions of individual pearl shell and baler shell ornaments (Akerman, 1973; 2018, Fig. 13.3; McAdam, 2008). Some of these marine shells have found their way into archeological sites and can be dated between 30 thousand and 18 thousand years ago in sites which were at least 100 km inland from the Indian Ocean (summarized in Akerman, 2018, 204).

The pierced shells from the archeological excavations at Mandu Mandu (Morse, 1993), now possibly dated to more than 40 thousand years ago, were an early indication of the sorts of signs that could have been used by early language users. Since language was a form of communication using symbols, and since symbols involved both arbitrariness and convention, in the earliest stages of language use it was important to distinguish between trustworthy people who had similar conventions (the in-group) and untrustworthy users of symbolic communication who did not share the conventions (the out-group).

When the Mandu Mandu shells were announced,[3] from excavation of a cave said to be 8–10 km from the shore (Morse, 1988), they were unusual in Australia, but now there are also marine scaphopod shells, said to be used as beads, from the site of Riwi which was 500 km from the shore when they were left there 30 thousand years ago (Balme & Morse, 2006). Since then, others have concentrated on pierced shells in Africa (Henshilwood et al., 2004) and the Mediterranean (Kuhn et al., 2001; Stiner, 1999) including earlier finds (Bar-Yosef Mayer, 2005; Bar-Yosef Mayer et al., 2009). Together with our indication of the way such beads might have been used among people speaking in symbols (Davidson & Noble, 1992), scholars now give due weight to previous studies of personal ornaments which date to the earliest presence of modern humans in Western Europe (White, 1989; 1992). Sometimes recognizing the presence and significance of art has required people to see what was already in front of their eyes.

Morphy (2007) described the early fatal interactions between the Yolngu and outsiders by illustrating the dead body of a Yolngu man who died as a result of the interaction with the white colonists on Flinders' 1803 circumnavigation. Another encounter more than 100 years later also left dead bodies. In both Yolngu cases, the bodies could be identified by the armbands they wore. In Morphy's words (2007, 31 and Fig. 3) "The armbands left behind marked the identities of those who had died. Identity, linking people with places and with ancestral beings lies at the heart of Yolgnu art…" It would be easy, if idle, to point to the fact that the 30 thousand year old female figurine from Willendorf (Weber et al., 2022) also marked her identity with decoration on the wrist (Fig. 8.10).

Many of the paintings of Dynamic Figures in the early rock art of Arnhem Land show individuals wearing items of attire, including necklaces and ornaments on the arm, in addition to skirts and headdresses (Fig. 9.6) (Lewis, 1988). Likewise, such headdresses were recorded in the *Gwion* figures of Australia's Kimberley as well as in Tierra del Fuego, Argentina (Fig. 9.7) (Fiore, 2020) and Levantine Rock art of eastern Iberia which in every case makes them stand out (Santos Da Rosa et al., 2021, Fig. 1).

[3] Permission to reproduce a photo was originally granted, but later withdrawn.

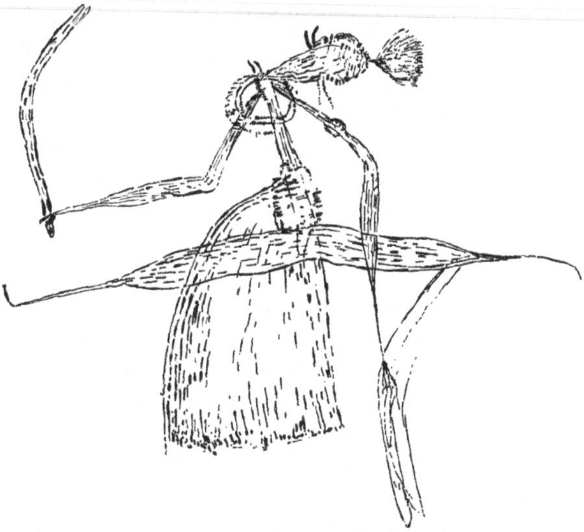

Fig. 9.6 Line drawing of a Dynamic figure from Boomerang Period in Deaf Adder Creek, Arnhem Land, Australia, with permission from Darrell Lewis. The figure has, at least, a "headdress, large 'dancing skirt', decoration on the [left] arm, and a 'necklet'." (Line drawing traced from photograph by Lewis, 1988, Fig. 8)

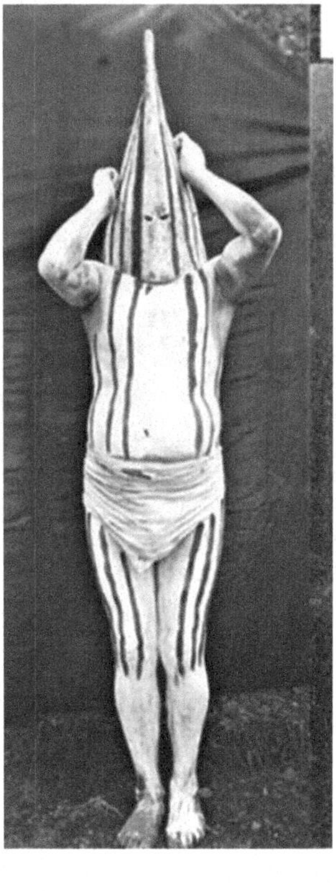

Fig. 9.7 Yamana/Yagan man from Tierra del Fuego painted and with headdress as a kina spirit. (Photo taken by Gusinde 1922, reproduced in Danae Fiore [2020, Fig. 6e] and Fiore secured permission of the Yagan community). Compare this to the Yulluna/Kalkadoon men in Fig. 9.25

The claim that the Mandu Mandu beads added to the understanding of the cognitive abilities of the first Sahulians has now been emphatically reinforced by a synthesis of the archeological evidence for "symbolic expression" in Sunda, Sahul and Wallacea (Langley et al., 2019, 17): "The abundant early traces of art, burial, and personal ornamentation found across Sunda, Sahul, and Wallacea now place the archaeological record of early modern humans in these regions on an equal footing with those of Eurasia and Africa." In addition, in Timor-L'Este, part of one of the Wallacean islands closest to Sahul (called Lesser Sunda islands), there were fragments of *Nautilus* shell at Jerimalai, from 42 thousand years ago which were drilled and marked with red ocher (Langley et al., 2016b). These fragments are similar to other finds, from another Wallacean island, Alor, of brightly colored and pierced shells which were probably sewn onto garments around 12 thousand years ago (Langley et al., 2023, 810–11) (Fig. 9.8), just as the beads with the Sunghir burials were, as shown in Chap. 8 (Fig. 8.9).

Finally, examination of a worked fragment of a kangaroo tibia dated to earlier than 45 thousand years ago at Carpenter's Gap in the southern Kimberley showed that it was probably a painted nose bone indicating direct human self-definition (Langley et al., 2016a).

There are still very few archeological sites with personal ornaments dated earlier than 30 thousand years ago, but these finds lend some support to the idea that ocher found at many early sites in Sunda and Sahul is an indication of body painting. Brumm et al. (2017, 4105–6) pointed out that there has been a systematic bias towards the much heavier sampling of sites in SW Europe than of those in Sunda, Wallacea and Sahul, explaining the greater abundance of evidence in Europe. We might argue that sample bias has been, in part, due to widespread past prejudices about Aboriginal archeology among non-Aboriginal people. These resemble the

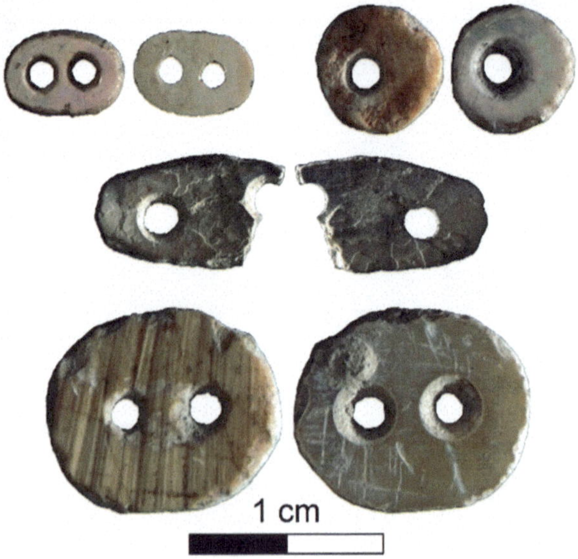

Fig. 9.8 Fragments of *Nautilus* shell pierced to be sewn on to garments, from Alor, Indonesia in Wallacea. (Langley et al., 2023. Photo and permission Michelle Langley)

prejudices about early cave art in Europe about 150 years ago, through the use of words such as "savage" and "primitive" for the people, and "craft" for their artistic activities. People have been known to judge art by their attitudes to the people who produced it, as I have suggested grazier Bill probably did.

It used to be easy to underestimate the significance of the finds of two burials at Lake Mungo in western NSW dated to around 40 thousand years ago because debates about burial in the archeological record have been focused on the Old World and whether any species other than modern humans buried the dead. The burials at Lake Mungo were certainly of modern humans. We can now recognize that the red ocher which surrounded the burial of Mungo III (Bowler & Thorne, 1976) would tend to support the idea of body painting in Sahul more than 40 thousand years ago. At the same time, we attach much greater importance to the fact that the earlier of the two burials (Mungo I) was not only one of the very first burials of a human known anywhere in the world, but certainly the first cremation (Bowler et al., 1970).

We can put all of these ideas together to indicate the cognitive significance of personal ornamentation, ceremonial burial and possibly body painting—all of which we should consider part of art—which was present in Australia very early, often earlier than any other known examples outside Sahul. Such is the nature of the archeological record, however, that at the same time as our recognition of the ways the first people used aspects of art in their first colonization of Sahul, discoveries elsewhere in the world have shown that some elements of art seem to have been appearing earlier than the appearance of modern humans, as I outlined in Chap. 7 (Sect. 7.6). The question is whether they represent the same process. I do not think so. I think this indicates that marking the environment was the start of people defining a relationship with places, and then marking their individuality was a way to define their own distinctiveness from others within that environment—among those places—allowing them to relate themselves symbolically with their environments. Of course, during the time when people were moving into a new continent such as Sahul one of their tasks would have been to make the environment theirs and they gave it a special significance even as they defined their own places in it. It was part of how they became Australians or New Guineans.

9.5 Ritual and Teaching

These are examples of the use of art to define individuality. The section that follows will show how, following the outline of ritual described in Chap. 5, art is engaged in ritual and story-telling—though many of the significant stories were actually sung. That is the context which allows people and groups to teach and to secure power for themselves. Learning stories and songs made people powerful. It is this dual expression—the thing itself and the idea associated with it, as I discussed it in Chap. 5—that defines how art is used in Australian Aboriginal societies, probably all societies. Following the insight of Kelly (2023), described in Chap. 5, that "the oral channel is doing almost all of the heavy lifting in terms of the information load," three things are essential for us to identify ritual in art in the absence of the people who drove it: (1) repetition; (2) indeterminacy without a story or song; (3) an oral channel that carried the story or song within the community.

There are two well-published examples of the use of art in ritual and teaching in the totemic societies of Sahul, though there are many more examples throughout Australia. The first is the series of studies by anthropologists studying the practices and behaviors of modern people who use art in initiations of people in northeastern Arnhem Land. The most detailed study was that by Morphy (1991) during his research among the Yolngu people following earlier studies among the same people by Warner, Thomson and the Berndts.

The second example is the use of ritual in maintaining linked ceremonies such as the Seven Sisters Dreaming across Australia, as was triumphantly revealed to the public in the *Songlines* exhibition at the National Museum of Australia (Neale, 2017b). We will deal with the *Songlines* example in the section on the tyranny of distance and place. These, of course, are modern examples, but they illustrate the complexity of what has arisen during the long history of the first occupation of Sahul up to the present day.

Before Morphy's study, Warner, working mostly around Milingimbi on the northern edge of Arnhem Land (not far from Ramangining where Fred Nanganarralil produced and sold the image in Fig. 9.2), described the social behavior of Yolngu people whom he called Murngin. They used materials we would call "art" during rituals and ceremonies, involving red feathers, red ocher body paint, and carved wooden spears, providing an example of large feathers from the crest of Sulphur-crested cockatoos as "used for the great ritual headdresses made by the men" (Warner, 1964 [1937], 146). Warner described the process by which people made bark paintings, hand stencils, and hollow cylindrical objects such as "trumpets" (which English speakers might call *didgeridoos*) and "coffins" (Warner, 1964 [1937), 503–505). He emphasized that the designs that carry ritual significance can be applied to many items of material culture and become sacred themselves through the association of the object, the design and its attached story or song.

I have previously expressed my preference to concentrate on Warner's way of looking at ritual aspects of behavior: "totemic societies recognize both a physical and a mental kinship with non-human groups" (Davidson, 2017). As Warner (1964, 409) put it after detailed analysis: in totemism "two aspects of society are being ritualized: the social relations of [people] to nature, and of [people] to [people]." These two types of interaction are those that all human societies need to get into balance. Australian Aboriginal people do it through the rituals associated with totemism, often involving art. The first people who came to Sahul could not have brought such practices and beliefs with them because they needed to know about the nature of Sahul before they could define their social relations to it.

Morphy's more detailed study of the artistic practices of the Yolngu in the 1970s followed such earlier studies: Warner in the 1920s (Hamby, 2008), Thomson in the 1930s and 1940s (Allen, 2011) and then the Berndts in 1947 (Pinchbeck, 2013b). They were all involved with art more or less explicitly. There were other collectors and non-Aboriginal scholars, particularly among the missionaries who worked in that region (as described by Morphy, 1998, 234–9) though their contribution was different (but important for the argument of this book) and I will discuss it in the penultimate section of this chapter. Such art had been part of Arnhem Land culture earlier: in the South Australian Museum, there are examples of bark paintings from the last quarter of the nineteenth century (Sutton, 1989, Figs. 53 and 72) and the first quarter of the twentieth century (his Figs. 55–57); Spencer collected bark paintings during that period for the National Museum of Victoria from Gunbalanya (once called Oenpelli) (Goldhahn et al., 2021; Mulvaney, 2008, 149). The Gunbalanya paintings do not seem to have had the ritual significance that was described by later anthropologists—perhaps because there does not seem to have been the repetition of imagery to accompany any storytelling, or perhaps because Spencer and others did not participate in the lives of the societies they collected from. Indeed, Spencer's own account described how he had seen the paintings on barks that formed the inside walls of Aboriginal dwellings and, with the single-mindedness of the colonist, had dismantled those spaces to obtain the paintings. You cannot get much less "participant" than that. The Aboriginal artist Namadbara described it in this way: "Mr Spencer saw the paintings in some of the bark humpies at Oenpelli and chopped them up" (Goldhahn et al., 2021, 54). This vandalism was a recognition that, for the colonial, the barks were art not anything else, though, by then perhaps Spencer was not sensitive to anything else. In the 1980s, there was a project to repaint some of the rock art images in northern Western Australia. The project provoked outrage among some non-Aboriginal people who claimed to value cultural heritage, and less outrage among involved Aboriginal people and non-Aboriginal people who worked with them (Mowaljarlai et al., 1988). The debate (and Spencer's earlier attitude to paintings in the dwellings of people in Arnhem Land) sounds like the equivalent of the saying: "No, Mr Van Gogh you cannot sell that painting because it belongs to all the world."

Tim Ingold (2000, 131) would have it that: "Hunters and gatherers of the past were painting and carving, but they were not producing art." This whole book is about how wrong that statement is. "Hunters and gatherers" were producing art, but it was not being understood as that, in part because those producers were not producing value through the art of a sort that critics themselves recognized. Fisher-gatherer-hunters have their own framework for recognizing value, but their product is no different from art that entered into the market. They were not selling it. The world has since moved seamlessly (apparently—though it may not seem like it to people who have been colonized) into a mode where Aboriginal paintings and carvings, accompanied by stories, are treasured elements of art—literally "treasured" because they are bought and sold. I suggest that Ingold would have had it that the product changed to enter the art market. Some of it has, but that is not the point. Rather, the art market, which includes the critics, has changed its attitude to people who once lived by fishing, gathering and hunting.

The people whom Warner was studying expressed their totemic relationships using designs painted on the bodies of initiates to be used in teaching about the inside knowledge of being a Yolngu (as illustrated in the photographs in Morphy, 2007, Figs. 2 & 7). Sixty years after Warner, that teaching occurred through painting similar designs on sheets of bark (Morphy, 1991), "originally a minor component of Yolngu material culture" (Morphy, 2007, 45). Warner collected much art including bark paintings but, apparently, although he wrote about "Murngin Artifacts" (Warner, 1964 [1937), Appendix II), he did not pay much attention to the material culture. Louise Hamby (2008) has suggested he was reluctant to single out particular items because he recognized that art objects were an integral part of the social system and did not really stand alone. They were, in fact, part of the four part art-system.

Ten years or so after Warner, Donald Thomson (1949, 1983) made substantial collections during his anthropological fieldwork around Yirrkala in eastern Arnhem Land, including bark paintings. As Nic Peterson (2005, 42) remarked, perhaps as a result of Thomson's inhibition in reaction to the publication of Warner's work at the time, these remained largely unknown until after Thomson's death when full analysis of the materials and Thomson's fieldnotes was possible. One detailed examination of the barks and his notes showed that sacred designs painted on bodies were subsequently painted on to sheets of bark: "In 1937 Thomson photographed men at Milingimbi on the final day of a *Ngarra* ceremony when the body paintings are revealed to others. These same designs appear to have been painted for Donald Thomson onto single sheets of bark" (Allen, 2011). In this way the designs that illustrated sacred songs ceased to be secret and became more recognizable as art open to everyone. It is difficult to deny that the designs painted on the bodies of young people could become art, but what is important is that both when painted on bodies or when similar forms were painted on barks, the acts that turned either or both paintings into art were the stories told about the pictures—sacred songs were secret, other stories were not. As elements of ritual, the stories were much more detailed, and both types of art had associated stories. They were both art.

After Thomson had worked in Arnhem Land, Catherine and Ronald Berndt collected both bark paintings and pictures on paper. In 1947 the Berndts obtained butchers' paper and crayons, for reasons connected with the ease of getting the images to ships anchored off shore, and asked 27 Yolngu men to use the special materials to illustrate "law, culture and knowledge" through pictures (Pinchbeck, 2013b). The drawings were presented 76 years later in the first exhibition and publication of "a significant element" of the paper-and-crayon works. These were described primarily as a "unique collection of artworks" in which the anthropology was "an unrivalled document of Yolngu knowledge and law" (Pinchbeck, 2013a)—the anthropology provided the stories that gave meaning to the art, just as it did through Thomson's fieldnotes. Commenting on the art collections made by the Berndts, Morphy (2007, 51) wrote that "The role of such collectors in *creating* [my emphasis] a market should not be underestimated," though, of course, the Yirrkala drawings remained mostly unseen until 1978 (Stanton, 2013, 19).[4] Art was always part of an art-system, but there has been a change in the way seers get to see it.

[4]As was the case with earlier examples of crayon and paper art collected by the Berndts at Birrundudu (see also Carty, 2021, 103–106). I am very grateful to Jason Gibson for alerting me to his work on these paintings and discussion https://adi.deakin.edu.au/project/collecting-at-the-

Morphy began his book *Ancestral Connections* (Morphy, 1991, 1–2) by pointing out the strange contrast between, on one hand, the scenario of two men painting the lid of a coffin while all other people were not involved, and, on the other, the fact that a week later one of the men was painting a sheet of bark in the same form but destined for sale. The first was secret and could not be seen by women and children; by contrast the painters of the commercial bark painting included women, and once completed, the bark was put on display in the shop and subsequently sold. "At the time of the Yolngu's first contact with Europeans, the same painting in an unmodified form would not have been seen by women in public." Morphy (1991, 2) went on to characterize the transition in the following way:

> In the years since first contact, some paintings have moved from the closed space of the men's ceremonial ground to the outside space of the craft shop, and beyond that to the art galleries of the world. … the shift to the outside has still left the inside apparently intact. Paintings occupy closed as well as open spaces, and though in some contexts they have become commodities, in others they remain part of the substance of ritual.

Some fully initiated people allow paintings to move from the sacred-and-closed/inside context to the public/outside domain. These were released with a political motive to inform governments about land ownership (Morphy & Layton, 1981, 61). This involved a succession of petitions from Yirrkala in 1963,[5] Barunga in 1988),[6] and the Uluru Statement from the Heart in 2017.[7] Art works were part of the system that included the text of the petition—what Wright called "ethnographic art overlaid with the desiderata of protest" (Wright, 2024, 5). The Petition shown in Fig. 9.9 is the original one given to Kim Beazley (senior), who had the idea to combine the petition with a Bark Painting (Wright, 2024, 314).

Thus, the symbols could be manipulated in two ways: first by changing the context in which the material objects were symbolic and hence their symbolism, and second by changing the meanings the symbols carried separately from their context. This has been a process probably applied many times before, in times for which we have no documentation, simply because any symbolic material has the qualities of arbitrariness and convention, and it is open to members of communities to alter the conventions in several different ways. Morphy (1977, 208–209) recognized this and commented on an example in which a bark painting was used by invoking mythology represented in the painting. The bark painting was used to control spiritual pollution that arose from a death and burial. This, Morphy wrote, "illustrates the dynamic response by the Yolngu to the changed conditions brought about by the increased contact with Europeans." The art was a catalyst for that change—the object could remain more or less unchanged, but the context changed around it. As

crossroads-anthropology-art-and-cultural-change-1939-85/. Mountford (1976, 94–5) had used paper and crayons to encourage drawing since the 1930s. His book was subsequently withdrawn.

[5] https://www.naa.gov.au/explore-collection/first-australians/aboriginal-petitions#1937

[6] https://aiatsis.gov.au/explore/barunga-statement

[7] https://ulurustatement.org/the-statement/view-the-statement/

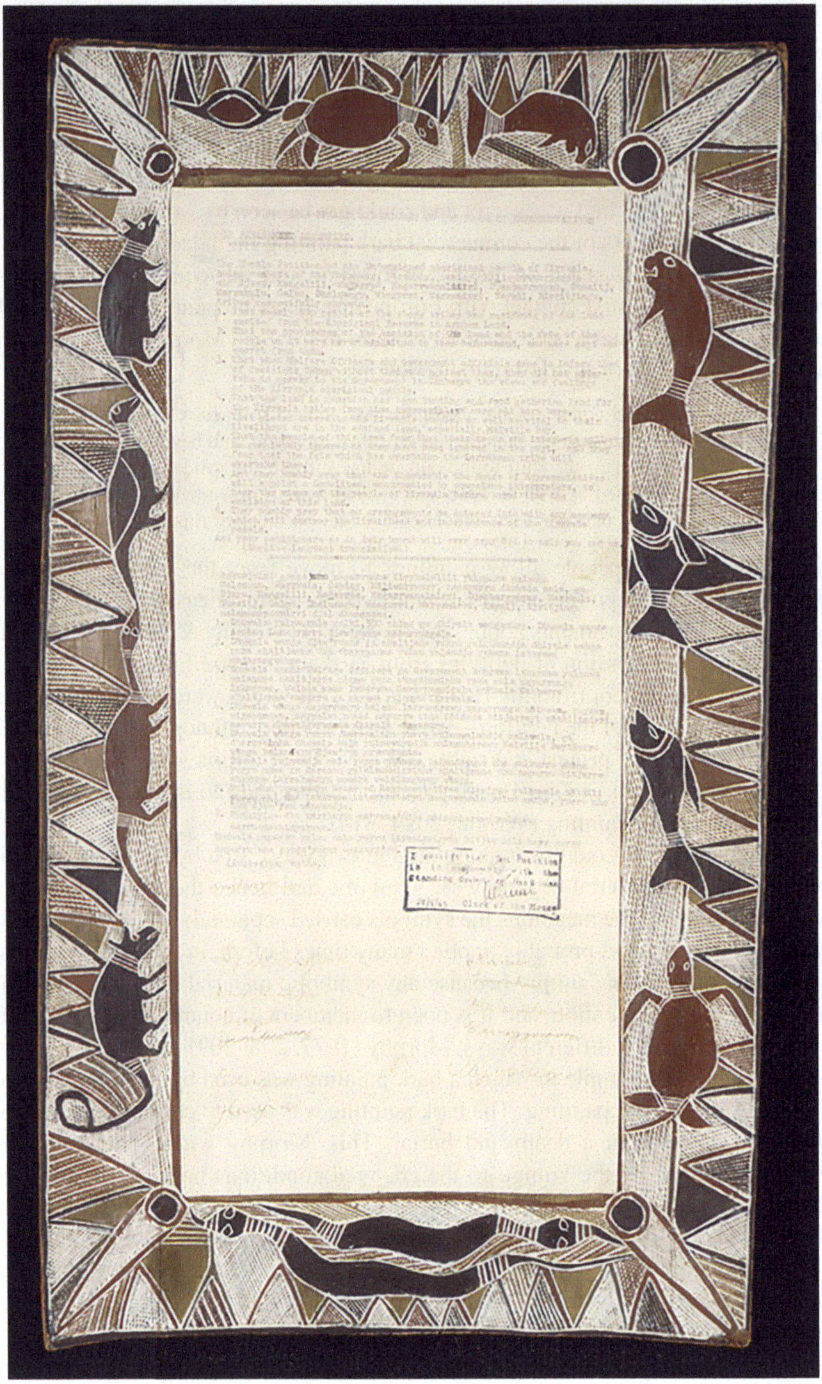

Fig. 9.9 Yirrkala Bark Petition (Yirritja) M02-0020.002. (It has a name in the non-Aboriginal world, but it is the name of white politicians.) "A visual text establishing the validity of Yolngu law" (Wright, 2024, 327). (With permission from Will Stubbs, Co-ordinator, Buku-Larrnggay Mulka Centre, Nhulunbuy)

with my comment about Ingold, it was not the art that changed, but the seers of it and the way in which they got to see it.

One of the features that emerged from Morphy's research with bark paintings in Yirrkala was that there are different levels of meaning for any painting: in particular, there are meanings in terms of myths and others in terms of aspects of the landscape real to anyone who cares to look. Morphy's informant Narritjin[8] said that "given a large enough bark on which he could represent everything, topography was the basis on which he would organize the painting" (Morphy, 1991, 251), though the interpretation in terms of such maps would only be accessible to the insiders. It is worth noting that one of the ways to represent scenes everywhere is through maps—but generally these have had to be very conventionalized representations of space (Utrilla et al., 2021). This view fits within the general belief expressed by the Yirrkala elders that "Country cannot speak for itself, so art must speak for Country" (Neale, 2020, 54) through the speech of those who are in the know. Neale is using the word Country as it is used by Aboriginal peoples of Australia to refer to their lands (Lewis & Rose, 1988, 54–62; Sutton, 1995), and that is not the same as the nation state(s) they live in.

Because Morphy's most detailed work took place after earlier work by other anthropologists, he was able to pay attention to the changes implicit over the previous 35 years. He has now discussed the further changes since he began to work with Yolngu at Yirrkala more than 25 years before (Morphy, 2007, 41–44): "There have … been significant shifts in the forms of art and contexts of ritual—but not a diminishing of them," including an expansion of the role of bark paintings and their inclusion among other examples of commercial "fine" art. The important point here is that variation is continually present in the cultural reproduction of the societies involved and that it is susceptible to the impact of commerce. It is particularly well-documented among the Yolngu but some of the flexibility that they have demonstrated has permitted the variation through time, not simply invention, in art traditions in many Aboriginal Countries in Australia.

We can understand something about this variation in cultural knowledge within Aboriginal societies by considering the example of the discovery of a Dreaming story among the Martu, in the Little Sandy Desert of Western Australia, documented by Tonkinson (1974, 85–86). In that case, Tonkinson witnessed the beginning of what he called a "dream-spirit ritual" when ten senior men exchanged notes to produce a single ritual. For this discussion to be productive, the original story must have had some connection to the other fragments of stories the several men knew. It is an important part of the way songlines were able to be constructed.

In most cases, such rituals did not survive for long, but in this case: "Over time, and with cultural transmission, the composers were quickly forgotten, and Aborigines of later generations came to regard them [the rituals—ID] as being Dreamtime creations" (Tonkinson, 1974, 85–86). One of the important things about

[8] Narritjin was one of the central figures surrounding the presentation of the Yirrkala Bark Petitions (Wright, 2024, e.g., 258–60).

forgetting the names of the people involved in recognition of the shared story is that it allowed for the appearance that beliefs were unchanging, even though particular individuals had the initial insight and discussed it with other known individuals so that they arrived collectively at a common story. It was regarded as not a "new" story but the successive revelation of something that had existed in earlier times and just been re-discovered. The sense of revelation was reinforced by forgetting the names of the group of individuals who agreed upon the story, and that, in turn, enhanced the sense that the song came from the Ancestors as the agents who did the creating.

In many places across Australia art sites with large numbers of engravings[9] or paintings (a small sample have been painted inside the grooves caused by engraving) distinguish those sites from the others with small numbers of engravings or paintings. One example of a large petroglyph site is Nganganterra in the Native Title Country of the Mulligan River People, dozens of dunes into the eastern side of the Simpson Desert (Fig. 9.10).

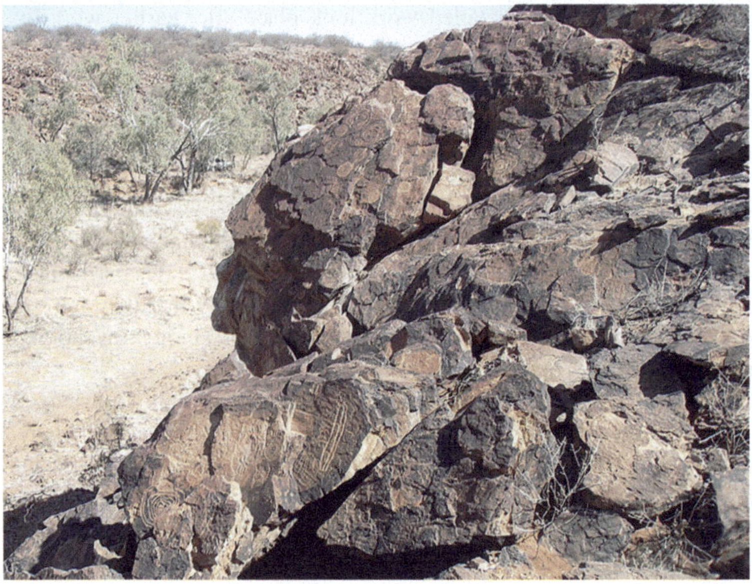

Fig. 9.10 General view of the engravings at a small part of Nganganterra rock art site on Mulligan River People Country in the Simpson Desert. Note the similarities and differences from Fig. 7.14. (Photo Iain Davidson with permission of the Mulligan River Aboriginal Corporation)

[9] Some prefer to call them "carvings" (Graeme Ward, 15 June 2024, personal communication) others petroglyphs. I just refer to two categories. (1) Those for which part of the original surface has been removed to make a new surface—engravings—though the same definition would apply to sculpting. (2) Those for which a new surface has been created by adding pigment—paintings.

The site is known for an image (Ross & Smith, 2009; Smith, 2013, Fig. 7.11) (No. 7 in Fig. 9.14) which has been compared to those called "Archaic Faces" in other parts of the country (David et al., 1992; Dix, 1977; Lewis & McCausland, 1987; McDonald, 2005; Mulvaney, 2015, 249–289; Smith, 2013, 252–254). Images called by that name have been said to be amongst the earliest images of human faces (Flood, 1997, 187–193), and thought (actually guessed, because none of them are well-dated) to be more than 14 thousand years old. Only Jalijbang 2 has a claim to a date, and the evidence suggests such faces are much more recent (David et al., 1992) (Fig. 9.11) and have characteristics (such as bodies) which make them unlike other "archaic faces". It is worth noting that the Jalijbang engraved human images are within the same group of shelters as paintings at Mennge-ya of more typical Victoria River Lightning Brothers painting style (Fig. 9.12) (David et al., 1994).

Mulvaney (2015, 249–289) argued cogently against giving all of the "Archaic Faces" images a young age on the grounds that many are in a heavily weathered state, and Smith (2013, 253) also argued against the recent age. Ross and Smith (2009, Fig. 1) compared the Nganganterra image with an image at the large rock engraving site of Carbine Creek, east of Dajarra (Isaacs, 1984) (Fig. 9.13).

More generally, Ross and Smith situated it on a map of other "Archaic Face" sites which are spread out across Australia from the Dampier Peninsula through Central Australia to these sites in the east of the distribution. Most of these sites lie in the arid zone of Australia on or slightly to the north of the Tropic of Capricorn.

Which motifs were in common among engraved images from Nganganterra and other sites? Mrs Hansen was very specific about the crossed circle (Fig. 9.14, 11) as a "Clever person" sign made when they "wanted to travel without being seen". Among the painted art of Arnhem Land (Brandl, 1973), identical signs occur in paintings of Rainbow Snakes at Magela Creek (p. 76, Fig. 161) and Deaf Adder Creek (p. 77, Fig. 164; p. 80, Fig. 170) where "all informants questioned about the possible meaning of the crossed circle—in some paintings it shows up as a spoked wheel—agreed at least by implication that this sign had some sacred-ritual significance" (p. 185). Love (1917) showed identical signs associated with two *Wanjina* figures in the Kimberley. We need to be careful, though, given the simplicity of the sign—Brandl warned "the step to mere speculation is short"—but the likelihood of special ritual significance is tantalizing.

Figure 9.14 top left shows a relatively recent sign of three lines superimposed on the circles on the engraved slab in Fig. 7.14. The sign is similar to another in Fig. 9.10 which is from a different part of the site, yet the two, much less-weathered, recent signs are very similar. The broader results of the selection of important images are shown in Fig. 9.14 and Table 9.1. Included in the Table are three sites in the Country of the Mulligan River People: Nganganterra, Mulligan River Site and Abudda Lakes. In addition, there are some of the images at the Dampier Peninsula sites of Murujuga (from my photographs but see Lorblanchet, 2018; McDonald & Mulvaney, 2023; Mulvaney, 2015); Roma Gorge in Central Australia (from images in Ross, 2003); Ewaninga in Central Australia (from my photographs); Flinders Ranges (Walsh, 1988); Mutawinji, NSW (Paakantyi Country from my photographs); Carbine Creek (from my photographs and Morwood, 1985); Selwyn Ranges,

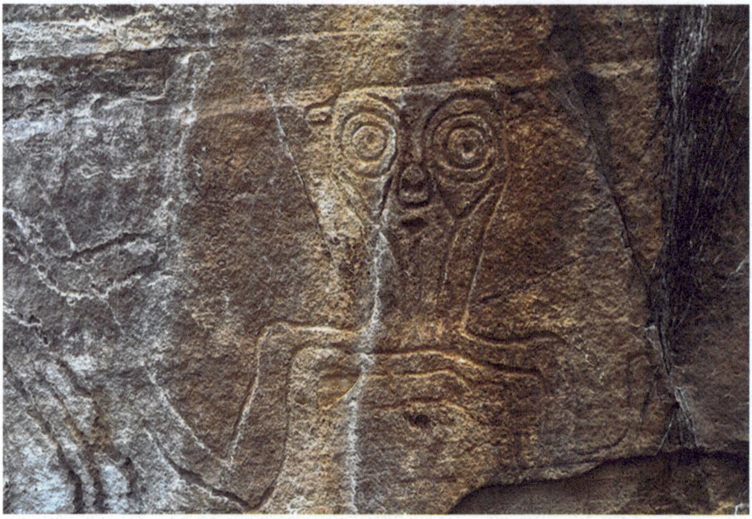

Fig. 9.11 Image of "Archaic Face" at Jalijbang, Northern Territory. (Photo by Iain Davidson, cropped. Permission from Traditional Owner Billy Harney)

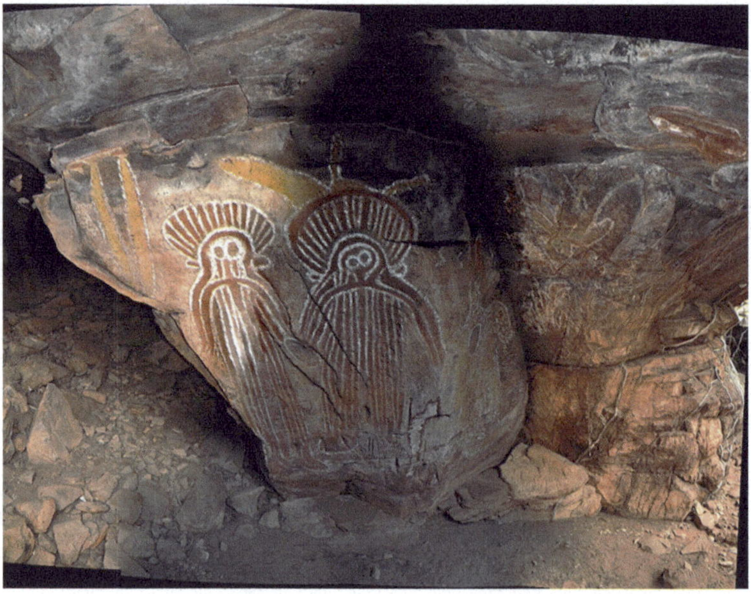

Fig. 9.12 Lightning Brothers and others in Mennge-ya rock shelter next to Jalijbang "Archaics". (Photo Iain Davidson reproduced with permission of Billy Harney, Traditional Owner)

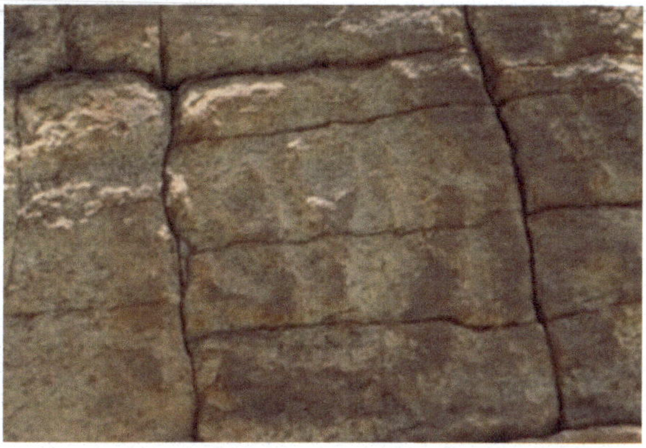

Fig. 9.13 Carbine Creek, near Dajarra, Queensland, showing the "Archaic Face" in the middle of the photograph. (Photograph Iain Davidson, January 2013)

Malbon Rock and the Cascades (Mitakoodi, Yulluna and Kalkadunga Country, from my photographs). As can be seen clearly in the images of multiple circles, there are several different weathering stages of the signs, and most people would take this as an indication that some of them are very ancient. The three lines which overlie the circles, though are quite recent.

There are other sites in Central Australia that include several large collections of engravings. Maynard (1979, 92) took the existence of those in South Australia and western New South Wales and their generally weathered appearance as an indication that such sites were the earliest ones in the sequence of Australian art (all over the continent). She called them "Panaramitee", after one of the sites in South Australia, and characterized them as being "dominated by macropod and bird tracks and circles, with a smaller number of crescents, groups of dots, human footprints, radiating lines, 'Tectiforms' or line mazes, and a tiny fraction of other nonfigurative designs." Maynard acknowledged that the identification of this set of sites as a distinctive style should be attributed to Edwards (e.g., Edwards, 1968) who found that the proportions of images of these types was remarkably similar between different sites. Mutawintji, a central ceremonial place in NSW (Beckett et al., 2008), has engravings generally attributed to the Panaramitee style.

The site I call Mulligan Art Site here[10] had fewer than 10% animal tracks and more than 90% non-figurative designs, mostly variations on a circle, so that the numbers put the site outside the range of Panaramitee sites discussed by Maynard (Davidson, 1983) (Fig. 9.15).

Galt-Smith (1997, 126–135) compared motif frequencies more generally between Panaramitee sites and Central Australian petroglyph sites and found that there was "no statistical basis for the Panaramitee style and no basis for including

[10] I called it Seymour Gap Site in 1983—reverting now to the original name is in recognition that the Native Title holders call themselves the Mulligan River Aboriginal Corporation https://www.mulliganriver.com.au/

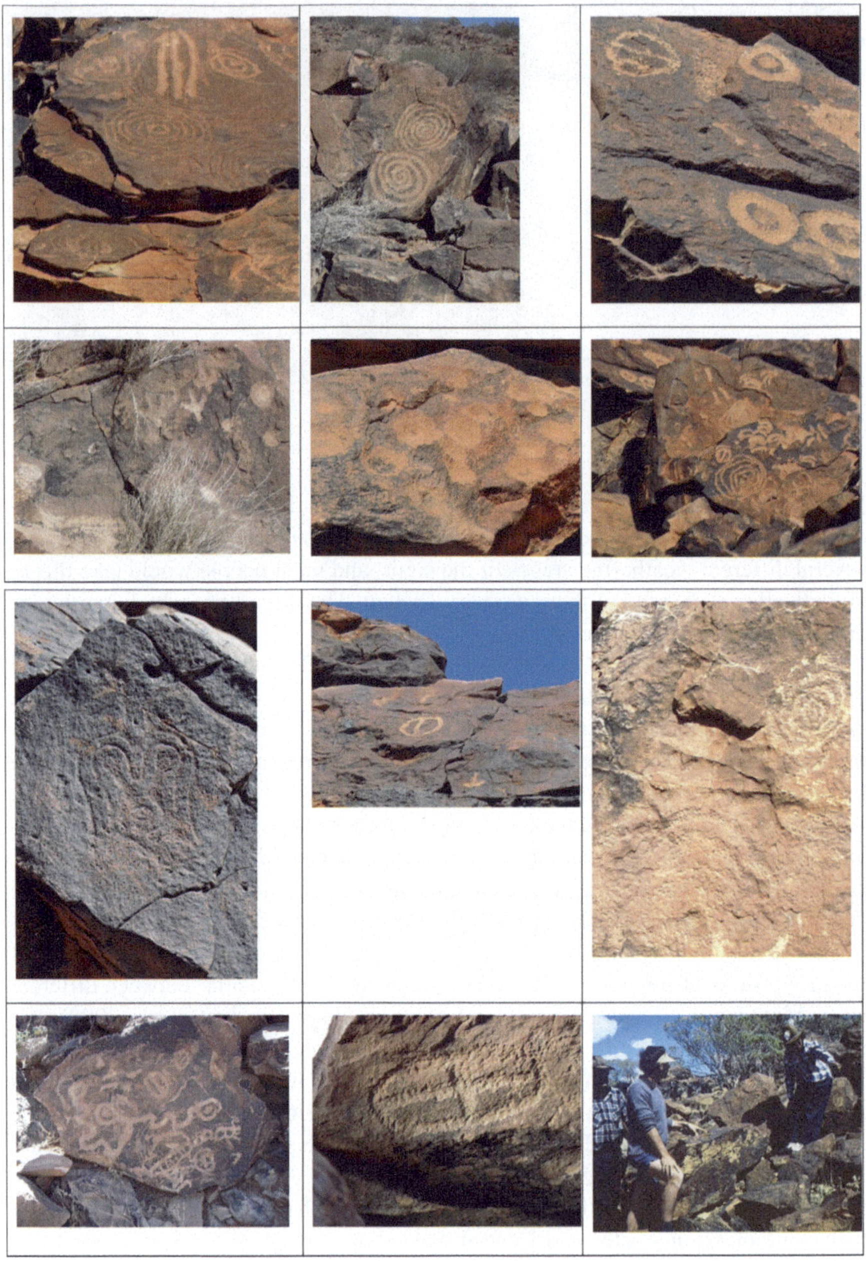

Fig. 9.14 Signs that occur at multiple sites across Australia along Seven Sisters Songlines (see Table 9.1). From the top left to right, then second, third and fourth rows also left to right: (1) Multiple circles (of at least three different ages); (2) Big spirals; (3) Circle and pit images, one alone at the top, two close together below, together with a much older singleton; (4) Trident/emu's feet, of different ages; (5) Pecked pits; (6) Small arcs and other images; (7) "Archaic faces";

the Central Australian sites within it." Despite some continuing to use the term (Franklin, 1991; Smith, 2013, 228–240), others have now expressed their skepticism that the label is useful in rock art studies, on the grounds that its definition is both too wide (for example, it does not make the distinctions I make here) and too narrow (it is restricted to one way of making art) (Rosenfeld, 1991). Rosenfeld argued that there are distinctive styles even in the earliest art, and Franklin's (2007) enumerations seem to confirm that judgment, because her statistical analysis groups together sites from related regions, even though it also separates those that are close together. That is to be expected unless the imagery was of the simplest form. In the case of one of the simplest forms, the circle, it could be restricted in the stories it could carry, but Morphy (1980) suggested that the meaning of the circle depends on the stories. The stories make the image complicated however simple its material form. I suspect that there were several different styles among the sites with engraved images, but they have not been teased out appropriately.

As I discussed in Chap. 6, Munn (1973) showed that (for Central Australia) graphic items such as circles could have a range of possible meanings (Fig. 6.5) which required either additional marks or songs/stories to differentiate them from their use in other songs/stories. Morphy (1980, 26) emphasized that in that context, the artist is speaking/singing about the iconic relation between the circle and the object involving the whole social context of, say, the domestic domain where circles may be (images of) vulvas, and straight lines may be penises. I have argued against any suggestion that the behavior of modern people is a simple guide to behavior in the past (Davidson, 1988), and that is not what I am proposing here and I will not go further about representations of sexuality. Rather, the images in Fig. 9.14, with the exception of those numbered 4 (tracks) and 7 (the possible "archaic face") are all marks that produce an expectation of meaning yet are indeterminate. Tracks are rare at the sites in the examples in Table 9.1, and "archaic faces" are quite abundant in Murujuga/the Burrup Peninsula (Mulvaney, 2015, 249–289) but are otherwise not abundant in Australia. The multiple circles in number 1 were made over many hundreds, maybe thousands, of years because they show several different episodes of weathering on the same surface, and the fact that in between there seem to be periods of no weathering may suggest that the production of multiple circles has a very long history.

The engraved images across Australia were either similar because the images were extremely simple—possible if the images were just single circles, but less

Fig. 9.14 (continued) (8) Barred circles; (9) Multiple nested arcs at entrance to Nganganterra; (10) Complex panels; (11) "Clever person sign" (Mulligan Art site); (12) Mrs Hansen explaining rock art to Ramón Tarragó and Jim Dwyer 1982 (Mulligan Art site). Unless otherwise indicated, all photos taken at Nganganterra Art Site with permission of Isabel and Avelina Tarragó, Traditional Owners for the Mulligan River People. Seven Sisters story was documented by Luise Hercus using the stories of Mick McLean. They include stories about the Wapiya girls who began their journey at the Paltyura/Pulchera Waterhole on Ethabuka Station which is part of the Mulligan River People's Native Title (Hercus, n.d.). (All photos by Iain Davidson)

Table 9.1 Shared signs at sites across Central Australia

	1 Multiple circles	2 Big spirals	3 Circle and pit	4 Emu's feet	5 Pecked pits	6 Small arcs	7 "Archaic faces"	8 Barred circles	9 Nested arcs	10 Complex panels	11 "Clever person sign"
Pilbara, WA (Murujuga)	Y	Y	Y	Y	Y		Y		Y		
Central Australia, NT (Roma Gorge)	Y	Y	Y		Y	Y	Y			Y	
Central Australia, NT (Ewaninga)	Y	Y	Y	Y	Y						
Flinders Ranges, SA	Y			Y					Y		
Mutawinji, NSW			Y	Y	Y				Y	Y	Y
Abudda Lakes, Qld	Y	Y	Y	Y				Y	Y	Y	Y
Mulligan Art Site (Qld)	Y	Y	Y	Y	Y	Y		Y		Y	Y
Nganganterra, Qld	Y	Y	Y	Y	Y	Y	Y	Y	Y	Y	
Carbine Creek, Qld	Y	Y	Y	Y	Y		Y	Y	Y	Y	
Selwyn Range, Qld	Y	Y	Y	Y		Y		Y	Y		
Malbon Rock, Qld				Y					Y	Y	
Cascades, Mount Isa, Qld			Y					Y		Y	

Numbers refer to those used in the caption of Fig. 9.14

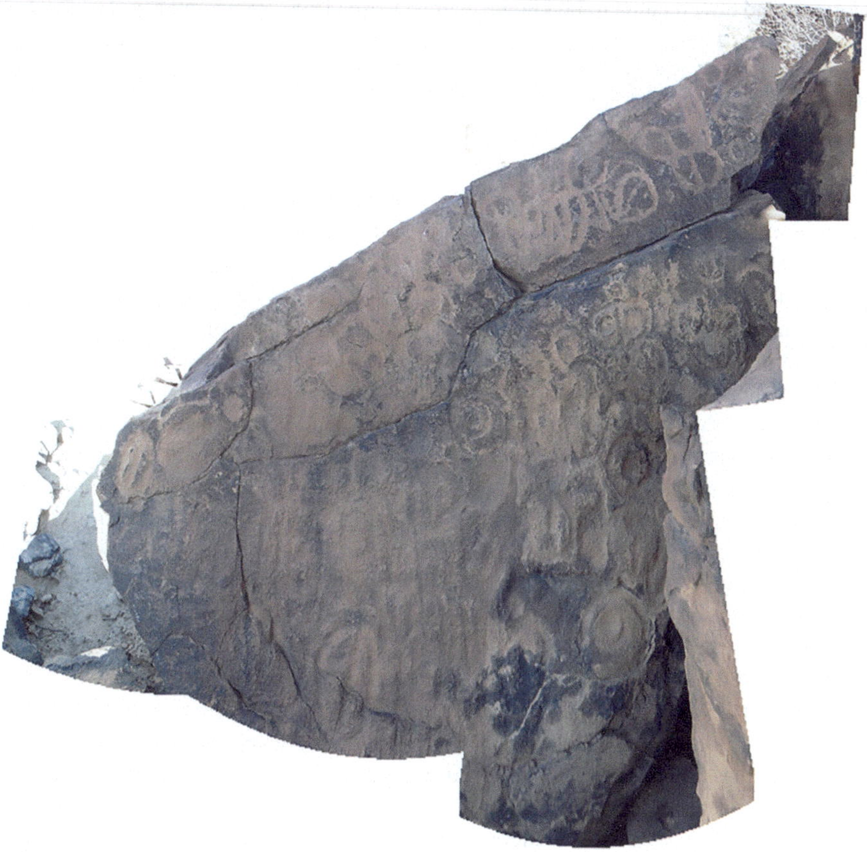

Fig. 9.15 Large panel beside the Mulligan River covered with multiple layers of engraved figures. There do not seem to be any images of animals or their tracks. (Photo Iain Davidson with permission of the Mulligan River People)

likely if they were, for example, barred circles—or, if they were more complex, the similarities are a result of shared meanings. Their complexity—multiple concentric circles or spirals—meant that there were linkages between the people who made them which were probably social. If the multiple circle panels are very old, that may mean that the linkages, too, were very old. Any meaning they may have had depended on people sharing a code of some sort which identified the art and the place. Such code would be transmitted among them in story and song, creating, as they did so, the songlines that are such an important part of the way Aboriginal people now and historically relate to each other and to Country.

We are constantly reminded in Australia (e.g., Jones & Russell, 2012) that the memory of places (Gibson, 2013; Hercus et al., 2009; Russell, 2012) is absolutely central to the construction of ideas about the landscape for the people who inhabit it (Hayward et al., 2018). Such songs and stories are essential to the understanding of

Country and how it came to be known as it is. Rose (1996, 7), an anthropologist who worked with Aboriginal people in the Northern Territory over many years, wrote about Country in the following terms: "People talk about country in the same way that they would talk about a person: they speak to country, sing to country, visit country, worry about country, feel sorry for country, and long for country. People say that country knows, hears, smells, takes notice, takes care, is sorry or happy". Rock art was part of that process by which people could structure their knowledge about the places most familiar to them and identify with it (e.g., Brady et al., 2023). Darrell Lewis and Deborah Bird Rose (1988, 65) put it: "…many art sites present particular difficulties with respect to consultation. Most of them are localizations of Dreaming power and information. In Dreaming sites (many of which are also art sites), numerous factors of cultural significance converge."

My suggestion is that people used the social connections stored in their Songlines, together with the cultural familiarity provided by production and identification in rock art of common non-representational images and the codes that allowed them to be given meaning. This common knowledge and coded meaning allowed people to move from regions of resource stress of one sort or another in their Country to regions where the hazards were different in the Country of others to whom the Songlines gave them connection. The shared meanings derived from the need for a common code to decipher indeterminate shared signs and were part of the social linkage between one place and another. In consequence people from different Countries were able to gather in groups who could engage in ceremony.

The central theoretical core of this argument is that the risks and hazards of extremely variable environments pose problems for survival and in societies without writing must be dealt with through information sharing (Minc, 1986) and ritual (Ross, 2013; Ross & Davidson, 2006). In particular, Aram Yengoyan (1968) argued long ago that some of the distinctive Australian rules about kinship (section and sub-section systems) were particularly suited to encouraging gatherings of people where they could find mates appropriate to those rules. Otherwise, in many cases, of people living in small, isolated groups in resource-poor regions there would not be enough others to find mates. Common songs and common codes about common artistic images brought people together for the continuation of society, and (for the European Upper Paleolithic) Conkey argued that such "aggregation sites" could be identified from the variation in the art on fragments of bone at a suite of sites (Conkey, 1980). We will return to the topic of aggregation sites in the section on Storytelling. These studies demonstrated that the location of art sites in the environment had an important impact on the social behavior of the groups who used it. Some of this variation arose because of the incorporation of stories associated with the art into the broader cosmology associated with Songlines that linked people in one part of the continent with those in another (Davidson et al., 2020). The power of the places at individual centers together with their linkage to other places through the coding of the images went some way to overcoming the tyranny of any individual place. The same stories and songs would be told in different parts of the Country. I suggest that this combination of shared songs/stories associated with coded imagery has something in common with the fixing of stories through writing and later printing.

9.6 Reducing the Indeterminacy

Morphy (1980, 30) argued "that a major distinction can be made between works of art that are intended to be interpreted on the basis of perceived formal resemblance between signifier and signified in a way that can be predicted by the producer, and works which are not intended to be interpreted on that basis." In doing so he was making the distinction I recognize here between marked places and those where producers could reduce the indeterminacy of the marking. As I described in Chap. 6, the currently earliest painted caves, those in the Maros in Sulawesi, such as Leang Bulu' Sipong 4 (Figs. 2.12 and 2.13) show a scene dated to more than 50 thousand years ago (Oktaviana et al., 2024). This seems to show some creatures in human guise, but with non-human heads—composite figures of animals and humans that get called therianthropes—hunting an *Anoa* (Aubert et al., 2019). Not only are these among the earliest paintings to survive (anywhere), but they are among the earliest scenes. Another scene has been dated to at least 51.2 thousand years ago at nearby Leang Karampuang (Oktaviana et al., 2024). I have argued before that the emergence of scenes made a critical difference to art because the art could then stand alone without the need for codes of meaning, or songs and stories of connection (Davidson, 2021). As I noted before, that does not mean that the songs and stories are informed by "knowledge that is untainted by time" (p. 18), and it does not mean that in figurative scenes there were no such codes or stories, just that the pictures could stand alone. Indeed, in these scenes from the Maros, there would need to be songs or stories about therianthropes. Kelly et al. (2021) go further and suggest that in the Wardaman Country of the Northern Territory, individual images did not stand alone, but were related to each other through the songs and stories.

Think of the difficulty of identifying the individuals, even those emphasized by gold circles around their heads. In Giotto's picture (Fig. 8.30) from Padua, if you were not part of the culture that knows the specific myth or even the type of myth represented there, you would not know. We take it for granted that the idea of a gold circle around the head has a meaning (some would say that it represents a halo and is an indication of holiness, though such things have never been observed and are only a convention of Christian art), but we have to acknowledge that it is a cultural convention we learn. Of course, if it is not a convention you, the reader, have learned, then my point is simply reinforced. Duccio, of course, recognized the difficulties by labelling some of his paintings (as I discussed in Chap. 3). The figures would stand alone as pictures of old men, but we can identify them as agents from the captions especially if we remember a particular bit of a story shared among people who used a place like that one. We can only link different sites in the landscape with each other, and with the landscape, if such figurative scenes are informed by stories—non-figurative signs required the stories to an even greater extent. This linkage of people, art and landscape seems to be the exemplification of the linkages Warner defined for totemic relationships among them. "Through this engagement, the mythology allows the rock art images to be agents in the landscape and not just in the rock art site" (Davidson & Nowell, 2021, 12), but that agency, we should know by now, was exercised through people and not as a result of the agency of any inanimate object.

There is another issue, though, and that is the question of figurative images in the art (which I have just smuggled in). So far, I have discussed non-figurative images, songs, codes and connections. Even with figurative images, such as the anthropomorphs of the Selwyn Ranges, there was a need for some sort of coding which would best be preserved and passed on in story or song. Why are some images more like people and some more like other animals (as discussed in, for example, Davidson et al., 2005)? Was there a move from non-figurative to figurative to remove the indeterminacy, and what is the implication of the emergence of a capacity to blur figurative and non-figurative which would maintain the mystery? There are two aspects of this which provide a way forward: one is the use of hand (and other) stencils as persisting marks on rocks (as I discussed in Chaps. 5 and 6) (Fig. 9.16); the other is the question of the transformation of signs, as I have discussed before (Davidson, 2013a, 40).

Stencils indicate that a person has marked a place either with a representation of themselves or another, or with a stencil of a familiar object, and some such marks might even be a way of memorializing particular places (Hayward et al., 2018). Any marking that is more than simple, or maybe any marking, still requires a song or story. But the important point here is that the stencil marks a place in a way in which not all of the meaning is indeterminate. Seers could still identify that a mark had been made, and they could identify that it was made in the shape of a (human) hand or other object. The indeterminate part is always the question why the mark was made.

There are three categories of marks of hands—(1) negative hand stencils, (2) prints of hands made by pressing the paint-covered hand against the rock (or other surface), and (3) the shapes of hands painted without using a hand as a template (sometimes with inaccurate results). Sometimes, negative stencils were infilled with designs (Brandl, 1973, 31, 169; Chazine, 2005; Walsh, 1988, 120–1). All three categories are found in some locations, as they are outside Australia at Chhattisgarh in India (Dubey-Pathak & Clottes, 2020). Tsang and colleagues argued that hand stencils "can have multiple meanings over time and space" (Tsang et al., 2023).

The stencils are shapes of human hands, most commonly made by placing a hand against the rock and blowing or otherwise applying paint to make a negative stencil. They are often among the very early markings in several continents (Sahul, Sunda, Europe). Warner (1964 (1937), 505) described how the Murngin[11] around Milingimbi in Arnhem Land made "favorite paintings" of the human hand: "first covering the palm of a man's hand with red ocher and then pressing it against the rock wall with the fingers spread. The painter then fills his mouth with a white or yellow wash and blows it around the hand to form a white outline around the red palm and fingers."

The early art from Lubang Jeriji Saléh in Borneo (then part of Sunda) included hand stencils at least 40 thousand years old and a painted figure of an animal at the same date (Aubert et al., 2018b). In all likelihood, the early stencils in Borneo are an indication that an expectation of meaning was created in Sunda. Hand stencils are found in many places in Sahul, including Australia (Walsh, 1979; 1988) where stencils were also used inventively to produce fragments of lines (Walsh, 1983) or other images, such as the "tridents" or "emu feet" composed by stencils of three

[11] Now better referred to as Yolngu (Sutton, 2009, 168).

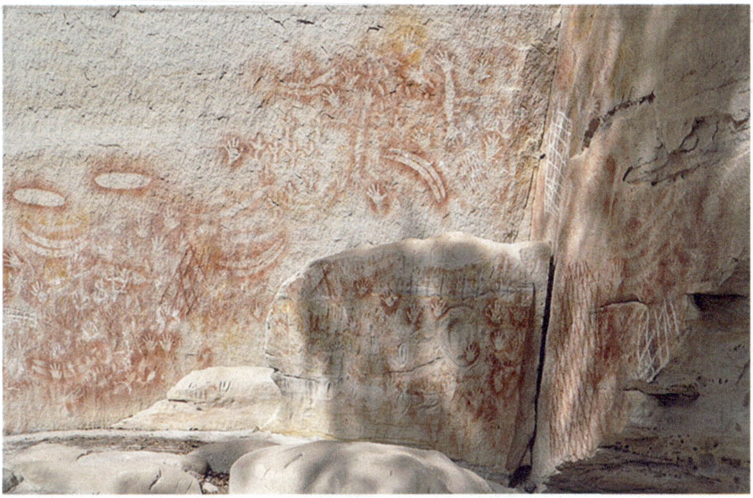

Fig. 9.16 Main panel at The Art Gallery, Carnarvon Gorge, Queensland, showing hand stencils, stencils of shields, boomerangs, "emu feet" composed of stencils of tips of boomerangs, grids or nets, and a human figure on the right. There are also incised engravings which have been called "vulvas". (Photograph Iain Davidson. Permission from Fred Conway, Traditional Owner)

boomerangs at Art Gallery in Carnarvon Gorge, Queensland (Fig. 9.16). Similar inventiveness was shown in Chauvet Cave in France where the palms of painted hands were pressed onto the cave wall to make a design of "large dots" (Baffier & Feruglio, 2001, Figs. 58–61; Clottes & Geneste, 2012, 588).

The technique of stenciling was used to produce images of human feet, boomerangs (of different sorts as shown here), shields (top left in Fig. 9.16), as well as those of models of animal tracks (Brady et al., 2020); rare objects (Hayward et al., 2018) such as dilly bags, armlets and spear-throwers; birds (Taçon et al., 2010); baler-shell pendants (Beaton & Walsh, 1977) (like that in Fig. 9.17); and an example of a bone dagger (Tsang et al., 2022a). Baler-shell pendants were identified in Carnarvon Gorge but have also been found in the Selwyn Ranges 900 km to the northwest in the Country of the Mitakoodi in whose Country they were reported by Roth (Beaton & Walsh, 1977; Roth, 1897, 112) (Fig. 9.17).

The amount of the arm is variable, and on occasion (as shown at The Tombs site and other sites in the Central Queensland Sandstone near Injune in Walsh, 1988, 118–9) a whole body was stenciled (Fig. 9.18). Mulvaney (2011, 109) described his sense of high emotion recording the Tombs site noting that of 12 handprints around the stenciled body 9 lacked a finger. Morwood (1984, 366) suggested that some hand stencil variations represented the way in which hands were used in sign languages across Australia (as shown in Roth, 1897, Ch. IV and Plates II–X). If that were true, some hand stencils carried meaning that was not indeterminate—provided the seers were familiar with the codes of hand gestural language.

Stencils are less common in Africa (mostly prints, not stencils, in SW Africa according to Manhire, 1998) though there are stencils in Algeria (Coulson & Campbell, 2001, Fig. 205). They occur in the Americas, and they are particularly

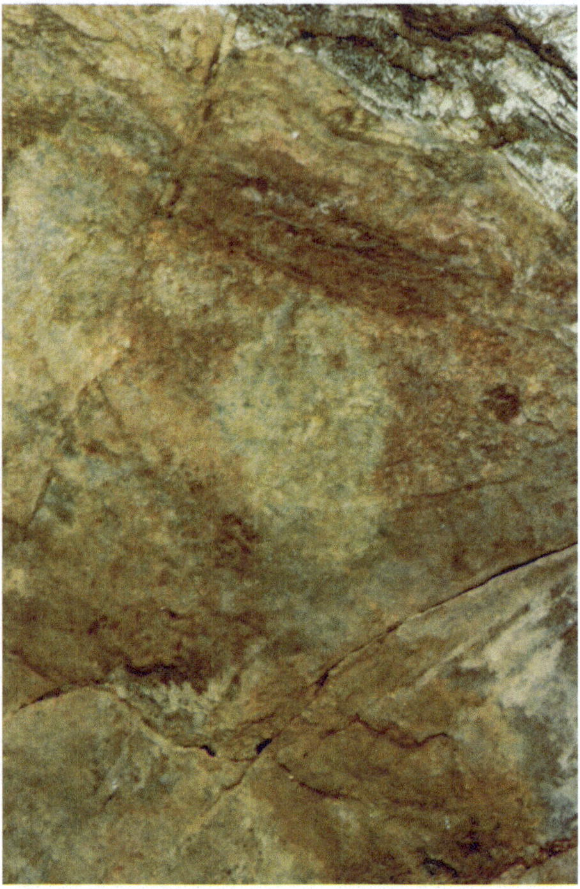

Fig. 9.17 Stencil of Pendant at Site 2 in the Selwyn Ranges (see also Fig. 2 Right). (Photo Iain Davidson, permission from Pearl Connelly and Fred Turner of the Mitakoodi Native Title organization)

well-represented at Cueva de las Manos, in southern Argentina (described by Aschero & Schneier, 2021) (Fig. 8.7). There are hand prints in Montana and Alberta (Keyser & Klassen, 2016, 151). The discontinuous distribution is probably an indication that there was no diffusion of the habit of making marks with meaning.

Anthony Forge (1991, 40) argued that hand stencils should not be considered as art at all. He was an anthropologist—a social anthropologist—of art. That is to say that he was someone who studied the social context of the art of modern people who were culturally different from him. His argument depended on the view that hand stencils "don't represent anything, they are an impression, mechanically reproduced from a natural part of a single cultural individual, or object." Forge objected that stencils "are not mediated through any symbolic system, they are not a part of culture," though he went on to point out that the variation in the times and places they were used, and whether hands or other objects were stenciled, meant that the marks were part of a culture (despite what he said). A recent study pointed out that hand stencils from the Auwim people of New Guinea were part of the process of constructing social narratives (Tsang et al., 2023). I believe that Forge's error here was because, as an anthropologist of art, he was primarily interested in whether the

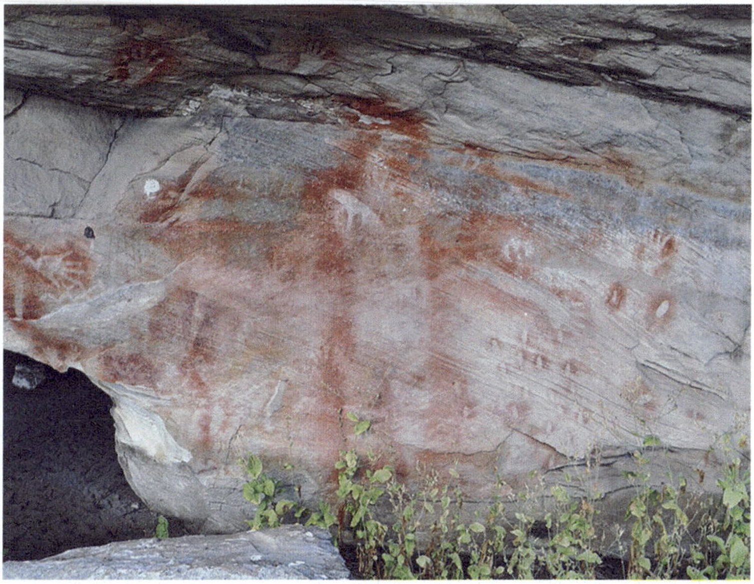

Fig. 9.18 Whole body stencil and other stencils in The Tombs site, Carnarvon Gorge (Mulvaney, 2011, 109). (Photo Iain Davidson. Permission from Fred Conway, Traditional Owner)

marks themselves should be considered as "art objects", as "works of art" rather than as part of the art-system which also involved people. It is easy to say that something is not part of the culture if you ignore the evidence that it is.

Forge might, instead, have taken his legitimate concern about the relationship between hand stencils and works of art to lead to a more original understanding of the systematic networks between people and material culture that define what art is. He could have been concerned about the artistic nature of the whole system of relationships between artist and seer, subject and object—including the location where the stencils were made as an intrinsic part of the "object". By the choices artists make about the relationships in the art-system, they make marks such as hand stencils intrinsically symbolic; in this case the choice is about where to place the mark, as well as how to do so and what it is.

More than 100 years ago, Charles Peirce, one of the principal philosophers of the concepts of symbols, icons and indexes, claimed that symbols are signs which "represent their objects, independently alike of any resemblance or any real connection, because dispositions or factitious habits of their interpreters insure their being so understood" (quoted in Nöth, 2010). Thus, on any interpretation, the stencil represents something more than just the hand (and arm) and generally that would include representing a person (synecdochally) by the stencil of part of their body—not necessarily a particular person (or even species of hominin), but a person, nevertheless. Seers understand that connection (simple and trivial though it may be) because of their "dispositions or factitious habits"—by which we can understand that those are part of their set of acquired cultural values. For that reason, I argue that negative stencils, hands or other things, are signs which are both arbitrary and conventional.

The artist made a choice to stencil a hand and an arbitrary length of the arm in a particular, chosen location.

In terms of our argument here, the point is that the stencil was a mark that could produce an expectation of meaning—any seer would understand that a human (or hominin, perhaps) was involved and may have had a reason for leaving a mark. Moreover, the meaning of most marks would actually be indeterminate. It might be possible through the ingenuity of an analysis to determine age or sex of the person who left the mark (as argued, for example, by Nelson et al., 2017; Snow, 2013) though there have recently been some suggestions of the need for caution (Fernández Navarro et al., 2024). For the average seer that would not be possible, and there would probably be no control against which to check identity of the person who left it—was it left by Lok or Liku, Ayla or Creb?[12] In the example from Maltravieso in Spain, prior to the dating (Hoffmann et al., 2018a), supposing it is to be believed (Aubert et al., 2018a; Hoffmann et al., 2018b), no one suspected that the stencil might have been made by a Neandertal rather than a person. If you cannot determine even the species responsible for a stencil of part of a hominin we can be certain that the meaning was indeterminate.

Hand stencils, then, are examples of the sorts of marks that humans (who had chosen to have a mouth full of a slurry of powdered ocher) might leave on a rock. The signs might not have been a work of art to anyone who saw them, but they showed that others had been there and had made a choice to leave a mark there. They could generate an expectation of meaning, yet the meanings of the marks themselves were generally indeterminate—unless there was an associated story.

On the evidence, and leaving aside the Neandertal stencil from Maltravieso, "the earliest known figurative depictions and hand stencils in the world" were made in Borneo and Sulawesi (Langley et al., 2019, Table 8) with dates earlier than 40 thousand years ago on both sides of the Makassar Strait that has always separated the two islands (there is one shown in Fig. 6.8). Relative to the number of rock art sites dated, there is a lower proportion of sites with hand stencils in Sahul (Langley et al., 2019, Table 3). There are, however, several hand-stencil sites in Tasmania, notably Ballawinne and Keyhole Cave, probably older than the 14-thousand-year-old rise of sea-level that isolated Tasmania (Lambeck & Chappell, 2001) as well as others that are probably more recent and made after the Tasmanians were separated from other people on the mainland (Dix, 2016).

The hand stencil is an image of a hand and the boomerang stencil one of a boomerang. They are, in the language of semiotics, indexical of a person in the first case, and of the material culture of a person in the second. It is a different matter why they are there. A song, story or code is still required to go beyond the superficial: why them, why then, why there? The important point is that by marking the rock with stencils of bits of people or of things that have a common familiarity, the process of transformation of signs could begin. The indexical mark, left in a way for seers to know that a mark had been left, became an iconic mark which allowed seers to know that a person with or without that material culture had been there. Single

[12] These are names of relevant characters in *The Inheritors* (Golding, 1950) and *Clan of the Cave Bear* (Auel, 1980), both novels about the late Paleolithic in Europe.

lines (e.g., the lines in Fig. 9.19 from a building in Salamanca, Spain, which echo similar examples of parallel lines in cave sites or on rocks elsewhere, which some call "abraded lines") might be indexical of human activity. Repetition of patterns, say of crossed lines as at Blombos, must be indexical of activity that can only be attributed to human agency. Once this transformation had happened, the option to make marks symbolic was open. This allowed the producers of the marks to spell out the meaning, which may not be iconic, but was indexical, with two meanings, one that people were there, and one which was the reason they made the marks. Seers had an expectation of meaning.

In the process of transformation of signs (Davidson, 2013a), non-figurative marks became indexical because of their common associations—they indicated that a particular person had made a mark of a particular type at a particular time in a particular place, so that there was a story to be told. The move to figurative marks could come about when such indexical signs became iconic (having some resemblance) through gestalt or pareidolic processes. Some account of the neurological processes involved has been given by Derek Hodgson (2013) and Robert Bednarik (2017, 101) wrote:

> Pareidolia is a form of apophenia …, the human tendency to perceive meaningful patterns within random data (Brugger 2001). The "abnormal meaningfulness" defining apophenia is neurologically rooted in the ability of the brain to sift through the mass of sensory information received to detect significant signals…

The Chauvet rhinoceros (Fig. 7.2) is an example of the need for gestalt cognitive processes to complete the figure. The Makapansgat cobble (Fig. 7.5) is an example of pareidolia in which the human mind, nowadays, recognizes the resemblance between the natural object and something such as a human face, for example in a tree trunk (Fig. 9.20). Australian "Archaic faces" (see Fig. 9.14—assuming the identification is correct) are also examples of pareidolia (Davidson, 2013a), except that they were undoubtedly produced by the artist, and it is the seers who experience pareidolia when looking at them. Bednarik's (2017) review found that modern humans record marks as art, although he could not similarly identify them.

In my previous treatment, the final stage of the production of pictures involved: "5) marking with accidental consequences, as in the dot-and-circle panels in Australia; 6) marking with ritual story telling; 7) marking without ritual but with meaning." In Fig. 9.14, the photo at top right gives an example of how dot-and-circle marks (number 3 in the caption) could accidentally come to have pareidolic properties. The top image would not have them by itself so the likelihood is that they were produced during ritual, but the two images at the bottom on the right (there is a much older example on the left) have been executed so close together that human cognition could recognize the pareidolic effect of iconic representation of eyes. Others have suggested that eyes are not the only anatomical elements that can be conjured from dots and circles occurring in pairs, but that is less frequently argued. The appearance of representation could, then, have been an accidental by-product of natural irregularities of shape.

In some sense this process mimics the comical conversation between Hamlet and Polonius discussing shapes in the clouds about which they could disagree, except that in these instances the shapes do not change with the wind (Davidson, 2013a,

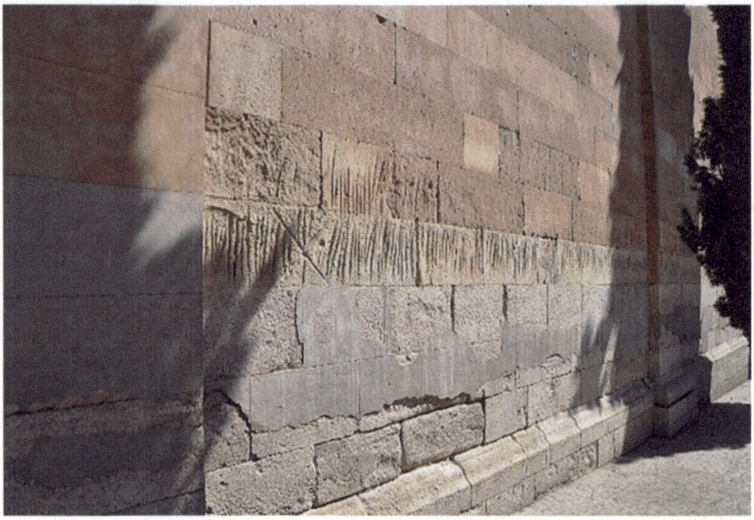

Fig. 9.19 Sandstone wall in Salamanca marked by parallel lines scored in two rows. Similar sequences of lines, which get called "abraded grooves," can be found in rock art in different parts of the world (as shown in the Victoria River District by Walsh, 1988). (Photograph Iain Davidson)

29–30). The pareidolic images remained fixed on the rock, and it is the persistence that allowed ritual storytelling to give them meanings that could be shared.

Pareidolia has now been discussed at length by others (e.g., Wisher et al., 2023) when talking about people producing images of animals in the painted caves of Cantabria. For the argument here, the important point in their account of pareidolia is that the artist has an important role. Even if it is possible for one person to see animals in the shapes of the cave, that resemblance does not guarantee that another person can see the same animals, see them when the artist was not there, or see them tomorrow. What was needed, always, was the storytelling. Because you cannot paint on clouds, but you can paint on cave walls, the producer needed to make marks that persist on the cave wall and then tell people about them so that the seer can see the animal the artist saw, perhaps guided by the story of the artist.

One of the reasons that much rock and cave art resonates with modern people in many places is that we expect certain sorts of resemblance. Chauvet Cave is well-known partly because some of the producers, nearly 40 thousand years ago, used modern-looking techniques of presenting visions of the world out there, and, of course, these were capable of triggering gestalt responses such as those to the disconnected marks in Fig. 7.2 which we shape up into the outline of the head of a rhinoceros. The outcome was that the "images of animals" seemed to be "living animals" which had agency towards other such animals, and seers could construct stories about that agency and interaction—whether or not they were the same as the stories of the producers.

You do need the image to be associated with a story, though. Morphy (1991, 155–158) described a limited experiment to test whether figurative images were recognizable as figurative when seen out of context. He redrew images from

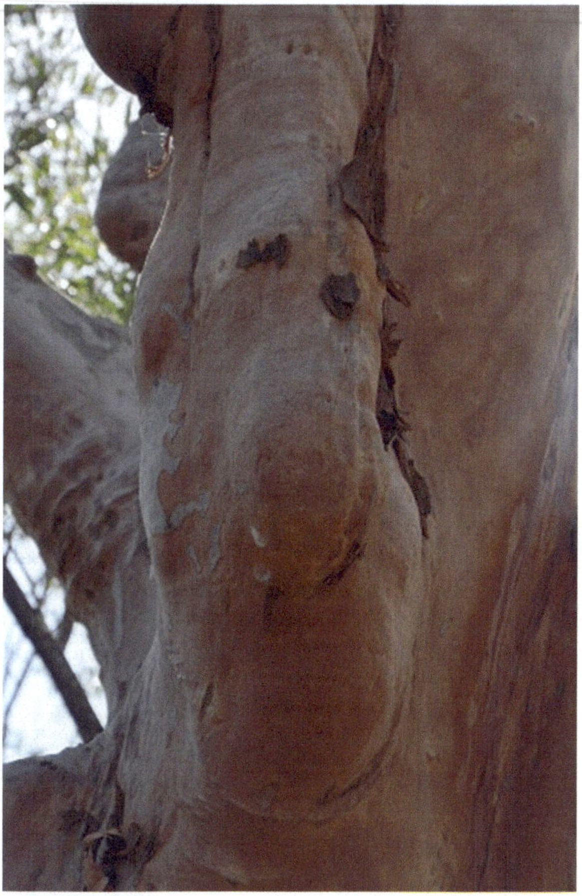

Fig. 9.20 I may be the only person who sees the marks on this tree as being like a human face, but it is an example both of pareidolia and of the need to have a story to account for the perception of resemblance. I told my companions my story about the importance of pareidolia in the history of art, particularly when pareidolia is accompanied by storytelling. (Photograph Iain Davidson)

photographs of bark paintings so that they were removed from all aspects of context. When he asked people about the animals shown in the drawings they identified about half of the species in the drawings correctly. We can suppose that in that case, naming otherwise indeterminate figures might have been enhanced by telling a story or singing a song. On the other side of Arnhem Land, Smith and others (2016, 1624) revisited the site of Doria Gudaluk, already well known for making archeologists think about the images at rock art sites, when the site was known as Beswick Creek (Macintosh, 1977), and concluded that "meaning cannot be obtained without the assistance of informed people." Lest we fall into the error of thinking that such a conclusion is exceptional within a body of art, consider for a moment our difficulty of getting any interpretation of the images in Breedon-on-the-Hill (Fig. 6.10) or the Giotto picture (Fig. 8.30), if you are not already/still part of the culture. The full elimination of indeterminacy requires interpretative songs or stories in cave art, Aboriginal art, church art, art of the last 300 years, or modern art. What was Goya's dog "thinking" as it sank in the quicksand (Fig. 9.21)? Was it even sinking, or would it climb out? Was it even thinking? We can tell Goya painted a dog, but not really much else without the context of the story.

Morphy and Layton (1981, 58–59), writing about the Yolngu in eastern Arnhem Land, detailed the processes by which the rights to produce or to sell particular bark paintings shifted over the years within a system of shared rights and responsibilities. That shift included slight changes in the content of the images which at one stage saw figurative art increase 20 times, presumably in response to the market. Equally, there were decisions to allow paintings to be made public, while withholding the complete set of stories. There were stories at different levels of society. Release of the images did not reveal all levels of the story to those who had not been initiated into them. Art is neither static, nor are the rules by which it is produced.

These two tendencies achieve the same objective. On one hand they reduced the indeterminacy associated with non-figurative representation, and on the other they expanded the range of the stories associated with art—some were exclusive and restricted, others were more secular and open to everyone. Different groups of people knew different stories depending on their status, and at the same time defined their status. This example shows that the indeterminacy may be eliminated by the artist, but the artist may still be sensitive to the seers who want to know the meaning.

Ingold (2000) has taken this further by pointing to two oppositions shown in bark paintings he illustrated. First, the images on a bark painting may have a superficial interpretation—in his example, a kangaroo and a human figure apparently spearing it. Second they may have a mythological interpretation—the kangaroo is an ancestral being which reveals the relationship between people and the landscape, and the

Fig. 9.21 Goya's dog. (From Museo del Prado, Madrid. Wikimedia Commons. Public Domain)

human figure would then be a *mimih* spirit figure. This is a recurrent theme both in bark paintings, at least one of which was collected by Baldwin Spencer (Carroll, 1977, Plate 2), and in rock art (Carroll, 1977, Plate 14). In the view of Ingold, the *mimih* is neither hunting nor butchering but inspecting the landscape as if it were a map of relationships (cf. Morphy, 1991, 221–2). Ingold's authority about the *mimih* is Luke Taylor, who elsewhere has written of the recent expansion of a market for sculptures of *mimih* (Taylor, 2005).

Ingold pointed out that in Central Australia the paintings of the Pintupi (best known by their use of dots and circles) (Perkins & Fink, 2000) did not (generally) include figurative images but concentrated on the imprint of the activities of Ancestors on the landscape (Ingold, 2000, 119–20). Smith (1992) interpreted the same differences in terms of environment, the need to bond to neighbors in harsh environments (Central Australia) and the desire to bound your country in fertile environments (Arnhem Land). In both cases, of course, the difference depended on what the artists and the community said to each other about their social environment and how they might use their belief system by representing it in the art.

The point is, though, that these works, whether Arnhem Land barks or Pintupi paintings, stand alone as works of art without our knowing the mythology they derive from nor the complex relationship between the mythology and the image—just as we do not need to be adherents to the same faith as Giotto to appreciate that his work was and is art, albeit from a time earlier than the mavens would recognize "Fine Art".

9.7 Overcoming the Tyranny of Distance and Place—The Importance of Persistence

In Chap. 6, I suggested that, with the emergence of writing and then of printing, people ceased to be tied to a particular place when they communicated by visual means. They could commit their communications to the written or printed word and they or others could carry them from there to another place (from the Ghent of van Eyck to the Aix of Cezanne, as it were) where the identical communication could take place without the storyteller needing to be present. This was possible when a form of writing was produced on clay tablets in Mesopotamia (Fig. 5.10). Nevertheless, some of the early surviving writing was produced on relatively immovable objects, such as the 4000-year-old Code of Hammurabi (Fig. 5.7) or the 2700-year-old Behistun inscription on a rock wall (Figs. 5.12 and 8.26).

The essence of art, though, involved a dual expression whereby, as I said in Chap. 6, "the producer can refer to a thought or subject—the literal surface—but the deeper cognitive meanings in the minds of both producer and seer might be different at different times, enhanced through the strength of the ritual and the cultural context." The seer required to have McLuhan's "actual process of thought" or Leonardo's "*cosa mentale*" and that, I suggest, was prompted by shared codes by which they could understand the images produced on the rocks. Rock art was a real example of McLuhan's "the medium is the message" because the seers had learned how to see the images they had committed to memory and the place where they are found. And in that way, they fixed the communication at a place for themselves and for anyone who was initiated into the coding. For any others, the communication was limited to the message: "this was an important place."

Dual expression in this sense is part of a process I have described previously (Davidson, 2023, 639) by which people constructed knowledge of places within Australia by naming them (Gibson, 2013; Hercus et al., 2009) so that they know what they call their Country (Hayward et al., 2018) and even came to know places they had never visited (Neale, 2017a; Russell, 2012). The individual place was memorable for the marking, but the memory invoked the story at that place. The connections between such marked places and the linked stories and rituals have come to be known as "Songlines" (see, for example Neale, 2017b; Neale & Kelly, 2020) because, in addition to the narrative of a story, the song enhances the memorability and creates links between locations in the landscape. The mnemonics, rhythm and structure of song are fundamental to the learning and transmission of songs in all sorts of societies and are a fundamental feature of oral cultures (Kelly, 2015, Ch. 3; Scalise Sugiyama, 2021). In Chap. 5, I cited scholars writing about Homer using standard phrases in "oral epic songs" to make stories more memorable, so this is a feature of oral cultures, not just those of fisher-gatherer-hunters.

One apparent manifestation of this connection to other regions is shown by the stories about Rainbows and Rainbow Serpents in Australia among the Waanyi of the southern margin of the Gulf of Carpentaria on the border of Queensland and the Northern Territory. Taçon (2008) showed through the stories of Aboriginal people and through the images of the rock art that there are connections between the Waanyi area and places to the north in Arnhem Land, to the Southwest in Alice Springs, and to the Southeast in Winton. My own work shows similar rock art images to the South in the Selwyn Ranges (Fig. 9.22): Yulluna[13] man, Tom Sullivan, said that this site was the "Wild Dog Dreaming". There are similar designs at the "Wild Dog Dreaming" rock art site in Boodjamulla/Lawn Hill National Park in Waanyi Country. A recent argument has also connected (much smaller) inverted-U petroglyphs with the Two Dog Dreaming (Franklin et al., 2021).

I look at these locations marked symbolically by similarities in the rock art and think of the massive ecological differences among the places represented. The differences in rainfall between Burketown just south of the Gulf of Carpentaria, in Waanyi Country, Cloncurry in the Selwyn Ranges, in Mitakoodi Country, and Boulia south of the Ranges, in Pitta Pitta Country, suggest that there would be an advantage to moving south from the floods or the devastation of cyclones when they occur around Burketown and Boodjamulla, or north from the droughts when they occur around Boulia. The existence of a Rainbow Serpent Songline, first told to me by Tom Sullivan in 1989, in that direction lends some credence to the connection (Davidson et al., 2004). Tom Sullivan took me to the Woodul Waterhole which is a site on the path of the Rainbow Serpent half way between Boulia and Cloncurry. When Tom and his family first took me there more than 10 years later, the waterhole

[13] This tribal grouping and language has several different spellings.

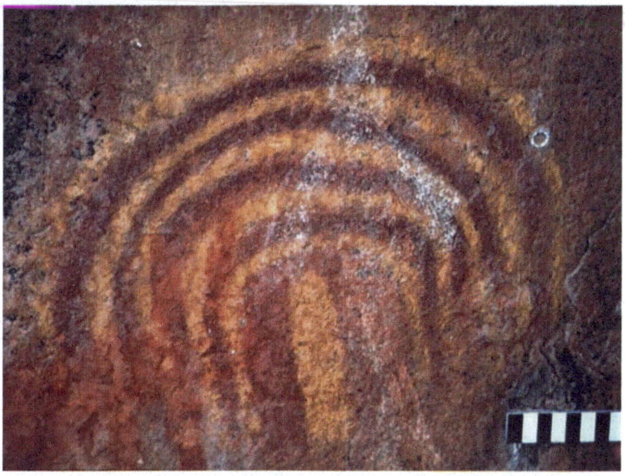

Fig. 9.22 "Rainbow" arcs from Selwyn Ranges south of Cloncurry, Australia. The red and yellow ocher arcs have a white human figure painted over them. (Photograph Iain Davidson. Permission from Yulluna man Lance Sullivan, Tom's Nephew)

was dry. He later sent me a photo he had taken of the waterhole when it was full after the drought broke (Fig. 9.23), so that we could use it in the paper we published together (Davidson et al., 2004).

In addition to the example of the Waanyi and the Rainbow Serpent, there are many other cases of art being used in social interactions. One of the results of the archeological research I led in the Selwyn Ranges of NW Queensland was that our analysis showed that ocher used for some of the paintings came from outside the region (Davidson et al., 2005). Trade in ocher was already known as a historic practice recorded by Roth (1897) in his ethnographic work while working as a doctor in the region and then from syntheses of trade or other movement of materials through the region (McBryde, 1987; McCarthy, 1939; Mulvaney, 1976). This was the first time our archeological work and analysis of ochers and paints could link the art with the movement of raw materials into a region within which there were known sources of ocher. The paintings were in rockshelters that lined the sides of a valley, and ochers were readily available at a quarry about 3 km away (Fig. 9.24), so this movement of ochers was not just about access to ocher but about the social relations of people who painted the rockshelter. I compare this with the linkages that can be established through the images themselves (Fig. 9.38). Together, trade in the means of production of the art and similarity in the imagery connect the art with the broader set of social relations for some of the other sites.

The first point is the arched features of the painted images from the Selwyn Ranges (see, for example, Figs. 9.1 and 2 Left) (Davidson et al., 2005). Figure 9.26 shows images from the site designated as Site 4 previously (Davidson et al., 2005, 114–115) which shows the arched shoulders seen in other figures. We do not know the "meaning" of the paint "under the shoulders" of the image, but it echoes, for me, the depressions interpreted as eyes in the "archaic face" at Nganganterra (Fig. 9.14). A comparison can also be made between the arched features of the "face" in Nganganterrra, the arched paint on the men dressed for ceremony (Fig. 9.25), and the arched "shoulders" on some of the anthropomorphic paintings from the Selwyn Ranges (Fig. 9.26).

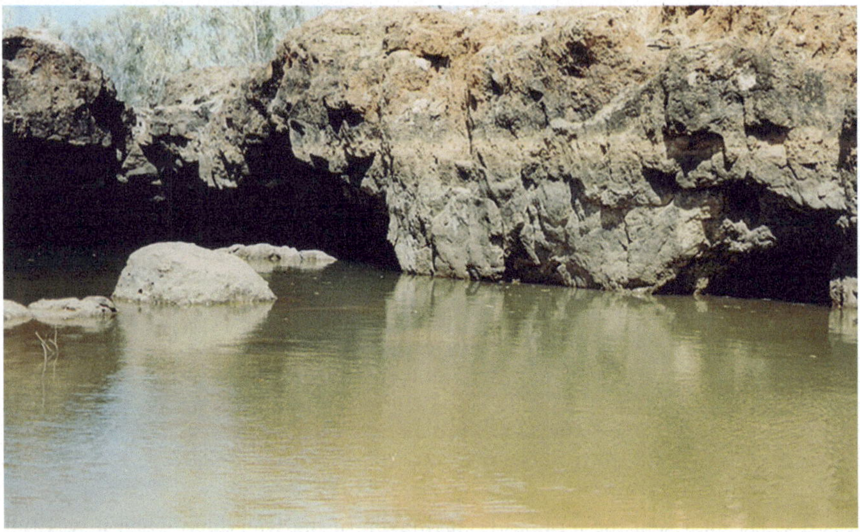

Fig. 9.23 Water deep in Woodul Waterhole, a place sacred to the Rainbow Serpent. (The photo was taken by Yulluna man, Tom Sullivan, for use in our publication [Davidson et al., 2004]). The waterhole is close to the ceremonial stone arrangement at Double Crossing Bore. I have never seen water in this hole, but the drought broke, and it filled

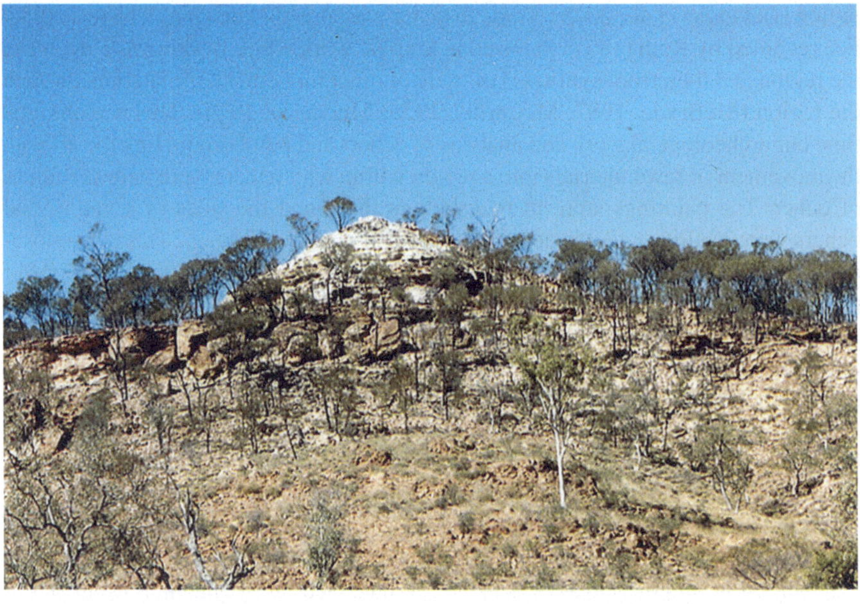

Fig. 9.24 Ocher Quarry in Anvil Creek, NW Queensland. The quarry is about 3 km south of the painted site we analyzed. Ocher from this quarry was used on some, but not all, of the local paintings. (Photo Iain Davidson)

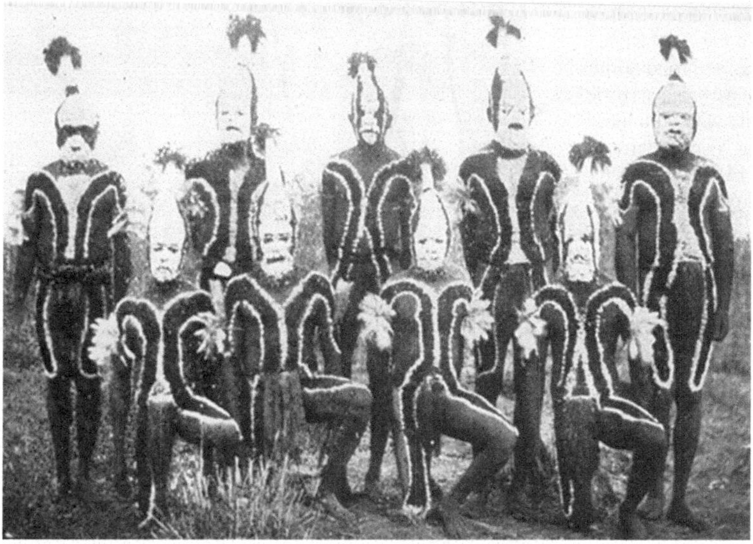

Fig. 9.25 Men dressed for the Boulia Corroboree. It seems likely that the man in the middle of the back row was Willie Sullivan, father of Tom Sullivan. These men were participating in the Molonga ceremony (called Mudlunga when it reached Lake Eyre in 1901/2 according to Hercus, 1980). (Previously published by Roth, 1910, Plate X. With permission from Lance Sullivan, Willie's grandson)

These figures are conventionally called "Anthropomorphs", meaning that they have the shape of humans. There are good reasons to suppose that some had a human shape and some were not human (as set out in Davidson et al., 2005, 112–6). Mitakoodi people, whose Country is around Cloncurry to the northeast of that of the Yulluna, refer to such images as hawks (Pearl and Gordon Connelly, 2013, personal communication), some appear to have crests, and others have the hands of mammals. Indeed, the Mitakoodi identification of some of these images as hawks would suggest that the arching can also be interpreted as the arching of wings of a bird. That indeterminacy and ambiguity was part of the way in which people had the option to ascribe meaning by uttering the associated songs and stories.

The second point is that such a comparison is compatible with the existence of songlines that are part of the network of mythological songs and stories linking the places of this region (Davidson et al., 2004) both through the transmitted songs that were told as Dreamings (as I heard both from Wankamadla/Mulligan River woman Mrs Hansen and from Yulluna man Tom Sullivan quite independently). This network was further enhanced by the movement of people both for ceremony and for the distribution of the many traded materials, including the narcotic *pituri* (Watson, 1983; Watson et al., 1983). Recent and historic estimates suggest very large quantities of the narcotic were involved (Silcock et al., 2012). John Mulvaney commented on the reach of such trading when he wrote: "In theory it was possible for a man who had brought *pituri* from the Mulligan River and ochre from Parachilna [to the south in the Flinders Range] to own a Cloncurry axe, a Boulia boomerang and wear

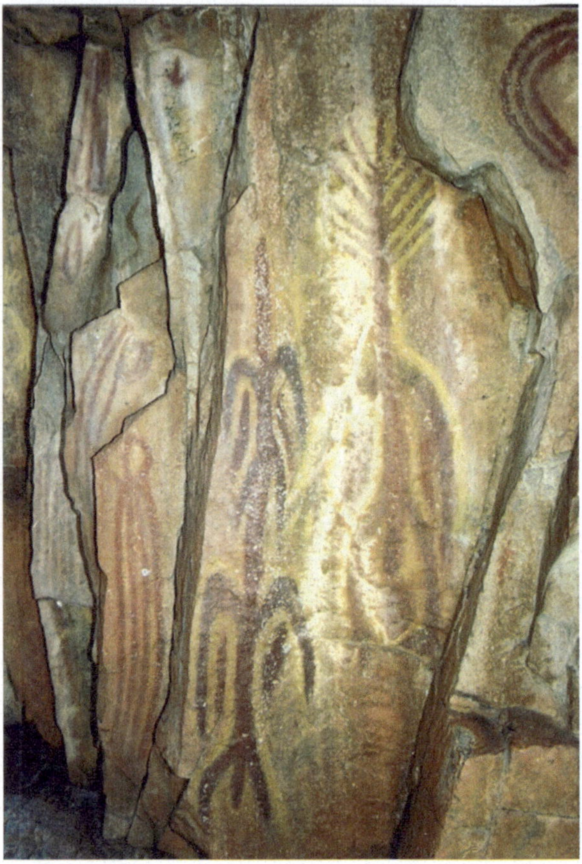

Fig. 9.26 Detail of Site 4 (see also Fig. 7.29 Right) showing "anthropomorphs" which have some stylistic similarity with both the image at Nganganterra (Fig. 9.14), and the men painted for ceremony in Fig. 9.25. There are nested arcs in the top right of the image. (Photograph Iain Davidson, permission from Pearl Connelly and Fred Turner of the Mitakoodi People)

shell pendants from Carpentaria and Kimberley" (Mulvaney, 1976, 80). Such connections required movements that were facilitated by use of message sticks which were themselves marked with distinctive designs (Kelly, 2024).

Modern Aboriginal behavior defines practices that limit the use of space according to moral rules defined within the kinship system (Munn, 1996)—you cannot go to particular places because they are "dangerous" unless you have been initiated to a level to do so. Mrs Hansen was at pains to point out the strict rules that governed the trade in *pituri* even in her time. Among other things, Nancy Munn stressed the importance of centers rather than boundaries—which corresponds well with marking places permanently by engraving them with the marks of the activity of ancestral people-who-went-before. To use the phrase she adopted, Aboriginal places were "center-oriented" and if people recognized bounds of their space, it did not seem to have had persisting markers on such a boundary. It seems that such an interpretation fits well with the old places marked by art, such as Nganganterra. The important point here is that because the signs were coded to have meaning, people had to commit the code to their phenomenal memory, a product of their living by oral tradition. They did not write words, but the combination of persisting signs with oral memory marked the distinctiveness and memorability of both the place and the message and, at the same time, allowed people to carry the memory of the meaning from one place to another. It was not the same as carrying the good news from one place to another in written form (even if the horse was dying), but it overcame some of the limitations of place and also structured the landscape. Boundaries were structured, but overlapping and were not devised to exclude people, but for crossing (Kelly & Brady, 2021, 723–725; Williams, 2019). By linking one place with another through the sharing of particular signs and the coded meanings they had, the structured meaning of the landscape was encoded in shared signs and initiation into their deeper meanings. As a result, people went some way towards overcoming the tyranny of distance.

In quite a different realm, writing about the concentration of images on plaquettes of stone at the Upper Paleolithic site of Parpalló in eastern Spain, I argued that such a concentration could arise if the plaquettes were used in a ritual at a central place for people who otherwise lived in scattered locations. Control of information can lead to centralization of the material signs of the activities of powerful people—as they appear to have done with that context for art (Davidson, 1989). In both cases, people could define the place by assigning it importance, both absolutely and through its social role in the landscape, putting it into context using stories that they told about it. Although the outcomes were similar—the concentration of power in particular places—one of the differences between the two examples is that people could see the Australian art in a fixed place, while the place, but not necessarily the art, was the main important outcome in Spain. In Australia, a knowledgeable woman like Mrs Hansen could explain the signs she saw (Fig. 9.14, bottom right) whereas the signs that concentrated power at Parpalló were not openly visible.

Two things stand out. First, if motifs share a common code to allow definition of meaning, there must have been some connection between places where similar or the same codes applied, as we know there was among a people travelling for their ordinary foraging, or among those travelling for ceremony (Galt-Smith, 1997). Second, we know from various studies that motif classes were both shared between sites across the country, and that some were not unlimited in their distribution. Distribution maps of image classes show that not all signs are found everywhere, so the maps give the appearance of bounded spaces (Ross, 2003, especially Fig. 7.7) yet there do not appear to be any boundaries. Ross showed that painted striped panels like those illustrated from the site known as Emily Gap (*Anthwerrke*) south of Alice Springs (*Mparntwe*) (Spencer & Gillen, 1899, 170–179) were found also at Jessie Gap, Kuyunba, Ooramina and Trephina Gorge, south and east of Alice Springs, but were not used as far east as Ngganganterra.[14] At the same time, signs she called "Oval/fern" are found at N'Dhala Gorge further east, and further south at Ewaninga and Rainbow Valley. People made oval/cross-hatched signs at Ngganganterra (Fig. 9.27), as well as at Abudda Lakes, Kyunba, Rainbow Valley and N'Dhala Gorge. The overlap in site names is enough to show that the distributions of coded signs were limited but not bounded.

[14] She has since found them at Mount Knuckey (18 September 2024, personal communication at a seminar in NERAM) about 30 km west of Ngganganterra. I suspect that they are part of a common tradition further east in the Selwyn Ranges, 400 km to the Northeast but the link is difficult to authenticate.

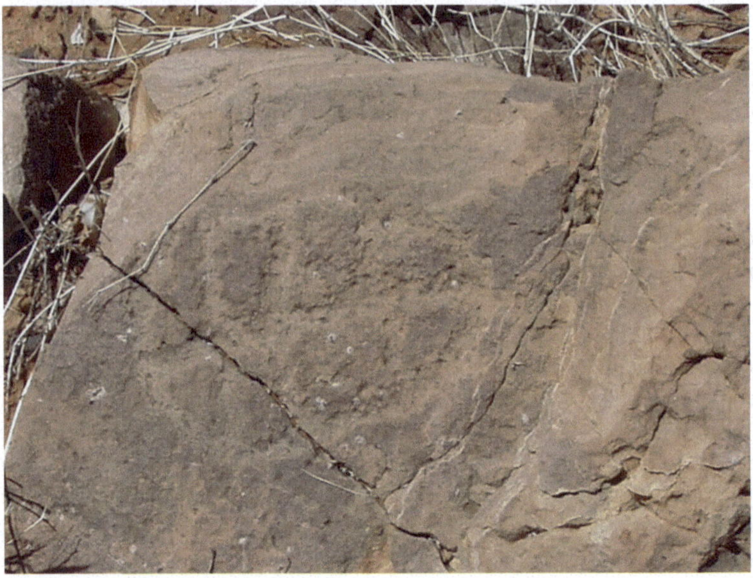

Fig. 9.27 Cross-hatched oval at Nganganterra August 2008. (Photograph Iain Davidson, with permission from the Mulligan River People)

One of the consequences of the combination of coded marks and story/songs is that people who are travelling around the country can mark the places that are important to them and recount the reasons. This is how they created the intersections that have come to be known as "Songlines" (Neale, 2017b; Neale & Kelly, 2020). People shared songs and stories along songlines and recounted Dreamings about the way the world came to be as it is. This process went a long way to overcoming the tyranny of distance. The art concreted the sense of belonging of one to the other—hence the practice of people knowing the nature of Country because they knew the stories even though they had never been there. By sharing the coded meanings of signs that persisted in different places, people could overcome the tyranny of place resulting from confining meanings to a particular site. It is difficult to think of any art which was confined to one single location either because the people who produced it did not move, or because they only produced art in one location, or because no one else produced anything else with the same properties at the time. This shows that the mental and material exemplification of rituals, such as those documented for the Yirrkala paintings, sets up a broader array of interconnections which often spread across enormous distances.

In Australia, the relevant comparison to writing could possibly be message sticks (there is a relevant discussion in Kelly, 2020; Kelly et al., 2024). At the time non-Aboriginal people disrupted and first described them, Australian Aboriginal societies did not use writing or printing to communicate with each other. They did not determine a fixed meaning of a message they communicated but committed it to

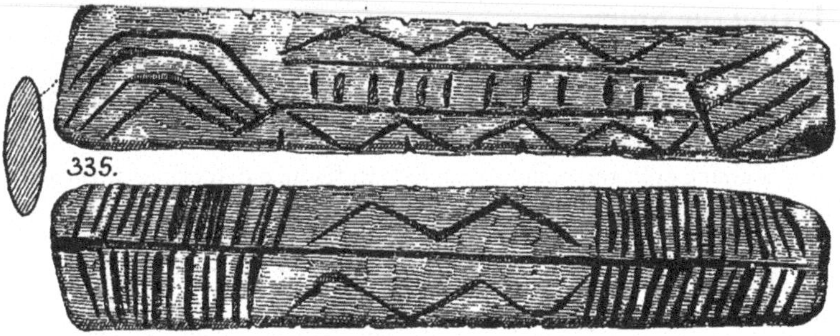

Fig. 9.28 Message stick drawing from Roth (1897, Plate XVIII). The description in Roth's text does not match the marks on the stick

memory that they could share with others and enhanced that memory with coded material signs carried on small wooden sticks (Fig. 9.28). Message sticks were marked with coded messages (Howitt, 1889) but the principal means of communication was the oral messages of the carriers (Kelly, 2020; 2024). As I indicated previously, message sticks "were used in contexts of interpersonal communication to facilitate friendly interaction" (Davidson, 1997, 133). In fact, because the message recorded as marked on the stick was not often of a repeatable code, the oral convention that there was a code depended equally on the "oral channel that was doing the heavy lifting" (in Kelly's words from Chap. 5).

Such sticks were used in Northwest Queensland to summon people to corroborees or to ask people to bring *pituri* (Davidson et al., 2004; Kimber, 2004). Roth (1897, 138) gave detailed description of one such message stick used by Mitakoodi to summon people to Cloncurry: "The message intended to be conveyed is to the effect that the sender wants his spears, boomerangs, and shields forwarded, as an initiation ceremony is about to be held at the river (Cloncurry), and the addressee is to come and join." One message-stick,[15] illustrated by Roth (Fig. 9.28) included images of nested arcs seen in rock art sites across the country (Roth, 1897, Fig. 335) (see Figs. 2 Right and 9.15).

This suggests a connection between some of the markings in rock art and some of the coded marks on message-sticks. The important messages were conveyed in the "oral channel" (Chap. 5) but the stick seems to have acted as a "passport" to corroborate the veracity of the message. Regions, such as Northwest Queensland, were connected by a network of trade and communications which linked the people in one place with other people and other places at a distance from them (Davidson et al., 2004; McBryde, 1987; Mulvaney, 1976).

[15] I am grateful to Dr. Piers Kelly for discussion of this particular message stick.

9.8 Fixing the Story

In my account of art's role in the changes in communication, I emphasized that, in societies with written texts, "fixing the story" was part of the process of making the story integral to the cultural values of society. Reference to those texts fixed by writing penetrated much of the art of those societies (see, for example, Fig. 4.8 Right). I suggested that this was, in part, a product of writing down a fixed text—everyone could tell the story of the picture because they could also refer to a written text. There is an important issue about who was able to read and how people who could not read could benefit from writing anything more than a text for a limited audience (Moss, 2021; 2023), Pursuing that issue could divert us from the study of visual art. We know that Australian Aboriginal societies did not have writing that permitted such fixity, so this could be an excuse to end this chapter here. As I pointed out previously, however, one of the effects of writing was to allow access to a single story at different times—something that became possible, in principle, with the persistence of rock (and other) art (Davidson, 2020, 765; 2024, 177).

At the same time, there seems to be some evidence of a wide distribution of some, but not all, cultural values in Australia through the expansion of knowledge of stories along the songlines associated with shared codes. This was exemplified at the widest level, as I have said, by the mapping of Songlines across Australia in the Seven Sisters Songline (Neale, 2017b) and other traditions (Davidson et al., 2004; Kerwin, 2012, 84–97). These reached into the traditional country of the Mulligan River People at Paltyura Waterhole (Hercus, n.d.) (Fig. 9.29). The Mulligan River channel ends in this Waterhole which is only about 40 km south of the Mulligan River Art Site.

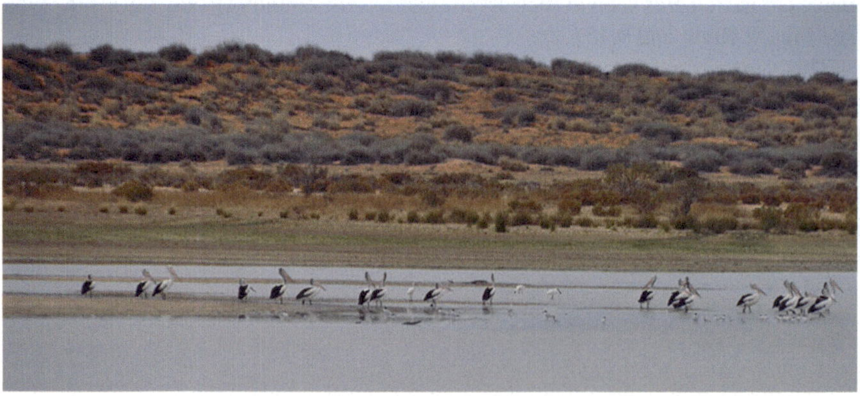

Fig. 9.29 Pelicans, Spoonbills and other birds in Paltyura/Pulchera Waterhole in the Simpson Desert, one of the northern and eastern endpoints of the Seven Sisters Songline (Hercus, n.d.). (Photo Iain Davidson September 2021, courtesy of Isabel and Avelina Tarragó)

The spread of cultural values is also shown at the more local or regional level both in the associations documented for modern paintings that have emerged from Papunya (Myers, 2002, 37–43) or Yirrkala (Morphy, 1991), and in the copious documentation of Songlines in Land and Native Title Claims. As Sutton (1995, 57) put it, there are:

> …localized Dreamings that travel hardly any distance; regional travellers with known starting and finishing points within the region and, thus, within the social and ritual ken of those with country on such tracks; and continental travellers which come into a region from places more or less unknown, pass through, and continue on to the countries of people little known, perhaps unknown. At certain points in the landscape, Dreamings of all these types may intersect, so that a single place has multiple connections to distinct but overlapping sets of people.

The Seven Sisters Songline was fixed precisely because some of those tracks, those songlines, were of continental scale dependent upon common codes for disambiguation. But there is a story that is repeated more often which emphasizes what would correspond to Sutton's "regional travellers" and was reflected in the rock art in particular regions. In several regions, there were distinctive rock art styles of limited distribution which were quite different from nearby distinctive rock art styles, despite some shared images. These would correspond with Sutton's "localized Dreamings". Ross (2013) summarized the arguments of that time concerning five of these in discussing what she called "A Continent of Nations." These were (1) the Kimberley of northwest Western Australia, (2) Arnhem Land, which has Yirrkala at its eastern edge, (3) South East Cape York, (4) Central Australia, of which Nganganterra (with its continental scale connections) is probably at the eastern edge, and (5) Northwest central Queensland, south of Cloncurry. Ross (2013, 169) concluded that regional styles emerged "when the relationship between people and place was under pressure, whether the pressure came from outside intruders, rising sea levels, population increase or the introduction of more intense interaction in the form of trade networks, or a combination of these factors." She went on to cite the views about the emergence of the apparently widespread belief system known as the Dreaming or Songlines (David, 2002; 2004) which we can, perhaps, trace back to the time of emergence of some of the rock art traditions more than 3.5 thousand years ago (*Wanjina* in the Kimberley, X-Ray art in Arnhem Land, Quinkans in Cape York) or their consolidation in the last 1000 years (Central Australia, NW Central Queensland). In this way, cosmological beliefs were tied in with the rock art that derived from them. The provision of ocher by trade was about social connection, because it involved movement of the goods into a region where there was already ocher, and that social connection can be identified through the movement of other materials, including stone axes. The axes can be traced to a source and from there into regions where they were traded despite the fact that similar products were made there (Davidson et al., 2005; 2021). Australia seems to have been crisscrossed by routes of travel which brought people together at the same time as marking their regional differences. In this, rock and other art was fundamental.

9.8.1 The Kimberley

The Kimberley is well known for two of the most distinctive painted art styles of Australia—the earlier *Gwion Gwion* figures and the later *Wanjina*[16] figures, but the story is a lot more complicated (e.g., Blundell & Woolagoodja, 2012; Gunn et al., 2022). *Gwion Gwion* figures show characteristic elongation and ornamentation of the human figure and other scholars refer to them as "finely executed". David (2017, 207–9) described them as "curious human shapes richly decorated with accoutrements such as feather or hair tassels, elaborate headgear, hand-propelled spears and boomerangs" and "decorated with tassels, arm bands and sporting elaborate head-dresses" (see Fig. 9.30). An Aboriginal account describes the earlier *Gwion* figures in the following way: "*Gwion* art began with the charismatic bird called *Gwion Gwion*. Explanations of the origin of the dark red paintings tell of a magic bird wiping fresh blood from the tip of its curved beak onto the walls of rock-shelters" (Ngarjno et al., 2000, 120).

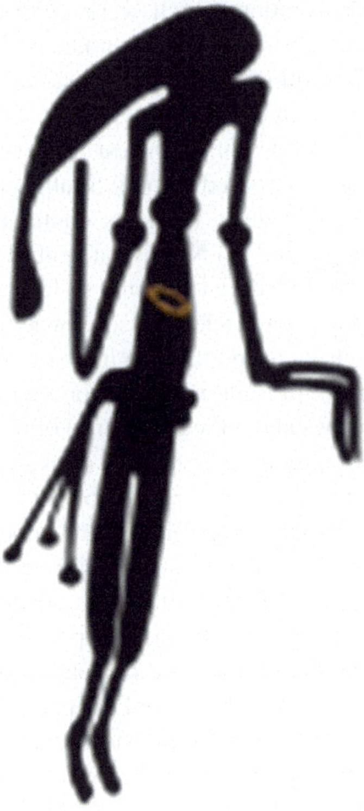

Fig. 9.30 Drawing of image of Gwion figure from the Kimberley, probably 12 thousand years old, showing evidence of arm ornaments, head-dress and waist ornament. (From Finch et al., 2020)

[16] There are two spellings of this word, one with and one without the "d". I have chosen to use this spelling because Donny Woolagoodja, who is part of that community, does not use the "d".

Tasselled *Gwion* figures are generally much smaller (~30 cm) than *Wanjinas*, but much more varied and one Aboriginal work described them as "Spirits from the creation epoch" and went on to describe *Wanjinas* as "Powerful ancestral beings whose images remain as cave drawings in the countries of the people of north Kimberley" (Jebb, 2008). Akerman (2016) summarized the state of knowledge of *Wanjinas*, following an earlier statement that "Images of *Wanjina* beings are characterised by halo-like headdresses and mouthless faces with large round eyes, fringed with eyelashes, set either side of an ovate nose. Most *Wanjina* images found as rock art are depicted without a mouth, although on some, more recent works on boards, bark or canvas the mouth and teeth is [sic] indicated" (Akerman, 2009). Despite these statements of obvious importance, the two classes of image are used today as symbols of Aboriginal organizations such as Mowanjum (Jebb, 2008) (Fig. 9.31) in telling the story of their festival to the world.

Many scholars have had their say about the styles of the imagery of the region and offered their version of a sequence from one style to another (e.g., Donaldson, 2007). Pursuit of the nuances of these versions would take us into the niceties of archeological analysis instead of the discussion that I intend of the broad pattern of communication using rock art. The original archeological classification into different styles was made without any understanding of the assumptions involved in claiming there was a connection between styles and chronology. The elongation and ornamentation of *Gwion Gwion* figures led some to question how to talk about the people who painted them (McNiven & Russell, 2005, 151). Andrée Rosenfeld, then one of the most respected archeologists of Australian art (Rosenfeld, 2000), was quoted as saying that one prominent argument by an authority of the time claimed that "the finely executed [*Gwion* or *Gwion Gwion*] paintings in the Kimberley are too fine to be Aboriginal and must be the work of earlier peoples…" but that these views "have no archeological basis and are informed by racist perceptions of what Aboriginal people are capable of" (McCulloch-Uehlin, 2001).[17] As I have suggested elsewhere, such racist judgments derived both from colonialist attitudes and from ignorance about the ways in which many peoples change the way they paint things as culture changes (Chap. 3).

Ian McNiven and Lynette Russell (2005, 149) showed that, historically, many scholars had discredited the argument that anyone other than Australian Aboriginal people were the producers of both *Gwion* and *Wanjinas*. An analysis of rock art images with some similarity to *Gwion* figures from all over the world showed that the Australian examples from the Kimberley were quite distinct, and that the closest images were those from Arnhem Land (White & Barry, 2004). Peter White and Michael Barry concluded that "archaeological evidence fails to establish a case for the exotic origin of *Gwion* images." The disagreement derives from the same social attitudes that drove the question of grazier Bill about "Art or scribbles" and the multiplicity of attitudes to art that are unreflective about critics' views of society and

[17] It is a curiosity that Rosenfeld's original remarks led to the pulping of the book in which they were published, so my quotation comes from a newspaper article about the events. Both antagonists are now dead.

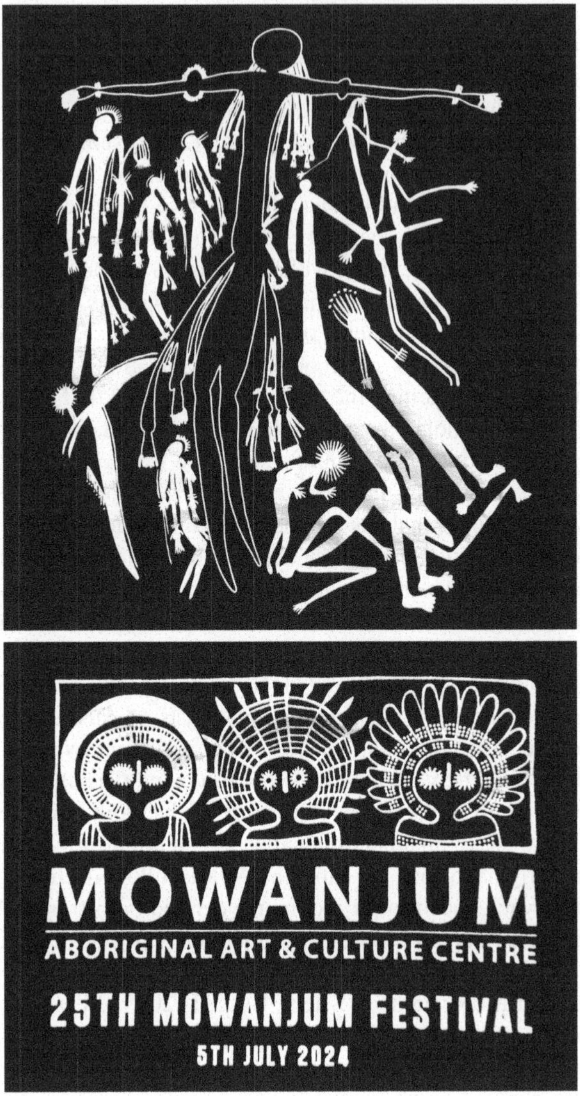

Fig. 9.31 Top Mowanjum Art and Cultural Centre use of Gwion rock art figures for branding (of a T-Shirt). Bottom Mowanjum Art and Cultural Centre use of Wanjina figures for branding

the art-system of producers, subject, product, and seers. In the context of all we have seen in this book, the discrepancies and disagreements can be attributed to the same sorts of contextual and cultural differences characteristic of all art particularly in the last 300 years. Archeologists and rock art specialists are, after all, creatures of our times in the same way as art historians. Aboriginal people generally have their own views about the role of art in the beliefs and history of their Country (e.g., Blundell & Woolagoodja, 2012).

In many previous studies published dates have been wishful thinking, partly based on the assumption that different styles were from different periods. For example, if there were no other means of getting a chronology, archeologists assumed that different styles belonged to different periods. More rigorous analysis and new methods have not only produced some credible first-estimates of the numbers (ages) associated with painting styles, but led to a need to evaluate the assumption that style equates to period by questioning the discreteness of the *Gwion* and *Wanjina* styles. At one stage, the Aboriginal people with responsibility for *Wanjinas* observed the *Gwion* figures and interpreted them "as the work of the capricious figures of the bush" (Layton, 1985, 448).

Meg Travers and June Ross (2016), following earlier authors where possible, classified a large number of images into 5 styles: 7 examples from the Irregular Infill Animal Period; 1249 from the *Gwion* Period; 795 from the Wararrajai *Gwion* Period; 375 from the Painted Hand Period; and 685 from the *Wanjina* Period. By using a statistical analysis of the attributes of this large number of motifs they showed that there was continuity between the several different styles, demonstrating the gradual and symbolic nature of changes. Statistical analysis showed that there seems to be a gradation between styles, but that does not demonstrate that those different styles were produced at different times. Although the analysis showed that the dates did not separate the styles in a simple way, the label "period" rather than "style" lingered into more recent publications, perhaps inappropriately. The conclusion could have been that these were overlapping styles, not representatives of different time periods. Such a conclusion might have led to much rethinking. This statement needs to be seen in the light of one by people who excavated at the Gunu Site Complex in the Kimberley (Moore et al., 2020):

> Although the relative stylistic sequence of the Kimberley rock art assemblage is generally accepted …, the degree of overlap between styles requires additional refinement and the chronology of the earlier styles remains problematic … . In addition, many motifs cannot be securely attributed to an identifiable stylistic period. [David et al. (2019b)] conclude that, based on the current evidence '*classic* Wanjina iconography probably dates to the last millennium' (our emphasis) while accepting the premise that incipient Wanjina design elements are likely to be older (Ross et al., 2016).

At one stage, a skeptical review of all previous attempts to provide Pleistocene dates from the Kimberley (Aubert, 2012) dismissed them all, though Richard Roberts has defended the 16 and 17 thousand year dates (David et al., 2013b, 6–7; Roberts et al., 1997). In more a recent project in the Kimberley, a luminescence age from a wasp nest overlying a yam figure provided a date of 16 thousand years ago (Ross et al., 2016, 16–18). Another dating project established that at least one figure from the Irregular Infill Animal style (already thought to be the earliest style in the region, pre-dating *Gwion Gwion* figures) could be assigned a radiocarbon date of about 17 thousand years ago (Finch et al., 2021, Figs. 5 and 6).

The manner in which the Irregular Infill Animals were represented had some similarity with the way in which the people of Sulawesi filled in pictures of animals 30 thousand years earlier (Fig. 9.32), though, of course, we might consider the number of options limited (without engaging in a detailed description of infill styles elsewhere). An earlier project in the Kimberley had used radiocarbon to date a similar figure to between 1 and 2 thousand years ago (Watchman et al., 1997 cited in

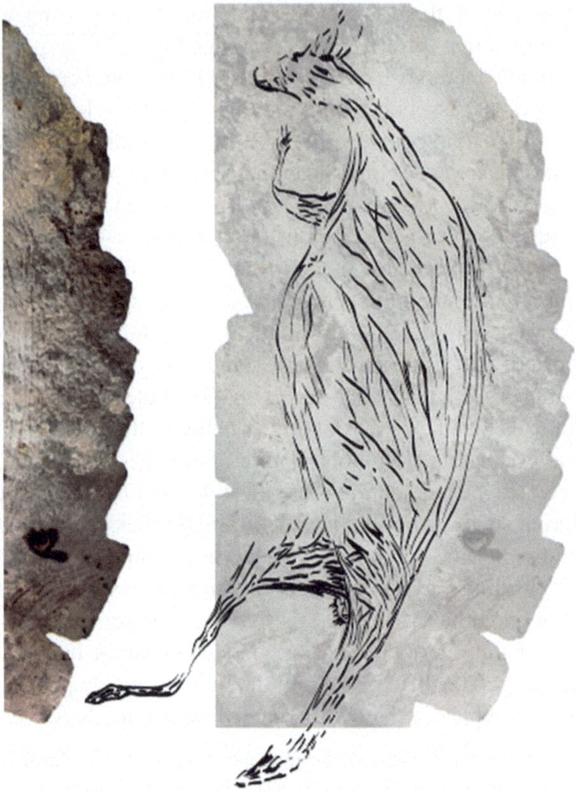

Fig. 9.32 Drawing of composite image of macropod with Irregular Infill, dated to 17 thousand years ago. (From Finch et al., 2021, Fig. 4)

Ross et al., 2016, Fig. 9). This could be taken to suggest that either, the technique of infill applied to these large animal images remained remarkably fixed over tens of thousands of years, or, no more than that the technique is simple.

Finch's team determined another date by measuring radiocarbon on samples from wasp nests both underlying and overlying *Gwion* figures in the Kimberley (Figs. 9.30 and 9.31 Top). The best example suggested that they cluster in a relatively narrow period around 12 thousand years ago (Finch et al., 2020).

In the earlier project, a series of absolute age determinations, using three different dating methods applied to styles of different types, showed that the speculative chronology archeologists expected (as in the quotation from Moore et al., 2020) was not supported straightforwardly (Ross et al., 2016). With the exception of the 16-thousand-year minimum date on the yam figure, all but one of the other determined ages had minimums of less than 3000 years—which means that they were older than 3000 years. The exception, from a large, infilled kangaroo figure said to be in *Wanjina* style was a millennium earlier than previous estimates and gave a minimum age of 5000 years (Ross et al., 2016, 29). The value of that date depends on the identification of the image as belonging to the *Wanjina* style.

The conclusion is that image painting has been happening in the caves of the Kimberley since 16–17 thousand years ago, though the use of ocher as paint probably dates back 40 thousand years (O'Connor & Fankhauser, 2001). The supposedly most recent style of paintings—*Wanjinas*—could be as old as 5000 years, though the recent art styles of the Kimberley region are more nuanced than I am suggesting (O'Connor et al., 2013). If the early date is confirmed, it needs to be put in context with the recent tradition of producing, and repainting (Layton, 1992, 37–40), *Wanjina*-style figures (Akerman, 2016). It would establish that the most recent style of Kimberley art has been fixed for a long time, longer, indeed than almost all the classic fixed texts derived from oral history mentioned in Chap. 8. Akerman (2009, 1) wrote:

> Wanjinas are cosmological Beings represented in the rock art of the central and northern Kimberley region of Western Australia ... [They] continue to be made and renewed today (e.g., Blundell & Woolagoodja, 2005). They are sometimes also painted and drawn on sheets of bark and are today common motifs in the contemporary art that emerges from the Kimberley Aboriginal centres at Mowanjum and Kalumburu.

Wanjinas, perhaps more than any other rock art image in Australia have penetrated the art world and not just as the logo for the Mowanjum Aboriginal organization. One *Wanjina* image was designed by Donny Woolagoodja, a Traditional Owner of *Wanjina* Country, for the Sydney Olympic Games opening ceremony (Fig. 9.33), when an estimated 3 billion people saw it on television (Blundell & Woolagoodja, 2012, 474).

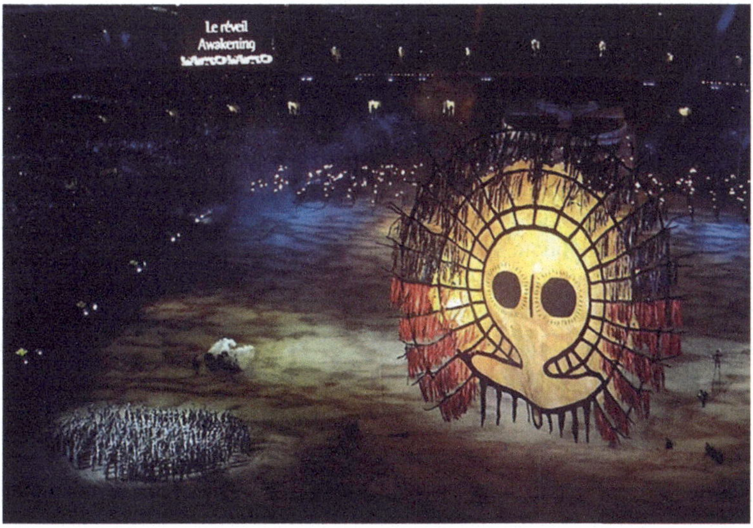

Fig. 9.33 Giant image of *Wanjina*, by Danny Woolagoodja, at the Sydney 2000 Olympic Games opening ceremony. https://nit.com.au/20-09-2022/3913/d-woolagoodja-whose-giant-wandjina-graced-the-sydney-olympics-opening-ceremony-dies-aged-75

9.8.2 Arnhem Land

Arnhem Land also has a long sequence of paintings on rocks—Chaloupka (1999) called it "the world's longest continuing art tradition"—full of complexity and nuance (some details are documented in Lewis, 1996, 13–14). David and colleagues (2013b) summarized the state of dating of Australian pictographs (i.e. not including various types of subtractive art—petroglyphs—which I generally refer to as engravings), concluding that many of the claims for early paintings are open to discussion, but must include a 26–28-thousand-year-old fragment of painted rock excavated from the Arnhem Land site of Nawarla Gabarnmang (David et al., 2013a). Early signs of art also include ocher crayons known to be as old as 65 thousand years at Majedbebe (Clarkson et al., 2017, 309).

There is some disagreement about the nuances. Pictures of animals, called "Large Naturalistic Figures," may have been produced early in the sequence (Chaloupka, 1999, 88–89, 99–103) (and show some similarity with Irregular Infill figures), but Lewis argued that they "may actually consist of a number of similar styles present throughout the whole sequence of Arnhem Land art" (Lewis, 1988, 70–72). Arguments that such paintings could be early because they include images of extinct fauna have always been problematic because some images may reflect the distortions of painting on rock surfaces; no one really knows when the particular species became extinct (Wroe et al., 2013); or no one really knows whether the animals looked like the modern reconstructions however imaginative they may be (Lewis, 1986; 2017). We have to remember, at all times, that artistic depictions are not necessarily accurate representations but result from cultural immersion in the contemporary symbolic conventions of representation.

Among the extinct animals painted in Arnhem Land are pictures of Thylacines (Chaloupka, 1999, 96–98)—the striped animal known as Tasmanian tigers, slightly dog-like, which David (2017, 210–11) suggests might even have been a domestic companion in Arnhem Land—extinct on the mainland of Australia for at least 3000 years, but present in Tasmania until 7 September 1936. Some hope was held out for a large painting of a large bird, thought to depict the extinct *Genyornis* (Gunn et al., 2011), but recent work using many different approaches on the site, the painting and other candidate species has eliminated that claim. It cannot be older than about 14 thousand years old because the split rock surface on which it was painted was not available until then and is older than 5 thousand years, long after the archeological last record of *Genyornis* (Barker et al., 2017; Cobden et al., 2017). One estimate has it that *Genyornis* did not survive to 40 thousand years ago (Johnson et al., 2016). That dating also fits with the claim that Large Naturalistic Figures are not necessarily early.

Lewis (1997) developed the argument, which had been hinted by many others, that before 12 thousand years ago low sea-levels ensured that the early art of Arnhem Land was part of a single range of variation with the early art of Kimberley. He suggested that the large number of human images from Arnhem Land he calls Dynamic Figures (such as that shown in Fig. 9.6) have many features in common with *Gwions* from the Kimberley, an overlap which has been suggested in a more recent analysis (White & Barry, 2004). Lewis suggested that the rise of sea-level then disrupted the human networks between the Kimberley and Arnhem Land and that the disappearance of boomerangs from the art in both places at about the same time is an indication of this.

The important point, here, is that there was an early tradition with Dynamic Figures (in the style of Fig. 9.6), and a later tradition known as X-Ray art. There has not yet been an analysis similar to that of Travers seeking to establish continuity between the ways in which different types of Anthropomorph were represented at different times, but there are suggestions. Morphy (1998, 55) addressed the question of reference to earlier styles when he wrote: "The same artist may produce an x-ray kangaroo, a sorcery figure or a battle scene with Mimi [Dynamic] figures. There is no reason why artists painting on rock surfaces should not also have drawn their inspiration from earlier representations." Ross (2013, 164–5) stated: "Complex X-Ray art emerged at about the same time as the Wandjina style. Complex X-Ray art developed gradually from an earlier prototype similar in overall form but lacking the distinctive internal detail of the more recent art…" citing Brandl (1973), Lewis (1988), Chaloupka (1999) (cited as the 1993 edition), Taçon (1993) (as well as a 1989 paper). We now know that X-Ray art emerged more than 5 thousand years ago (Jones et al., 2017), and Morphy has demonstrated that X-Ray art continued into the period of contact with Macassans from Sulawesi and beyond (Morphy, 1998, 61).

The work among the Yolngu of Arnhem Land showed some of the processes which have led to change in the society of art producers and consumers (Morphy & Layton, 1981, 71). In particular, Morphy was able to show that "In Arnhem Land certain clans who revealed their sacred paintings lost control over them and were unable to prevent their dissemination in ways which ran counter to the traditional value system." I do not want to draw a direct analogy between this modern study and the processes that might have applied in the past, rather I just want to show that such processes exist when societies do not fix their stories in writing, or do not otherwise provide a way back from social change. In their comparative study, Morphy and Layton (1981) were able to show that while elders in Arnhem Land lost control of their art, dairy farmers in France were able to step back from social changes because they did retain important controls.

9.8.3 Central Australia

I have suggested that Nganganterra is an eastern extension of the Central Australian tradition, so I will put other sites into the context of behavior at that site. Brett Galt-Smith (1997, Ch. 8), used ethnographic and rock-art data from Central Australia, to show that there are petroglyph (engraving) sites alongside pictograph (painting) sites, predominantly geometric images and without hand-stencils. Ross (2003) used that set of sites and their rock art as the basis for her argument that much rock art in Central Australia involved ritual. She selected seven variables synthesized by

Rappaport (1999) from his life-long survey of rituals (mostly ignoring the evidence of rock art[18]) and showed how they applied to the Central Australia rock art. Ross and Davidson (2006) summarized the evidence for these characteristics in the Central Australian rock art.

Rappaport's (1999, 24) first two criteria—invariance and repetition—are fundamental parts of his definition and closely aligned. I referred to them in Chap. 5: "the performance of more or less invariant sequences of formal acts and utterances not entirely encoded by the performers." It is immediately apparent from Figs. 9.10 and 9.14 that there is substantial repetition of the images of multiple circles through the site. Indeed, such images also seem to be made at different times and subsequently weathered. This was a long-standing tradition and one shared to the east where there are sites at the Mulligan River with multiple circles (Fig. 9.34) and with repeated simpler images of circles, barred circles, and "Clever person signs" (Fig. 9.35). The extension of these invariant and repeated signs is indicated in Table 9.1. Similar signs are also found at Nganganterra, and several other Central Australian sites (Table 9.1).

The evidence for a specialized time for ceremonies is difficult to gather for archeological sites. In general, it depends on the weather, or rather the availability of food for ceremony as a product of well-timed rains leaving water and growing plants. In the Simpson Desert this is not likely to have been a fixed time, but the gorge within which Nganganterra sits would always have been a good place for the concentration of water after it rained, and grasses would have been abundant there at that time. Likewise, the Mulligan River would have flowed when rain fell in the

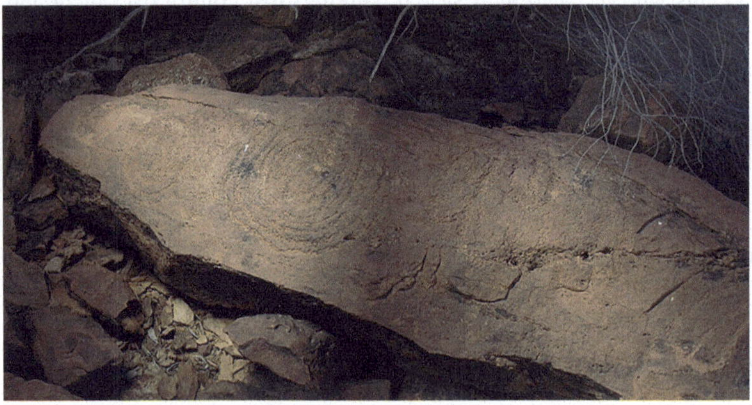

Fig. 9.34 Repeated and invariant multiple circles on slab at Mulligan River Site. Note also that there is a barred circle with more or less the same weathering at the right-hand end of the slab. (Photograph Iain Davidson with permission of Isabel and Avelina Tarragó)

[18] This is not surprising given that Rappaport mostly worked in New Guinea (as a cultural anthropologist) where the evidence of rock art is less well known than in Australia. That, in itself, is an interesting phenomenon given that New Guinea was part of Sahul at a time when we know that rock art was produced in Australia.

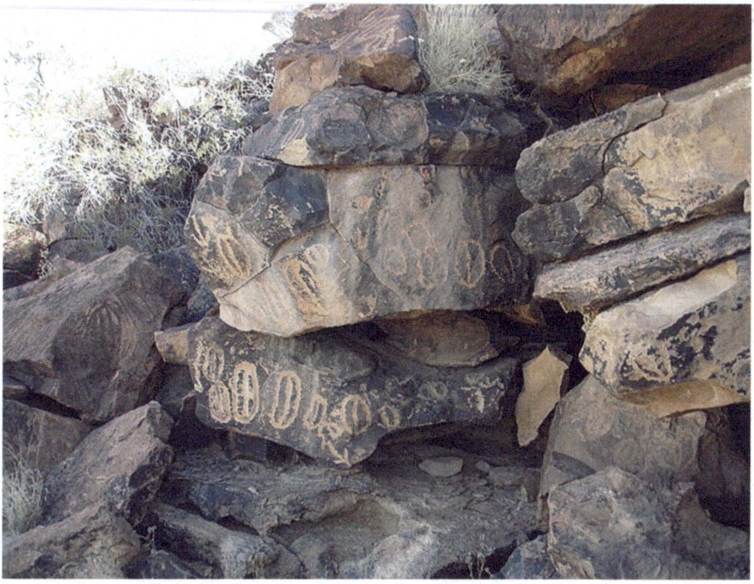

Fig. 9.35 Repeated images of barred circles at Mulligan River Site. Note that the weathering is very different from that in Fig. 9.34. But these are vertical not horizontal surfaces so the conditions for weathering were different. There are some tridents, some old images of multiple barred circles (to the left of the image) and one image which may be a "Clever person sign". (Photograph Iain Davidson with permission of Isabel and Avelina Tarragó)

region (there is a study of its reliable springs by Silcock et al., 2020), and it would have filled the Abudda Lakes. The specialized time for ceremonies would have been a result not of a predictable season, but of predictable environmental responses to perceivable events.

Having said that, people do seem to have concentrated the art in specialized places in the two other sites associated with Nganganterra, the Mulligan River Site and Abudda Lakes. Following the principle established by Ross (2003, 269) the significance of the place was also established at both Nganganterra and Mulligan River by the repeated addition of images to the original ones at different times, such that the sites include standardized, repeated images at different stages of weathering. As we stated previously: "The repeated production of rock art within a restricted area indicates that once a place is marked with rock art, it is drawn into the cultural framework of those who produce and view the rock art and in this way often became the locus for future rock art production" (Ross & Davidson, 2006, 322).

My research in the Toko Ranges (Davidson, 1983) suggested that there was little evidence (but not none) for rock art in those hills, and it was the trio of Nganganterra, Mulligan River and Abudda Lakes which seem to have concentrated large amounts of activity for rock art production. There are small patches of rock art at the entrance to Nganganterra (Fig. 9.36 also in Fig. 9.14) and around Mulligan River but not in

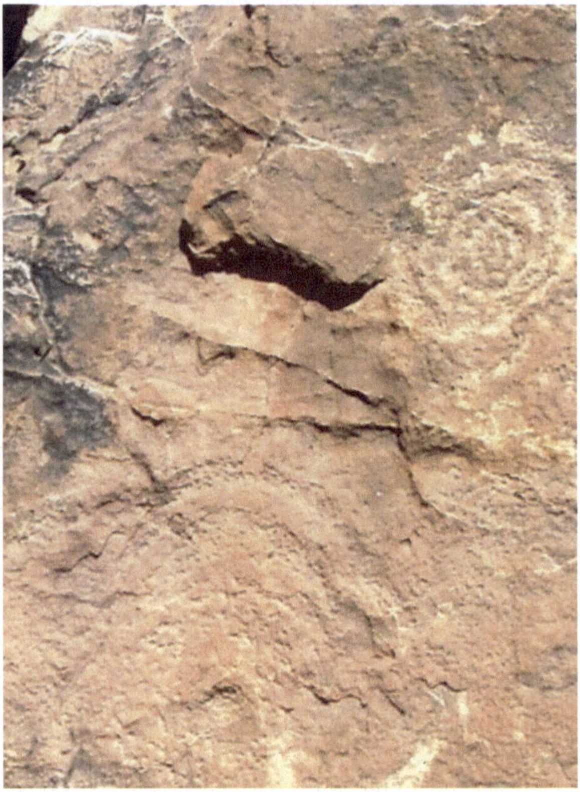

Fig. 9.36 Images from the rocks outside the main site of Nganganterra with a relatively recent set of concentric circles and below that some nested arcs. This suggests that this was not a place with the main ritual, since it was not a specialized place where standardized, repeated images were produced. (Photograph Iain Davidson, with permission of Isabel and Avelina Tarragó)

the main site, as well as in some other places—but these isolated cases would not necessarily have involved ritual.

In considering stylized behavior or stylized form, as features of Rappaport's characteristics of rituals, Ross identified two main types: those which derive from iconic reference but are so stylized as to lack the markers of resemblance; those which are principally indexical and acquire their distinctiveness through the repeated association between the image type and the meaning. We will see the stylized iconic reference in discussing the Northwest Central Queensland images, and we referred to the haloes around some human figures in religious art as examples of the stylized indexicality of images (Fig. 8.30). The point I have been emphasizing, though, is that for an art of circles and other geometric signs to be part of the systematic knowledge of anyone, it requires stylized behavior and teaching—it requires a code, such as that described by Munn (1973), or Green (2014) for Central Australia, or Morphy (1980) for Arnhem Land. I would argue that despite Munn's protestation that such art was "representational", it is meaningless without such stylization, but the coded stylization allowed the possibility of different levels of meaning.

Rappaport (1999, Ch. 4) showed that ritual could only be identified if there was performance and participation. Again, we have problems with identifying that for anything other than eye-witness accounts. Ross was able to identify places where there were grindstones associated with art, and others where images seemed to have been repeatedly rubbed. I would argue that the spaces adjacent to and around the art sites of Nganganterra, Mulligan River Site and Abudda Lakes allowed for both performance and participation.

So, finally, we are left with the criterion that Rappaport identified of rituals having a form which can hold and transfer a canonical message. I have been emphasizing throughout this book that storytelling is an intrinsic part of all art-systems. It seems an appropriate conclusion to the investigation of Rappaport's criteria for ritual that once we have identified that a site fulfils all but the last of the criteria, we might reasonably conclude that there was in addition a canonical message transmitted through initiation into the coding of the images.

9.8.4 Northwest Central Queensland, South of Cloncurry

Study of the painted images from Northwest Central Queensland has shown that the anthropomorphic images themselves are very variable, some more like images of people, others more like images of other animals, and to some extent they reflect a set of beliefs which emphasize(d) the connections (Davidson et al., 2005, 112–116). The art is probably mostly less than 1000 years old (Ridges et al., 2000). Ross (1997) showed that the rock art of the region is stylistically distinctive with an anthropomorphic motif that is regionally restricted and that there were two forms of the motif: Basic (images in Fig. 9.37 are basic)—generally monochrome, small and without details and found in rock shelters and less accessible areas; and Detailed (Figs. 2 Left, 9.1, and 9.26 all show Detailed images)—generally bichrome or polychrome and have elaborate details such as headdresses. I suspect that some of the monochrome examples may have had some yellow details which transformed to red on the surface of the rock through exposure of finely ground ocher to strong sunlight (Cook et al., 1990). Detailed anthropomorphs are highly visible and found at reliable water sources (for details of this analysis by Ross see Morwood, 2002, 197–8)—but they are embedded in assemblages of art which have variable content across the region (Ridges, 2006). This is a result of groups of painters, who were closely related to each other culturally, nevertheless differentiating one group from the other at the same time. In light of the discussion of ritual for Central Australia, this seems to be a good example of images that began with a high level of iconicity, but which were used in ways in which that resemblance was a problem. These were likely highly symbolic.

When Mal Ridges analyzed the paintings at the different sites across the region from Mount Isa to McKinlay, he found two sorts of pattern (Fig. 9.38 Left for Motifs and Right for Design Elements of Anthopomorphs). Using a database derived from our work in Northwest Queensland, Ridges (2003) showed more detail of how these different interactions worked by calculating all the linkages between sites using the design elements of the anthropomorphs (Fig. 9.38 Right).

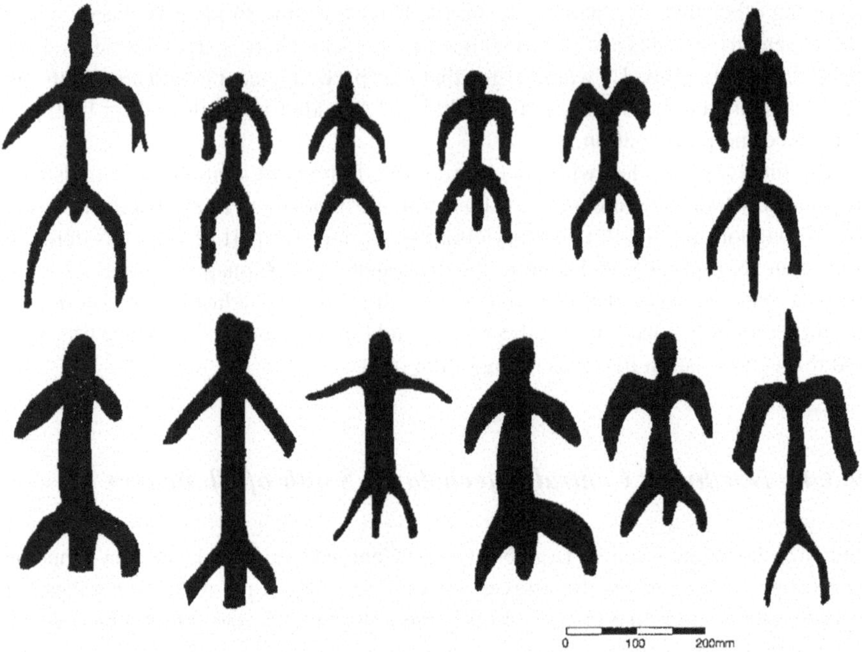

Fig. 9.37 Collection of images of painted figures from the Selwyn Ranges and Northwest Central Queensland, drawn by June Ross. (Permission from June Ross, but see also Davidson et al., 2005)

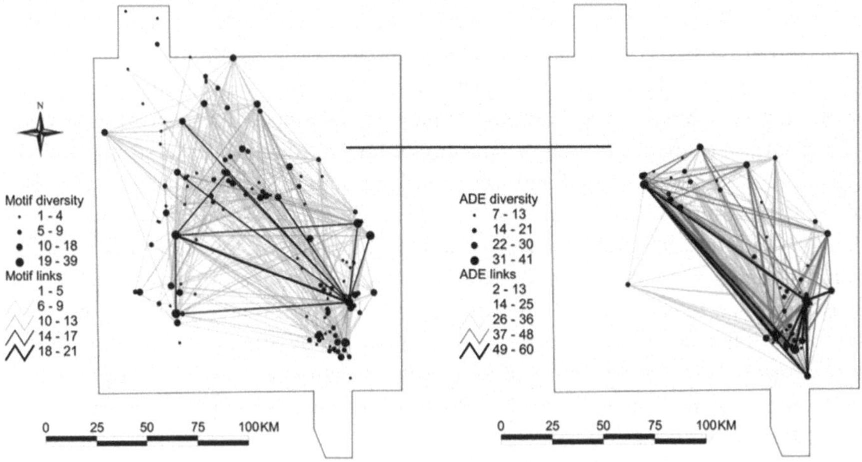

Fig. 9.38 Linkages between art sites in NW Queensland calculated by Ridges (2006) showing the patterning of links between sites for Anatomical Design Elements (ADE) of anthropomorphs on the right, and links for Motifs on the left. The red line (added by the author) marks the northernmost point for the anthropomorphs showing that the Motifs common to the Anthropomorph region spread strongly to the northwest and to the west

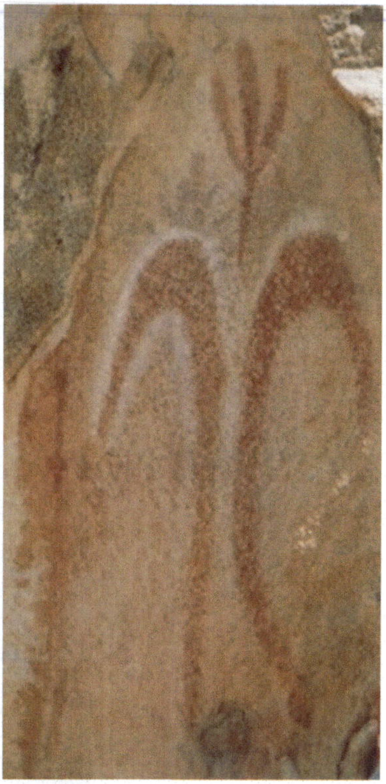

Fig. 9.39 Anthropomorphic figure from Saxby Waterhole. (Photograph Iain Davidson, permission from Pearl Connelly and Fred Turner of the Mitakoodi People)

The heavier the lines the more the links, and the larger the dots show greater the diversity. For the anthropomorphs, there is a very strong pattern of linkage along the axis north-west and south-east (Fig. 9.39).

Note that the detail of this analysis depends only on the recognition of similarity between the design elements of the anthropomorphs. In the second analysis, Ridges showed the distribution of particular non-anthropomorphic image categories (for example, quadrupedal animals such as those in Fig. 9.40 and nested arcs as in Fig. 2 Right). Putting the two maps side by side shows a much less dominant set of linkages for the Motif categories, which, importantly, are not on the same bearings as those for the Anatomical Design Elements.

These analyses show the distinctiveness of particular assemblages of rock art and the patterns of relationship between different groups. They can be taken to show something like the local (the anthropomorphs) and the regional (the other motifs) levels of connection, which tie in, also, to the continental travels of songlines represented by the Rainbow Serpent and the Wild Dog Dreamings (Taçon, 2008) (see Fig. 9.22). Indeed, while recording one site where there are anthropomorphs, we found an engraved spiral high on the rocks above the paintings which seems to connect to the spirals at Nganganterra (Fig. 9.41) 350 km away, giving us further insight into the possibility that there were continental travels associated with the beliefs.

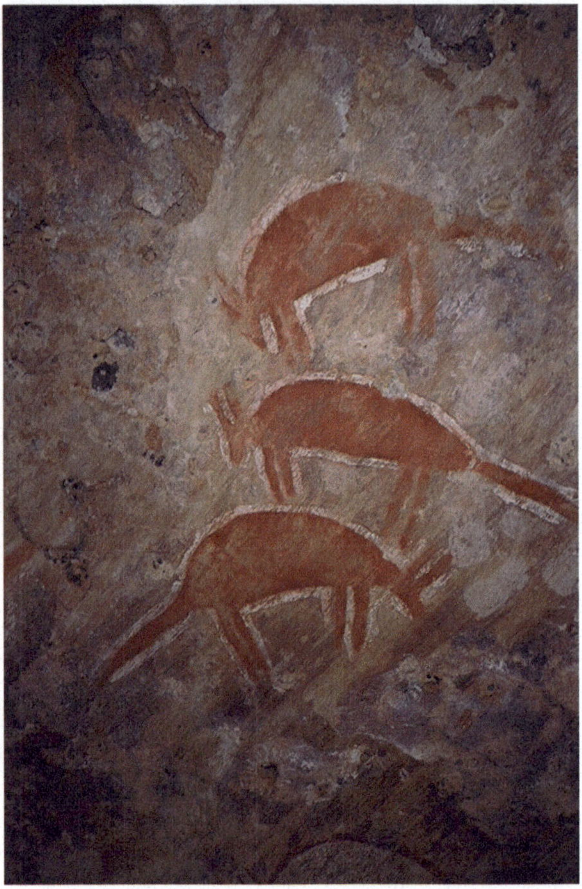

Fig. 9.40 Animal motifs from the site Lance Sullivan calls *Ngura buli majumba*. (Photograph Iain Davidson. Permission from Yulluna man Lance Sullivan, Tom's Nephew)

The pattern of linkages was much weaker and spread much further to the northwest (as predicted by the linkages in the Rainbow and Wild Dog Dreamings, I discussed previously with reference to Fig. 9.22) and also to the southwestern region where there were almost no anthropomorphic figures.

This variation within the region can be taken as an indication of what Sutton called "localized Dreamings"; the evidence that people brought ocher into the region even though there were ocher sources within a short distance of the sites is evidence of "regional travellers" as is the evidence for the trade of materials, such as axe heads, through the region (Davidson et al., 2005, 104–8); and, in all likelihood, the conjunction of an ongoing tradition of the Rainbow and Wild Dog Dreamings both in the art and in the oral stories should be considered as evidence for Sutton's (1995, 57) "continental travellers."

Northwest Central Queensland, then, seems to epitomize the different scales at which art provided linkages between groups of people, allowing those in one region to maintain connections elsewhere through a trade in ochers beyond the utilitarian,

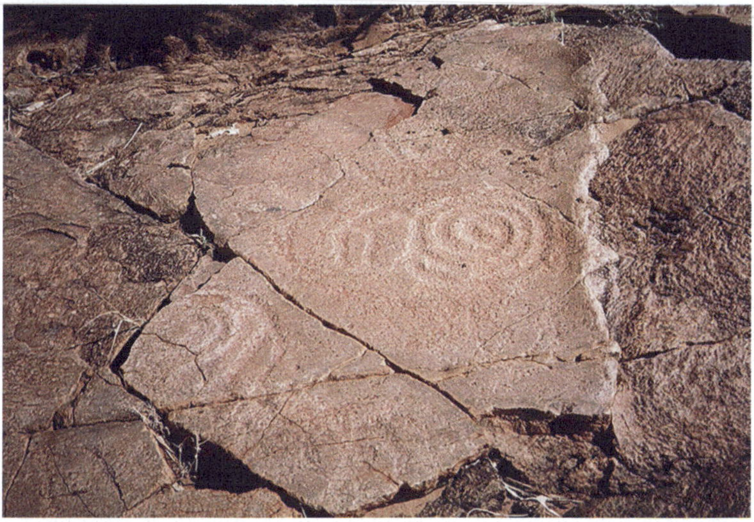

Fig. 9.41 Spirals engraved above a rock hole where anthropomorphs were painted. The spirals show some similarity to those at the Mulligan Art Site and Nganganterra. (Photograph Iain Davidson. Permission from Yulluna man Lance Sullivan)

and a maintenance of stories or songs which linked to north or south. As we pointed out previously, there are also well-known stone arrangements in the region at which people from different parts of the region met for ceremony, certainly within the last 85 years (Davidson, 2008).

9.8.5 Storytelling, Aggregation and Dispersal

The question of the role of rock art in the movement of people in the regions of their activity may seem like an esoteric bit of archeology, but it is not really. It demonstrates the importance of storytelling and singing songs in both the rituals associated with rock art and the lives of the people who practiced them. The evidence of songlines in Australia seems to suggest that the songs and stories which accompanied the visual images were an intrinsic part of both the nature of the images as art and of the patterns of movement of people. And the persistence of images was an important part of how people knew there was a story or a song about the place, with associated rituals.

The differentiation of art styles relatively early in the northern part of the continent makes the point that symbolism was used to mark some aspect of identity over areas much wider than the sporadic occurrences of evidence that people produced personal ornaments or fragments of ocher. There was also not necessarily a clean break between acknowledged styles, as was once thought. This is more a recognition of the tyranny of former archeological habits than anything else. Once upon a time, archeologists had no other option than to distinguish art styles as belonging to sharply differentiated cultures; now we have independent means of defining chronology, those same styles seem to merge into each other, and are part of the way in which people define cultural values. Understanding that is the real art history. Cultures are not fixed entities—they never have been—but they change as a part of continual processes of coming in to being (discussed in relation to stone tools in Davidson, 2016).

The patterns of similarity and difference within a set of artistic images seems, therefore, to be a product of the patterns of movement among groups of people who defined themselves, in part, through that production of art. Archeological studies of art have explored this set of relationships. Conkey (1980) looked at the supposedly decorative marks on bone artefacts from northern Spain around the famous site of Altamira (see Fig. 2.10) and found that there was a restricted range of design elements (see Fig. 9.42) shown in two images from her paper.

Further analysis showed that these were usually combined in particular groups of elements (as shown in Fig. 3 reproduced from her paper on the right of Fig. 9.42). We could say that this was one way in which scribbles might become art. Conkey argued that people lived in dispersed locations at different times where they marked bone tools in ways we might regard as scribbles, but when those people got together at a particular site there were distinctive characteristics of the patterning of their scribbles into combinations with definable principles of construction. We could argue that the social arrangements at that time and that place were made manifest in the art produced. McDonald and Veth (2012) have applied this language of aggregation and dispersal to a situation in the Western Desert of Western Australia.

Galt-Smith (1997, Ch. 8) showed for Central Australia that the situation was not as simple as Conkey suggested. Central Australian ethnography shows two different types of aggregation. One was represented by places of major ceremonial aggregation of people from different regions, and in Central Australia these were only represented by petroglyph (engraving) sites (as it might be Nganganterra). The other

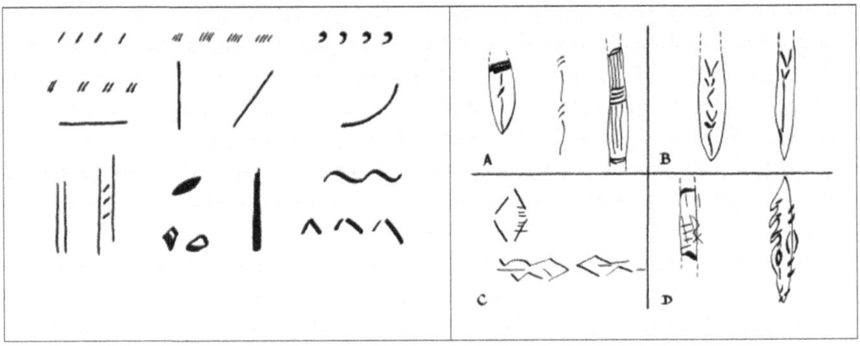

Fig. 9.42 Ranges of marks at sites around Altamira in northern Spain, taken to show that individual design elements at sites to which people dispersed (her Fig. 4 on Left) might be combined when they aggregated at Altamira (her Fig. 3, in A–D). (From Conkey, 1980. Permission from Meg Conkey)

involved more local totemic groupings and these were only present among the pictograph (painting) sites, predominantly geometric images. These would correspond with Sutton's (1995, 57) "regional travellers" on one hand and localized travelers on the other. In addition, there were dispersion sites, which included either petroglyphs or pictographs, and were characterized by hand-stencils. In this way, Galt-Smith was able to show the variation among the rock art characteristics of ethnographically identifiable aggregation sites and dispersal sites, and also show that the patterning for engraving (petroglyph) sites was different from that for painting (pictograph) sites.

The art-system, therefore, seems to have contained variations which permitted the spread of ideologies at these three different scales by using art as a means of encoding information in different ways. There were no independent checks on the information or the codes, except within the system by which knowledge was passed on from "people-who-knew" to "others-who-did-not". When you or I encounter written texts, we can, in principle, read them without having been taught anything by another person except how to read: the encoded information in rock art provided persistence of the images at rock art sites, and people who were taught the code learned to decipher them. People could only know the meaning of the images at sites if they had been initiated into the codes involved with them. Individuals in positions of power could use their socially-determined influence, and only the system of knowledge exchange and persistence at different scales within any community provided checks and balances.

In discussing writing (Chap. 5) I said:

> the 'story' included in the written text could be seen and read by several others, so that this was getting close to one-to-many communication, and with it, characteristic asynchronous communication which released the storytelling from some of its tethering to time. The power of the insiders shifted to include those who controlled access to writing and reading, such as teachers in schools or their equivalents.

It seems reasonable to state that rock art involved a system of a relatively unvarying initiation of younger people by elders, involving visual images that allowed asynchronous communication once younger people were educated into the meanings. Provided people were permitted to decipher persistent signs because there was relatively little change in the codes, rock art among a mobile people "could be referred to many times without reference to the original acts of communication." But it depended on those with power being able to teach a consistent message so that the code would be passed on faithfully.

9.9 Freeing the Story from Institutions of Control

I have (self-consciously) stretched the framework I established in Chap. 8 by likening processes unleashed by the emergence of writing to processes in Australian Aboriginal art despite the fact that Aboriginal people did not have writing but depended on the conventions within commonly held oral traditions. We will find it is an even bigger stretch to understand processes in these societies that provided the same access to change as that provided by printing on the one hand—which allowed people in different places to know the same things, with independent validation—and communication at a distance on the other—which enabled people to know things instantaneously in different places.

Morphy and Layton (1981) told a story[19] about the expansion of the range of paintings so that the Yolngu elders allowed uninitiated people to put paintings on sale. The missionaries based at Milingimbi and at Yirrkala just before Warner's study both encouraged a trade in Yolngu art and artifacts "as a source of income for the mission…it taught people the value of labour…showed…that Yolngu had skills as artists and craftsmen which demonstrated their equality as human beings" (Morphy, 2007, 32). Of course, we can take it for granted that the Yolngu themselves started off with no doubt about the value of their labor, their skills and their equality as human beings. Perhaps the emphasis on these things—or, at least, the acts of making them visible—reflected the attitudes of missionaries, whose work profited from the monetary income from the art. The process allowed Yolngu to exercise their own agency in production of art, though the images were bound up with rights, particularly rights to land, and there were strict protocols about who could dispose of paintings and how (Morphy & Layton, 1981, 58). Taylor (2015) has documented how the imagery of bark paintings has responded to tradition, the agency of the artists, and the demands of the market.

Participants in ceremony can see images painted on the bodies of young people due to be initiated. If those images are transferred onto sheets of bark they can be seen both by them and by others. This has some of the effect of fixing the story about the images. But at the same time it has the potential to place the emphasis onto the image on the bark, rather than to the orally-transmitted story that generated the initiation.

One response to commercial demand was to be selective about which designs could enter the market, in a context where there was great uniformity among the traditional designs. By selective release, clans retained their control of the designs that could be seen outside the secret/sacred context—including the design on the coffin described at the beginning of Morphy's book (1991). Another response was to increase the amount of figurative representation in the bark paintings. That increase sometimes involved the substitution of an identifiable—determinate—animal figure on the bark paintings for the indeterminate abstract or symbolic figure used in secret/sacred contexts. "What [the artist] has in effect created is a mock sacred art, which allows him to maintain full control over the clan's closed sacred art" (Morphy & Layton, 1981, 60).

We could say that this had exactly the effect that Socrates predicted in the *Phaedrus* (see Chap. 8): seers (in this case purchasers of bark paintings) seemed to know many things (the species represented) but remained mostly ignorant of the

[19] This part of the story seems most likely to be due to Morphy.

interpretive depth (the social implications of those animals) and the secret/sacred context. It would seem to be the case that the problem of the validation of the message was provided by a different mechanism: the full message was/is only conveyed in initiation into different levels of society, and the contents of the information were restricted to those who had already been initiated. That system of secrecy and restricted membership of those-who-know acted/acts as a powerful means of corroboration. Those codes were fixed, though perhaps not as much as the powerful would like to think, but there was another level at which the same material marks could be interpreted even in the absence of the codes—just as people can see the innovative strength of Picasso's development of Cubism and his retreat from Abstraction (Mallen, 2022), but some others still hanker after a meaning, and others can concentrate on the fierce way he treated and depicted women (Mechling, 2023).

My representation of this art is that it seems to be remarkably fixed and depended on social stability to allow the initiated to fulfil their ritual obligations using the learned codes. But Aboriginal life has been disrupted by outsiders—non-Aboriginal people and society. These included visiting traders from Indonesia (before Europeans), missionaries and long-term pastoralists or miners from Europe, and other invaders of Australia. Some of the interactions included unsought violence, marginalization, or paid work. Ritual obligations were sometimes fulfilled using quite novel means of producing images because of circumstances such as lack of access to traditional ochers. These circumstances sometimes involved production of art which looks like a quite different tradition (O'Connor et al., 2013).

As a result of all the disruption, the traditions of painting among the Yolngu (and elsewhere) allowed the emergence of novelty, particularly when commerce was attached to particular practices. It was not novelty on a grand scale like printing in Europe—or perhaps it was. The different scales of relationship and the confrontation with other pasts created opportunities for social change—but to a large extent the stimulus was provided by commerce.

A similar process applied in the Central Desert of Australia where the acrylic paintings of Pintupi people began a movement that has dominated many other groups of Aboriginal painting. These constituted "a new form, objects made for sale and not for their own ritual or practical use" which the Aboriginal painters tried to "define … in terms meaningful to them" (Myers, 1991, 27). As Myers documented throughout his account of the emergence of an Aboriginal "high art" in Central Australia: "Papunya people could express themselves, demonstrate their creativity, and develop their own sense of self. That is, they could find confidence and money, rather than demoralization and despair, in activity linked to their own values" (Myers, 2002, 131). The paintings produced derived from the painters' deeply internalized principles of art, just as the more-figurative barks of the Yolngu did, but the product was not anything that they had produced before. The outcome was a previously-unseen combination of those principles with the demands of commercialization.

9.10 Overcoming the Tyranny of Time

Kimber (2004) and Foster (1998) have both recounted the use of Aboriginal messengers to carry messages written on paper, or "yabber," and carried in a "cleft stick". Although the individual messengers covered prodigious distances, this just illustrates the need that Europeans in Australia felt for more rapid communications, as well as the likely travel of individual messengers on such paths for other reasons. The practice is surrounded by anecdotes of the way the messages were regarded by carriers who were unable to read. Anecdotes of theft by messengers in this sort of system use identical words whether in India 400 years ago, or Australia 150 years ago, suggesting that the anecdotes were driven by common racist attitudes among colonists rather than by any certain aspect of the behavior of the messengers, or even repeated cases of it. Nevertheless, veracity of the message was an issue.

To some extent the question of veracity was solved for Aboriginal rock art not by confirmation of any claim but by people embedding their activities in a common set of life opportunities. There was an ongoing tradition of people being summoned to ceremonies by messages sent from the ceremonial place to distant places. Message sticks were involved in that process including their use as a passport (Davidson et al., 2004), and the well-documented use of "paper yabber" was just plugging in to the same system of people carrying messages over long distances by foot. The results were neither instantaneous nor anywhere near it in the way that the semaphore, the telegraph, and subsequent digital communication are. Yet the framework was there.

First, there were massive continental-scale continuities in beliefs exemplified by songlines (Neale, 2017b; Neale & Kelly, 2020) which could only have existed in the presence of common stories and common codes for the imagery that accompanied the traditions (Hercus, n.d.). Some of that system is visible in the rock art. That long travels were not only possible because of such traditions but actually happened is shown by the documentation of the trade in a ceremony called *Molonga* in North West Queensland (Roth, 1897, 117–125) and *Mudlunga* around Lake Eyre (Hercus, 1980). Roth recorded the dates at which it was first performed in Carandotta, the place of his first mention in 1893, and at other places among the European settlements as it travelled from the north to the south. He encountered it in Boulia in 1895 nearly 250 km further south. In a letter of 1898, Gillen (Mulvaney et al., 1997, 206) talked about the arrival of the ceremony in Alice Springs (600 km to the west of Boulia). He called it *Chitching-alla*, but claimed that the ceremony was identical with *Molonga*, and said that it had begun in the saltwater country (which would imply a journey of at least 1000 km). Mick Maclean reported his excitement as a child at the arrival of the same ceremony, which he called *Mudlunga*, at a station near Maree 1000 km south of Boulia in 1901 (Hercus, 1980).

Kimber (1990) suggested that:

> …the Mulunga ceremony illustrates the dynamism of Aboriginal cultures, and their flexibility – their proselytising, in fact, of new ideas so long as there is a basic framework of understanding. This dynamic aspect of the various Aboriginal cultures within Australia is as important (one might well argue is more important) to understand as those which reflect conservatism.

This openness to new ideas is well-illustrated by the incorporation into the Dreaming beliefs of elements of ceremony from non-Aboriginal events, such as military events on the south coast (White, 1980). Together with the ideology of the Dreaming exemplified by the Tonkinson story, successive revelation could be depersonalized so that it became attributed to the Ancestors and generated an ideology of timelessness—what Stanner called "everywhen"—which overcame the tyranny of time. In this context, the standardization of images in some of the art sites allowed the rapid spread of information, sacred and secular, across vast distances. One of the performances of the Mulunga was supervised by a Wangkamadla man (Hercus, 1980, note 4), and Wangkamadla own the site of Nganganterra which we have been discussing. People accepted invitations to participate in ceremonies across huge areas which is an indication of the extent of relatedness of people and their stories.

None of the Australian Aboriginal peoples at their various contacts with Europeans or others had the use of writing, printing, or instantaneous communication across distance, but they were able to incorporate the conservatism that writing gave through the engagement of their communities with their rituals. The existence of a hierarchy of successively revealed meanings meant that individual elements of the set of meanings could be used in other contexts (such as by being used in commerce) and take on new importance. Those who controlled the commerce could acquire power through their control of the commercialization of art. Finally, sharing over large distances of the emerging practices involved in producing art did not free people from the tyranny of time, but reduced the impact of the lack of restrictions.

9.11 Essay 9 Art and Archeohistory (LzBe1, Selwyn Ranges and Cathedral Cave, Carnarvon Gorge)

This painstaking analysis leads us to the situation where art before the Art 1 of the modern world derived its full force from the interactions between the producer and the cultural knowledge of people in the rest of the community. Cultural context shared within the community, such as that for Fig. 4.8 Right, allowed much greater understanding—it illustrates part of the song of Odysseus—but could be interpreted at some level because it represented a scene—it is a picture of a sailor on a boat with rowers. Scenes allowed individual images to tell a story familiar to anyone (Fig. 4.8 Left), but specific images were even more meaningful when the cultural context of the society was understood (as in Figs. 4.8 Right and 5.22 Right). This is the difference in cognition between the implicational system that depends on inputs from external sources and the propositional system that depends on the cultural knowledge in the mind, as Barnard and colleagues would suggest (Barnard et al., 2016). In the scene of a man and a woman in Fig. 4.8 Left, the fact that the man anticipates something is a result of the external inputs to the implicational system. That would be true of the scene of a man being rowed in a boat (Fig. 4.8 Right) but by using the propositional system (the cultural knowledge) such a scene has a much richer meaning because the mind remembers the cultural context of the story of the Sirens.

Early Australians were already making their distinctive marks by symbolic structuring of their relationships with the environment and hence each other. The use of beads (Balme & O'Connor, 2019; Langley et al., 2019) relatively early in Australia gives some indication of the antiquity of a structuring of interpersonal relations that was also symbolic in nature. That symbolic structure of personal relations evolved within Australia and ended up with the system of Songlines recorded in the recent past. Among other things, Songlines show how the First Australians used the symbolism of art to define their relationships to each other and to the land (Neale & Kelly, 2020).

In Australia, the country is crisscrossed with Songlines or Dreaming tracks which create the cultural context of First Peoples (e.g., Neale & Kelly, 2020). Cultural stories of people who live in particular places have been linked: each location where there is a Dreaming story became a node along a Songline, giving importance to place and to attachment to Country. Each place is important for the stories that link it to other parts of the landscape through the songs, whether the community know the distant places or not—though sometimes people even have knowledge of places they have never visited (Russell, 2012). Research to investigate the age of such belief systems has shown that they reached their modern form after a process of change (David, 2002; Hiscock & Faulkner, 2006). This is only to be expected, as the cosmology related to the formation of the landscape, so could only have come into existence once the landscape was known. The issue becomes how a symbolic system evolved and impacted behavior across a whole country.

The first step has been to understand the way in which the first colonists of Sahul adapted to the new continent (Crabtree et al., 2021; Salles et al., 2024). The well-documented art sequences across northern Australia (Veth et al., 2011), with different regional traditions back to the Pleistocene, indicate an early symbolic differentiation of populations probably before the last period of low sea-level and maximum cold. Chronologies are not always anchored but long sequences can be shown, despite some uncertainty, for Arnhem Land (David et al., 2017), Kimberley (David et al., 2019a; Finch et al., 2020; Travers & Ross, 2016), Murujuga (McDonald et al., 2018; Mulvaney, 2015), and Cape York (Cole & Watchman, 2005; Morwood & Hobbs, 1995). In other places, it would seem that sequences were not so long, and this almost certainly has implications for the ways in which different regions were used (Brady & Bradley, 2014; Davidson et al., 2005). Other places seem to have discontinuities within an underlying phased art sequence (McDonald, 2017), but in different regions people reacted to their local circumstances with distinctive outcomes (which may have included environmental change, population numbers, migration of people, and consequent social change including variation in religious system) (Ross, 2013). The outcome was diversity of material culture, languages, ritual and social organization (Davidson, 2014a). People interacted and differentiated their art styles in different ways and the way each group chose played its part in the evolution of diversity in other aspects of society.

People developed relationships with each other and may have demonstrated relationships ritually and with material markers (Balme & O'Connor, 2019; Balme et al., 2018; McAdam & Davidson, 2018). They also marked places as they moved, and we see those places as rock art (Davidson, 2012). The ritual and other significance of such places made a difference to foraging decisions (Davidson, 2013b, 16): the significance defined attachment to places, but the ranks of resources differ depending on the location of the people in relation to particular resources. In this way, marked places, such as rock art sites, affected foraging decisions—the mesa in Fig. 9.43 is marked in this way and overlooks a large archeological site—potentially making cultural behavior inefficient (Byrne et al., 2004). For example, if places acquired ritual significance through marking, the cultural choices about using or avoiding such places defined options about access to food sources. Locations became important not just because of the access they gave to resources, but through

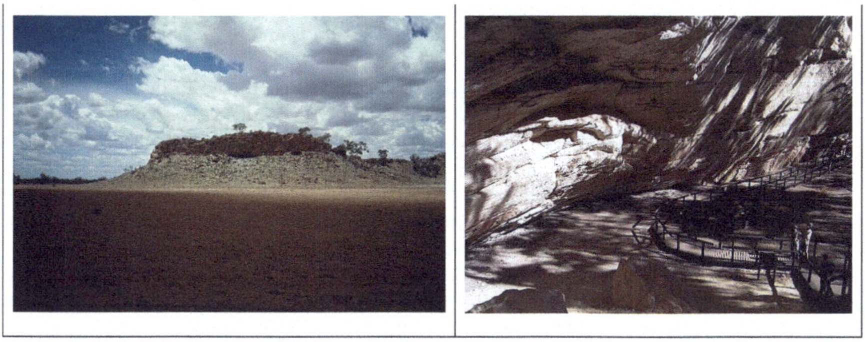

Fig. 9.43 Left: LzBe1 Mesa, Selwyn Ranges, south of Cloncurry, Queensland. (Photo Iain Davidson). Right: Cathedral Cave, Carnarvon Gorge, Queensland showing the boardwalk for visitation (now burned in bushfires) 2005. (Photo Iain Davidson)

the ritual knowledge about them. In such a way, art sites that seem principally important, to archeologists and others, because of the content of the images, can be seen as significant to foragers as a result of their location in relation to resources. It may, however, be more difficult to argue for the same sort of implication for art deep in caves (wherever it occurs, but particularly) in Western Europe because, even though it was probably used ritually, it would only have been seen by people making special purpose visits inside the cave—and for that they would already have needed to know the art was there.

The place of the art itself became one of the central aspects of communities, just as Cathedral Cave in the Central Queensland Highlands did (Fig. 9.43 Right) (Giorgi & Taçon, 2019). One of the clichés about rock art emphasizes that it is all produced on rocks, but that is not a trivial observation. As a result of it being made on the rock, once teller, story and (original) seers are all gone, the material sign is still present on the rock, and the rock is still in the same place it was originally. The materiality of the image still communicates something even if the producer and the initial seer(s) have gone. The archeologist is a new agent and that agent's task is to understand the meanings that attach to the rock art now.

To some extent, the site with the rock art is the equivalent of the art museum, in which art from a variety of producers, each with different stories told to different seers, is concentrated. The big difference is still about the transferability of the material image, the finances that make that possible, and the fact that all of the art in museums was produced after it became possible to print images of it.

References

Adeleye, M. A., Hopf, F., Haberle, S. G., Stannard, G. L., Mcwethy, D. B., Harris, S., & Bowman, D. M. J. S. (2024). Landscape burning facilitated Aboriginal migration into Lutruwita/Tasmania 41,600 years ago. *Science Advances, 10*(46), eadp6579. https://doi.org/10.1126/sciadv.adp6579

Akerman, K. (1973). Aboriginal baler shell objects in Western Australia. *Mankind, 9*(2), 124–125.

Akerman, K. (2009). *Wanjinas--a synopsis of knowledge to date. A Kimberley Foundation scoping paper*. Kimberley Foundation of Australia.

Akerman, K. (2016). *Wanjina: Notes on some iconic ancestral beings of the Northern Kimberley*. Hesperian Press.

Akerman, K. (2018). The esoteric and decorative use of bone, shell, and teeth in Australia. In M. Langley, M. Litster, D. Wright, & S. K. May (Eds.), *The archaeology of portable art: Southeast Asian, Pacific, and Australian perspectives* (pp. 199–219). Routledge.

Allen, L. (2011). *Ancestral powers and the aesthetic: Arnhem land paintings and objects from the Donald Thomson collection*. Museum Victoria Publishing.

Allen, J., & O'Connell, J. F. (2020). A different paradigm for the initial colonisation of Sahul. *Archaeology in Oceania, 55*(1), 1–14. https://doi.org/10.1002/arco.5207

Aschero, C. A., & Schneier, P. (2021). 20. The "black series" in the hunting scenes of Cueva de las Manos, Río Pinturas, Patagonia, Argentina. In I. Davidson & A. Nowell (Eds.), *Making scenes: Global perspectives on scenes in rock Art* (pp. 310–326). Berghahn Books.

Atkinson, A. (1997). *The Europeans in Australia. A history* (Vol. 1). Oxford University Press.

Aubert, M. (2012). A review of rock art dating in the Kimberley, Western Australia. *Journal of Archaeological Science, 39*, 573–577.

Aubert, M., Brumm, A., Ramli, M., Sutikna, T., Saptomo, E. W., Hakim, B., et al. (2014). Pleistocene cave art from Sulawesi, Indonesia. *Nature, 514*(7521), 223–227. https://doi.org/10.1038/nature13422

Aubert, M., Brumm, A., & Huntley, J. (2018a). Early dates for 'Neanderthal cave art' may be wrong. *Journal of Human Evolution, 125*, 215–217.

Aubert, M., Setiawan, P., Oktaviana, A. A., Brumm, A., Sulistyarto, P. H., Saptomo, E. W., et al. (2018b). Palaeolithic cave art in Borneo. *Nature, 564*(7735), 254–257. https://doi.org/10.1038/s41586-018-0679-9

Aubert, M., Lebe, R., Oktaviana, A. A., Tang, M., Burhan, B., Hamrullah, et al. (2019). Earliest hunting scene in prehistoric art. *Nature, 576*(7787), 442–445. https://doi.org/10.1038/s41586-019-1806-y

Auel, J. M. (1980). *The clan of the cave bear*. Hodder and Stoughton.

Baffier, D., & Feruglio, V. (2001). La salle Brunel. In J. Clottes (Ed.), *La Grotte Chauvet. L'art des origines* (pp. 64–73). Seuil.

Bahn, P. G. (1998). *The Cambridge illustrated history of prehistoric art*. Cambridge University Press.

Balme, J. (2013). Of boats and string: The maritime colonisation of Australia. *Quaternary International, 285*, 68–75. https://doi.org/10.1016/j.quaint.2011.02.029

Balme, J., & Morse, K. (2006). Shell beads and social behaviour in Pleistocene Australia. *Antiquity, 80*(310), 799–811.

Balme, J., & O'Connor, S. (2019). Personal ornaments in early prehistory bead making in Aboriginal Australia from the deep past to European arrival: Materials, methods, and meanings. *PaleoAnthropology, 177*, 195.

Balme, J., Davidson, I., McDonald, J., Stern, N., & Veth, P. (2009). Symbolic behaviour and the peopling of the southern arc route to Australia. *Quaternary International, 202*(1–2), 59–68.

Balme, J., O'Connor, S., & Langley, M. C. (2018). Marine shell ornaments in northwestern Australian archaeological sites: Different meanings over time and space. In M. Langley, M. Litster, D. Wright, & S. K. May (Eds.), *The archaeology of portable art: Southeast Asian, Pacific and Australian perspectives* (pp. 258–273). Routledge.

Bardon, G. (1991). *Papunya Tula. Art of the Western Desert*. McPhee Gribble.

Barker, B., Lamb, L., Delannoy, J.-J., David, B., Gunn, R. G., Chalmin, E., et al. (2017). Archaeology of JSARN–124 site 3, central-western Arnhem Land: Determining the age of the so-called 'Genyornis' painting. In B. David, P. S. C. Taçon, J. J. Delannoy, & J. M. Geneste (Eds.), *The archaeology of rock art in western Arnhem Land, Australia* (Vol. Terra Australis 47, pp. 423–496). ANU Press.

Barnard, P. J., Davidson, I., & Byrne, R. W. (2016). Toward a richer theoretical scaffolding for interpreting archaeological evidence concerning cognitive evolution. In T. Wynn & F. Coolidge (Eds.), *Cognitive models in Palaeolithic archaeology* (pp. 45–67). Oxford University Press.

Bar-Yosef Mayer, D. E. (2005). The exploitation of shells as beads in the Palaeolithic and Neolithic of the Levant. *Paléorient, 31*(1), 176–185.

Bar-Yosef Mayer, D. E., Vandermeersch, B., & Bar-Yosef, O. (2009). Shells and ochre in Middle Paleolithic Qafzeh Cave, Israel: Indications for modern behavior. *Journal of Human Evolution, 56*(3), 307–314.

Beaton, J. M., & Walsh, G. L. (1977). Che-ka-ra. *Mankind, 11*(1), 46–48.

Beckett, J., Hercus, L., Martin, S., & Colyer, C. (2008). *Mutawintji: Aboriginal cultural association with Mutawintji National Park*. Office of the Registrar, Aboriginal Land Rights Act 1983.

Bednarik, R. G. (2017). Pareidolia and rock art interpretation. *Anthropologie (Brno), 55*(1/2), 101–117.

Bird, M. I., Condie, S. A., O'Connor, S., O'Grady, D., Reepmeyer, C., Ulm, S., et al. (2019). Early human settlement of Sahul was not an accident. *Scientific Reports, 9*(1), 8220. https://doi.org/10.1038/s41598-019-42946-9

Blundell, V., & Woolagoodja, D. (2005). *Keeping the Wanjinas fresh: Sam Woolagoodja and the enduring power of Lalai*. Fremantle Press.

Blundell, V., & Woolagoodja, D. (2012). Rock art, aboriginal culture, and identity: The Wandjina paintings of northwest Australia. In J. McDonald & P. Veth (Eds.), *A companion to rock art* (pp. 472–487). Wiley-Blackwell.

Bowern, C. (2012). The riddle of Tasmanian languages. *Proceedings of the Royal Society B: Biological Sciences, 279*, 4590–4595. https://doi.org/10.1098/rspb.2012.1842

Bowern, C. (Ed.). (2023). *The Oxford guide to Australian languages*. Oxford University Press.

Bowler, J. M., & Thorne, A. (1976). Human remains from Lake Mungo: Discovery and excavation of Lake Mungo III. In R. L. Kirk & A. G. Thorne (Eds.), *The origin of the Australians* (pp. 127–138). Australian Institute of Aboriginal Studies.

Bowler, J. M., Jones, R., Allen, H., & Thorne, A. G. (1970). Pleistocene human remains from Australia: A living site and human cremation from Lake Mungo, western New South Wales. *World Archaeology, 2*, 39–60.

Bradshaw, C. J. A., Ulm, S., Williams, A. N., Bird, M. I., Roberts, R. G., Jacobs, Z., et al. (2019). Minimum founding populations for the first peopling of Sahul. *Nature Ecology & Evolution, 3*(7), 1057–1063. https://doi.org/10.1038/s41559-019-0902-6

Bradshaw, C. J. A., Reepmeyer, C., Saltré, F., Agapiou, A., Kassianidou, V., Demesticha, S., et al. (2024). Demographic models predict end-Pleistocene arrival and rapid expansion of pre-agropastoralist humans in Cyprus. *Proceedings of the National Academy of Sciences, 121*(21), e2318293121. https://doi.org/10.1073/pnas.2318293121

Brady, L. M., & Bradley, J. J. (2014). Reconsidering regional rock art styles: Exploring cultural and relational understandings in northern Australia's Gulf country. *Journal of Social Archaeology, 14*(3), 361–382.

Brady, L. M., Bradley, J. J., Kearney, A., & Wesley, D. (2020). A rare miniature and small-scale stencil assemblage from the Gulf of Carpentaria: Replication and meaning in Australian rock art. *Antiquity, 94*(375), 780–796.

Brady, L. M., Bradley, J., & Kearney, A. (2023). *Jakarda Wuka (too many stories): Narratives of rock art from Yanyuwa Country in northern Australia's Gulf of Carpentaria*. Sydney University Press.

Brandl, E. J. (1973). *Australian Aboriginal paintings in western and central Arnhem Land: Temporal sequences and elements of style in Cadell River and Deaf Adder Creek art*. Australian Institute of Aboriginal Studies.

Brumm, A., Langley, M. C., Moore, M. W., Hakim, B., Ramli, M., Sumantri, I., et al. (2017). Early human symbolic behavior in the Late Pleistocene of Wallacea. *Proceedings of the National Academy of Sciences, 114*(16), 4105–4110. https://doi.org/10.1073/pnas.1619013114

Brumm, A., Oktaviana, A. A., Burhan, B., Hakim, B., Lebe, R., Zhao, J.-X., et al. (2021). Oldest cave art found in Sulawesi. *Science Advances, 7*(3), eabd4648. https://www.science.org/doi/10.1126/sciadv.abd4648

Byrne, R. W., Barnard, P. J., Davidson, I., Janik, V. M., McGrew, W. C., Miklósi, Á., & Wiessner, P. (2004). Understanding culture across species. *Trends in Cognitive Sciences, 8*(8), 341–346.

Carroll, P. J. (1977). Mimi from western Arnhem Land. In P. J. Ucko (Ed.), *Form in indigenous art; schematisation in the art of Aboriginal Australia and Prehistoric Europe* (pp. 119–130). Australian Institute of Aboriginal Studies.

Cartailhac, E., & Breuil, H. (1906). *La caverne d'Altamira à Santillana, près Santander (Espagne)*. Monaco.

Carty, J. (2021). *Balgo: Creating country*. UWA Publishing.

Chaloupka, G. (1999). *Journey in time. The world's longest continuing art tradition. The 50, 000 year story of Australian Aboriginal rock art of Arnhem Land*. Reed New Holland.

Chazine, J.-M. (2005). Rock art, burials, and habitations: Caves in East Kalimantan. *Asian perspectives, 44*(1), 219–230.

Clarkson, C., Jacobs, Z., Marwick, B., Fullagar, R., Wallis, L., Smith, M. A., et al. (2017). Human occupation of northern Australia by 65,000 years ago. *Nature, 547*(7663), 306–310. https://doi.org/10.1038/nature22968

Clottes, J., & Geneste, J. M. (2012). Twelve years of research in Chauvet Cave: Methodology and main results. In J. McDonald & P. Veth (Eds.), *A companion to rock art* (pp. 583–604). Blackwell.

Cobden, R., Clarkson, C., Price, G. J., David, B., Geneste, J.-M., Delannoy, J.-J., et al. (2017). The identification of extinct megafauna in rock art using geometric morphometrics: A Genyornis newtoni painting in Arnhem Land, northern Australia? *Journal of Archaeological Science, 87*(Supplement C), 95–107. https://doi.org/10.1016/j.jas.2017.09.013

Cole, N., & Watchman, A. (2005). AMS dating of rock art in the Laura region, Cape York Peninsula, Australia–protocols and results of recent research. *Antiquity, 79*, 661–678.

Conkey, M. W. (1980). The identification of prehistoric hunter-gatherer aggregation sites: The case of Altamira. *Current Anthropology, 21*(5), 609–630.

Cook, N. D. J., Davidson, I., & Sutton, S. A. (1990). Why are so many ancient rock paintings red? *Australian Aboriginal Studies, 1990*(1), 30–32.

Coulson, D., & Campbell, A. (2001). *African rock art*. Harry N. Abrams. Inc.

Crabtree, S. A., White, D. A., Bradshaw, C. J. A., Saltré, F., Williams, A. N., Beaman, R. J., et al. (2021). Landscape rules predict optimal superhighways for the first peopling of Sahul. *Nature Human Behaviour, 5*(10), 1303–1313. https://doi.org/10.1038/s41562-021-01106-8

David, B. (2002). *Landscapes, rock-art and the dreaming. An archaeology of preunderstanding*. Leicester University Press.

David, B. (2004). Rock-art and the experienced landscape: The emergence of late Holocene symbolism in north-east Australia. In C. Chippindale & G. Nash (Eds.), *The figured landscapes of rock-art. Looking at pictures in place* (pp. 153–181). Cambridge University Press.

David, B. (2017). *Cave art*. Thames & Hudson.

David, B., Chant, D., & Flood, J. (1992). Jalijbang 2 and the distribution of pecked faces in Australia. *Memoirs of the Queensland Museum, 32*(1), 61–77.

David, B., McNiven, I. J., Attenbrow, V., Flood, J., & Collins, J. (1994). Of Lightning Brothers and White Cockatoos: Dating the antiquity of signifying systems in the Northern Territory, Australia. *Antiquity, 68*(259), 241–251.

David, B., Barker, B., Petchey, F., Delannoy, J.-J., Geneste, J.-M., Rowe, C., et al. (2013a). A 28,000 year old excavated painted rock from Nawarla Gabarnmang, northern Australia. *Journal of Archaeological Science, 40*(5), 2493–2501. https://doi.org/10.1016/j.jas.2012.08.015

David, B., Geneste, J.-M., Petchey, F., Delannoy, J.-J., Barker, B., & Eccleston, M. (2013b). How old are Australia's pictographs? A review of rock art dating. *Journal of Archaeological Science, 40*(1), 3–10. https://doi.org/10.1016/j.jas.2012.08.019

David, B., Taçon, P. S. C., Delannoy, J.-J., & Geneste, J.-M. (Eds.). (2017). *The archaeology of rock art in western Arnhem Land, Australia*. Australian National University Press.

David, B., Delannoy, J.-J., Mialanes, J., Clarkson, C., Petchey, F., Geneste, J.-M., et al. (2019a). 45,610–52,160 years of site and landscape occupation at Nawarla Gabarnmang, Arnhem Land plateau (northern Australia). *Quaternary Science Reviews, 215*, 64–85. https://doi.org/10.1016/j.quascirev.2019.04.027

David, B., Delannoy, J.-J., Petchey, F., Gunn, R. G., Huntley, J., Veth, P., et al. (2019b). Dating painting events through by-products of ochre processing: Borologa 1 Rockshelter, Kimberley, Australia. *Australian Archaeology, 85*(1), 57–94.

Davidson, D. S. (1937). A preliminary consideration of Aboriginal Australian decorative art. *American Philosophical Society Memoirs, 9*, 147.

Davidson, I. (1983). On the edge of the Simpson: Recent additions to understanding of prehistoric artefact distributions in arid Australia. *Australian Archaeology, 17*, 27–37.

Davidson, I. (1988). The naming of parts: Ethnography and the interpretation of Australian prehistory. In B. Meehan & R. Jones (Eds.), *Archaeology with ethnography: An Australian perspective* (pp. 17–32). Department ofPrehistory, ANU.

Davidson, I. (1989). Freedom of information: Aspects of art and society in western Europe during the last ice age. In H. Morphy (Ed.), *Animals into art* (pp. 440–456). Unwin Hyman.

Davidson, I. (1997). The power of pictures. In M. Conkey, O. Soffer, D. Stratmann, & N. G. Jablonski (Eds.), *Beyond art: Pleistocene image and symbol* (Vol. 23, pp. 128–158). The California Academy of Sciences.

Davidson, I. (2008). Chapter 7. Ethnological studies and archaeology of North West Central Queensland. In R. McDougall & I. Davidson (Eds.), *The Roth family, anthropology and colonial administration* (pp. 121–132). Left Coast Press.

Davidson, I. (2010a). Australian archaeology as a historical science. *Journal of Australian Studies, 34*(3), 377–398.

Davidson, I. (2010b). The colonization of Australia and its adjacent islands and the evolution of modern cognition. *Current Anthropology, 51*(s1), s177–s189. https://doi.org/10.1086/650694

Davidson, I. (2012). Four questions about foraging models and the process of colonisation. *Australian Archaeology, 74*, 19–20.

Davidson, I. (2013a). Origins of pictures: An argument for transformation of signs. In K. Sachs-Hombach & J. R. J. Schirra (Eds.), *Origins of pictures. Anthropological discourses in image science* (pp. 15–45). Halem.

Davidson, I. (2013b). Peopling the last new worlds: The first colonisation of Sahul and the Americas. *Quaternary International, 285*, 1–29. https://doi.org/10.1016/j.quaint.2012.09.023

Davidson, I. (2014a). Hunter gatherers in Australia: Deep histories of continuity and change. In V. Cummings, P. Jordan, & M. Zvelebil (Eds.), *Oxford handbook of the archaeology and anthropology of hunter gatherers* (pp. 368–404). Oxford University Press.

Davidson, I. (2014b). It's the thought that counts: Unpacking the "package" of behaviour of the first people of Sahul and its adjacent islands. In M. Porr & R. Dennell (Eds.), *East of Africa. Southern Asia, Australia and human origins* (pp. 243–256). Cambridge University Press.

Davidson, I. (2016). Stone tools: Evidence of something in between culture and cumulative culture? In M. N. Haidle, N. J. Conard, & M. Bolus (Eds.), *The nature of culture* (pp. 99–120). Springer Netherlands.

Davidson, I. (2017). Images of animals in rock art: Not just "good to think". In B. David & I. J. McNiven (Eds.), *Oxford handbook of the archaeology and anthropology of rock art* (pp. 435–467). Oxford University Press.

Davidson, I. (2020). Marks, pictures and art: Their contribution to revolutions in communication. *Journal of Archaeological Method and Theory, 27*(3), 745–770. https://doi.org/10.1007/s10816-020-09472-9

Davidson, I. (2021). 1. Scenes and non-scenes in rock art. In I. Davidson & A. Nowell (Eds.), *Making scenes: Global perspectives on scenes in rock art* (pp. 16–31). Berghahn Books.

Davidson, I. (2023). Humans making history through continuities and discontinuities in art. *Cambridge Archaeological Journal, 33*, 637–654. https://doi.org/10.1017/S0959774323000057

Davidson, I. (2024). Arte Paleolítico en la Península Ibérica y más allá/Art and communication in Iberia and beyond. In A. B. Marín-Arroyo, O. Moro-Abadía, & A. Diez-Castillo (Eds.), *Arte, pensamiento simbólico, modos de vida en la Prehistoria* (pp. 173–188). Editorial Universidad de Cantabria.

Davidson, I., & Noble, W. (1992). Why the first colonisation of the Australian region is the earliest evidence of modern human behaviour. *Archaeology in Oceania, 27*, 135–142.

Davidson, I., & Nowell, A. (2021). Introduction. Behind the scenes—did scenes in rock art create new ways of seeing the world? In I. Davidson & A. Nowell (Eds.), *Making scenes: Global perspectives on scenes in rock art* (pp. 1–15). Berghahn Books.

Davidson, I., Tarragó, I., & Sullivan, T. (2004). Market forces. In V. Donovan & C. Wall (Eds.), *Making connections: A journey along Central Australian Aboriginal trading routes* (pp. 12–23). Arts Queensland.

Davidson, I., Cook, N. D. J., Fischer, M., Ridges, M., Ross, J., & Sutton, S. A. (2005). Archaeology in another country: Exchange and symbols in North West Central Queensland. In I. Macfarlane, M.-J. Mountain, & R. Paton (Eds.), *Many exchanges: Archaeology, history, community and the work of Isabel McBryde* (pp. 101–128). Aboriginal History Inc.

Davidson, I., Burke, H., Sullivan, L., Wallis, L. A., Artym, U., & Barker, B. (2020). Cultural conflict in text and materiality: The impact of words and lead on the northwest Queensland colonial frontier, Australia. *World Archaeology, 51*(5), 724–740. https://doi.org/10.1080/00438243.2020.1711153

Davidson, I., Burke, H., Connelly, P., Porter, S., Sullivan, H., Sullivan, L., et al. (2021). Oral tradition, history, and archaeohistory of Indigenous Australia. In I. J. McNiven & B. David (Eds.), *The Oxford handbook of the archaeology of Indigenous Australia and New Guinea* (pp. 157–186). Oxford University Press.

Dix, W. (1977). Facial representations in Pilbara rock engravings. In P. J. Ucko (Ed.), *Form in Indigenous Art. Schematisation in the art of Aboriginal Australia and prehistoric Europe* (pp. 277–285). Australian Institute of Aboriginal Studies.

Dix, S. (2016). New Aboriginal hand stencil discoveries at the Freycinet National Park, Tasmania. *Australian Archaeology, 82*(1), 76–79. https://doi.org/10.1080/03122417.2016.1165927

Domingo Sanz, I. (2021). 16. New insights into the analysis of Levantine Rock Art scenes informed by observations on Western Arnhem Land Rock Art. In I. Davidson & A. Nowell (Eds.), *Making scenes: Global perspectives on scenes in rock art* (pp. 240–258). Berghahn Books.

Donald, M. (1991). *Origins of the modern mind. Three stages in the evolution of culture and cognition*. Harvard University Press.

Donaldson, M. (2007). Introduction and overview of Kimberley rock art. In M. Donaldson & K. Keneally (Eds.), *Rock art of the Kimberley* (pp. 1–24). Kimberley Society.

Dubey-Pathak, M., & Clottes, J. (2020). Hands in Chhattisgarh rock art. *Adoranten*, 1–21.

Edwards, R. (1968). Prehistoric rock engravings at Thomas Reservoir, Cleland Hills, western central Australia. *Records of the South Australian Museum, 15*(4), 647–670.

Evans, N., & McConvell, P. (1998). The enigma of Pama-Nyungan expansion in Australia. In R. Blench & M. Spriggs (Eds.), *Archaeology and Language II. Correlating archaeological and linguistic hypotheses*. Routledge.

Fernández Navarro, V., Godinho, R. M., García Martínez, D., & Garate Maidagan, D. (2024). Exploring the utility of Geometric Morphometrics to analyse prehistoric hand stencils. *Scientific Reports, 14*(1), 6336. https://doi.org/10.1038/s41598-024-56889-3

Finch, D., Gleadow, A., Hergt, J., Levchenko, V. A., Heaney, P., Veth, P., et al. (2020). 12,000-Year-old Aboriginal rock art from the Kimberley region, Western Australia. *Science Advances, 6*(6), eaay3922. https://doi.org/10.1126/sciadv.aay3922

Finch, D., Gleadow, A., Hergt, J., Heaney, P., Green, H., Myers, C., et al. (2021). Ages for Australia's oldest rock paintings. *Nature Human Behaviour, 5*(3), 310–318. https://doi.org/10.1038/s41562-020-01041-0

Fiore, D. (2020). The art of making images: Technological affordance, design variability and labour organization in the production of engraved artefacts and body paintings in Tierra del Fuego (Southern South America). *Journal of Archaeological Method and Theory., 27*, 481–510. https://doi.org/10.1007/s10816-020-09474-7

Flood, J. (1997). *Rock art of the dreamtime: Images of ancient Australia*. Angus & Robertson.

Forge, J. A. (1991). Handstencils: Rock art or not art. In A. Rosenfeld & P. G. Bahn (Eds.), *Rock art and prehistory* (pp. 39–44). Oxbow Press.

Foster, R. (1998). Paper yabber: The messenger and the message. *Aboriginal History, 22*, 105–115.

Franklin, N. R. (1991). Explorations of the Panaramitee style. In P. Bahn & A. Rosenfeld (Eds.), *Rock art and prehistory* (pp. 120–135). Oxbow Books.

Franklin, N. R. (2007). Discontinuous dreaming networks: Analyses of variability in Australian pre-historic petroglyphs. *Rock Art Research, 24*(1), 79–103.

Franklin, N. R., Giorgi, M., Habgood, P. J., Wright, N., Gorringe, J., Gorringe, B., et al. (2021). Gilparrka Almira, a rock art site in Mithaka Country, southwest Queensland: Cultural connections, dreaming tracks and trade routes. *Archaeology in Oceania, 1*, 284–303. https://doi.org/10.1002/arco.5244

Gabriel, J., & Gorecki, P. (2014). *The 'Karawari Caves Precinct' of the Sepik River Basin, Papua New Guinea*. Cairns Institute, James Cook University.

Galt-Smith, B. (1997). *Motives for motifs: Identifying aggregation and dispersion settlement patterns in the rock art assemblages of Central Australia. Unpublished Bachelor of Arts (Honours) thesis*. University of New England, Armidale.

Gibson, J. (2013). Addressing the Arrernte: FJ Gillen's 1896 Engwura speech. *Australian Aboriginal Studies, 2013*(1), 57–72.

Giorgi, M., & Taçon, P. S. C. (2019). Carnarvon Gorge: Safekeeping a place and Indigenous agency within rock art research and management. *Australian Archaeology, 85*(2), 184–195. https://doi.org/10.1080/03122417.2019.1647508

Goldhahn, J., Taylor, L., Tacon, P. S. C., May, S. K., & Maralngurra, G. (2021). Paddy Compass Namadbara and Baldwin Spencer: An artist's recollection of the first commissioned Aboriginal bark paintings in Oenpelli, 1912. *Australian Aboriginal Studies*, (2), 46–65.

Golding, W. (1950). *The inheritors*. Faber and Faber.

Gould, R. (1969). Puntutjarpa rockshelter: A reply to Messrs. Glover and Lampert. *Archaeology and Physical Anthropology in Oceania, 4*(3), 229–237.

Green, J. (2014). *Drawn from the ground. Sound, sign and inscription in Central Australian sand stories*. Cambridge University Press.

Gunn, R. G., Douglas, L. C., & Whear, R. L. (2011). What bird is that? Identifying a probable painting of Genyornis newtoni in Western Arnhem Land. *Australian Archaeology, 73*, 1–12.

Gunn, R. G., David, B., Delannoy, J.-J., Smith, B., Unghangho, A., Waina, I., et al. (2022). Superpositions and superimpositions in rock art studies: Reading the rock face at Pundawar Manbur, Kimberley, northwest Australia. *Journal of Anthropological Archaeology, 67*, 101442. https://doi.org/10.1016/j.jaa.2022.101442

Hamby, L. (2008). The reluctant collector: Lloyd Warner. In N. Peterson, L. Allen, & L. Hamby (Eds.), *The makers and making of Indigenous Australian museum collections* (pp. 355–385). Melbourne University Press.

Hayes, E. H., Fullagar, R., Field, J. H., Coster, A. C. F., Matheson, C., Nango, M., et al. (2022). 65,000-years of continuous grinding stone use at Madjedbebe, Northern Australia. *Scientific Reports, 12*(1), 11747. https://doi.org/10.1038/s41598-022-15174-x

Hayward, J. A., Johnston, I. G., May, S. K., & Taçon, P. S. C. (2018). Memorialization and the stencilled rock art of Mirarr Country, northern Australia. *Cambridge Archaeological Journal, 28*(3), 361–378. https://doi.org/10.1017/S095977431800015X

Henshilwood, C. S., d'Errico, F., Vanhaeren, M., van Niekerk, K., & Jacobs, Z. (2004). Middle Stone Age shell beads from South Africa. *Science, 304*(5669), 404.

Henshilwood, C. S., d'Errico, F., Van Niekerk, K. L., Coquinot, Y., Jacobs, Z., Lauritzen, S.-E., et al. (2011). A 100,000-year-old ochre-processing workshop at Blombos Cave, South Africa. *Science, 334*, 219–222.

Hercus, L. (1980). 'How we danced the Mudlunga': Memories of 1901 and 1902. *Aboriginal History, 4*(1), 4–32.

Hercus, L. (n.d.). *The journey of the Seven Sisters through the Lake Eyre region*.

Hercus, L., Hodges, F., & Simpson, J. (2009). *The land is a map: Placenames of indigenous origin in Australia*. ANU Press.

Hiscock, P., & Faulkner, P. (2006). Dating the dreaming? Creation of myths and rituals for mounds along the Northern Australian coastline. *Cambridge Archaeological Journal, 16*(2), 209–222.

Hodgson, D. (2013). Ambiguity, perception, and the first representations. In K. Sachs-Hombach & J. R. J. Schirra (Eds.), *Origins of pictures. Anthropological discourses in image science* (pp. 401–423). Halem.

Hoffmann, D. L., Standish, C. D., García-Diez, M., Pettitt, P. B., Milton, J. A., Zilhão, J., et al. (2018a). U-Th dating of carbonate crusts reveals Neandertal origin of Iberian cave art. *Science, 359*(6378), 912–915. https://doi.org/10.1126/science.aap7778

Hoffmann, D. L., Standish, C. D., Pike, A. W. G., García-Diez, M., Pettitt, P. B., Angelucci, D. E., et al. (2018b). Dates for Neanderthal art and symbolic behaviour are reliable. *Nature Ecology & Evolution, 2*, 1044–1045. https://doi.org/10.1038/s41559-018-0598-z

Howitt, A. W. (1889). Notes on Australian message sticks and messengers. *The Journal of the Anthropological Institute of Great Britain and Ireland, 18*, 314–332.

Ingold, T. (2000). Chapter 7. Totemism, animism and the depiction of animals. In T. Ingold (Ed.), *The perception of the environment: Essays on livelihood, dwelling and skill* (pp. 111–131). Routledge.

Iriki, A., & Sakura, O. (2008). The neuroscience of primate intellectual evolution: Natural selection and passive and intentional niche construction. *Philosophical Transactions of the Royal Society B: Biological Sciences, 363*(1500), 2229–2241.

Isaacs, J. (1984). *Australia's living heritage*. Lansdowne.

Jebb, M. A. (Ed.). (2008). *Mowanjum: 50 years community history*. Mowanjum Aboriginal Community and Mowanjum Artists.

Johnson, C. N., Alroy, J., Beeton, N. J., Bird, M. I., Brook, B. W., Cooper, A., et al. (2016). What caused extinction of the Pleistocene megafauna of Sahul? *Proceedings of the Royal Society of London B: Biological Sciences, 283*(1824), 20152399. https://doi.org/10.1098/rspb.2015.2399

Jones, S., & Russell, L. (2012). Archaeology, memory and oral tradition: An introduction. *International Journal of Historical Archaeology, 16*(2), 267–283. https://doi.org/10.1007/s10761-012-0177-y

Jones, T., Levchenko, V., & Wesley, D. (2017). How old is X-ray art? Minimum age determinations for early X-ray rock art from the 'Red Lily' (Wulk) Lagoon rock art precinct, western Arnhem Land. In B. David, P. S. C. Taçon, J. J. Delannoy, & J.-M. Geneste (Eds.), *The archaeology of rock art in western Arnhem Land, Australia* (pp. 129–143). ANU Press.

Kelly, L. (2015). *Knowledge and power in prehistoric societies: Orality, memory, and the transmission of culture*. Cambridge University Press.

Kelly, P. (2020). Australian message sticks: Old questions, new directions. *Journal of Material Culture, 25*(2), 133–152. https://doi.org/10.1177/1359183519858375

Kelly, P. (2023). *Thoughts on 'proto-writing' in the Upper Palaeolithic*. Retrieved from https://bravenewwords.info/2023/01/08/thoughts-on-proto-writing-in-the-upper-palaeolithic/?utm_source=pocket_saves

Kelly, P. (2024). Comparing Australian message sticks and sequentially marked objects of the Upper Palaeolithic: Problems and opportunities. *Topics in Cognitive Science*. https://doi.org/10.1111/tops.12762

Kelly, M., & Brady, L. M. (2021). Boundaries, relationality, and style provinces in Australian rock art. In I. J. McNiven & B. David (Eds.), *The Oxford handbook of the archaeology of Indigenous Australia and New Guinea* (pp. 717–742). Oxford University Press.

Kelly, M., David, B., & Flood, J. (2021). 11. Scene but not heard. Seeing scenes in a Northern Australian Aboriginal site. In I. Davidson & A. Nowell (Eds.), *Making scenes: Global perspectives on scenes in rock art* (pp. 162–178). Berghahn Books.

Kelly, P., Lei, J., Bibiko, H.-J., & Barker, L. (2024). AMSD: The Australian Message Stick Database. *PLoS One, 19*(4), e0299712. https://doi.org/10.1371/journal.pone.0299712

Kerwin, D. (2012). *Aboriginal dreaming paths and trading routes. The colonisation of the Australian economic landscape*. Sussex Academic Press.

Keyser, J. D., & Klassen, M. A. (2016). *Plains Indian rock art*. University of Washington Press.

Kimber, R. G. (1990). Mulunga old Mulunga. 'Good corroboree', they reckon. In P. Austin, R. M. W. Dixon, T. Dutton, & I. White (Eds.), *Language and history: Essays in honour of Luise A. Hercus* (Vol. C-116, pp. 175–191). Pacific Linguistics.

Kimber, R. G. (2004). Message-sticks and 'Yabber' Mail. In V. Donovan & C. Wall (Eds.), *Making connections: A journey along Central Australian Aboriginal trading routes* (pp. 27–32). Arts Queensland.

Kuhn, S. L., Stiner, M. C., Reese, D. S., & Gulec, E. (2001). Ornaments of the earliest Upper Paleolithic: New insights from the Levant. *Proceedings of the National Academy of Sciences of the United States of America, 98*(13), 7641–7646.

Lambeck, K., & Chappell, J. (2001). Sea level change through the Last Glacial cycle. *Science, 292*(5517), 679–686. https://doi.org/10.1126/science.1059549

Langley, M. C., O'Connor, S., & Aplin, K. (2016a). A 46,000-year-old kangaroo bone implement from Carpenter's Gap 1 (Kimberley, northwest Australia). *Quaternary Science Reviews, 154*, 199–213. https://doi.org/10.1016/j.quascirev.2016.11.006

Langley, M. C., O'Connor, S., & Piotto, E. (2016b). 42,000-year-old worked and pigment-stained Nautilus shell from Jerimalai (Timor-Leste): Evidence for an early coastal adaptation in ISEA. *Journal of Human Evolution, 97*, 1–16.

Langley, M. C., Litster, M., Wright, D., & May, S. K. (Eds.). (2018). *The archaeology of portable art: Southeast Asian, Pacific, and Australian perspectives*. Routledge.

Langley, M. C., Clarkson, C., & Ulm, S. (2019). Symbolic expression in Pleistocene Sahul, Sunda, and Wallacea. *Quaternary Science Reviews, 221*, 105883. https://doi.org/10.1016/j.quascirev.2019.105883

Langley, M. C., Hakim, B., Agus Oktaviana, A., Burhan, B., Sumantri, I., Hadi Sulistyarto, P., et al. (2020). Portable art from Pleistocene Sulawesi. *Nature Human Behaviour, 4*(6), 597–602. https://doi.org/10.1038/s41562-020-0837-6

Langley, M. C., Kealy, S., & Mahirta, u., & O'Connor, S. (2023). Sequins from the sea: Nautilus shell bead technology at Makpan, Alor Island, Indonesia. *Antiquity, 97*(394), 810–828. https://doi.org/10.15184/aqy.2023.97

Layton, R. H. (1985). The cultural context of hunter-gatherer rock art. *Man: Journal of the Royal Anthropological Institute, 20*(3), 434–453.

Layton, R. H. (1992). *Australian rock art. A new synthesis*. Cambridge University Press.

Lewis, D. (1986). The Dreamtime animals: A reply. *Archaeology in Oceania, 21*, 140–143.

Lewis, D. (1988). *The rock paintings of Arnhem Land, Australia. Social, ecological and material culture change in the Post-Glacial period*. BAR Internationals Series 45.

Lewis, D. (1996). In defence of Arnhem Land rock art research. *Australian Archaeology, 43*(1), 12–20.

Lewis, D. (1997). Bradshaws: The view from Arnhem Land. *Australian Archaeology, 44*(1), 1–16.

Lewis, D. (2017). Megafauna identification for dummies: Arnhem Land and Kimberley 'megafauna' paintings. *Rock Art Research, 34*(1), 82–100.

Lewis, D., & McCausland, B. (1987). Engraved human figures and faces from Wardaman country, eastern Victoria River district, Northern Territory. *Australian Aboriginal Studies*, (1), 67–79.

Lewis, D., & Rose, D. B. (1988). *The shape of the dreaming. The cultural significance of Victoria River rock art*. Aboriginal Studies Press.

Lorblanchet, M. (2018). *Archaeology and petroglyphs of Dampier (Western Australia). An archaeological investigation of Skew Valley and Gum Tree Valley* (Vol. 27). Technical Reports of the Australian Museum Online.

Love, J. R. B. (1917). Notes on the Worora tribe of north-western Australia. *Transactions and Proceedings of the Royal Society of South Australia, XLI*, 21–38.

Macintosh, N. W. G. (1977). Beswick Creek cave two decades later: A reappraisal. In P. J. Ucko (Ed.), *Form in Indigenous Art. Schematisation in the art of Aboriginal Australia and prehistoric Europe* (pp. 191–197). Australian Institute of Aboriginal Studies.

Malafouris, L. (2013). *How things shape the mind. A theory of material engagement*. MIT Press.

Malaspinas, A.-S., Westaway, M. C., Muller, C., Sousa, V. C., Lao, O., Alves, I., et al. (2016). A genomic history of Aboriginal Australia. *Nature, Advance Online Publication, 538*, 207–214. https://doi.org/10.1038/nature18299

Mallen, E. (2022). The embodiment of artistic objects in Pablo Picasso's Cubism. *Arts, 11*(1), 32. Retrieved from https://www.mdpi.com/2076-0752/11/1/32

Manhire, A. (1998). The role of hand prints in the rock art of the south-western Cape. *The South African Archaeological Bulletin, 53*(168), 98–108.

Maynard, L. (1979). The archaeology of Australian Aboriginal art. In S. M. Mead, I. Brymer, & S. Martich (Eds.), *Exploring visual art of Oceania: Australia, Melanesia, Micronesia and Polynesia* (pp. 82–110). The University of Hawaii Press.

McAdam, L. E. S. (2008). *Beads across Australia: An ethnographic and archaeological view of the patterning of Aboriginal ornaments* (Unpublished Ph.D. thesis). University of New England, Armidale.

McAdam, L. E. S., & Davidson, I. (2018). Beads and boundaries. In M. Langley, M. Litster, D. Wright, & S. K. May (Eds.), *The archaeology of portable art: Southeast Asian, Pacific, and Australian perspectives* (pp. 220–240). Routledge.

McBryde, I. (1987). Goods from another country: Exchange networks and the people of the Lake Eyre basin. In D. J. Mulvaney & J. P. White (Eds.), *Australians to 1788* (pp. 252–273). Fairfax, Syme and Weldon.

McCarthy, F. D. (1939). "Trade" in Aboriginal Australia and "trade" relationships with Torres Strait, New Guinea and Malaya. *Oceania, 9*, 405–438.

McConvell, P., & Bowern, C. (2011). The prehistory and internal relationships of Australian languages. *Language and Linguistics Compass, 5*(1), 19–32. https://doi.org/10.1111/j.1749-818X.2010.00257.x

McCulloch-Uehlin, S. (2001) Rock the house, a row mixing reputation, rock art and race has led a publisher to quietly rewrite an entry in an illustrious publication, Inquirer. *The Weekend Australian*, 20–21 October 2002. Retrieved from www.theaustralian.com.au

McDonald, J. (2005). Archaic faces to headdresses. The changing role of rock art across the arid zone. In P. Veth, M. A. Smith, & P. D. Hiscock (Eds.), *Desert peoples: Archaeological perspectives* (pp. 116–141). Blackwell Publishing.

McDonald, J. (2017). Discontinuities in arid zone rock art: Graphic indicators for changing social complexity across space and through time. *Journal of Anthropological Archaeology, 46*, 53–67. https://doi.org/10.1016/j.jaa.2016.08.005

McDonald, J., & Mulvaney, K. (Eds.). (2023). *Murujuga: Dynamics of the dreaming*. UWA Publishing.

McDonald, J., & Veth, P. (2012). The social dynamics of aggregation and dispersal in the Western Desert. In J. McDonald & P. Veth (Eds.), *A companion to rock art* (pp. 90–102). Blackwell Publishing Ltd.

McDonald, J., Reynen, W., Ditchfield, K., Dortch, J., Leopold, M., Stephenson, B., et al. (2018). Murujuga Rockshelter: First evidence for Pleistocene occupation on the Burrup Peninsula. *Quaternary Science Reviews, 193*, 266–287. https://doi.org/10.1016/j.quascirev.2018.06.002

McNiven, I. J., & Russell, L. (2005). *Appropriated pasts: Indigenous peoples and the colonial culture of archaeology*. Altamira Press.

Mechling, L. (2023, June 7). "This is an experiment": Is Hannah Gadsby's Picasso exhibition really that bad? *The Guardian*.

Merlan, F. (1989). The interpretive framework of Wardaman rock art: A preliminary report. *Australian Aboriginal Studies, 1989*(2), 14–24.

Mimura, N. (2013). Sea-level rise caused by climate change and its implications for society. *Proceedings of the Japan Academy, Series B, 89*(7), 281–301.

Minc, L. D. (1986). Scarcity and survival: The role of oral tradition in mediating subsistence crises. *Journal of Anthropological Archaeology, 5*, 39–113.

Moore, M. W., Westaway, K., Ross, J., Newman, K., Perston, Y., Huntley, J., et al. (2020). Archaeology and art in context: Excavations at the Gunu Site Complex, Northwest Kimberley, Western Australia. *PLoS One, 15*(2), e0226628. https://doi.org/10.1371/journal.pone.0226628

Morphy, H. (1977). Yingapungapu: Ground sculpture as bark painting. In P. Ucko (Ed.), *Form in indigenous art* (pp. 205–209). Australian Institute of Aboriginal Studies.

Morphy, H. (1980). What circles look like. *Canberra Anthropology, 3*(1), 17–36.

Morphy, H. (1991). *Ancestral connections: Art and an Aboriginal system of knowledge*. University of Chicago Press.

Morphy, H. (1998). *Aboriginal art*. Phaidon Press.

Morphy, H. (2007). *Becoming art: Exploring cross-cultural categories*. Berg.

Morphy, H., & Layton, R. H. (1981). Choosing among alternatives: Cultural transformation and social change in Aboriginal Australia and the French Jura. *Mankind, 13*(1), 56–73.

Morphy, F., & Morphy, H. (2009). The Blue Mud Bay case: Refractions through saltwater country. *Dialogue, 28*(1), 15–25.

Morse, K. (1988). Mandu Mandu rockshelter: Pleistocene human coastal occupation of North West Cape, Western Australia. *Archaeology in Oceania, 23*, 81–88.

Morse, K. (1993). Shell beads from Mandu Mandu rockshelter, Cape Range Peninsula, Western Australia, before 30,000 bp. *Antiquity, 67*, 877–883.

Morwood, M. J. (1984). The prehistory of the Central Queensland Highlands. In F. Wendorf & A. E. Close (Eds.), *Advances in world archaeology* (Vol. 3, pp. 325–380). Academic.

Morwood, M. J. (1985). Facts and figures: Notes on rock art in the Mt. Isa area, northwestern Queensland. *Rock Art Research, 2*(2), 140–145.

Morwood, M. J. (2002). *Visions from the past. The archaeology of Australian Aboriginal art*. Allen and Unwin.

Morwood, M. J., & Hobbs, D. R. (Eds.). (1995). *Quinkan prehistory. The archaeology of Aboriginal art in S.E. Cape York Peninsula, Australia* (Vol. 3). Anthropology Museum, University of Queensland.

Moss, C. R. (2021). Between the lines: Looking for the contributions of enslaved literate laborers in a second-century text (P. Berol. 11632). *Studies in Late Antiquity, 5*(3), 432–452.

Moss, C. R. (2023). The secretary: Enslaved workers, stenography, and the production of early Christian literature. *The Journal of Theological Studies, 74*(1), 20–56.

Mountford, C. P. (1976). *Nomads of the Australian desert*. Rigby.

Mowaljarlai, D., Vinnicombe, P., Ward, G. K., & Chippindale, C. (1988). Repainting of images on rock in Australia and the maintenance of Aboriginal culture. *Antiquity, 62*(237), 690–696.

Mulvaney, D. J. (1976). 'The chain of connection': The material evidence. In N. Peterson (Ed.), *Tribes and boundaries in Australia* (pp. 72–94). Australian Institute of Aboriginal Studies.

Mulvaney, K. (2003). Transformations–rock walls to canvas: Representations of the totemic geography in Aboriginal Australia. *Before Farming, 5*, 297–308.

Mulvaney, D. J. (2008). "Annexing all I can lay hands on" Baldwin Spencer as ethnographic collector. In N. Peterson, L. Allen, & L. Hambly (Eds.), *The makers and making of Indigenous Australian museum collections* (pp. 141–162). Melbourne University Press.

Mulvaney, D. J. (2011). *Digging up a past*. UNSW Press.

Mulvaney, K. (2015). *Murujuga Marni. Rock art of the macropod hunters and mollusc harvesters*. UWA Publishing.

Mulvaney, D. J., Morphy, H., & Petch, A. (Eds.). (1997). *"My dear Spencer": The letters of F.J. Gillen to Baldwin Spencer*. Hyland House.

Munn, N. (1973). *Walbiri iconography. Graphic representation and cultural symbolism in a Central Australian Society*. Cornell University Press.

Munn, N. (1996). Excluded spaces: The figure in the Australian Aboriginal landscape. *Critical Inquiry, 22*(3), 446–465.

Myers, F. (1991). Representing culture: The production of discourse (s) for Aboriginal acrylic paintings. *Cultural Anthropology, 6*(1), 26–62.

Myers, F. (2002). *Painting culture: The making of an Aboriginal high art*. Duke University Press.

Neale, M. (2017a). White man got no dreaming. In M. Neale (Ed.), *Songlines: Tracking the seven sisters* (pp. 202–209). National Museum of Australia Press.

Neale, M. (Ed.). (2017b). *Songlines: Tracking the seven sisters*. National Museum of Australia Press.

Neale, M. (2020). 3. Knowledge in country and the third archive. In M. Neale & L. Kelly (Eds.), *Songlines. The power and the promise* (pp. 45–66). Thames and Hudson.

Neale, M., & Kelly, L. (Eds.). (2020). *Songlines. The power and the promise*. Thames and Hudson.

Nelson, E., Hall, J., Randolph-Quinney, P., & Sinclair, A. (2017). Beyond size: The potential of a geometric morphometric analysis of shape and form for the assessment of sex in hand stencils in rock art. *Journal of Archaeological Science, 78*, 202–213. https://doi.org/10.1016/j.jas.2016.11.001

Newsome, A. E. (1980). The eco-mythology of the red kangaroo in Central Australia. *Mankind, 12*(4), 327–333.

Ngarjno, Ungudman, Banggal, & Nyawarra. (2000). *Gwion Gwion: Secret and sacred pathways of the Ngarinyin Aboriginal people of Australia (with parallel texts in French and German) (J. Doring, editor)*. Könemann.

Norman, K., O'Connor, S., & Bird, M. (2021). Island hopping to Sahul. In I. J. McNiven & B. David (Eds.), *The Oxford handbook of the archaeology of Indigenous Australia and New Guinea*. Oxford University Press.

Norman, K., Bradshaw, C. J. A., Saltré, F., Clarkson, C., Cohen, T. J., Hiscock, P., et al. (2024). Sea level rise drowned a vast habitable area of north-western Australia driving long-term cultural change. *Quaternary Science Reviews, 324*, 108418. https://doi.org/10.1016/j.quascirev.2023.108418

Nöth, W. (2010). The criterion of habit in Peirce's definitions of the symbol. *Transactions of the Charles S. Peirce Society, 46*(1), 82–93.

O'Connell, J. F., Allen, J., Williams, M. A., Williams, A. N., Turney, C. S., Spooner, N. A., et al. (2018). When did Homo sapiens first reach Southeast Asia and Sahul? *Proceedings of the National Academy of Sciences, 115*(34), 8482–8490.

O'Connor, S., & Arrow, S. (2008). Boat images in the rock art of northern Australia with particular reference to the Kimberley, Western Australia. In G. Clark, B. F. Leach, & S. O'Connor (Eds.), *Islands of inquiry. Colonisation, seafaring and the archaeology of maritime landscapes* (pp. 397–409). ANU Press.

O'Connor, S., & Fankhauser, B. (2001). Art at 40, 000 BP? One step closer: An ochre covered rock from Carpenter's Gap Shelter 1, Kimberley region, Western Australia. In A. Anderson, I. Lilley, & S. O'Connor (Eds.), *Histories of old ages. Essays in honour of Rhys Jones* (pp. 287–300). Pandanus Books, Australian National University.

O'Connor, S., Balme, J., Fyfe, J., Oscar, J., Oscar, M., Davis, J., et al. (2013). Marking resistance? Change and continuity in the recent rock art of the southern Kimberley, Australia. *Antiquity, 87*(336), 539–554.

O'Connor, S., Kealy, S., Reepmeyer, C., Samper Carro, S. C., & Shipton, C. (2023). Terminal Pleistocene emergence of maritime interaction networks across Wallacea. *World Archaeology, 54*(2), 244–263. https://doi.org/10.1080/00438243.2023.2172072

Oktaviana, A. A. (2018). Hand stencils and boats in the painted rock art of the karst region of Muna Island, Southeast Sulawesi. In *The archaeology of Sulawesi: Current research on the Pleistocene to the historic period* (Vol. 2015, pp. 61–77). ANU Press.

Oktaviana, A. A., Joannes-Boyau, R., Hakim, B., Burhan, B., Sardi, R., Adhityatama, S., et al. (2024). Narrative cave art in Indonesia by 51,200 years ago. *Nature, 631*(8022), 814–818. https://doi.org/10.1038/s41586-024-07541-7

Pawley, A. (2005). The chequered career of the Trans New Guinea Phylum: Recent historical research and its implications. In A. Pawley, R. Attenborough, J. Golson, & R. Hide (Eds.), *Papuan pasts: Investigations into the cultural, linguistic and biological history of the Papuan speaking peoples* (pp. 67–108). Pacific Linguistics, Australian National University.

Perkins, H., & Fink, H. (2000). *Papunya Tula: Genesis and genius*. Art Gallery of NSW.

Peterson, N. (2005). Thomson's place in Australian anthropology. In B. Rigsby & N. Peterson (Eds.), *Donald Thomsons, the man and scholar* (pp. 29–44). The Academy of the Social Sciences in Australia.

Pike, A. W. G., Hoffmann, D. L., García-Diez, M., Pettitt, P. B., Alcolea, J., De Balbín, R., et al. (2012). U-series dating of Paleolithic art in 11 caves in Spain. *Science, 336*(6087), 1409–1413. https://doi.org/10.1126/science.1219957

Pinchbeck, C. (2013a). Introduction. In C. Pinchbeck (Ed.), *Yirrkala drawings* (pp. 14–15). Art Gallery of NSW.

Pinchbeck, C. (Ed.). (2013b). *Yirrkala drawings*. Art Gallery of NSW.

Rappaport, R. A. (1999). *Ritual and religion in the making of humanity*. Cambridge University Press.

Ridges, M. (2003). *Numerous indications. The archaeology of regional hunter-gatherer behaviour in northwest central Queensland, Australia* (Unpublished Ph.D.). University of New England, Armidale.

Ridges, M. (2006). Scale and its effects on understanding regional behavioural systems: An Australian case study. In G. Lock & B. L. Molyneaux (Eds.), *Confronting scale in archaeology* (pp. 145–161). Springer.

Ridges, M., Davidson, I., & Tucker, D. (2000). The organic environment of paintings on rock. In G. K. Ward & C. Tuniz (Eds.), *Advances in dating Australian rock-markings. Papers from the first Australian rock-picture dating workshop* (pp. 61–70). Australian Rock Art Association.

Roberts, R. G., Walsh, G. L., Murray, A., Olley, J., Jones, R., Morwood, M. J., et al. (1997). Luminescence dating of rock art and past environments using mud-wasp nests in northern Australia. *Nature, 387*, 696–699.

Rose, D. B. (1996). *Nourishing terrains: Australian Aboriginal views of landscape and wilderness*. Australian Heritage Commission.

Rosenfeld, A. (1991). Panaramitee: Dead or alive? In P. Bahn & A. Rosenfeld (Eds.), *Rock art and prehistory* (pp. 136–144). Oxbow Books.

Rosenfeld, A. (2000). 5.1 Rock art: A multifaceted heritage. In S. Kleinert & M. Neale (Eds.), *The Oxford companion to Aboriginal art and culture* (pp. 103–109). Oxford University Press.

Ross, J. (1997). *Painted relationships: An archaeological analysis of a distinctive anthropomorphic rock art motif in northwest central Queensland* (B.A. Honours). Department of Archaeology and Palaeoanthropology, University of New England, Armidale.

Ross, J. (2003). *Rock art, ritual and relationships. An archaeological analysis of rock art from the central Australian arid zone* (Ph.D.). University of New England, Armidale.

Ross, J. (2013). A continent of nations: The emergence of new regionally distinct rock art styles across Australia. *Quaternary International, 285*, 161–171. https://doi.org/10.1016/j.quaint.2012.01.007

Ross, J., & Davidson, I. (2006). Rock art and ritual: An archaeological analysis of rock art in arid Central Australia. *Journal of Archaeological Method and Theory, 13*(4), 305–341. https://doi.org/10.1007/s10816-006-9021-1

Ross, J., & Smith, M. A. (2009). An engraved 'archaic face' in the northeastern Simpson Desert. *Australian Archaeology, 69*(1), 68–70.

Ross, J., Westaway, K., Travers, M., Morwood, M. J. J., & Hayward, J. (2016). Into the past: A step towards a Robust Kimberley rock art chronology. *PLoS One, 11*(8), e0161726. https://doi.org/10.1371/journal.pone.0161726

Roth, W. E. (1897). *Ethnological studies among the North-West-Central Queensland Aborigines (Facsimile Edition)*. Government Printer.

Roth, W. E. (1910). North Queensland Ethnography. Bulletin No. 15. Decoration, deformation, and clothing. *Records of the Australian Museum, 8*, 20–54.

Russell, L. (2012). Remembering places never visited: Connections and context in imagined and imaginary landscapes. *International Journal of Historical Archaeology, 16*(2), 401–417. Retrieved from http://www.jstor.org/stable/23258949

Salles, T., Joannes-Boyau, R., Moffat, I., Husson, L., & Lorcery, M. (2024). Physiography, foraging mobility, and the first peopling of Sahul. *Nature Communications, 15*(1), 3430. https://doi.org/10.1038/s41467-024-47662-1

Salvador, R. J. (1997). *Maize. The Encyclopedia of Mexico: history, culture, and society*. Fitzroy Dearborn Publishers.

Santos Da Rosa, N., Fernández-Macías, L., Mattioli, T., & Díaz-Andreu, M. (2021). Dance scenes in Levantine rock art (Spain): A critical review. *Oxford Journal of Archaeology, 40*(4), 342–366.

Satterthwaite, L., & Arthur, B. (2005). Technology and material culture. In B. Arthur & F. Morphy (Eds.), *Macquarie atlas of Indigenous Australia* (pp. 48–65). Macquarie University, Macquarie Dictionary.

Scalise Sugiyama, M. (2021). Co-occurrence of ostensive communication and generalizable knowledge in forager storytelling. *Human Nature, 32*(1), 279–300. https://doi.org/10.1007/s12110-021-09385-w

Sehasseh, E. M., Fernandez, P., Kuhn, S., Stiner, M. C., Mentzer, S., Colarossi, D., et al. (2021). Early Middle Stone Age personal ornaments from Bizmoune Cave, Essaouira, Morocco. *Science Advances, 7*(39), eabi8620. https://doi.org/10.1126/sciadv.abi8620

Silcock, J. L., Tischler, M., & Smith, M. A. (2012). Quantifying the Mulligan River Pituri Trade of Central Australia. *Ethnobotany Research and Applications, 10*, 037–044.

Silcock, J. L., Tischler, M. K., & Fensham, R. J. (2020). *Oases at the gates of hell: Hydrogeology, cultural history and ecology of the Mulligan River Springs, Far Western Queensland*. Paper presented at the proceedings of the Royal Society of Queensland.

Sillitoe, P. (1988). From head-dresses to head-messages: The art of self-decoration in the highlands of Papua New Guinea. *Man, 23*, 298–318.

Smith, C. (1992). Colonising with style: Reviewing the nexus between rock art, territorially and the colonisation and occupation of Sahul. *Australian Archaeology, 34*, 34–42. Retrieved from http://www.jstor.org/stable/40287092

Smith, M. A. (2013). *The archaeology of Australia's deserts*. Cambridge University Press.

Smith, C., Domingo Sanz, I., & Jackson, G. (2016). Beswick Creek Cave six decades later: Change and continuity in the rock art of Doria Gudaluk. *Antiquity, 90*(354), 1613–1628.

Snow, D. R. (2013). Sexual dimorphism in European Upper Paleolithic cave art. *American Antiquity, 78*(4), 746–761. Retrieved from http://www.jstor.org/stable/43184971

Specht, J., Torrence, R., & Mulvaney, K. (2021). Petroglyphs and place: Complex histories at four sites in New Britain. *Archaeology in Oceania, 56*(3), 196–228.

Spencer, B. (1928). *Wanderings in wild Australia* (Vol. 2). Macmillan.
Spencer, B., & Gillen, F. J. (1899). *Native tribes of Central Australia*. Macmillan.
Spriggs, M. (2007). The Neolithic and Austronesian expansion within Island Southeast Asia and into the Pacific. In S. Chiu & C. Sand (Eds.), *From Southeast Asia to the Pacific. Archaeological perspectives on the Austronesian expansion and the Lapita cultural complex* (pp. 104–140). Academia Sinica.
Standish, C. D., Pettitt, P., Collado, H., Aguilar, J. C., Milton, J. A., García-Diez, M., et al. (2025). The age of hand stencils in Maltravieso cave (Extremadura, Spain) established by U-Th dating, and its implications for the early development of art. *Journal of Archaeological Science: Reports, 61*, 104891. https://doi.org/10.1016/j.jasrep.2024.104891
Stanton, J. E. (2013). The crown jewels of our collection: The Yirrkala crayon drawings. In C. Pinchbeck (Ed.), *Yirrkala drawings* (pp. 17–20). Art Gallery of NSW.
Stiner, M. C. (1999). Palaeolithic mollusc exploitation at Riparo Mochi (Balzi Rossi, Italy): Food and ornaments from the Aurignacian through Epigravettian. *Antiquity, 73*(282), 735–754.
Stokes, J. L. (1969). *Discoveries in Australia: With an account of the coasts and rivers explored and surveyed during the voyage of the H.M.S. Beagle (Facsimile Edition)*. Libraries Board of South Australia.
Sutton, P. (Ed.). (1989). *Dreamings: The art of Aboriginal Australia*. Viking.
Sutton, P. (1995). *Country. Aboriginal boundaries and land ownership in Australia*. Aboriginal History Inc.
Sutton, P. (2009). *The politics of suffering. Indigenous Australia and the end of the liberal consensus*. Melbourne University Press.
Taçon, P. S. C. (1993). Regionalism in the recent rock art of western Arnhem Land, Northern Territory. *Archaeology in Oceania, 28*, 112–120.
Taçon, P. S. C. (2000). Rock-art sites. In S. Kleinert & M. Neale (Eds.), *The Oxford companion to Aboriginal art and culture* (pp. 690–691). Oxford University Press.
Taçon, P. S. C. (2008). Rainbow colour and power among the Waanyi of Northwest Queensland. *Cambridge Archaeological Journal, 18*(2), 163–176. https://doi.org/10.1017/s0959774308000231
Taçon, P. S. C., Langley, M. C., May, S. K., Lamilami, R., Brennan, W., & Guse, D. (2010). Ancient bird stencils in Arnhem Land, Northern Territory, Australia. *Antiquity, 84*(324), 416–427.
Taylor, L. (2005). Chapter 13. Manifestations of the *Mimih*. In *The power of knowledge: The resonance of tradition* (pp. 182–198). Aboriginal Studies Press.
Taylor, L. (2015). Categories of 'old' and 'new' in western Arnhem Land bark painting. In *Long history deep time* (pp. 101–117). ANU Press.
Thomson, D. F. (1949). *Economic structure and the ceremonial exchange cycle in Arnhem Land*. Macmillan.
Thomson, D. F. (1983). *Donald Thomson in Arnhem Land. Compiled and introduced by N.Peterson*. Currey O'Neil.
Tonkinson, R. (1974). *The Jigalong Mob: Aboriginal victors of the desert crusade*. Cummings.
Travers, M., & Ross, J. (2016). Continuity and change in the anthropomorphic figures of Australia's northwest Kimberley. *Australian Archaeology, 82*(2), 148–167. https://doi.org/10.1080/03122417.2016.1210757
Tsang, R., Brady, L. M., Katuk, S., Taçon, P. S. C., Ricaut, F.-X., & Leavesley, M. G. (2021). Agency, affect and archaeologists: Transforming place with rock art in Auwim, upper Karawariarafundi region, East Sepik, Papua New Guinea. *Rock Art Research, 38*(2), 183–194.
Tsang, R., Katuk, S., May, S. K., Taçon, P. S. C., Ricaut, F.-X., & Leavesley, M. G. (2022a). Rock art and (re) production of narratives: A cassowary bone dagger stencil perspective from Auwim, East Sepik, Papua New Guinea. *Cambridge Archaeological Journal, 32*(4), 547–565.
Tsang, R., Pleiber, W., Kariwiga, J., Plutniak, S., Forestier, H., Taçon, P. S. C., et al. (2022b). Rock art and long-distance prehistoric exchange behavior: A case study from Auwim, East Sepik, Papua New Guinea. *The Journal of Island and Coastal Archaeology, 17*(3), 432–444.

Tsang, R., Katuk, S., May, S. K., Taçon, P. S. C., Ricaut, F. X., & Leavesley, M. G. (2023). Hand stencils and communal history: A case study from Auwim, East Sepik, Papua New Guinea. *Archaeology in Oceania, 58*(1), 115–130.

Twain, M. (1989 [1897]). *Following the Equator: A journey around the world.* Dover Publications.

Utrilla, P., Mazo, C., Domingo, R., & Bea, M. (2021). 14. Maps in prehistoric art. In I. Davidson & A. Nowell (Eds.), *Making scenes: Global perspectives on scenes in rock art* (pp. 207–222). Berghahn Books.

van den Bergh, G. D., Li, B., Brumm, A., Grün, R., Yurnaldi, D., Moore, M. W., et al. (2016). Earliest hominin occupation of Sulawesi, Indonesia. *Nature, 529*(7585), 208–211. https://doi.org/10.1038/nature16448. http://www.nature.com/nature/journal/v529/n7585/abs/nature16448.html#supplementary-information

Veth, P., Stern, N., McDonald, J., Balme, J., & Davidson, I. (2011). Ch. 9. The role of information exchange in the colonisation of Sahul. In R. Whallon, W. A. Lovis, & R. K. Hitchcock (Eds.), *Information and its role in hunter-gatherer bands* (pp. 203–220). Cotsen Institute of Archaeology Press.

Walsh, G. L. (1979). Mutilated hands or signal stencils? A consideration of irregular hand stencils from central Queensland. *Australian Archaeology, 9*(1), 33–41.

Walsh, G. L. (1983). Composite stencil art: Elemental or specialised? *Australian Aboriginal Studies, 1983*(2), 34–44.

Walsh, G. L. (1988). *Australia's greatest rock art.* E.J. Brill/Robert Brown & Associates.

Warner, W. L. (1964 [1937]). *A Black Civilization. A social study of an Australian Tribe.* Harper Torchbooks.

Watson, P. L. (1983). *This precious foliage* (Vol. 26). Oceania Monographs, University of Sydney.

Watson, P. L., Luanratana, O., & Griffin, W. J. (1983). The ethnopharmacology of pituri. *Journal of Ethnopharmacology, 8*(3), 303–311. https://doi.org/10.1016/0378-8741(83)90067-3

Weber, G. W., Lukeneder, A., Harzhauser, M., Mitteroecker, P., Wurm, L., Hollaus, L.-M., et al. (2022). The microstructure and the origin of the Venus from Willendorf. *Scientific Reports, 12*(1), 1–10.

White, I. M. (1980). The birth and death of a ceremony. *Aboriginal History, 4*(1), 32–41.

White, R. (1989). Production complexity and standardization in early Aurignacian bead and pendant manufacture: Evolutionary implications. In P. A. Mellars & C. Stringer (Eds.), *The human revolution: Behavioural and biological perspectives on the origin of modern humans* (pp. 366–390). Princeton University Press.

White, R. (1992). Beyond art: Toward an understanding of the origins of material representation in Europe. *Annual Review of Anthropology, 21*(1), 537–564.

White, J. P., & Barry, M. (2004). 'Exotic Bradshaws' or Australian 'Gwion': An archaeological test. *Australian Aboriginal Studies, 2004*(1), 37–44.

Wickham, C. (1842). Notes on Depuch Island. *Journal of the Royal Geographical Society of London, 12*(ArticleType: Research-article/Full publication date: 1842), 79–83. Retrieved from http://www.jstor.org/stable/1797981

Williams, N. M. (2019). A boundary is to cross: Observations on Yolngu boundaries and permission. In N. M. Williams & E. S. Hunn (Eds.), *Resource managers: North American and Australian hunter-gatherers* (pp. 131–153). Routledge.

Wisher, I., Pettitt, P., & Kentridge, R. (2023). Conversations with caves: The role of pareidolia in the Upper Palaeolithic figurative art of Las Monedas and La Pasiega (Cantabria, Spain). *Cambridge Archaeological Journal, 1-24*, 315–338. https://doi.org/10.1017/S0959774323000288

Wright, C. (2024). *Naku Dharuk the Bark Petitions. How the people of Yirrkala changed the course of Australian Democracy*. Text Publishing.

Wroe, S., Field, J. H., Archer, M., Grayson, D. K., Price, G. J., Louys, J., et al. (2013). Climate change frames debate over the extinction of megafauna in Sahul (Pleistocene Australia-New Guinea). *Proceedings of the National Academy of Sciences, 110*(22), 8777–8781. https://doi.org/10.1073/pnas.1302698110

Yengoyan, A. A. (1968). Demographic and ecological influences on Aboriginal marriage sections. In R. B. Lee & I. DeVore (Eds.), *Man the hunter* (pp. 185–199). Aldine.

Chapter 10
... A Thousand Words

Abstract This chapter draws together the different threads of the argument both in terms of what art might be, and why there might be one set of ideas for people who have narrow interests in "Fine Art" and a different set for people who have broader visions of what art is. I present a case for the relationship between these two visions, and how, in considering them, we can understand much more about art in the history of humans. Importantly, people who are interested in art (of any period or style) should consider the roles art has had throughout, at least, the last 50,000 years.

What has been the main argument of this book? If a picture is worth a thousand words, could I have used just a few pictures instead of my lengthy argument? The whole point is that I could not. Everyone needs at least a thousand words to explain why they have used a particular picture. The words are built in to the art-system as I have argued it. They are part of it, but the picture is not it.

I have cited several anecdotes that have influenced the way I have been thinking about art throughout a long(ish) career. In discussing them I hope I have got you to think about art in ways that will make some of you see it anew. Much of what I say has also been said by someone else, but I think the whole story is original. Recognizing the pressures on the concept of art, and using criteria we have applied to other examples, the simple takeaway is this: it should be uncontroversial for people who are interested in art (of any period or style) to include art of the past or present and consider the roles art has had throughout, at least, the last 50,000 years. Rothko is said to have looked at Pompeii, Picasso at Iron Age sculptures. We should be able to look at art of any period past or present through the lens of the lessons of this book to understand more about what we are looking at. We do not need to know those facts about Rothko and Picasso to appreciate their art, rather we need to know how the culture we live in affects the art we make and see, and that all of us dredge our culture—our understanding—through our experience of previous works. Those works of very old art were part of the culture that Rothko (Fig. 10.1) and Picasso lived in. Our culture has been formed, in part, by the work of Rothko and Picasso,

I. Davidson, *Art or Scribbles? In the Eye of the Beholder: The Evolutionary Emergence of Visual Communication*, Interdisciplinary Evolution Research 11,
https://doi.org/10.1007/978-3-031-90458-5_10

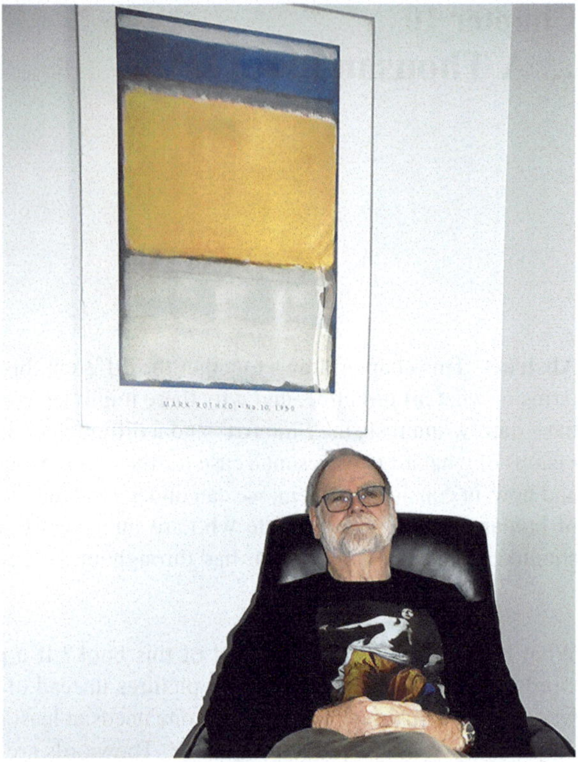

Fig. 10.1 The author seated in a motel wearing a t-shirt printed with an image of a painting of Lucretia by the 17th century artist Artemisia Gentileschi, exhibited in Hamilton, Victoria, April 2024. I am sitting below a print of a painting by Rothko (perhaps, as an example of motel art). Photo Helen Arthurson (with permission)

and the way we see any art today has been "dredged" through those works by someone, maybe us.

I began with an anecdote about a grazier, Bill, who disparaged the rock art made by the Aboriginal people whose Native Title rights have now been recognized. That land is leased to his family, and the Mitakoodi used art from a site on the property during the Determination of that Native Title. I arrived at a position that was nicely described by Pasztory as "it is society that decides what is and what isn't art." Grazier Bill's social context and mine were very different while he was alive—though we were both men and white. I have tried to set out the decision-making by society in this book and recognize that all people look at art through the ideas we get from our class, gender and race as well as our attitudes to colonialism and indigeneity, and they usually reflect our position on the scale from privilege to disadvantage.

I have been influenced in this by the Australian artist Brett Whitely writing about "… pictures …[that] filtered and sometimes dredged the whole of Europe through them." All art builds on our experience, and our expectations derive from traditions established by our knowledge of earlier art. One point of the book is that people who are interested in art of the last three or four hundred years can profit from understanding the art of the three or four hundred years before that, which would take them back to Van Eyck (Fig. 8.21). They might then continue backwards to realize how art got there—from understanding about the archeohistory of art and looking at a range of traditions that stretch back perhaps 2000 generations. But it also works the other way around. Stephanie Moser (1996) has shown that attitudes to our very distant forebears, known only from archeology—such as those who produced the images from Sulawesi (Figs. 2.12 and 4.7) or from Iberia (Fig. 4.6)—have been determined by the imagery produced to illustrate the scientific conclusions, but that those images themselves were influenced by the dominant ideology of the time.

As an archeologist, I have believed for a long time that we should call many things art—and I have puzzled about the boundaries of "pattern", "design", "decoration" and art (Davidson, 2013). Are there criteria by which we distinguish motel art from museum art, or abstract expressionism from imaginative wrapping paper? Why is it that Gainsborough's *Blue Boy* (Fig. 7.7) is one of the great paintings, while the same image printed on playing cards is not even art? We experience this puzzle in many different ways, but, for example, some people refer to street murals as "graffiti" (e.g., the Adnate art destroyed by developers in Fig. 3.19), others recognize that it is art (e.g., a portrait by Adnate has just been selected for the 2024 Packing Room Prize in the Archibald Portrait Prize at the Art Gallery of NSW). The solution lies in our social attitudes to art in particular contexts.

Through discussions of a definition of art in Chap. 3, I have sought to provide an understanding that there is an art-system involving all four elements of producer (artist) and seer (viewer) as well as the subject of the art and the object the producer makes (the "art object"). My emphasis has been on the people—artists and seers—and their agency in establishing the characteristics of the relationship between the subject and the object in the art-system. Morphy (2007, xi) began with a definition that differed from the approach I have used here: "art objects are ones with aesthetic and or semantic attributes (but in most cases both), that are used for representational or presentational purposes." Morphy's emphasis was firmly on the "art object" and hence he was driven into the camp of discussing "Fine Art" as opposed to something else. On the other hand, Morphy acknowledged that "Fine Art" "has been defined by people in the West" and "Indigenous Australians produced art for thousands of years before they encountered Europeans…" who introduced them to the "select category of fine art" and the concept of aesthetics. As I pointed out in the previous chapter, Ingold (2000, 131) would deny the judgment that what the Yolngu were producing was art, and this book is, incidentally, an argument about why he was wrong. The problem arose partly from the concept of aesthetics, which is either entirely subjective as I have suggested, or at best might differ between different cultural systems. It is difficult to see how Aboriginal "art objects" can have aesthetic attributes within the same framework as aesthetic criteria for any other context—unless the qualities of aesthetics are properties of the seer and not of the "art object". But the "art objects" can be considered "art" by my criteria of providing a systemic relationship between idea, artist, object, and seer informed by stories or songs and contributing to a set of cultural values—which may, recursively, contribute to the mental attitude with which the artist considers the next idea (Fig. 10.2).

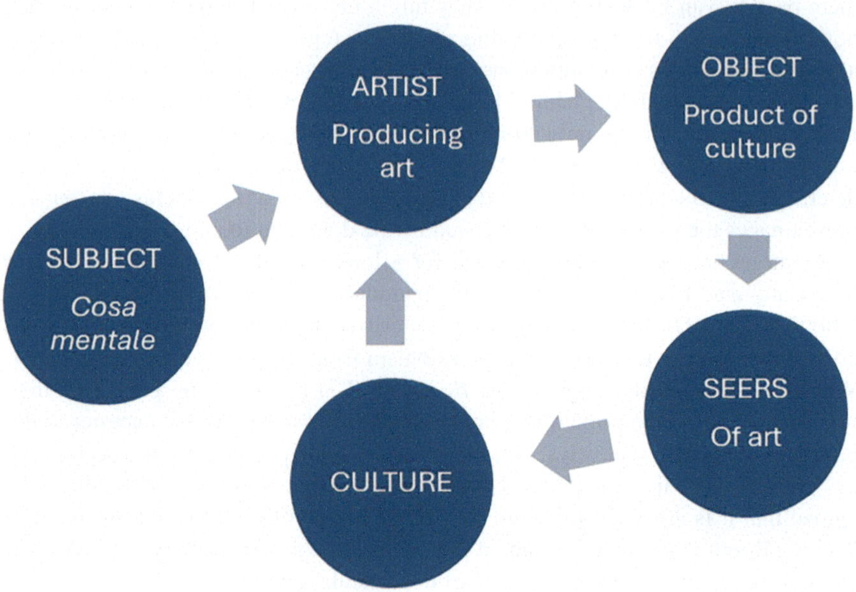

Fig. 10.2 Model of the art-system and culture. © Iain Davidson

We have been discussing art, and the elements of identifying it and analyzing how it has been used. I have tried to emphasize that any "aesthetic" judgment is generally solipsistic: "I know what I like" without consideration of why you should like it too (except that I might tell you to). So "aesthetics" becomes the property of things that I like, not of "art objects". Our encounter with grazier Bill is one way to emphasize that one person's aesthetic art is another person's unbearable scribble. Once a market for art emerged in 18th century Europe (after Artemisia Gentileschi), it was absolutely necessary for that market to invent a concept to define what was good and what was bad as I have described in Chap. 3. Just as importantly, by analyzing the field of art in Australia, using the same methods as applied to histories of other art, we can show what elements of the art-system are in common in two completely independent bodies of work, without invoking concepts such as "aesthetics" or the market. We can avoid the subjectivity of the judgment between art and scribbles by not passing judgment in either way.

From my slight acquaintance with books of art history, many of those things I appreciate might not be called art by the important people who stand guard at its portals. As a young man growing up in Chicago, Shiner (2001) saw that there was art in both the Art Institute and the Field Museum, and he wondered why they were kept separate. The Art Institute declares that its mission is: "We collect, care for, and interpret works of art across time, cultures, geographies, and identities, centering the vision of artists and makers".[1] The Field Museum declares that its purpose is:

[1] https://www.artic.edu/about-us/mission-and-history

"The Field Museum connects all of us to the natural world and the human story".[2] You can find art in two sorts of museums. In one we exhibit the products of non-western people, which were often stolen or acquired by unequal interactions, as in the well-known case of the marble sculptures from the Parthenon in Athens (Hamilakis, 1999), the contents of the Summer Palace in Beijing (Gillman, 2019), or the Benin Bronzes (Hicks, 2020). In the others we exhibit works of art by named artists, often produced as "Fine Art". It would be an over-simplification, but not by much, to say that in one we exhibit the things produced by people who lost wars to the museum builders, and in the other—the National Galleries of Art—we exhibit works paid for by privilege and wealth generated by the evils associated with such colonial aspirations. Sometimes this strange opposition emerged into the art: a small number of pictures show interactions between putatively enslaved people and those who might be their enslavers in the 18th and 19th centuries (Procter, 2020, Chs 4, 10).

Shiner (2001, 18) gave something of a clue to the answer to his own question by emphasizing the importance of funding art production before the 17th century through patronage and commissions, certainly in Europe but probably elsewhere. He then pointed out that the 17th and 18th centuries saw the rise of market economies and, arguably, it was the commerce in art that made all the difference to the emergence of what he (and others) call "Fine Art". There is much more stuff that should be called art than just the rarified category that those who are specialists in appreciation of "fine" art after the middle of the eighteenth century (sometimes) call "fine". One of the issues here is that devotees of "fine" art have concentrated on the art object itself (as in the definition by the Art Institute) and not on the whole art-system of relationships between the artist, the subject, the object and themselves, bound together by stories and songs. By ignoring that set of relationships in the art-system, "fine" art seems to be special when the object can be bought and sold as if it existed without the stories of its creator and creation. Of course, the commerce in such "fine" art also led to a market in archeological artefacts after devotees decided that they too belonged with other art. There is an ongoing close relationship between archeology (and archeologists) and art of all sorts (Megaw, 2003).

I still believe and have been arguing here (especially in Chap. 3), that much of the earliest stuff (Art 3), and most archeological art (Art 2), deserves to be called art and to be considered on a par with those things in Art Institutes (Art 1). Many objects in Ethnographic Museums should also be considered as art (Art 1), but they were produced without commerce (at least they were produced without money). There may, of course, have been systems of payment for production without money, but generally speaking in ways that the guardians of Art 1 did not consider, such as in rights and obligations. The biggest difference, to which Shiner was referring, lay in the commercial structure of the system within which Fine Art was produced. We now expect statements about the price of pictures. In Chap. 3, I laid out the details of the financial transactions associated with a painting by Arthur Streeton (Fig. 3.16). Churcher (2014) gives very little other detail about the "meaning" of the

[2] https://www.fieldmuseum.org/

Beloved French Film Star Alain Delon's Art Collection Fetched $8.8 Million at a Paris Auction, Doubling Its Low Estimate
The top lot was a painting by Raoul Dufy, 'The Bay at Sainte-Adresse' (1906), which sold for $1.1 million.

Klimt's 'Lady With a Fan' Sets a New Auction Record in Europe, Making $108.4 Million at Sotheby's Mini-Marathon of London Sales
The artist's final portrait—it was still on the easel in Klimt's studio when he died in 1918—is now the most expensive artwork ever sold in Europe.

By the Numbers: A Breakdown of Results From Sotheby's London Auctions of Modern and Contemporary Art, June 2023
Get the stats behind the spin.

Artnet News, June 2023

Fig. 10.3 Three headlines from *Artnet News* showing the importance of the price of paintings

painting. That may have been an aberration but, in many contexts, it is the norm. Three headlines of items in a recent edition of the online journal *Artnet News* (Fig. 10.3) show that for many the main preoccupation is still the price (but not the value) of paintings.

Reading Shiner allowed me to reach a fundamental conclusion: what really changed in early 18th century Europe (because it did not happen, for example, in early 18th century Australia) was the social conditions of production of (fine) art. These involved replacing the system of patrons of artists by a market in art. Previously artists were commissioned by the church and/or the rich or royalty; now they sell their works. To maintain the boundaries of privilege and entitlement, private collections and art museums emerged, and academies of art were created. Churches, royalty and the rich continued to patronize the arts, but now they purchased paintings not painters, provided they had the approval of new gatekeepers. These novelties were enmeshed with the rise of commerce such that artists (and others) made a living from the fact that people began to sell and buy art. One of my University teachers told me of a senior colleague who had written that the appalling behavior of the Borgias and the Medicis of Medieval Tuscany could be justified because the great art of Renaissance Italy arose out of their nastiness. There was a bit more to it than that, of course, in terms of the technology of production, in terms of paints and supports, as well as revolutions in perspective and shading. It turns out that the senior colleague never wrote such a thing, but the person who told me was making their own comment about the importance of inequality and patronage in the production of art.

We divorce the "work of art" in a museum from the relationships between artists and the ideas they had when creating the work (Leonardo's *cosa mentale*) and exhibiting it in a museum with an interpretive sign that is generally minimally informative, too small and at the wrong height. Part of the ritual of the museum is the bow to the painting to peer at the tiny interpretive sign. The seer is excluded from the system in which the work was produced. And there are other separations: between the work and the story artists tell about it; between the four elements of the art-system made up of seer of the work and the artist, the idea and the work itself; between the work of art and the art-system; between the functions the artists have in mind for their work and the functions to which their work is put. There are at least four things in any such relationship, and they are held together by stories and songs that originate with artists and the culture of their society. Separating the parts, to give emphasis on the work, forces a view that causes the confusions I have been trying to clear up. In reality the production of art necessitates the stories that the producers could tell. The objects they produce impact on the culture within which producers and seers live (Fig. 10.2) and become a new part of it. We do not need to hear the particular story because we grow up in a culture with expectations that certain sorts of pictures will have certain sorts of story. We know that pictures of people with gold circles around their heads are indicative of a particular sort of person—just as we think that St Hardulph must have been important for some reason. We all have different cultural expectations because we have seen different visual images from other people, and all heard different stories about them. It is another question what this knowledge was like at a time when there were no photographs and little ability to convey any sense of color of paintings elsewhere. How much did anyone know about the Sistine Chapel before photography? As an example, in Toledo, the Transparente (Fig. 1.5) was described in 1890 as follows (Ford, 1890, 138–9): "In this fricassée of marble the plunging figure of San Rafael is prominent with his legs kicking out above him in the air. He holds in his rt [*sic*] hand a huge gilt fish." This does not give much idea of either the work or the associated stories. This imperfect knowledge more than anything else determines why culture can never be considered as a monolithic entity—it is made up from our experiences and they are all different. By-and-large we and the people around us will be exposed to similar things because we do not experience everything there is, and the boundaries are never fixed.

I recounted how I heard some people in the Museum of Modern Art say that they liked an exhibition of Cubist art but wanted to know what it meant. In the case where people make marks to represent something, it is an achievement to turn marks into that something else—a picture—and it is not something that we can take as built-in to the human brain. People generate an expectation of meaning by making marks, and I discussed hand-stencils as an example. All animals make marks on the ground or other parts of their environment. What was significant was that humans made marks and noticed that, so that they could make other marks and generate an

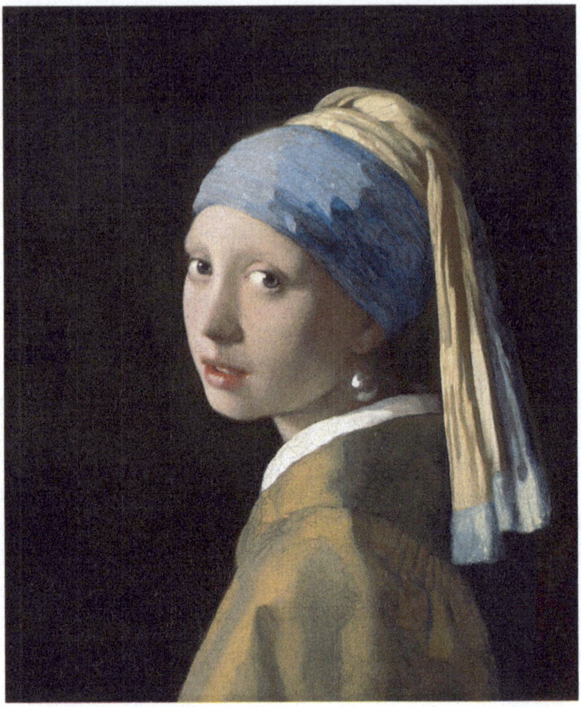

Fig. 10.4 Johannes Vermeer, 1665, The Girl with the Pearl Earring. Public domain, via Wikimedia Commons

expectation of meaning. When we see marks made by others, we have an expectation that the artist meant something by them—we expect that they stenciled hands and other objects for a reason. The reality is that any meaning is generally indeterminate—that is one reason why there are "so many meanings". "Meaning" is generally determined by the words of the artist, and by the songs and stories that created the cultural expectations with which we have learned to see it. It is that indeterminacy that enables novelist (Chevalier, 1999) and the film-maker to construct a fictitious story about *The Girl with the Pearl Earring* with little more than Vermeer's picture to go on (Fig. 10.4).

One way in which people resolve the indeterminacy in art is through the use of scenes, but not all art includes scenes, and we are dependent on stories or songs that go with it. Picasso and Braque turned marks into pictures OF something (though the seers in my story of the museum were not sure what), talking to each other about how they were going about it. In doing so they made us all know that there were many ways to see the world and they wanted us to realize that if we would go beyond our culture's view there was more than what we had already got. Representations of life are like a box of chocolates[3]—not just photographs of pretty scenes. I struggle to reconcile the views that have been expressed to me that one painter was so good his pictures were just like a photograph (as was said to me about

[3] This deliberately echoes the popular movie *Forrest Gump*.

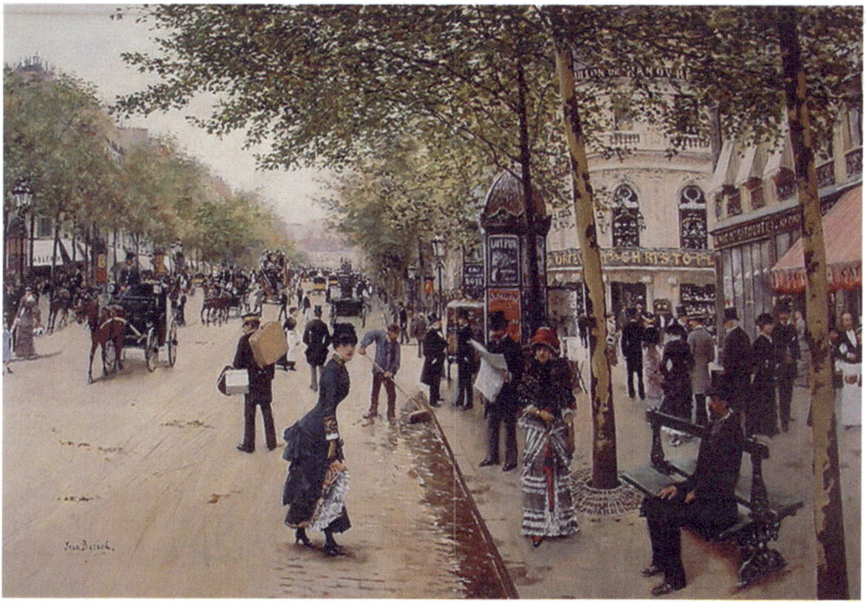

Fig. 10.5 *Boulevard des Capucines* Jean Béraud 1889. In 1873–4 Monet showed his impression of the same street at the first show by the Impressionists (Fig. 7.27). (Public domain, via Wikimedia Commons)

Fig. 10.5), or that some of my photographs are so good they look just like a painting. Neither statement is true, but it says something about public attitudes to art. Somewhere in there is art, but it is not necessarily a property of the pictures themselves, but of the view the commenters have arising from their cultural values.

What Picasso and Braque achieved was to get seers to look at the pictures they had painted in order to work out how we see any subject, not just as a pretty thing. The impact of the photograph was the need to understand the consequences of the way we see things (Berger, 2008). There may, in fact, have been more than one "meaning" and more than one way to perceive things. As it were, you never know what you are going to get.

It is true of well-known studies of modern Aboriginal art that the work does not have "meaning" by itself but is part of the Aboriginal cultural world of Songlines which connect places together and create a more-or-less common ideology right across the country. That is true of my more limited experience of Aboriginal traditions, and I have tried to show the linkages to the sites I know best which were part of and thus created a network of cultural values across almost all of Australia. There is a sense in which meaning derives from the network of relationships for all art—not in terms of Songlines, perhaps, but in terms of the cultural world of the producer, and our own cultural worlds, our experiences. So, the story I told about the Tim Storrier horse at NERAM (Fig. 7.12) encapsulates this as it includes the sculpture and artistic equipment, his Archibald Prize winning portrait (with similar artistic equipment) and the Bosch Vagabond (Fig. 7.11), the John Sell Cotman picture of the nineteenth century horse (Fig. 7.13), and the fact that the NERAM fundraising exhibition (at which I purchased the Rees print from Fig. 5.22) is called "Packsaddle". The whole cultural context is part of the reduction of indeterminacy of the mark-making. Once I realized that, I understood much more about art in general, and recognized the significance of what I used to say rather intemperately: "Art is too important to be left to people who are interested in art." They need to understand cultural context and stories about the past too. As the Aboriginal poet Oodgeroo (Oodgeroo Noonuccal [Kath Walker], 1970) wrote "The past is all around us and within." In all societies, we internalize the conventions that art establishes, and they become part of our set of cultural values.

As I have outlined, mostly in Chap. 9, the market had a really important role in the development of First Nations art in two well-studied regions of Australia, Yirrkala and Papunya—and, of course, subsequently with its development all over Australia. Missionaries in Northeast Arnhem Land fostered the Yolngu art movement in the 1930s as a way to provide financial independence for the community (Morphy, 1998, 236–7), but subsequent developments, described by Morphy and Layton (1981), led to the erosion of the exclusive rights of senior men and the acquisition of new power by less senior men through the sale of art. Those rights of senior men, though, involved educating young people into the cultural values of Yolngu society and the conventions and codes of artistic production. That some of those artistic conventions were also seen, and that many people went through initiations meant that they became part of the whole cultural context of the society.

In a different context, the fact that much of the cave art of Western Europe was produced inside caves which were often difficult to access must mean that some people controlled access to the art, whatever other purposes that access may have had. This tradition, which continued in some places during 30 thousand years, must have required two things: (1) that people knew that there was art in caves; and (2) that society was divided into those who could tell a story about it and others to whom they would tell the story when they went into the caves together. These are the conditions for an expectation of meaning and the reduction of the indeterminacy through that process of generating cultural value. It does not really matter whether that story was about fertility, sex, the nuances of the hunt, or the importance of trance visions of other worlds. What was important was that there was social inequity between those who did the paintings, those who knew the stories and those who had been told them.

In Papunya in the early 1970s, one of the strong motivations for the emphasis on art was the possibility of selling pictures in Alice Springs (Bardon, 1991; Johnson, 2010, Ch 2), but it was not always a simple process and required careful nurturing (Myers, 2000). This was seen as an opportunity for the Aboriginal community to generate an income independent of the settlement managers (Myers, 2002, 120–146) but painters and marketing organizations had to balance Aboriginal cultural traditions and the pressures of commercialization (Myers, 2000). It was evident that the marks in the paintings had meanings, though some of those could not be divulged to people who were not already initiated. To that extent, the meaning was indeterminate. The boards and canvases, therefore, had value and hence power for the painters and storytellers.

I argue that all forms of communication, even those that do not give that appearance, have a material element—a visual element—which persists. The persistence of the material thing-that-can-be-seen is a fundamental feature of human culture from the beginning of art. It turns out that persistence has its own importance, particularly when meanings are not generally known, as it signifies that the art is powerful—not that the art itself has agency, but that people have power through knowing about the meanings of the persistent works.

Reading Morphy (2007), who has dedicated himself to the study of the anthropology of art, I discovered that he too has been chasing some of the same paradox about the dual expression of the art—the "art object" and the idea associated with it—but approached it rather differently. To some extent, this book has been consistent with the way Morphy (1999) concluded:

> This gap between the promise of results and their achievement, between the potential to provide immediate access to symbolic systems of the past and the difficult and limited nature of that access, makes the archaeology of art appear simultaneously an exciting and at times frustrating discipline. Perhaps more than any other form of archaeological data it hints at possible interpretative richness, but this richness can only be gained through interpretative restraint and analytic rigour, leavened by the imagination of the archaeological agent.

In addition, there are two points which show the importance of the restraint and analysis I have presented here. First, Morphy (2007) bemoaned that anthropologists had tended to neglect art, while art historians had disdained discussion of anthropological art. This is more or less what I pointed out at the beginning of the preface. Second, his book (Morphy, 1991) was a prolonged discussion of how one particular example of Northern Australian Aboriginal art had become "Fine Art", just as Myers (2002) discussed another such process for Central Australia.

Throughout this book I have had at the forefront the attitude to art expressed by one of the great synthesizers, Ernest Gombrich, that art involves unquestioned beliefs. I pointed out in Chap. 2 that he omitted the cultural context within which he was making that judgment—what were his own "unquestioned beliefs"? Gombrich (1995, 39–43) emphasized that "within the prescribed rites and customs of our [western, white, male] lives"... "the whole story of art is not a story of progress in technical proficiency, but a story of changing ideas and requirements." I go further and ask the question about the importance of those changing ideas and requirements in art—its role in culture and cultural change. And I pointed out that the concept of "culture" has many different meanings. In Gombrich's terms, what are the processes that prescribe rites and customs? The answer to that can be found in a broad understanding of ritual as I described in Chap. 5. Rituals, formal or informal, are concerned with teaching and reinforcing the values by which people belong to a particular society or its subdivisions of class, gender or race. Rituals constitute the culture of the Warlpiri and pervade the whole society because ritual meanings ensure that everything stands both for itself and for something else. As I said "rituals constitute their culture."

"Culture" is never a fixed entity, as I discussed in Chap. 3 and have reiterated in different forms throughout. Most obviously, if there is anything in this book which is new to you, then that is an indication that the culture of your community was different from mine, but now they are more closely similar. All of the examples I cite demonstrate that "culture" in any definition is additive and subjective. Bourdieu (1968) understood that our definitions of artistic matters depend on the cultural values any individual accepts and displays because we are all immersed in our own version of them. But they also contribute to ongoing cultural values.

I have been hinting at the idea that the interplay between art and culture is central to the way people understand what they mean by art—it determines what reaches the eye of the beholder (Fig. 10.6). Those cultural values may involve class, gender, and race. The question becomes, then, how people get to belong to or be identified with particular categories of class, gender or race, particularly given that these are social categories not biological ones. Again, the clue can be seen in the way we all acquire credentials, including education and training, and that process will be broadly similar within a society but there are experiences that define differences which may be large or small.

All of which leaves us with the conundrum of why the distribution of rock art in the world is as it is. I cited Paul Taçon (2000) as saying that in Australia: "Rock paintings and engravings can be found almost everywhere suitable surfaces occur."

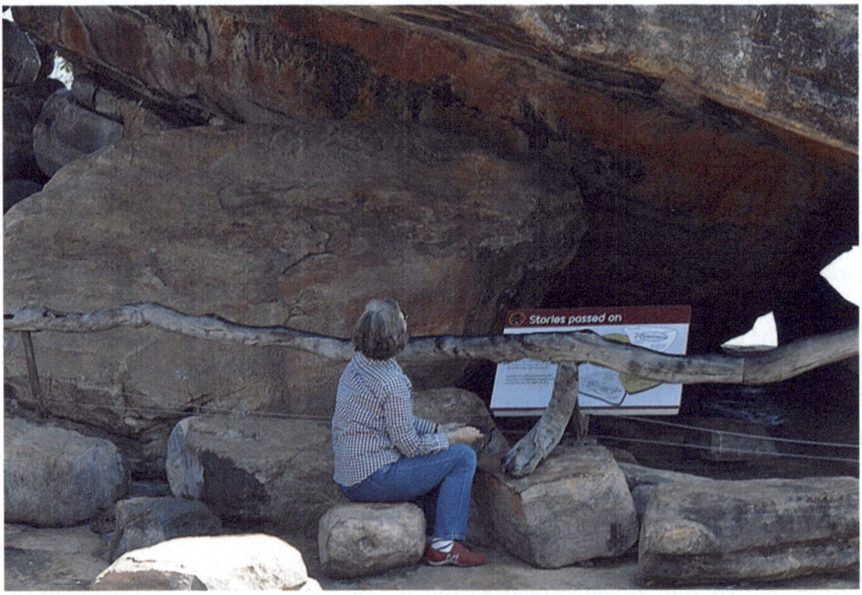

Fig. 10.6 The beholder looking at art at Ubirr Rock, Kakadu National Park. As the interpretation board indicates, the nature of the art depends on the stories passed on. (Photo Iain Davidson)

My experience in many different regions in Australia, Europe, Southwest Asia and the Americas, although not comprehensive, suggests that First Nations people used rock art in the ways I have described rather more than people in some regions outside Sahul. I would attribute this to the sorts of social contexts of production I described in the previous chapter. Thus, Central Australia "contained variations which permitted the spread of ideologies" at local, regional and continental scales "by using art as a means of encoding information in different ways." There was tight control of the ways knowledge was passed on within the system at least in part through "the encoded information in rock art [that] provided persistence of the images at rock art sites."

One of my earlier definitions of art was: "Art is the making and marking of surfaces where the images have implicational meaning for the observer whether or not they know the full propositional meaning of the artist" (Davidson, 2013, 42). We can now see that the definition can be put into the context of the expectation of meaning and indeterminacy, and of the semantic roles which involve a mental process and whole series of elements that make up the process of producing art as I described it in Table 8.1. This is the difference in cognition between the implicational system which depends on inputs from external sources and the propositional system which depends on the cultural knowledge stored in the mind, as Barnard and colleagues would suggest (Barnard et al., 2016).

Any art is composed of a set of marks which we see as something more than just marks. All other animals make some sort of mark, but not all of them behave as if they can work out how the marks got there. Humans can. And we can go further and make marks that signify much more than the fact that we have made marks. Our ancestors made marks, worked out that they were signs of the activity that caused them, then realized that they could make similar marks for the sake of signaling something in addition and tell stories about them. In the case where people make marks to represent something, the pictures might be worth a thousand words, but they will be your thousand or my thousand but not anyone else's thousand. We will each have a different idea of the propositional meaning of the marks (Barnard et al., 2016) because we have absorbed different experiences during our lives such that we each have slightly different cultures. What Morphy would call the "inside" meaning comes from our education in and immersion in the cultural background we bring when viewing marks. It is the full propositional meaning that makes art art. Art gained its importance by its positioning to be seen and enjoyed by anyone within a broader cultural context. The point may be that marks which cannot be put into such a cultural context get called scribbles.

10.1 Essay 10 A Thousand Final Words … (Carved Phalange from La Garma, Iberia and the Xian Terracotta Warriors, China)

There is no doubt that great beauty is one of the characteristics of much art, but beauty can occur at any time, not just in "Fine Art." The first image for this Essay (Fig. 10.7 Left) is an image of a cow carved onto the toe bone that was originally in the back foot of a cow. It was found in the cave of La Garma, in northern Spain, which had been sealed from the outside world for 16.5 thousand years (Arias et al., 2007–2008). The object has a hole drilled through it lengthwise so that it seems to have been suspended in a way that, rotated clockwise, the whole image could be seen. This was an image of great accomplishment in which the carver paid considerable attention not only to their image of the cow they were familiar with but to the way that it would be seen by the seers. It has now been transferred to the collections of the Museo Arqueológico de Santander, where it is exhibited as if it were a work of art. In all likelihood it is difficult for viewers to appreciate it in its context of examples of other objects from Upper Paleolithic art that could be rotated on a string or of its ability to reveal a more interesting story about the role of children within the society of the time (Nowell, 2015, 896–7).

Culley (2021) pointed out that scenes were more prevalent in the art of Upper Paleolithic Europe on portable pieces of stone or bone, and they are rare—or at least differently constructed— in the art on the walls inside caves. Archeologists and others have argued endlessly about the meaning of cave art (e.g., David, 2017) in Western Europe, but the important point is that there seem to have been two different contexts. In portable art, there were scenes that could be interpreted by almost

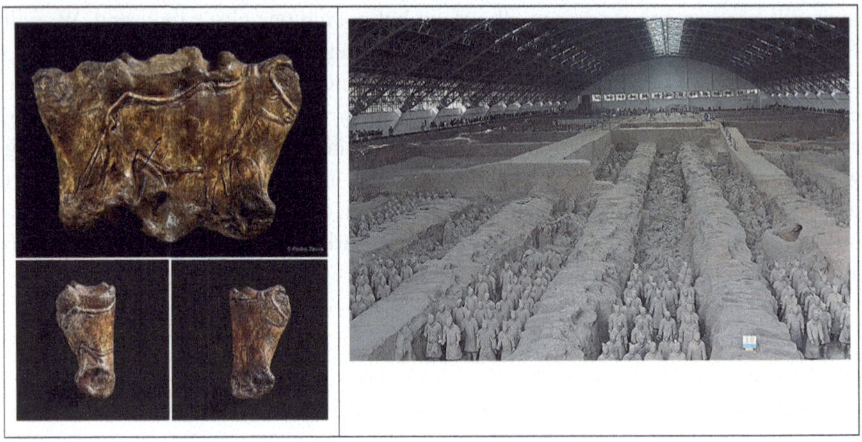

Fig. 10.7 Left: Carved bovid phalange, La Garma (Iberia) (photo Pedro Saura. Permission from D. Roberto Ontañón Peredo, director del Museo de Prehistoria y Arqueología de Cantabria). Right: Terracotta warriors, excavated at Xian, China. (Photo Iain Davidson)

anyone who sees them; on the rock walls inside the caves, the nature of the art was different, and, almost without argument, (young) people had to be taught the cultural context and hence the propositional meanings of art in their society—the stories that went with that scene. The La Garma bone was inside the cave, but may have had a cultural context that needed to be taught in order to understand the rituals involving the art on the cave walls.

The second image (Fig. 10.7 Right) is from the Xian Terracotta Warriors. It became possible to look down on the art of Africa, Asia, or the Americas because it was "only" ritual. Art objects should be "appreciated" for their beauty alone. The Xian Terracotta Warriors become objects of beauty in the museum where some of them were housed and in various touring exhibits of a few specimens (I saw three completely out of context in Barcelona in 2004). The point was that there were several thousand of the models of warriors, which were deposited ritually and then buried to mark the death of the Emperor Qin Shi Huang 2200 years ago (Fig. 10.7 Right). Selection of individual specimens enables us to appreciate them as works of "art" by themselves, ignoring the context that makes the whole experience much more meaningful. The context of these remarkable objects was provided by the text written within a hundred years of the burial. A picture may be worth a thousand words, but a picture with a thousand words places more emphasis on context than on the subjective seductions of the picture as Fine Art.

These images remind me, and you the reader, that I have barely mentioned Africa, the Americas or Asia in this book. I seem, at one level, to have made a major omission, and I need to say a few things about that in the context of the argument in this book.

First, the examples demonstrate that the portion of the whole art world that is partitioned off to be called "Fine Art" by those so inclined represents an amazing bias when we, the seers, contemplate the art of the world. After all, Picasso is said to have begun a revolution in the body of Art 1 when he included an African mask in place of the face of one of the women in his *Les Demoiselles d'Avignon, 1907.* The artist almost certainly saw such a mask in the ethnographic museums of Paris (Murrell 2000) and then made the face of his *Portrait of Gertrude Stein* (Fig. 3.23) in a mask-like form. Picasso, Matisse and Modigliani were all influenced in what they produced in the early twentieth century by the forms of African art. Like the Benin Bronzes, of which the *Brass Rooster* (Fig. 8.29) is an example that has been returned to Nigeria, many of the examples of art from Africa had been looted or acquired in dubious circumstances (Hicks, 2020; McDougall, 2008). For our purposes here, as with the marbles from the Parthenon in Athens and the Xian Warriors, the African objects, by divorcing them, often forcefully, from their cultural contexts became no more than art objects, but are now considered very much in terms of the story of their acquisition by museums outside that context, which can now be given importance it would not have had when they were removed. In those museums, they were exhibited as if they were "Primitive" art (the term used by those discussing the use of such style by Picasso and others), with all of the deliberate or implied pejoratives of the word primitive. In fact, they have all of the characteristics of art being discussed here: producer, seers, story, subject, object and transferability. They also have both an implicational or surface meaning in terms of what they represent. They are beautiful. But they also had an original context such that knowledge of them and their stories imbued cultural values. That gave them a propositional meaning too—"actual processes of thought".

Shiner (2001, 5–8) was right to point to a revolution of art in the eighteenth century through which "Fine Art" was taken to exist as an aesthetic ideal, but his more important point was about the social transformation that brought this about in which European men, the upper classes and the wealthy were deemed to set the standards of beauty required for works to belong to *Civilisation* (as the BBC would have had it). But for Africa and the Americas, and to some extent Asia, it was almost as if the "works of art" had to be removed from their cultural context before those in the west could regard them as art. Whatever the ideas of that small, self-selected band of brothers, earlier art and other art have been all around, involved men and women, high and low, rich and poor and are still art. Indeed, there was more than money would buy.

Likewise, Bourdieu (1987) saw the point that art always involved a contextualized and politicized process. We should view things the other way around. Context and politics have been determining what should be regarded as art. People create something that they present to an audience with a story, and in doing so their act enters an art-system. The producer and the observer(s) are the essential elements that turn something into a "work of art", but the work is not something that exists outside the system of social relationships of the people concerned. There is no essence called "art."

Paradoxically, this leads to the conclusion that "Fine Art"—Constable's *Hay Wain*, or one of the *Haystacks* by Monet—for all the skill and beauty of the finished product, is impoverished compared with the social, political and ritual context of some of the art dubbed "primitive". That is because much of "fine" art has lost that link between Art 1 and context precisely because it is made for transfer and exhibit in museums as Fine Art. "Scribbles," such as the curves in Fig. 2 Right, need explanation provided by the storytelling of the painters but to people like me or grazier Bill they can be opaque without the stories. We both approached the same marks and came to different conclusions because our whole cultural histories differed—for me the rock art of the Selwyn Ranges is art but for grazier Bill it was scribbles. But that means the rock art might be the true "art" with propositional meaning in which they required the *cosa mentale* that imbued the culture of the society in which they were painted, even though they have no obvious implicational meaning. Whereas, however incorrect we may have been about the "anthropomorphic figure" of Fig. 2 Left, the very fact that we can give it an implicational meaning (however incomplete at a propositional level) makes our initial interpretation impoverished. Maybe Bill had everything the wrong way around—those things that look like scribbles can ONLY be interpreted if we know how to interpret them; we think we know what other things are, but we often do not really know. Our minds are made up by knowing the keys to everything.

References

Arias, P., Ontañon, R., Álvarez, E., Cueto, M., García, C., & Teira, L. C. (2007–2008). Falange grabada de la galería inferior de la Garma: Aportación al estudio del arte mobiliar del Magdaleniense Medio. *Veleia, 24–25*, 97–129.

Bardon, G. (1991). *Papunya Tula. Art of the Western Desert*. McPhee Gribble.

Barnard, P. J., Davidson, I., & Byrne, R. W. (2016). Toward a richer theoretical scaffolding for interpreting archaeological evidence concerning cognitive evolution. In T. Wynn & F. Coolidge (Eds.), *Cognitive models in Palaeolithic archaeology* (pp. 45–67). Oxford University Press.

Berger, J. (2008). *Ways of seeing*. Penguin UK.

Bourdieu, P. (1968). Outline of a sociological theory of art perception. *International Social Science Journal, 20*(4), 589–612.

Bourdieu, P. (1987). The historical genesis of a pure aesthetic. *The Journal of Aesthetics and Art Criticism, 46*, 201–210.

Chevalier, T. (1999). *Girl with a pearl earring*. Harper Collins.

Churcher, B. (2014). *Australian notebooks*. The Miegunyah Press.

Culley, E. V. (2021). 12. A comparison of "scenes" in parietal and non-parietal upper paleolithic imagery: formal differences and ontological implications. In I. Davidson & A. Nowell (Eds.), *Making scenes: global perspectives on scenes in rock art*. Berghahn Books.

David, B. (2017). *Cave art*. Thames & Hudson.

Davidson, I. (2013). Origins of pictures: an argument for transformation of signs. In K. Sachs-Hombach & J. R. J. Schirra (Eds.), *Origins of pictures. Anthropological discourses in image science* (pp. 15–45). Halem.

Ford, R. (1890). *A handbook for travellers in Spain* (7th ed.). John Murray.

Gillman, D. (2019). The old summer palace and the rhetoric of national treasures. *Santander Art and Culture Law Review, 5*(2), 233–244.

Gombrich, E. H. (1995). *The story of art* (16th ed.). Phaidon.

Hamilakis, Y. (1999). Stories from exile: Fragments from the cultural biography of the Parthenon (or 'Elgin') marbles. *World Archaeology, 31*(2), 303–320.

Hicks, D. (2020). *The brutish museums: the Benin bronzes, colonial violence and cultural restitution*. Pluto Press.

Ingold, T. (2000). Chapter 7. Totemism, animism and the depiction of animals. In T. Ingold (Ed.), *The perception of the environment: essays on livelihood, dwelling and skill* (pp. 111–131). Routledge.

Johnson, V. (2010). *Once upon a time in Papunya*. UNSW Press.

McDougall, R. (2008). Chapter 4. The making of *Great Benin*: Felix Norman Roth and Henry Ling Roth. In R. McDougall & I. Davidson (Eds.), *The Roth family, anthropology and colonial administration* (pp. 73–92). Left Coast Press.

Megaw, J. (2003). Figuring it out: What are we? Where do we come from? The parallel visions of artists and archaeologists By Colin Renfrew, p. 224, 175 illustrations. ISBN 0 500 05114 3. Thames & Hudson, London, 2003. Price£ 32 (hb).

Morphy, H. (1991). *Ancestral connections: art and an aboriginal system of knowledge*. University of Chicago Press.

Morphy, H. (1998). *Aboriginal art*. Phaidon Press.

Morphy, H. (2007). *Becoming art: exploring cross-cultural categories*. Berg.

Morphy, H., & Layton, R. H. (1981). Choosing among alternatives: cultural transformation and social change in Aboriginal Australia and the French Jura. *Mankind, 13*(1), 56–73.

Moser, S. (1996). Visual representation in archaeology: depicting the missing-link in human origins. In *Picturing knowledge: Historical and philosophical problems concerning the use of art in science* (pp. 184–214).

Myers, F. (2000). In sacred trust. Building the Papunya Tula market. In H. Perkins & H. Fink (Eds.), *Papunya Tula. Genesis and Genius* (pp. 235–245). Art Gallery of New South Wales in Association with Papunya Tula Artists.

Myers, F. (2002). *Painting culture: the making of an aboriginal high art*. Duke University Press.
Nowell, A. (2015). Learning to see and seeing to learn: children, communities of practice and pleistocene visual cultures. *Cambridge Archaeological Journal, 25*(4), 889–899. https://doi.org/10.1017/S0959774315000360
Oodgeroo Noonuccal [Kath Walker]. (1970). The past. In *My people* (p. 93). Jacaranda Books.
Procter, A. (2020). *The whole picture: the colonial story of the art in our museums & why we need to talk about it*. Cassell.
Shiner, L. (2001). *The invention of art: a cultural history*. The University of Chicago Press.
Taçon, P. S. C. (2000). Rock-art sites. In S. Kleinert & M. Neale (Eds.), *The Oxford companion to aboriginal art and culture* (pp. 690–691). Oxford University Press.